CONVERSION FACTORS BETWEEN U.S. CUSTOMARY SYSTEM (USCS) AND THE STANDARD INTERNATIONAL (SI) SYSTEM

Quantity	USCS to SI	SI to USCS
Length	1 in = 25.400 mm	1 m = 39.37 in
	1 ft = 0.3048 m	1 m = 3.281 ft
Area	$1 \text{ in}^2 = 645.2 \text{ mm}^2$	$1 \text{ mm}^2 = 1.550(10^{-3}) \text{ in}^2$
	$1 \text{ ft}^2 = 0.0929 \text{ m}^2$	$1 \text{ m}^2 = 10.76 \text{ ft}^2$
Volume	$1 \text{ in}^3 = 16.39(10^3) \text{ mm}^3$	$1 \text{ mm}^3 = 61.02(10^{-6}) \text{ in}^3$
	$1 \text{ ft}^3 = 0.028 \text{ m}^3$	$1 \text{ m}^3 = 35.31 \text{ ft}^3$
Area moment of inertia	$1 \text{ in}^4 = 0.4162(10^6) \text{ mm}^4$	$1 \text{ m}^4 = 2.402(10^{-6}) \text{ in}^4$
Mass	1 slug = 14.59 kg	1 kg = 0.06852 slug
Force	1 lb = 4.448 N	1 N = 0.2248 lb
	1 kip = 4.448 kN	1 kN = 0.2248 kip
Moment	1 in·lb = 0.1130 N·m	1 N·m = 8.851 in·lb
	1 ft·lb = 1.356 N·m	1 N·m = 0.7376 ft·lb
Force per unit length	1 lb/ft = 14.59 N/m	1 N/m = 0.06852 lb/ft
Pressure; Stress	1 psi = 6.895 kPa	1 kPa = 0.1450 psi
	1 ksi = 6.895 MPa	1 MPa = 0.1450 ksi
	$1 \text{ lb/ft}^2 = 47.88 \text{ Pa}$	$1 \text{ kPa} = 20.89 \text{ lb/ft}^2$
Work; Energy	1 lb·ft = 1.356 J	1 J = 0.7376 lb·ft
Power	1 lb·ft/s = 1.356 W	1 W = 0.7376 lb·ft/sec
	1 hp = 745.7 W	1 kW = 1.341 hp

INTERMEDIATE
MECHANICS OF MATERIALS

INTERMEDIATE
MECHANICS OF MATERIALS

Madhukar Vable
Michigan Technological University

New York   Oxford
OXFORD UNIVERSITY PRESS
2008

Oxford University Press, Inc., publishes works that further Oxford University's
objective of excellence in research, scholarship, and education.

Oxford New York
Auckland Cape Town Dar es Salaam Hong Kong Karachi
Kuala Lumpur Madrid Melbourne Mexico City Nairobi
New Delhi Shanghai Taipei Toronto

With offices in
Argentina Austria Brazil Chile Czech Republic France Greece
Guatemala Hungary Italy Japan Poland Portugal Singapore
South Korea Switzerland Thailand Turkey Ukraine Vietnam

Published by Oxford University Press, Inc.
198 Madison Avenue, New York, New York 10016
http://www.oup.com

Oxford is a registered trademark of Oxford University Press

ISBN 978-0-19-518855-4

Printing number: 9 8 7 6 5 4 3 2 1

Printed in the United States of America
on acid-free paper

To my friend and my love
Professor Pushpalatha Murthy

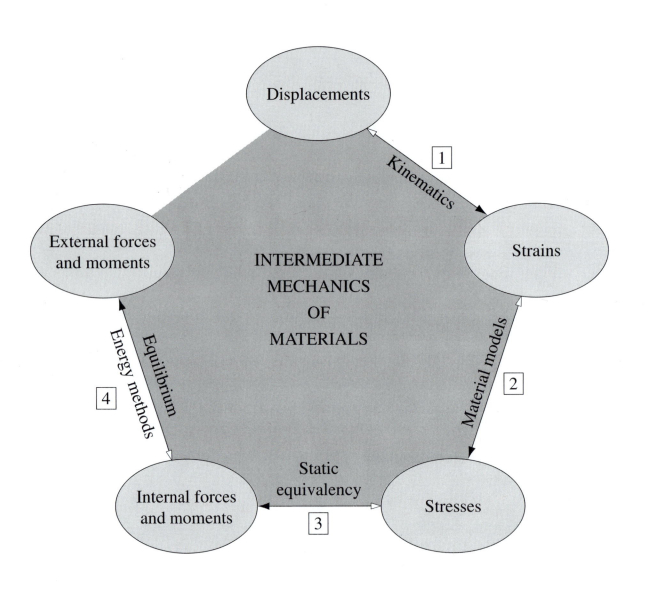

CONTENTS

Chapter 4   Composite Structural Members

Chapter 5   Inelastic Structural Behavior

Chapter 6 Thin-Walled Structural Members

Chapter 7 Energy Methods

Chapter 8   Elasticity and the Mechanics of Materials

Chapter 9 Finite Element Method

Appendix A Statics and Mechanics of Materials Review

Appendix B   Basic Matrix Algebra

Appendix C   Information Charts and Tables

PREFACE

Plastics engineering is becoming a popular new minor, and biomedical engineering is a growing new discipline; in both fields, stress analysis requires an understanding of the mechanics of inelastic and nonlinear material behavior. Applications of inelastic and nonlinear material are also growing as metals compete with new materials in engineering design by operating in the plastic region through prestressing. The growing use of metal matrix composites, polymer composites, reinforced concrete, and wooden beams stiffened with steel strips and other laminated structures emphasizes the need to expose students to the analysis of one-dimensional composite structural members. The ubiquity of the finite element method in engineering analysis and design emphasizes the educational importance of energy principles and the concepts of the finite element method. Equally significantly, the evaluation and use the results produced by commercial finite element computer software requires an understanding of stress and strain transformation in three dimensions and of failure theories. Capstone senior design projects have added to the importance of understanding unsymmetric bending and the concept and use of shear centers in design. Beam and shaft vibrations, beams on elastic foundations, Timoshenko beams, and so on are among the many topics in existing aerospace, civil, and mechanical engineering courses that use the principles of mechanics of materials. If a student is to be taught the mechanics of the topics described without being overwhelmed by all their inherent complexities, then the presentation of the material must have coherence and compactness that consolidates what he or she has already learned in the introductory course and builds on it. This is the underlying design of this book, and the Note to the Instructor elaborates on it.

An introductory course on the mechanics of materials is required in most engineering disciplines. The educational philosophy behind the requirement is to teach common mechanics concepts and principles in a single course and to present the extensions and applications of the mechanics concepts and principles in individual disciplines. This educational philosophy addresses the need of interdisciplinary education while realizing curriculum efficiency by reducing duplication of course content—an important consideration as educators attempt to modernize engineering education by incorporating research into a burgeoning curriculum. The introductory course on the mechanics of materials has served the engineering community well, but the tremendous growth in the applications of mechanics of materials is bringing added importance to a second course, Intermediate Mechanics of Materials. The intermediate course is often taught by instructors who use their own notes in conjunction with either optional topics from a textbook designed primarily to teach the introductory course or selected topics from a book primarily designed for a graduate course on advanced mechanics of materials. The pedagogical needs of the students with respect to theoretical details, numerical examples, and posttext problems are difficult to meet and put undue burden on the instructor teaching the course—such has been my experience in the past twenty years. This book is designed to provide educational material for a second course on the mechanics of materials taught to juniors or seniors.

Although consistent in its design and notation with my introductory mechanics of materials book, this book does not depend upon the book used for an introductory course. There are many pedagogical features to help students meet the educational objectives. The Note to the Student describes some of the features that address their pedagogical needs.

I welcome any comments, suggestions, concerns, or corrections that will help me improve the book. Readers may relay their input to the publisher or to me. My e-mail address is mavable@mtu.edu.

ACKNOWLEDGMENTS

I am indebted to Professor I. Miskioglu for his friendship and for reviewing parts of the manuscript, even on very short notice.

Thanks to Professor G. Jayaraman for using this book in note form and providing valuable input and corrections.

Thanks to my graduate student Mr. JaiHind Maddi for his help with the solutions manual and to Michigan Technological University for their support.

Thanks to the following and other anonymous reviewers whose comments have significantly improved this book:

Professor Lawrence Agbezuge of Rochester Institute of Technology
Professor Ron Averill of Michigan State University
Professor Paul E. Barbone of Boston University
Professor Mark E. Barkey of University of Alabama
Professor David M. Barnett of Stanford University
Professor Donald M. Blackketter of University of Idaho
Professor Aaron S. Budge of Minnesota State University
Professor Jack Chessa of University of Texas
Professor Mark Garnich of University of Wyoming
Professor Ronald U. Goulet of University of Tennessee
Professor Abhijit Gupta of Northern Illinois University
Professor Stephen M. Heinrich of Marquette University
Professor Mohammad Mahinfalah of North Dakota State University
Professor Robert Rizza of Milwaukee School of Engineering
Professor Laster W. Schmerr Jr. of Iowa State University
Professor Scott Short of Northern Illinois University
Professor Ziheng Yao of West Virginia University

Thanks to the following whose work makes this book beautiful and professional:

Karen Shapiro, Managing Editor, Editorial/Design/Production
Annika Sarin, Designer
Rachel Perkins, Design Associate
Brenda Griffing, Copyeditor
Trent Haywood, Senior Copywriter

Dawn Stapleton, Associate Editor, Engineering
Adriana Hurtado, Editorial Assistant, Engineering

Thanks to Danielle Christensen, sponsoring editor of this book for her friendship, help, and support throughout the various stages of the book publication process.

To my children, Anusha and Adhiraj, who are such a joy, I am sorry I could not incorporate your suggestion for the title of this book 'An exciting sequel.'

NOTE TO THE STUDENT

The following features, in particular, should help you meet the learning objectives of the book.

- Appendix A briefly reviews some of the concepts from the prerequisite course on introductory mechanics of materials. Other brief reviews, particularly in the first three chapters, are introduced in the text before new concepts are built on the introductory material.

- Appendix B reviews basic matrix algebra that is needed in the book.

- All internal forces and moments are in ***bold italics,*** emphasizing their difference from external forces and moments.

- Every chapter starts with the section titled *Overview,* which describes the motivation for studying the chapter and the major learning objective(s) in the chapter.

- Every chapter ends with the section titled *Closure,* which highlights the important points and concepts studied in the chapter.

- Every example statement is followed by a section called *Plan* and ends with *Comments.* Developing a *plan* before solving a problem is essential for the development of analysis skills. The *comments* are observations deduced from the example, highlighting concepts discussed in the text that precedes the example.

- On the inside back cover of the textbook is a *formula sheet* for easy reference. To give your instructor the option of permitting the use of the formula sheet in an exam, there is no explanation of the variables or the equations.

NOTE TO THE INSTRUCTOR

The best way of showing how the presentation in this book meets the objectives stated in the Preface is to draw your attention to specific features. This note also gives my own instructor's perspective on topics covered in each chapter.

In the introductory course on the mechanics of materials, the students learn the theories for axial rods, torsion of circular shafts, and symmetric bending of beams. The derivation of all three theories is presented in a consolidated form as a synopsis in Table 3.3 (on pages 113–115), which highlights the commonality in the three theories and the modular character that is depicted in Figure 3.2 (on page 104). The four links connecting the five variables shown in Figure 3.2 are kinematic equations relating displacements and strains, the constitutive equations relating strains and stresses, the equivalencies between stresses and internal forces, and the equilibrium equations relating internal and external forces. Any changes to the assumptions in one module affect only the equations in that module; the equations in other modules remain unchanged. With this view, the beam vibration equations are a simple modification of equilibrium equations as demonstrated in Example 3.6 on page 122. Similarly, inclusion of dynamic terms in axial members (Problem 3.25) and the torsion of shafts (Problem 3.28) and the foundation effects in beams on elastic foundations (Problem 3.23) are simple modifications of equilibrium equations and are given as posttext problems. In Example 3.7 the derivation of equations governing the deformation of the Timoshenko beam is demonstrated as a change in kinematics, while all other equations remain the same; although we carry a new set of variables, the process of moving from one step to the next remains the same as in the derivation of the elementary theories highlighted in Table 3.3. Having demonstrated the modularity in the derivation of theories and how complexities are incorporated, the theories on composite structural members (Chapter 4), inelastic structural behavior (Chapter 5), and thin-walled structural members (Chapter 6) can be obtained by modifications as described in the Overview of each chapter.

Based on student performance and feedback, I believe that the foregoing presentation consolidates what the student learned in the introductory course, and that the repetitive and the compact character of the derivations both helps in the understanding and retention of the key ideas and exposes the students to a vast array of complexities in the derivation of theories of one-dimensional structural members. It is not my intention to convey the impression that students understand all the implications of all the complexities they see in this book. For greater understanding of complexities, time has to be spent with the application of the solution to the equations, as is done in the courses that will use these theories. I believe, however, that students learn to appreciate the mechanics of incorporating complexities into the elementary theory of one-dimensional structures they learned in the introductory course of mechanics of materials.

Chapters 1 through 3 briefly review introductory mechanics of materials, introduce notation, and then introduce new concepts that build on what the student already knows. Students in the introductory course have seen stress and strain transformation in two dimensions. Familiar conclusions from two dimensions are derived by using the matrix method, and the matrix method approach is generalized for use in stress and strain transformation in three dimensions. The familiar generalized Hooke's law is discussed as a subclass of linear material models in Chapter 2. One of the conclusions that is observed is that the principal direction of stresses and strains is the same only for isotropic materials; for materials of other types, such as orthotropic materials, the principal directions for stresses and strains are different. Stress concentration factors and stress intensity factors are introduced as a means of extrapolating nominal stress results from elementary structural theories into regions of stress concentration and vicinity of cracks. Failure theories and fatigue are introduced, and are used along with stress concentration factors and stress intensity factors during a review of axial, torsion, and bending problems in Chapter 3. Shear stresses in bending are covered in more detail than is needed in a review, as this topic is critical in the determination of shear centers and because many students in the introductory course struggle with it. Discontinuity functions, introduced in Chapter 3, are used throughout the book for statically determinate and indeterminate axial, torsion, and bending problems.

Chapter 7 covers the classical energy methods in detail. Once more, concepts are elaborated by using one-dimensional structural members. Chapter 8 introduces the basic equations in elasticity and its relationship to mechanics of materials. Elasticity is also used to systematically obtain results for thick cylinders, thin rotating disks, and torsion of noncircular shafts, all within the capability of an undergraduate student. In Chapter 9, the Rayleigh-Ritz method introduced in Chapter 7 is used to formulate a procedure for applying the finite element method. The terminology used in the finite element method is introduced, and the method is applied to solve simple axial, torsion, and bending problems for which analytical solutions can be obtained.

The book requires elementary knowledge of partial derivatives, some matrix algebra, and simple calculus, which most undergraduate students have been exposed to in their curriculum. Though the mathematics is not involved, the algebra can be tedious. Upper-end calculators and symbolic manipulators can reduce the tedium, but each instructor will have to define what is permissible.

The book has more material than can be covered in a fifteen-week, three-credit course to accommodate the pace and choice of topics of individual instructions. A sample syllabus, lecture slides, and sample exams that I use with this book are posted as pdf files for downloading on my personal web page.[1] All these, along with a solution manual, will also be available to the instructors through the publisher.

---

[1] University network changes sometimes change web addresses. I will maintain a personal web page with a link for the educational material for *Intermediate Mechanics of Materials.* The current web address is http://www.me.mtu.edu/%7Emavable/

1 Stress and Strain

1.0 OVERVIEW

The roof of a building collapses because it was not designed to be strong enough to withstand the snow loads in northern climates, as shown in Figure 1.1(a). The lid of a sauce bottle cannot be opened because the bond between the bottle and the lid is too strong. A crankshaft vibrates and causes noise because it is not stiff enough. The diver in Figure 1.1(b) cannot jump high enough to execute her dives because the diving board was made too stiff. Strength and stiffness are both important in the design of structural elements. Stress, a measure of the intensity of internal forces, and strain, a measure of the intensity of deformation, are two fundamental variables in the mechanics of materials that are used in assessing of strength and stiffness.

(a)   (b)

Figure 1.1 Examples of inadequacy in (a) strength and (b) stiffness. [*Photos courtesy of NOAA (a) and U.S. Navy (b): Photographer Mate Second Class Jsoloh Sellers III.*]

1

In this chapter we start by briefly reviewing the concepts of stress and strain at a point. The notation of double subscripts on stresses and strain is described. Of particular importance is the use of the subscripts in determining the direction of a stress component on a surface. Strains, which are derivatives of displacements, can be approximately evaluated by means of finite difference approximation, a numerical technique discussed in this chapter.

Analysis of forces and deformation, hence of stresses and strains, is conducted in a coordinate system that is chosen to simplify analysis. But the maximum stresses and strains that may cause failure in a material may exist on planes and in directions that are different from the chosen coordinate system. Stress and strain transformation equations provide a means of moving from one coordinate system to another. In this chapter we start with the equations of stress and strain transformation in two dimensions that were studied in the introductory course on the mechanics of materials. We will cast these familiar equations in matrix form and use the matrix method to derive familiar conclusions in two dimensions. We will then generalize the matrix method approach for use with stress and strain transformation in three dimensions.

The learning objectives in this chapter are:

1. To understand the concept of stress and the use of double subscripts in determining the direction of stress components on a surface.
2. To understand the concept of strain and the use of small-strain and finite difference approximations.
3. To understand stress and strain transformations in three dimensions.

1.1 STRESS ON A SURFACE

The forces of attraction and repulsion between two particles (atoms or molecules) in a body are assumed to act along the line that joins the two particles.[1] The forces vary inversely as an exponent of the radial distance separating the two particles. Thus, every particle exerts a force on every other particle, as shown symbolically on an imaginary surface of a body in Figure 1.2(a). These forces between the particles hold the body together and are referred to as the internal forces. The shape of the body changes when we apply external forces on the body. The change in shape implies that the distance between the particles must change, which further implies that the forces between the particles (internal forces) must change. When the change in the internal forces exceeds some characteristic material value, the body will break. Thus, the strength of the material can be characterized by the measure of *change in the intensity* of internal forces. This measure of change in the intensity of internal forces is what we call stress.

We can replace all the forces that are exerted on any single particle in Figure 1.2(a) by the resultant of these forces on that particle as shown in Figure 1.2(b). The magnitude

---

[1] Forces that act along the line joining two particles are called *central forces*. The concept of central forces started with Newton's universal gravitation law: The force between two particles is inversely proportional to square of the radial distance between two particles and acts along the line joining the two particles. At atomic levels, the central forces vary not with the square of the radial distance but with an exponent, which is a power of 8 or 10.

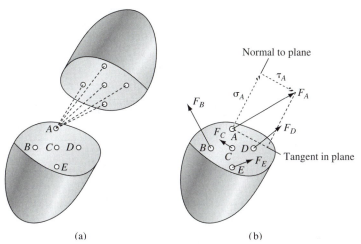

Figure 1.2 Internal forces between particles on the two sides of an imaginary cut. (a) Forces between particles in a body shown on particle A. (b) The resultant force on each particle.

and direction of these resultant forces will vary with the location of the particle (point), as shown in Figure 1.2(b). In other words, when external forces are applied, an internal distributed force system is generated in the material. It is the intensity of this internal distributed force system that we are trying to measure as stress. Furthermore, we note that force is a vector; hence the internal distributed forces (stress on a surface) can be resolved into normal (perpendicular to the surface) and tangential (parallel to the surface) distributed forces as shown in Figure 1.2(b). We summarize the description in this paragraph with the following definition.

Definition 1 The internal distributed forces on an imaginary cut surface of a body is called the *stress on a surface*. The internal distributed force that is normal to the surface of an imaginary cut is called the *normal stress* on a surface. The internal distributed force that is parallel to the surface of an imaginary cut surface is called the *shear stress* on the surface.

Table 1.1 shows the various units of stress used in this book. It should be noted that, *one psi* is equal to 6.95 kPa or *approximately 7 kPa*. Alternatively, 1 kPa is equal to 0.145 psi or *approximately 0.15 psi*.

TABLE 1.1 Units of Stress

Units	Description	Basic Units
psi	Pounds per square inch	lb/in^2
ksi	Kilopounds (kips) per square inch	$10^3 \, lb/in^2$
Pa	Pascals	N/m^2
kPa	Kilopascals	$10^3 \, N/m^2$
MPa	Megapascals	$10^6 \, N/m^2$
GPa	Gigapascals	$10^9 \, N/m^2$

Normal stress on a surface may be viewed as the internal forces that are developed owing to the material resistance to the *pulling apart or pushing together of two adjoining planes* of an imaginary cut. Like pressure, normal stress is always perpendicular to the surface of the imaginary cut. But unlike pressure, which can only be compressive, normal stress can be tensile.

> **Definition 2** A normal stress that pulls the surface away from a body is called a *tensile stress*. A normal stress that pushes the surface into a body is called a *compressive stress*.

In other words, tensile stress acts in the direction of the outward normal to the surface, while compressive stress is opposite to the direction of the outward normal to the surface. Normal stress is usually reported as tensile (T) or compressive (C), not as positive or negative. Thus, $\sigma = 100$ MPa (T) and $\sigma = 10$ ksi (C) are the conventional ways of reporting tensile and compressive normal stresses, respectively.

Shear stress on a surface may be viewed as the internal forces that are developed owing to the resistance of the material to the *sliding of two adjoining planes* along the imaginary cut. Like friction, shear stresses act tangentially to the plane in the direction opposite the impending motion of the surface. But unlike friction, shear stress is not related to the normal forces (stresses).

1.2   STRESS AT A POINT

The breaking of a structure starts at a point where the internal force intensity (i.e., stress) exceeds some material characteristic value. This implies that we need to refine our definition of "stress on a surface" to "stress at a point." But an infinite number of planes (surfaces) can pass through a point. Which imaginary surface do we shrink to zero? When we shrink the surface area to zero, which internal force component should we use? Both difficulties can be addressed by assigning directions to the orientation of the imaginary surface and to the direction of the internal force on this surface and then carrying the description of the directions as the subscripts of the stress components just as we carried x, y, and z as subscripts to describe the components of a vector.

Figure 1.3 shows a body cut by an imaginary plane that has an outward normal in the i direction. On this surface we have a differential area ΔA_i on which a resultant

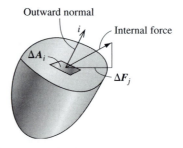

Figure 1.3 Stress at a point.

$$[\sigma] = \begin{bmatrix} \sigma_{xx} & \tau_{xy} & \tau_{xz} \\ \tau_{yx} & \sigma_{yy} & \tau_{yz} \\ \tau_{zx} & \tau_{zy} & \sigma_{zz} \end{bmatrix}$$

Figure 1.4 Stress matrix in three dimensions.

force[2] acts. The component of the force in the j direction is ΔF_j. A component of average stress is $(\Delta F_j/\Delta A_i)$. If we shrink ΔA_i to zero, we get the definition of a stress component at a point as shown by Equation (1.1).

$$\sigma_{ij} = \lim_{\Delta A_i \to 0} \left(\frac{\Delta F_j}{\Delta A_i} \right)$$

Direction of outward normal to the imaginary cut surface

Direction of the internal force component

(1.1)

Now when we look at a stress component, the first subscripts tells us the orientation of the imaginary surface and the second subscript tells us the direction of the internal force.

In three dimensions, each subscript i and j can refer to the x, y, or z direction. In other words, there are nine possible combinations of the two subscripts, as shown in the stress matrix in Figure 1.4. The diagonal elements in the stress matrix are normal stresses, and all off-diagonal elements in the stress matrix represent shear stresses.

Note that to specify the stress at a point, we must have a magnitude and two directions. Table 1.2 shows the number of components needed to specify a scalar, a vector, and a stress. Now force, moment, velocity, and acceleration are all different quantities, but all are called vectors. In a similar manner, stress belongs to a category called *tensors*,[3] or more specifically, *stress is a second-order tensor*,[4] where "second order" refers to the exponent in the last row. In this terminology, a vector is a tensor of order one, and a scalar is a tensor of order zero.

TABLE 1.2   Comparison of Number of Components

Quantity	Dimensions		
	One	Two	Three
Scalar	$1 = 1^0$	$1 = 2^0$	$1 = 3^0$
Vector	$1 = 1^1$	$2 = 2^1$	$3 = 3^1$
Stress	$1 = 1^2$	$4 = 2^2$	$9 = 3^2$

[2] If a resultant moment is included, then the stress is referred to as couple stress. Couple stress is important if stress analysis is conducted at such a small scale that the moment transmitted by the bonds between molecules must be incorporated. See Fung [1965] in Appendix D for additional details.

[3] Tensor calculus like vector arithmetic is a branch of mathematics. See Synge and Schild [1978] in Appendix D for additional details.

[4] To be labeled as tensor, a quantity must also satisfy certain coordinate transformation properties, discussed briefly in Section 1.3.

In the introductory course on the mechanics of materials, it was explained that shear stress is symmetric, as shown by Equations (1.2a), (1.2b), and (1.2c).

$$\tau_{xy} = \tau_{yx} \tag{1.2a}$$

$$\tau_{yz} = \tau_{zy} \tag{1.2b}$$

$$\tau_{zx} = \tau_{xz} \tag{1.2c}$$

The symmetry of shear stress given by Equations (1.2) implies that *stress at a point has nine stress components, but only six are independent*. Similarly, in two dimensions stress at a point has four nonzero components in general, but only three are independent.

1.2.1 Sign Convention for Stress

Definition 3 We will consider ΔA_i to be positive if the outward normal to the surface is in the positive *i* direction. If the outward normal is in the negative *i* direction, then ΔA_i will be considered to be negative.

With this convention in mind, we can immediately deduce the sign for stress. A stress component can be positive in two ways: both the numerator and denominator are positive in Equation (1.1), or both the numerator and denominator are negative in Equation (1.1). This leads to the following set of definitions.

Definition 4 A *positive* stress component multiplied by a surface area that has an outward normal in the positive direction produces an internal force in the positive direction.

or

A *positive* stress component multiplied by a surface area that has an outward normal in the negative direction produces an internal force in the negative direction.

We conclude this section with the following points to remember.

- Stress is an internal quantity that has units of force per unit area.
- A stress component at a *point* is specified by magnitude and two directions (i.e., stress is a second-order tensor). The first subscript on stress gives the direction of the outward normal of the imaginary cut surface. The second subscript gives the direction of the internal force.
- The sign of a stress component is determined from the direction of the internal force and the direction of the outward normal to the imaginary cut surface.
- Stress on a *surface* needs a magnitude and only one direction to specify it (i.e., stress on a surface is a vector).

1.2.2 Stress Elements

The discussion in the preceding section shows that stress at a point is an abstract quantity, and developing an intuitive feel for it is difficult. Stress on a surface, however, is easier to visualize as a distributed force on a surface.

> **Definition 5** A stress element is an imaginary object that helps us visualize stress at a point by allowing us to construct surfaces that have outward normals in the directions of the coordinates.

It shall be seen in the discussion that follows that a stress element is a cube in Cartesian coordinates, a fragment of a cylinder in cylindrical coordinates, and a fragment of a sphere in spherical coordinates. We start our discussion with the construction of a stress cube to emphasize the basic principles of the construction of stress elements. We can use a similar process to draw stress elements in cylindrical and spherical coordinate systems.[5]

1.2.3 Construction of a Stress Cube

Consider the point at which we want to describe stress. Around this point, imagine an object that has sides parallel to the coordinate system. A cube has six surfaces with outward normals that are either in the positive or negative coordinate direction. Thus, we have accounted for the first subscript in our stress definition. We know that force is in the positive or negative direction of the second subscript. We use our sign convention to show the stress in the direction of the force on each of the six surfaces.

To demonstrate the construction just described, we will assume that all nine stress components shown in the stress matrix in Figure 1.4 are positive.

Let us consider the first row in the stress matrix in Figure 1.4. The first subscript gives us the direction of the outward normal, which is the *x* direction. Surfaces *A* and *B* in Figure 1.5 have the outward normal in the *x* direction, and it is on these surfaces that the stress component of the first row will be shown.

The outward normal on surface *A* is in the positive *x* direction [denominator is positive in Equation (1.1)]. For the stress component to be positive on surface *A*, the force must be in the positive direction (numerator must be positive), as shown in Figure 1.5.

The outward normal on surface *B* is in the negative *x* direction [denominator is negative in Equation (1.1)]. For the stress component to be positive on surface *B*, the force must be in the negative direction (numerator must be negative), as shown in Figure 1.5.

Let us now consider the second row in the stress matrix in Figure 1.4. From the first subscript, we know that the normal to the surface is in the *y* direction. Surface *C* has an outward normal in the positive *y* direction; therefore all forces on surface *C* are in the positive direction of the second subscript, as shown in Figure 1.5. Surface *D* has an outward normal in the negative *y* direction. Therefore, all forces on surface *D* are in the negative direction of the second subscript, as shown in Figure 1.5.

---

[5] See Example 1.2 for treatment of a stress element in spherical coordinates.

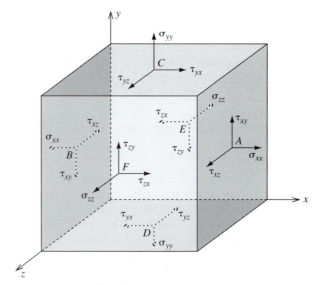

Figure 1.5 Stress cube showing all positive stress components.

The components of the third row in the stress matrix are shown on surfaces E and F in Figure 1.5, in accordance with the foregoing logic.

The positive normal stress components (e.g., σ_{xx}) are pulling the cube in opposite directions; that is, the cube is in tension owing to a positive normal stress component. As mentioned earlier, normal stresses are reported as tension or compression, not as positive or negative.

Figure 1.5 shows that if we consider the symmetric pair of shear stress components, then this pair either points toward a corner or away from it—this observation can be used in drawing shear stresses on the three surfaces of the stress cube once the shear stress on one of its surfaces has been drawn.

1.2.4 Plane Stress

Plane stress is one of the two types of two-dimensional simplification in the mechanics of materials. In Section 1.6.1, we will study plane strain, the other type of two-dimensional simplification, and in Section 2.2, we will see the difference between the two types of simplification. By "two-dimensional" we imply that one of the coordinates does not play a role in the description of the problem. If we choose z to be the coordinate, we set all stresses with subscript z to be zero. Figure 1.6(a) shows the stress matrix in plane stress, Figure 1.6(b) the associated stress cube for the point, and Figure 1.6(c) the simplified two-dimensional representation of the stress cube as viewed from the z axis. We note that the plane with outward normal in the z direction is stress free. *Stress-free* surfaces are also called *free surfaces*.

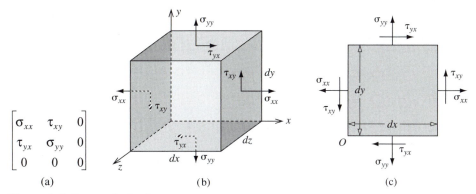

$$\begin{bmatrix} \sigma_{xx} & \tau_{xy} & 0 \\ \tau_{yx} & \sigma_{yy} & 0 \\ 0 & 0 & 0 \end{bmatrix}$$

(a)   (b)   (c)

Figure 1.6 Stress cube in plane stress.

EXAMPLE 1.1

Show the nonzero stress components on surfaces *A*, *B*, and *C* of two cubes represented in Figure 1.7 for the following coordinate systems:

$$\begin{bmatrix} \sigma_{xx} = 80 \text{ MPa (T)} & \tau_{xy} = 30 \text{ MPa} & \tau_{xz} = -70 \text{ MPa} \\ \tau_{yx} = 30 \text{ MPa} & \sigma_{yy} = 0 & \tau_{yz} = 0 \\ \tau_{zx} = -70 \text{ MPa} & \tau_{zy} = 0 & \sigma_{zz} = 40 \text{ MPa (C)} \end{bmatrix}$$

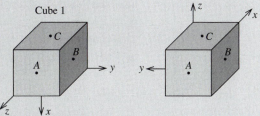

Figure 1.7 Cubes in different coordinate systems.

PLAN

We can identify the surface with the outward normal in the direction of the first subscript. If the outward normal is in the positive coordinate direction, then the denominator in Equation (1.1) is positive; otherwise it is negative. Knowing the sign of the denominator and the sign of stress component, we can determine the sign of the internal force. We can show the internal force on the identified surface in the positive coordinate direction of the second subscript if the sign of the internal force is positive and in the negative coordinate direction if sign of the internal force is negative.

SOLUTION (Figure 1.8)

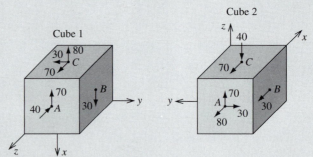

Figure 1.8 Solution of Example 1.1.

Cube 1 The first subscript of σ_{xx}, τ_{xy}, and τ_{xz} shows that the outward normal is in the x direction; hence these components will be shown on surface C. The outward normal on surface C is in the *negative x* direction; hence the denominator in Equation (1.1) is negative. Therefore:

- the internal force must be in the *negative x* direction to produce a *positive* (tensile) σ_{xx},
- the internal force must be in the *negative y* direction to produce a *positive* τ_{xy},
- the internal force must be in the *positive z* direction to produce a *negative* τ_{xz}.

The first subscript of τ_{yx} shows that the outward normal is in the y direction, and hence this component will be shown on surface B. The outward normal on surface B is in the *positive y* direction; hence the denominator in Equation (1.1) is positive. Therefore:

- the internal force must be in the *positive x* direction to produce a *positive* τ_{yx}.

The first subscript of τ_{zx}, σ_{zz} shows that the outward normal is in the z direction, and hence these components, will be as shown on surface A. The outward normal on surface A is in the *positive z* direction; hence the denominator in Equation (1.1) is positive. Therefore:

- the internal force must be in the *negative x* direction to produce a *negative* τ_{zx},
- the internal force must be in the *negative z* direction to produce a *negative* (compressive) σ_{zz}.

Cube 2 The first subscript of σ_{xx}, τ_{xy}, and τ_{xz} shows that the outward normal is in the x direction, and hence these components will be shown on surface A. The outward normal on surface A is in the *negative x* direction, and hence the denominator in Equation (1.1) is negative. Therefore:

- the internal force must be in the *negative x* direction to produce a *positive* (tensile) σ_{xx},
- the internal force must be in the *negative y* direction to produce a *positive* τ_{xy},
- the internal force must be in the *positive z* direction to produce a *negative* τ_{xz}.

The first subscript of τ_{yx} shows that the outward normal is in the y direction, and hence this component will be as shown on surface B. The outward normal on surface B is in the *negative y* direction; hence the denominator in Equation (1.1) is negative. Therefore:

- the internal force must be in the *negative y* direction to produce a *positive* τ_{yx}.

The first subscript of τ_{zx}, σ_{zz} shows that the outward normal is in the z direction, and hence these components will be as shown on surface C. The outward normal on surface C is in the *positive z* direction, and hence the denominator in Equation (1.1) is positive. Therefore:

- the internal force must be in the *negative x* direction to produce a *negative* τ_{zx},
- the internal force must be in the *negative z* direction to produce a *negative* (compressive) σ_{zz}.

COMMENTS

1. In drawing the normal stresses, we could have made use of the fact that σ_{xx} is tensile, hence pulls the surface outward, and σ_{zz} is compressive, hence pushes the surface inward, which is a quicker way of getting the direction of these stress components than the arguments based on signs and subscripts.

2. In drawing the shear stresses, we could have made use of the fact that a symmetric pair of shear stress components points either toward a corner or away from it. Thus, having drawn τ_{xz} on surface C in cube 1 pointing toward the corner edge, we could have drawn τ_{zx} on surface A upward, as then it too would point toward the corner edge. Similarly, having drawn τ_{xy} on surface C in cube 1 pointing away from the corner edge, we can draw τ_{yx} on surface B going downward, as then it too would point away from the corner edge.

EXAMPLE 1.2

Figure 1.9 shows the positive stress components on a stress element drawn in the following spherical coordinate system:

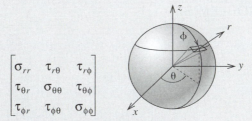

$$\begin{bmatrix} \sigma_{rr} & \tau_{r\theta} & \tau_{r\phi} \\ \tau_{\theta r} & \sigma_{\theta\theta} & \tau_{\theta\phi} \\ \tau_{\phi r} & \tau_{\phi\theta} & \sigma_{\phi\phi} \end{bmatrix}$$

Figure 1.9 Stresses in spherical coordinates.

PLAN

We can construct a stress element with surfaces that have outward normal in the r, θ, and ϕ direction. The first subscript will identify the surface on which the row of stress components is to be shown. The second subscript then will show the direction of the stress component on the surface.

SOLUTION

We draw the stress element with lines in the directions of r, θ, and ϕ as shown in Figure 1.10(a).

- The stresses σ_{rr}, $\tau_{r\theta}$, and $\tau_{r\phi}$ will be on surface A in Figure 1.10(b).
- The outward normal on surface A is in the *positive r* direction, thus the forces must be in the *positive r*, θ, and ϕ directions to result in *positive* σ_{rr}, $\tau_{r\theta}$, and $\tau_{r\phi}$.
- The stresses $\tau_{\theta r}$, $\sigma_{\theta\theta}$, and $\tau_{\theta\phi}$ will be on surface B in Figure 1.10(b).
- The outward normal on surface B is in the *negative* θ direction. Thus the force must be in the *negative r*, θ, and ϕ directions to result in positive $\tau_{\theta r}$, $\sigma_{\theta\theta}$, and $\tau_{\theta\phi}$.
- The stresses $\tau_{\phi r}$, $\tau_{\phi\theta}$, and $\sigma_{\phi\phi}$ will be on surface C in Figure 1.10(b).
- The outward normal on surface C is in the *positive* ϕ direction. Thus, the force must be in the *positive r*, θ, and ϕ directions to result in positive $\tau_{\phi r}$, $\tau_{\phi\theta}$, and $\sigma_{\phi\phi}$.

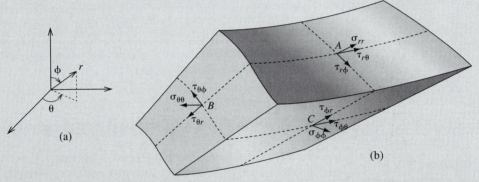

(a)

(b)

Figure 1.10 Stress element in spherical coordinates.

PROBLEM SET 1.1

1.1 Show the nonzero stress components on the A, B, and C faces of the cube shown in Figure P1.1.

$$\begin{bmatrix} \sigma_{xx} = 100 \text{ MPa (T)} & \tau_{xy} = 200 \text{ MPa} & \tau_{xz} = -125 \text{ MPa} \\ \tau_{yx} = 200 \text{ MPa} & \sigma_{yy} = 175 \text{ MPa (C)} & \tau_{yz} = 225 \text{ MPa} \\ \tau_{zx} = -125 \text{ MPa} & \tau_{zy} = 225 \text{ MPa} & \sigma_{zz} = 150 \text{ MPa (C)} \end{bmatrix}$$

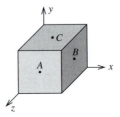

Figure P1.1

1.2 Show the nonzero stress components on the *A*, *B*, and *C* faces of the cube shown in Figure P1.2.

$$\begin{bmatrix} \sigma_{xx} = 90 \text{ MPa (C)} & \tau_{xy} = -200 \text{ MPa} & \tau_{xz} = 0 \\ \tau_{yx} = -200 \text{ MPa} & \sigma_{yy} = 175 \text{ MPa (C)} & \tau_{yz} = 225 \text{ MPa} \\ \tau_{zx} = 0 & \tau_{zy} = 225 \text{ MPa} & \sigma_{zz} = 150 \text{ MPa (C)} \end{bmatrix}$$

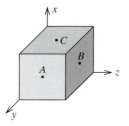

Figure P1.2

1.3 Show the nonzero stress components on the *A*, *B*, and *C* faces of the cube shown in Figure P1.3

$$\begin{bmatrix} \sigma_{xx} = 0 & \tau_{xy} = -15 \text{ ksi} & \tau_{xz} = 0 \\ \tau_{yx} = -15 \text{ ksi} & \sigma_{yy} = 10 \text{ ksi (C)} & \tau_{yz} = 25 \text{ ksi} \\ \tau_{zx} = 0 & \tau_{zy} = 25 \text{ ksi} & \sigma_{zz} = 20 \text{ ksi (T)} \end{bmatrix}$$

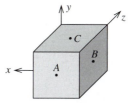

Figure P1.3

1.4 Show the nonzero stress components on the *A*, *B*, and *C* faces of the cube shown in Figure P1.4.

$$\begin{bmatrix} \sigma_{xx} = 0 & \tau_{xy} = -15 \text{ ksi} & \tau_{xz} = 0 \\ \tau_{yx} = -15 \text{ ksi} & \sigma_{yy} = 10 \text{ ksi (C)} & \tau_{yz} = 25 \text{ ksi} \\ \tau_{zx} = 0 & \tau_{zy} = 25 \text{ ksi} & \sigma_{zz} = 20 \text{ ksi (T)} \end{bmatrix}$$

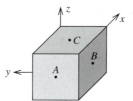

Figure P1.4

1.5 Show the nonzero stress components on the *A*, *B*, and *C* faces of the cube shown in Figure P1.5.

$$\begin{bmatrix} \sigma_{xx} = 70 \text{ MPa (T)} & \tau_{xy} = -40 \text{ MPa} & \tau_{xz} = 0 \\ \tau_{yx} = -40 \text{ MPa} & \sigma_{yy} = 85 \text{ MPa (C)} & \tau_{yz} = 0 \\ \tau_{zx} = 0 & \tau_{zy} = 0 & \sigma_{zz} = 0 \end{bmatrix}$$

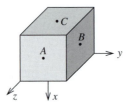

Figure P1.5

1.6 Show the nonzero stress components on the *A*, *B*, and *C* faces of the cube shown in Figure P1.6.

$$\begin{bmatrix} \sigma_{xx} = 70 \text{ MPa (T)} & \tau_{xy} = -40 \text{ MPa} & \tau_{xz} = 0 \\ \tau_{yx} = -40 \text{ MPa} & \sigma_{yy} = 85 \text{ MPa (C)} & \tau_{yz} = 0 \\ \tau_{zx} = 0 & \tau_{zy} = 0 & \sigma_{zz} = 0 \end{bmatrix}$$

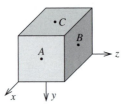

Figure P1.6

1.7 Show the stress components of a point in plane stress on the square in Figure P1.7.

$$\begin{bmatrix} \sigma_{xx} = 100 \text{ MPa (T)} & \tau_{xy} = -75 \text{ MPa} \\ \tau_{yx} = -75 \text{ MPa} & \sigma_{yy} = 85 \text{ MPa (T)} \end{bmatrix}$$

Figure P1.7

1.8 Show the stress components of a point in plane stress on the square in Figure P1.8.

$$\begin{bmatrix} \sigma_{xx} = 100 \text{ MPa (T)} & \tau_{xy} = -75 \text{ MPa} \\ \tau_{yx} = -75 \text{ MPa} & \sigma_{yy} = 85 \text{ MPa (T)} \end{bmatrix}$$

Figure P1.8

1.9 Show the stress components of a point in plane stress on the square in Figure P1.9.

$$\begin{bmatrix} \sigma_{xx} = 27 \text{ ksi (C)} & \tau_{xy} = 18 \text{ ksi} \\ \tau_{yx} = 18 \text{ ksi} & \sigma_{yy} = 85 \text{ ksi (T)} \end{bmatrix}$$

Figure P1.9

1.10 Show the stress components of a point in plane stress on the square in Figure P1.10.

$$\begin{bmatrix} \sigma_{xx} = 27 \text{ ksi (C)} & \tau_{xy} = 18 \text{ ksi} \\ \tau_{yx} = 18 \text{ ksi} & \sigma_{yy} = 85 \text{ ksi (T)} \end{bmatrix}$$

Figure P1.10

1.11 Show the nonzero stress components in the r-θ-x cylindrical coordinate system on the A, B, and C faces of the stress element shown in Figure P1.11.

$$\begin{bmatrix} \sigma_{rr} = 145 \text{ MPa (C)} & \tau_{r\theta} = 100 \text{ MPa} & \tau_{rx} = -125 \text{ MPa} \\ \tau_{\theta r} = 100 \text{ MPa} & \sigma_{\theta\theta} = 160 \text{ MPa (T)} & \tau_{\theta x} = 165 \text{ MPa} \\ \tau_{xr} = -125 \text{ MPa} & \tau_{x\theta} = 165 \text{ MPa} & \sigma_{xx} = 150 \text{ MPa (T)} \end{bmatrix}$$

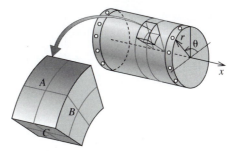

Figure P1.11

1.12 Show the nonzero stress components in the r-θ-x cylindrical coordinate system on the A, B, and C faces of the stress element shown in Figure P1.12.

$$\begin{bmatrix} \sigma_{rr} = 10 \text{ ksi (C)} & \tau_{r\theta} = 22 \text{ ksi} & \tau_{rx} = 32 \text{ ksi} \\ \tau_{\theta r} = 22 \text{ ksi} & \sigma_{\theta\theta} = 0 & \tau_{\theta x} = 25 \text{ ksi} \\ \tau_{xr} = 32 \text{ ksi} & \tau_{x\theta} = 25 \text{ ksi} & \sigma_{xx} = 20 \text{ ksi (T)} \end{bmatrix}$$

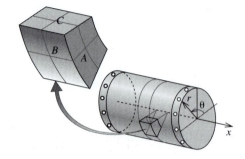

Figure P1.12

1.13 Show the stress components of a point in plane stress on the stress element shown in Figure P1.13 in polar coordinates.

$$\begin{bmatrix} \sigma_{rr} = 125 \text{ MPa (T)} & \tau_{r\theta} = -65 \text{ MPa} \\ \tau_{\theta r} = -65 \text{ MPa} & \sigma_{\theta\theta} = 90 \text{ MPa (C)} \end{bmatrix}$$

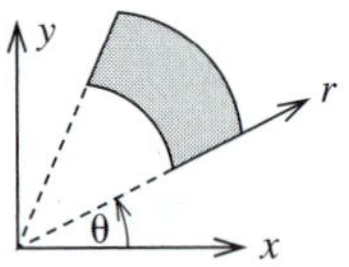

Figure P1.13

1.14 Show the stress components of a point in plane stress on the stress element shown in Figure P1.14 in polar coordinates.

$$\begin{bmatrix} \sigma_{rr} = 125 \text{ MPa (T)} & \tau_{r\theta} = -65 \text{ MPa} \\ \tau_{\theta r} = -65 \text{ MPa} & \sigma_{\theta\theta} = 90 \text{ MPa (C)} \end{bmatrix}$$

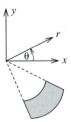

Figure P1.14

1.15 Show the stress components of a point in plane stress on the stress element shown in Figure P1.15 in polar coordinates.

$$\begin{bmatrix} \sigma_{rr} = 25 \text{ ksi (C)} & \tau_{r\theta} = 12 \text{ ksi} \\ \tau_{\theta r} = 12 \text{ ksi} & \sigma_{\theta\theta} = 18 \text{ ksi (T)} \end{bmatrix}$$

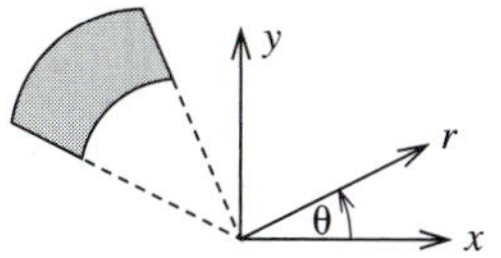

Figure P1.15

1.16 Show the stress components of a point in plane stress on the stress element shown in Figure P1.16 in polar coordinates.

$$\begin{bmatrix} \sigma_{rr} = 25 \text{ ksi (C)} & \tau_{r\theta} = 12 \text{ ksi} \\ \tau_{\theta r} = 12 \text{ ksi} & \sigma_{\theta\theta} = 18 \text{ ksi (T)} \end{bmatrix}$$

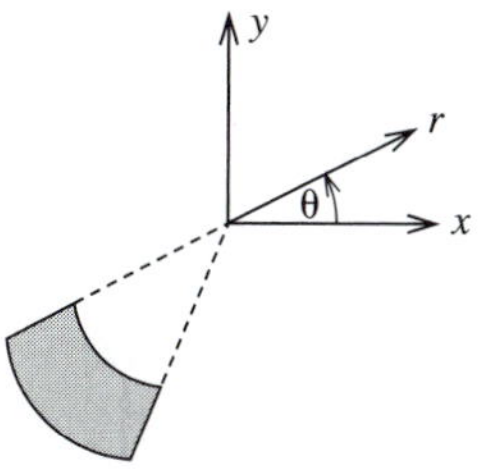

Figure P1.16

1.17 Show the nonzero stress components in the r-θ-ϕ spherical coordinate system on the A, B, and C faces of the stress-element shown in Figure P1.17.

$$\begin{bmatrix} \sigma_{rr} = 135 \text{ MPa (C)} & \tau_{r\theta} = 100 \text{ MPa} & \tau_{r\phi} = -125 \text{ MPa} \\ \tau_{\theta r} = 100 \text{ MPa} & \sigma_{\theta\theta} = 160 \text{ MPa (C)} & \tau_{\theta\phi} = 175 \text{ MPa} \\ \tau_{\phi r} = -125 \text{ MPa} & \tau_{\phi\theta} = 175 \text{ MPa} & \sigma_{\phi\phi} = 150 \text{ MPa (T)} \end{bmatrix}$$

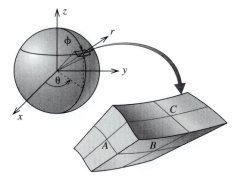

Figure P1.17

1.18 Show the nonzero stress components in the r-θ-ϕ spherical coordinate system on the A, B, and C faces of the stress element shown in Figure P1.18.

$$\begin{bmatrix} \sigma_{rr} = 0 & \tau_{r\theta} = -18 \text{ ksi} & \tau_{r\phi} = 0 \text{ ksi} \\ \tau_{\theta r} = -18 \text{ ksi} & \sigma_{\theta\theta} = 10 \text{ ksi (C)} & \tau_{\theta\phi} = 25 \text{ ksi} \\ \tau_{\phi r} = 0 & \tau_{\phi\theta} = 25 \text{ ksi} & \sigma_{\phi\phi} = 20 \text{ ksi (T)} \end{bmatrix}$$

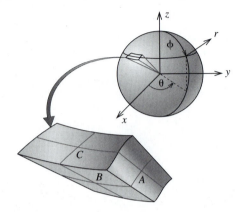

Figure P1.18

1.19 Show that the normal stress σ_{xx} on a surface can be replaced by equivalent internal normal force N and internal bending moments M_y and M_z as given in Equations (1.3) and shown in Figure P1.19.

$$N = \int_A \sigma_{xx} dA \tag{1.3a}$$

$$M_y = -\int_A z\sigma_{xx} dA \tag{1.3b}$$

$$M_z = -\int_A y\sigma_{xx} dA \tag{1.3c}$$

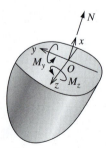

Figure P1.19

1.20 The normal stress on a cross section is given by $\sigma_{xx} = a + by$, where y is measured from the centroid of the cross section. If A is the cross-sectional area, I_{zz} is the area moment of inertia about the z axis, N and M_z, respectively, are the internal axial force and internal bending moment given by Equations (1.3a) and (1.3c), show the following:

$$\sigma_{xx} = \frac{N}{A} - \left(\frac{M_z}{I_{zz}}\right)y \tag{1.4}$$

We will see Equation (1.4) in combined axial and symmetric bending problems later in the book.

1.21 The normal stress on a cross section is given by $\sigma_{xx} = a + by + cz$, where y and z are measured from the centroid of the cross section. Use Equations (1.3a) and (1.3c) to show:

$$\sigma_{xx} = \frac{N}{A} - \left(\frac{M_z I_{yy} - M_y I_{yz}}{I_{yy} I_{zz} - I_{yz}^2} \right) y - \left(\frac{M_y I_{zz} - M_z I_{yz}}{I_{yy} I_{zz} - I_{yz}^2} \right) z \tag{1.5}$$

Equation (1.5) is used in the unsymmetric bending of beams. Note that if either y or z is an axis of symmetry, then $I_{yz} = 0$. In such a case, Equation (1.5) simplifies considerably.

1.22 In Figure P1.22, which shows an infinitesimal element in plane stress, F_x and F_y are the body forces acting at the point and have the dimensions of force per unit volume. By converting stresses shown in Figure P1.22 into forces and writing equilibrium equations, show

$$\frac{\partial \sigma_{xx}}{\partial x} + \frac{\partial \tau_{yx}}{\partial y} + F_x = 0 \tag{1.6a}$$

$$\frac{\partial \tau_{xy}}{\partial x} + \frac{\partial \sigma_{yy}}{\partial y} + F_y = 0 \tag{1.6b}$$

$$\tau_{xy} = \tau_{yx} \tag{1.6c}$$

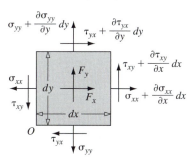

Figure P1.22

1.3   STRESS TRANSFORMATION IN TWO DIMENSIONS

In this section the stress transformation equations from the introductory course on the mechanics of materials are briefly reviewed and cast in matrix form. This gives another perspective on familiar concepts. In Section 1.4, this new perspective will be extended to stress transformation in three dimensions.

The relationship of stresses at *a point* in different coordinate systems, that is, transformation of stress components with coordinate systems, is called stress transformation. Stress transformation can also be viewed as relating stresses on different planes that pass through *a point*. The normals of the plane define the axis of a coordinate system to which we are transforming the stress components. We define two coordinate systems that we will use in this chapter.

Definition 6 The fixed reference coordinate system in which the entire problem is described is called the *global coordinate system*.

Definition 7 A coordinate system that can be fixed at any point on the body and has an orientation that is defined with respect to the global coordinate system is called the *local coordinate system*.

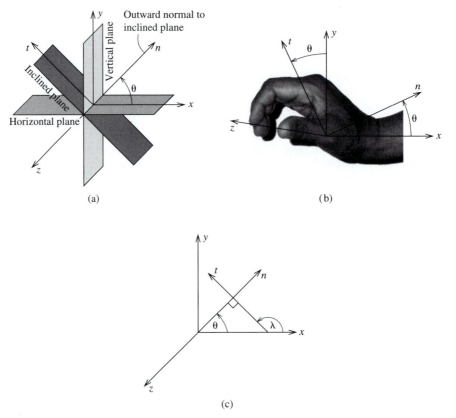

Figure 1.11 Local and global coordinate systems.

In this book the global coordinate system will most often be the *x-y-z* Cartesian coordinate system. Relating internal forces and moments to external forces and moments is usually done in a global coordinate system. The internal quantities are then used to obtain stresses in the global coordinate system.

We assume that the point is in plane stress, and we know the stresses in the *x-y-z* coordinate system, that is, on the horizontal and vertical planes shown in Figure 1.11(a). We seek to find stresses on inclined planes, which can be obtained by rotation about the *z* axis. Alternatively, we seek to find stresses in the *n-t-z* coordinate system, where *n* is the normal direction to a plane through that point as shown in Figure 1.11(a) and (c), and *t* is the tangent direction to the plane such that *n-t-z* is a right-handed coordinate system as shown in Figure 1.11(b).

In the introductory course on the mechanics of materials, the stresses in the *n-t-z* coordinate system were related to stresses in the Cartesian coordinate system by means of the wedge method, which is briefly elaborated here. Figure 1.12(a) shows the stresses acting on the vertical, horizontal, and inclined planes that compose the stress wedge. The stresses are converted to forces by multiplying by the areas of each plane on which the stresses act, as shown on the force wedge in Figure 1.12(b). By equilibrium of forces in the *n* and *t* directions, we obtain Equations (1.7a) and (1.7b).

$$\sigma_{nn} = \sigma_{xx}\cos^2\theta + \sigma_{yy}\sin^2\theta + 2\tau_{xy}\sin\theta\cos\theta \qquad (1.7a)$$

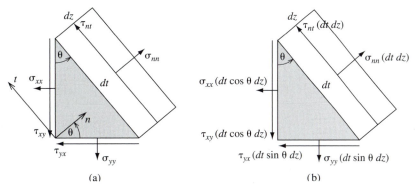

Figure 1.12 (a) Stress wedge. (b) Force wedge.

$$\tau_{nt} = -\sigma_{xx}\cos\theta\,\sin\theta + \sigma_{yy}\sin\theta\,\cos\theta + \tau_{xy}(\cos^2\theta - \sin^2\theta) \tag{1.7b}$$

By substituting $90 + \theta$ for θ in Equation (1.7a), we obtain the normal stress in the t direction as shown in Equation 1.7c.

$$\sigma_{tt} = \sigma_{xx}\sin^2\theta + \sigma_{yy}\cos^2\theta - 2\tau_{xy}\cos\theta\,\sin\theta \tag{1.7c}$$

On adding Equations (1.7a) and (1.7c), we obtain Equation (1.8).

$$\sigma_{nn} + \sigma_{tt} = \sigma_{xx} + \sigma_{yy} \tag{1.8}$$

Equation (1.8) implies that the sum of normal stresses in an orthogonal system is independent of coordinate transformation.

1.3.1 Matrix Method in Two Dimensions

Figure 1.11(c) shows the normal (n) and tangent (t) directions of a plane. The direction cosines of the unit vectors in the n and t directions can be written as in Equation (1.9a).

$$n_x = \cos\theta \qquad n_y = \sin\theta \qquad t_x = \cos\lambda \qquad t_y = \sin\lambda \tag{1.9a}$$

Noting that $\lambda = 90 + \theta$, we obtain Equation (1.9b).

$$t_x = -n_y \qquad t_y = n_x \tag{1.9b}$$

To develop stress transformation equations in matrix form, we introduce the matrix notation given in Equation (1.9c).

$$\{n\} = \begin{Bmatrix} n_x \\ n_y \end{Bmatrix} \qquad \{t\} = \begin{Bmatrix} t_x \\ t_y \end{Bmatrix} \qquad [\sigma] = \begin{bmatrix} \sigma_{xx} & \tau_{xy} \\ \tau_{yx} & \sigma_{yy} \end{bmatrix} \tag{1.9c}$$

The symmetry of shear stresses implies that the stress matrix is symmetric. In matrix notation, this symmetry is expressed by Equation (1.10),

$$[\sigma]^T = [\sigma] \tag{1.10}$$

where $[\ ]^T$ implies the transpose of the matrix.

In matrix notation, Equations (1.7a), (1.7b), and (1.7c) can be written (see Problems 1.23 and 1.24) as Equations (1.11a), (1.11b), and (1.11c),

$$\boxed{\sigma_{nn} = \{n\}^T [\sigma]\{n\}} \tag{1.11a}$$

$$\boxed{\tau_{nt} = \{t\}^T [\sigma]\{n\}} \tag{1.11b}$$

$$\boxed{\sigma_{tt} = \{t\}^T [\sigma]\{t\}} \tag{1.11c}$$

where $\{\ \}^T$ implies the transpose of the column. Equations (1.11a) through (1.11c) show that we can obtain a stress component by pre- and postmultiplying the stress matrix by the unit vectors in the direction of the subscripts. Because the shear stress is symmetric (see Problem 1.28), we could have premultiplied by the unit normal and postmultiplied by the unit tangent. In other words, the order of multiplication is immaterial in the calculation of shear stress.

Definition 8 Stress on a surface is called *traction* or a *stress vector.*

Note that stress at a point is a second-order tensor. When we specify a surface, the orientation of the surface is defined; hence the stress on the surface needs only one direction to specify, which implies that stress on a surface is a vector quantity. Mathematically the stress vector $\{S\}$ is defined as in Equation (1.12).

$$\boxed{\{S\} = [\sigma]\{n\}} \tag{1.12}$$

In long notation Equation (1.12) can be written as in Equations (1.13a) and (1.13b).

$$S_x = \sigma_{xx} n_x + \tau_{xy} n_y \tag{1.13a}$$

$$S_y = \tau_{yx} n_x + \sigma_{yy} n_y \tag{1.13b}$$

Since direction cosines do not have units, the stress vector has the units of stress. Thus, we now know three different quantities that have units of force per unit area:

- pressure, which is a scalar quantity,
- traction, which is a vector quantity,
- stress, which is a second-order tensor.

To better appreciate the concept of traction, we once more consider the force wedge shown in Figure 1.12(b). We represent the incline area $dt\,dz = dA$ and use the direction cosine notation of Equation (1.9a) to show the two-dimensional representation of the force wedge, as in Figure 1.13(a). The forces shown on the inclined plane could be resolved into forces in the x and y coordinates, as shown in Figure 1.13(b). By equilibrium of forces in Figure 1.13(b), we obtain Equations (1.13a) and (1.13b).

Alternatively, the stress vector $\{S\}$ could be written as in Equation (1.14).

$$\{S\} = \sigma_{nn}\{n\} + \tau_{nt}\{t\} \tag{1.14}$$

Equations (1.12) and (1.14) are representations of the vector $\{S\}$ in two different coordinate systems, as shown in Figure 1.14.

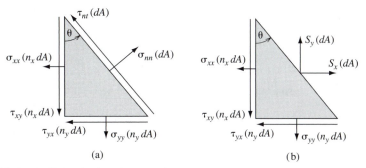

Figure 1.13 Statically equivalent force wedges.

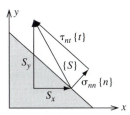

Figure 1.14 Stress vectors in different coordinate systems.

1.3.2 Principal Stresses

Recall the following definitions from the introductory course on the mechanics of materials.

Definition 9 Planes on which the shear stresses are zero are called the *principal planes*.

Definition 10 The normal direction to the principal planes is referred to as the *principal direction*, or the principal axis for stresses.

Definition 11 The angles the principal axis makes with the global coordinate system are called *principal angles*.

Definition 12 Normal stress on a principal plane is called *principal stress*.

Definition 13 The greatest principal stress is called *principal stress* 1.

In plane stress, the plane with outward normal in the z direction has zero shear stress. Therefore, by Definition 9, the z direction is a principal direction. Since there are three principal planes, there are three principal stresses at a point. The three principal stresses are labeled σ_1, σ_2, and σ_3.

Let the unit normal vector of a principal plane be given by $\{p\}$ and the corresponding principal stress be σ_p. By Definitions 9 and 12, the shear stress on the principal plane is zero and the normal stress is the principal stress. If $\{p\}$ is the normal direction in Equations (1.12) and (1.14), the vector $\{S\}$ can be written as Equation (1.15).

$$\{S\} = [\sigma]\{p\} = \sigma_p\{p\} \tag{1.15}$$

In long notation, Equation (1.15) can be written as follows:

$$\{S\} = \begin{bmatrix} \sigma_{xx} & \tau_{xy} \\ \tau_{yx} & \sigma_{yy} \end{bmatrix} \begin{Bmatrix} p_x \\ p_y \end{Bmatrix} = \begin{bmatrix} \sigma_p & 0 \\ 0 & \sigma_p \end{bmatrix} \begin{Bmatrix} p_x \\ p_y \end{Bmatrix} \tag{1.16}$$

Alternatively, Equation (1.16) can be written as Equation (1.17).

$$\begin{bmatrix} (\sigma_{xx} - \sigma_p) & \tau_{xy} \\ \tau_{yx} & (\sigma_{yy} - \sigma_p) \end{bmatrix} \begin{Bmatrix} p_x \\ p_y \end{Bmatrix} = 0 \tag{1.17}$$

Equation (1.17) represents two equations in the two unknowns p_x and p_y. For a nontrivial (nonzero) solution to exist, the determinant of the matrix must be zero. This is a classic statement of an eigenvalue problem. To show that the eigenvalues of the stress matrix are the principal stresses, we set the determinant of the matrix to be zero to obtain Equation (1.18).

$$\sigma_p^2 - \sigma_p(\sigma_{xx} + \sigma_{yy}) + (\sigma_{xx}\sigma_{yy} - \tau_{xy}^2) = 0 \tag{1.18}$$

Equation (1.18) is called the *characteristic equation* in an eigenvalue problem. We can find the roots of the quadratic equation as given in Equation (1.19a).

$$\sigma_{1,2} = [(\sigma_{xx} + \sigma_{yy}) \pm \sqrt{(\sigma_{xx} + \sigma_{yy})^2 - 4(\sigma_{xx}\sigma_{yy} - \tau_{xy}^2)}]/2 \tag{1.19a}$$

The terms in Equation (1.19a) can be rearranged to obtain Equation (1.19b).

$$\sigma_{1,2} = \left[\left(\frac{\sigma_{xx} + \sigma_{yy}}{2} \right) \pm \sqrt{\left(\frac{\sigma_{xx} - \sigma_{yy}}{2} \right)^2 + \tau_{xy}^2} \right] \tag{1.19b}$$

Equation (1.19b) gives the formulas for principal stresses we saw in the introductory course on the mechanics of materials. We conclude the discussion with the following observations.

> **Definition 14** The *eigenvalues* of the stress matrix are the principal stresses.
>
> **Definition 15** The *eigenvectors* of the stress matrix are the principal directions.

To determine the eigenvectors, we can use *either* of the two equations represented in (1.17), along with the fact that the sum of the square of direction cosines is 1. Example 1.3 demonstrates the process of determining the eigenvectors.

1.3.3 Maximum Shear Stress

In the introductory course on the mechanics of materials, we saw that the maximum shear stress exists on a plane that is 45° to the principal planes. The magnitude of the maximum shear stress is given by

$$\tau_{max} = \left| \max\left(\frac{\sigma_1 - \sigma_2}{2}, \frac{\sigma_2 - \sigma_3}{2}, \frac{\sigma_3 - \sigma_1}{2} \right) \right| \tag{1.20}$$

where the term on the right-hand side is chosen to give the maximum magnitude of difference between the two principal stresses. Figures 1.15 through 1.17 show the various planes on which the maximum shear stress will exist, depending upon the relative values of the principal stresses.

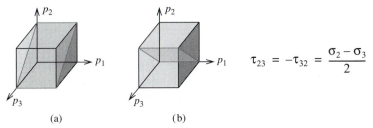

Figure 1.15 Planes of maximum shear stress that are 45° to principal planes 2 and 3.

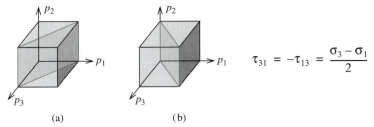

Figure 1.16 Planes of maximum shear stress that are 45° to principal planes 1 and 3.

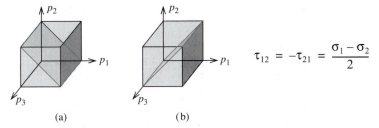

Figure 1.17 Planes of maximum shear stress that are 45° to principal planes 1 and 2.

EXAMPLE 1.3

The following stresses were found at a point in plane stress on a fuselage of an aircraft:

$$\sigma_{xx} = 30 \text{ MPa (T)} \qquad \sigma_{yy} = 60 \text{ MPa (C)} \qquad \tau_{xy} = 50 \text{ MPa}$$

Use the matrix method to determine (a) the normal and shear stresses on a plane passing through the point that is at 30° counterclockwise to the x axis, (b) the principal stresses, and (c) principal direction 1.

PLAN

(a) We can find the angle made by the normal and the tangent of the plane with the x axis. By using Equations (1.11a) and (1.11b), we can obtain the normal and shear stresses on the inclined plane. (b) After subtracting σ_p from the diagonal term in the stress matrix, we can set the determinant to zero to find the eigenvalues, which are the principal stresses. (c) After finding the eigenvalues in part (b), we can find eigenvector 1, which gives us principal direction 1.

SOLUTION

(a) The inclined plane AA is at an angle of 30° from the x axis, hence direction n of the outward normal is $\theta = 120°$, as shown in Figure 1.18. Direction t is determined by the requirement that n-t-z be a right-handed coordinate system. Thus, the tangent direction is at an angle of $(180 + 30)$ degrees from the x axis. In matrix form, we can write the information as shown in Equation (E1).

$$[\sigma] = \begin{bmatrix} 30 & 50 \\ 50 & -60 \end{bmatrix} \quad \{n\} = \begin{Bmatrix} \cos\ 120 \\ \sin\ 120 \end{Bmatrix} = \begin{Bmatrix} -0.5 \\ 0.866 \end{Bmatrix}$$

$$\{t\} = \begin{Bmatrix} \cos\ 210 \\ \sin\ 210 \end{Bmatrix} = \begin{Bmatrix} -0.866 \\ -0.5 \end{Bmatrix} \tag{E1}$$

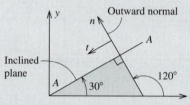

Figure 1.18 Outward normal to the plane in Example 1.3.

We can find the stress vector from Equation (1.15) as follows:

$$\{S\} = [\sigma]\{n\} = \begin{bmatrix} 30 & 50 \\ 50 & -60 \end{bmatrix} \begin{Bmatrix} -0.5 \\ 0.866 \end{Bmatrix} = \begin{Bmatrix} 28.30 \\ -76.96 \end{Bmatrix} \tag{E2}$$

The normal stress can be found from Equation (1.11a) as shown in Equation (E3).

$$\sigma_{nn} = \{n\}^T[\sigma]\{n\} = \begin{Bmatrix} -0.5 \\ 0.866 \end{Bmatrix}^T \begin{Bmatrix} 28.30 \\ -76.96 \end{Bmatrix} = -14.15 - 66.65 \qquad \text{or} \tag{E3}$$

ANS. $\sigma_{nn} = 80.8$ MPa (C)

The shear stress can be found from Equation (1.11b) as shown in Equation (E4).

$$\tau_{nt} = \{t\}^T[\sigma]\{n\} = \begin{Bmatrix} -0.866 \\ -0.5 \end{Bmatrix}^T \begin{Bmatrix} 28.30 \\ -76.96 \end{Bmatrix} = -24.5 + 38.48 \qquad \text{or} \tag{E4}$$

ANS. $\tau_{nt} = 14.0$ MPa

(b) We can subtract σ_p from the diagonal term in the stress matrix and set the determinant to zero as shown in Equation (E5).

$$\begin{bmatrix} 30 - \sigma_p & 50 \\ 50 & -60 - \sigma_p \end{bmatrix} = 0 \tag{E5}$$

Expanding the determinant in Equation (E5), we obtain the following characteristic equation:

$$[(30 - \sigma_p)(-60 - \sigma_p) - 2500] = 0 \quad \text{or} \quad \sigma_p^2 + 30\sigma_p - 4300 = 0 \tag{E6}$$

By solving for the two roots in Equation (E6), we obtain $\sigma_p = -82.3$ and $\sigma_p = 52.26$. Following the convention that $\sigma_1 > \sigma_2$, we report the results for principal stresses as follows:

ANS. $\quad \sigma_1 = 52.3$ MPa (T) $\qquad \sigma_2 = 82.3$ MPa (C)

Checking Results: We note that $\sigma_1 + \sigma_2 = -30$ and $\sigma_{xx} + \sigma_{yy} = -30$, which validates the results in accordance with Equation (1.8).

(c) We can substitute σ_1 in the first row of Equation (1.17) to obtain Equation (E7).

$$-22.26 p_x + 50 p_y = 0 \quad \text{or} \quad p_x = 2.246 p_y \tag{E7}$$

We note that the square of direction cosines is 1 and obtain Equation (E8).

$$(2.246 p_y)^2 + (p_y)^2 = 1 \quad \text{or} \quad p_y = \pm 0.4067 \tag{E8}$$

From Equation (E7) we can find two solutions for p_x as $p_x = \pm 0.9134$. Thus, the following possible solutions exist for principal direction 1:

ANS. $\quad \begin{Bmatrix} \theta_x = 24° \\ \theta_y = 66° \end{Bmatrix} \quad \text{or} \quad \begin{Bmatrix} \theta_x = 156° \\ \theta_y = 114° \end{Bmatrix}$

Checking Results: If we use Equation (1.11a) to find the normal stress in principal direction 1, we can obtain principal stress 1 as follows:

$$\sigma_1 = \{p\}^T \begin{bmatrix} 30 & 50 \\ 50 & -60 \end{bmatrix} \begin{Bmatrix} 0.9134 \\ 0.4067 \end{Bmatrix} = \begin{Bmatrix} 0.9134 \\ 0.4067 \end{Bmatrix}^T \begin{Bmatrix} 47.7370 \\ 21.2680 \end{Bmatrix} = 52.25 \qquad \text{Checks.}$$

COMMENT

The two solutions for principal direction 1 are 180° apart. However, once principal direction 1 has been chosen, principal direction 2 is fixed because principal direction 3 is the z axis and according to our convention, principal directions 1-2-3 form a right-handed coordinate system.

EXAMPLE 1.4

Prove that the coefficients of the characteristic equations are independent of the coordinates.

PLAN

We can write the characteristic equation in an n-t coordinate system and then show that the coefficients in the n-t coordinate system have the same form and are equal to the coefficients in the x-y coordinate system.

SOLUTION

The characteristic equation in the n-t coordinate system can be written as follows:

$$\sigma^2 - \sigma(\sigma_{nn} + \sigma_{tt}) + (\sigma_{nn}\sigma_{tt} - \tau_{nt}^2) = 0 \tag{E1}$$

From Equation (1.8), we have $\sigma_{nn} + \sigma_{tt} = \sigma_{xx} + \sigma_{yy}$; that is, the sum of normal stresses is independent of coordinate transformation. Thus, we would accomplish our task if we could show that the last term is independent of coordinate transformation. We can write the last term as Equation (E2).

$$(\sigma_{nn}\sigma_{tt} - \tau_{nt}^2) = \left(\frac{\sigma_{nn} + \sigma_{tt}}{2}\right)^2 - \left(\frac{\sigma_{nn} - \sigma_{tt}}{2}\right)^2 - \tau_{nt}^2 \tag{E2}$$

We can write Equations (1.7a) and (1.7c) in terms of a double angle and obtain Equation (E3).

$$\frac{\sigma_{nn} - \sigma_{tt}}{2} = \left(\frac{\sigma_{xx} - \sigma_{yy}}{2}\right)\cos 2\theta + \tau_{xy} \sin 2\theta \tag{E3}$$

Substituting Equations (E3) and (1.7b) in the last two terms of Equation (E2), we obtain Equation (E4).

$$\left(\frac{\sigma_{nn} - \sigma_{tt}}{2}\right)^2 + \tau_{nt}^2 = \left(\frac{\sigma_{xx} - \sigma_{yy}}{2}\right)^2 + \tau_{xy}^2 \tag{E4}$$

Substituting Equations (E4) and (1.8) into Equation (E2), we obtain Equation (E5).

$$\left(\frac{\sigma_{nn} + \sigma_{tt}}{2}\right)^2 - \left(\frac{\sigma_{nn} - \sigma_{tt}}{2}\right)^2 - \tau_{nt}^2 = \left(\frac{\sigma_{xx} + \sigma_{yy}}{2}\right)^2 - \left(\frac{\sigma_{xx} - \sigma_{yy}}{2}\right)^2 - \tau_{xy}^2$$

$$(\sigma_{nn}\sigma_{tt} - \tau_{nt}^2) = (\sigma_{xx}\sigma_{yy} - \tau_{xy}^2) \tag{E5}$$

The result of Equation (E4) proves that all coefficients of the characteristic equations are independent of coordinate transformation.

COMMENTS

1. If the coefficients of the characteristic equation do not change with the coordinate system, then the roots of the characteristic equations (i.e., the principal stresses) do not change with the coordinate system. In other words, the *principal stresses at a point are unique*.

2. The characteristic equation can be rewritten in the form $\sigma^2 - \sigma I_1 + I_2 = 0$, where

$$I_1 = \sigma_{xx} + \sigma_{yy}, \quad \text{and} \quad I_2 = \begin{vmatrix} \sigma_{xx} & \tau_{xy} \\ \tau_{yx} & \sigma_{yy} \end{vmatrix}$$

We call I_1 and I_2 stress invariants to emphasize that these coefficients are invariant with coordinate transformation. We will see these quantities again in the next section.

1.4 STRESS TRANSFORMATION IN THREE DIMENSIONS

The concepts and formulas developed in two dimensions can be extended to three dimensions by noting that the matrix relationships described in Section 1.3 do not depend upon the size of the matrix. Figure 1.19(a) shows the direction cosines of the unit normal in three dimensions, which can be written as follows:

$$n_x = \cos \theta_x \qquad n_y = \cos \theta_y \qquad n_z = \cos \theta_z$$

We define the quantities in Figure 1.19(b) by means of

$$\{n\} = \begin{Bmatrix} n_x \\ n_y \\ n_z \end{Bmatrix} \qquad \{S\} = \begin{Bmatrix} S_x \\ S_y \\ S_z \end{Bmatrix} \qquad [\sigma] = \begin{bmatrix} \sigma_{xx} & \tau_{xy} & \tau_{xz} \\ \tau_{yx} & \sigma_{yy} & \tau_{yz} \\ \tau_{zx} & \tau_{zy} & \sigma_{zz} \end{bmatrix} \qquad (1.21)$$

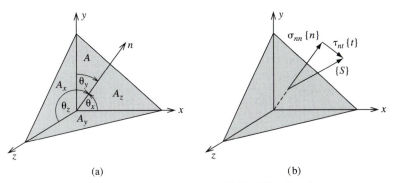

(a) (b)

Figure 1.19 (a) Direction cosines of a unit normal. (b) Equilibrating shear stress.

The normal stress can be found from Equation (1.11a). In two dimensions, the normal and tangent directions were related. In three dimensions, however, there are an infinite number of tangents in the inclined plane shown in Figure 1.19(a). The calculation of shear stress in three dimensions entails problems of two types.

1. The tangent direction is known. The shear stress in that direction may be found from Equation (1.11b). Shear stress found from Equation (1.11b) represents the component of the stress vector in the given tangent direction; this may not be the shear stress necessary for equilibrium of the wedge shown in Figure 1.19(a).

2. The shear stress that is necessary for equilibrium may be found from Equation (1.14), as shown in Figure 1.19(b). We can find the stress vector $\{S\}$ from Equation (1.12) and σ_{nn} from Equation (1.11a). From Equation (1.14) we obtain $\tau_{nt}\{t\} = \{S\} - \sigma_{nn}\{n\}$. The magnitude and direction of the shear stress can be determined from the vector $\tau_{nt}\{t\}$.

EXAMPLE 1.5

The stresses at a point on a frame of a car are given by the following stress matrix:

$$\begin{bmatrix} 30 & 0 & 20 \\ 0 & 30 & -10 \\ 20 & -10 & 0 \end{bmatrix} \text{MPa}$$

Determine (a) the normal stress on a plane that has outward normals at 60°, 60°, and 45° to the x, y, and z directions, respectively; (b) the shear stress magnitude and direction on the plane in part (a) that is in equilibrium with the given state of stress; and (c) the value of the shear stress on the plane with outward normals given in part (a) and tangent direction that makes the angles 45°, 135°, and 90° to the x, y, and z directions, respectively.

PLAN

(a) We can find the direction cosines of the outward normal from the given angles. We can then find the stress vector $\{S\}$. By premultiplying with the unit normal vector, we can find the normal stress. (b) From Equation (1.14) we can find the shear stress vector, from which we can find the magnitude and direction of the shear stress. (c) We can find the direction cosines of the tangent direction from the given angles. By premultiplying the stress vector by the unit tangent vector, we can find the shear stress.

SOLUTION

(a) We can find the direction cosine from the given angles as shown in Equation (E1).

$$n_x = \cos 60 = 0.5 \qquad n_y = \cos 60 = 0.5 \qquad n_z = \cos 45 = 0.7071 \qquad \text{(E1)}$$

We can find the stress vector from Equation (1.15) as shown in Equation (E2).

$$\{S\} = [\sigma]\{n\} = \begin{bmatrix} 30 & 0 & 20 \\ 0 & 30 & -10 \\ 20 & -10 & 0 \end{bmatrix} \begin{Bmatrix} 0.5 \\ 0.5 \\ 0.7071 \end{Bmatrix} = \begin{Bmatrix} 29.142 \\ 7.929 \\ 5.0 \end{Bmatrix} \tag{E2}$$

From Equation (1.11a) the normal stress can be found as shown in Equation (E3).

$$\sigma_{nn} = \{n\}^T[\sigma]\{n\} = \begin{Bmatrix} 0.5 \\ 0.5 \\ 0.7071 \end{Bmatrix}^T \begin{Bmatrix} 29.14 \\ 7.929 \\ 5.0 \end{Bmatrix} = 14.571 + 3.965 + 3.536 \tag{E3}$$

ANS. $\sigma_{nn} = 22.1$ MPa (T)

(b) From Equation (1.14) we obtain Equation (E4).

$$\tau_{nt}\{t_e\} = \{S\} - \sigma_{nn}\{n\} = \begin{Bmatrix} 29.142 \\ 7.929 \\ 5.0 \end{Bmatrix} - 22.071 \begin{Bmatrix} 0.5 \\ 0.5 \\ 0.7071 \end{Bmatrix} = \begin{Bmatrix} 18.107 \\ -3.107 \\ -10.607 \end{Bmatrix} \tag{E4}$$

The magnitude of the vector in Equation (E4) gives us the magnitude of shear stress, as shown in Equation (E5).

$$\tau_{nt} = \sqrt{18.107^2 + 3.107^2 + 10.607^2} \tag{E5}$$

ANS. $\tau_{nt} = 21.21$ MPa

The direction of the shear stress is the unit vector in Equation (E2) and can be found as shown in Equation (E6).

$$\{t_e\} = \frac{1}{21.21} \begin{Bmatrix} 18.107 \\ -3.107 \\ -10.607 \end{Bmatrix} \tag{E6}$$

ANS. $\{t_e\} = \begin{Bmatrix} 0.8537 \\ -0.1465 \\ -0.5 \end{Bmatrix}$

(c) The direction cosines of the given tangent directions are shown in Equation (E7).

$$t_x = \cos 45 = 0.7071 \qquad t_y = \cos 135 = -0.7071 \qquad t_z = \cos 90 = 0 \tag{E7}$$

From Equation (1.11b) we have: $\tau_{nt} = \{t\}^T[\sigma]\{n\} = \{t\}^T\{S\}$. Substituting the stress vector from Equation (E2), we obtain the shear stress as shown in Equation (E8).

$$\tau_{nt} = \left\{ \begin{array}{c} 0.7071 \\ -0.7071 \\ 0 \end{array} \right\}^T \left\{ \begin{array}{c} 29.142 \\ 7.929 \\ 5.0 \end{array} \right\} = 20.606 - 5.607 \qquad \text{(E8)}$$

ANS. $\tau_{nt} = 15$ MPa

COMMENTS

1. The problem highlights how Equations (1.14) and (1.7b) are used differently in the calculation of shear stress. There are an infinite number of tangent directions on a plane in three dimensions. Depending upon the tangent direction, the value of τ_{nt} calculated from Equation (1.7b) will change. This is not surprising, since this value represents a component of the stress vector in a given direction. Of the infinite directions in the tangent plane, there is only one direction in which the shear stress value will be in equilibrium with the given stress state. This unique equilibrating direction is calculated from Equation (1.14).

2. The magnitude of shear stress in part (b) can also be calculated from the knowledge of the stress vector components in the n and t directions are the normal and shear stress, hence

$$\tau_{nt} = \sqrt{|S|^2 - \sigma_{nn}^2} = \sqrt{29.142^2 + 7.929^2 + 5.0^2 - 22.072^2} = 21.21$$

which is the same value as in part (b).

1.4.1 Principal Stresses

The equivalent form of Equations (1.16) and (1.17) in three dimensions can be written as follows:

$$\begin{bmatrix} \sigma_{xx} & \tau_{xy} & \tau_{xz} \\ \tau_{yx} & \sigma_{yy} & \tau_{yz} \\ \tau_{zx} & \tau_{zy} & \sigma_{zz} \end{bmatrix} \left\{ \begin{array}{c} p_x \\ p_y \\ p_z \end{array} \right\} = \begin{bmatrix} \sigma_p & 0 & 0 \\ 0 & \sigma_p & 0 \\ 0 & 0 & \sigma_p \end{bmatrix} \left\{ \begin{array}{c} p_x \\ p_y \\ p_z \end{array} \right\} \qquad \text{or}$$

$$\begin{bmatrix} (\sigma_{xx} - \sigma_p) & \tau_{xy} & \tau_{xz} \\ \tau_{yx} & (\sigma_{yy} - \sigma_p) & \tau_{yz} \\ \tau_{zx} & \tau_{zy} & (\sigma_{zz} - \sigma_p) \end{bmatrix} \left\{ \begin{array}{c} p_x \\ p_y \\ p_z \end{array} \right\} = 0 \qquad \text{(1.22)}$$

The matrix Equation (1.22) represents three equations in the three unknown direction cosines. For a nontrivial solution, the determinant of the matrix must be zero. Once more we observe that the eigenvalues of the stress matrix are the principal stresses, and the eigenvectors are the principal directions. To determine the eigenvectors corresponding to an eigenvalue, we can use any two of the three equations in Equation (1.22), along with the knowledge that the square of the direction cosines is 1, as shown in Equation (1.23).

$$\boxed{p_x^2 + p_y^2 + p_z^2 = 1} \tag{1.23}$$

1.4.2 Principal Stress Convention

Principal stresses may be ordered or unordered. Generally speaking, the following convention is observed for ordering of principal stresses. In three dimensions the principal stresses are reported such that $\sigma_1 > \sigma_2 > \sigma_3$. In two dimensions the principal stresses are reported such that $\sigma_1 > \sigma_2$, but σ_3 is not governed by any order. This is because σ_3 is equal to σ_{zz}, which depends upon whether the state of stress is plane stress or plane strain. By convention, calculated principal stresses are reported in ordered form. In discussion of failure theories (see, e.g., Section 2.3), the principal stresses are usually considered to be unordered, so that the failure criterion is unaffected by the ordering convention.

The principal angles are calculated from the direction cosines of the principal directions. By convention, the angles reported are between zero and 180°, that is,

$$0° \leq \theta_x, \theta_y, \theta_z \leq 180° \tag{1.24}$$

EXAMPLE 1.6

Show that the principal directions are orthogonal if the principal stresses are not equal.

SOLUTION

Let $\{p_n\}$ and $\{p_m\}$ be the principal directions corresponding to any two principal stresses σ_n and σ_m. We can use Equation (1.15) to write Equations (E1) and (E2).

$$[\sigma]\{p_n\} = \sigma_n\{p_n\} \tag{E1}$$

$$[\sigma]\{p_m\} = \sigma_m\{p_m\} \tag{E2}$$

We can premultiply Equation (E1) by $\{p_m\}^T$ and Equation (E2) by $\{p_n\}^T$ to obtain Equations (E3) and (E4).

$$\{p_m\}^T[\sigma]\{p_n\} = \sigma_n\{p_m\}^T\{p_n\} \tag{E3}$$

$$\{p_n\}^T[\sigma]\{p_m\} = \sigma_m\{p_n\}^T\{p_m\} \tag{E4}$$

The left- and right-hand sides of Equation (E3) are scalar quantities, and hence we can take the transpose of both sides and write Equations (E5) and (E6) by using matrix identities (see Appendix B for brief review of matrix arithmetic):

$$[\{p_m\}^T[\sigma]\{p_n\}]^T = \sigma_n[\{p_m\}^T\{p_n\}]^T \qquad \text{or} \tag{E5}$$

$$\{p_n\}^T[\sigma]^T\{p_m\} = \sigma_n\{p_n\}^T\{p_m\} \tag{E6}$$

By noting that the stress matrix is symmetric as given by Equation (1.10), we can rewrite Equation (E6) as Equation (E7).

$$\{p_n\}^T[\sigma]\{p_m\} = \sigma_n\{p_n\}^T\{p_m\} \tag{E7}$$

Subtracting Equation (E7) from Equation (E4), we obtain Equation (E8).

$$[\sigma_m - \sigma_n]\{p_n\}^T\{p_m\} = 0 \tag{E8}$$

Since $\sigma_m \neq \sigma_n$, Equation (E8) implies that $\{p_n\}^T\{p_m\} = 0$, which is the condition of orthogonality of two vectors. Thus we have shown that any two principal directions are orthogonal if the principal stresses are not equal.

COMMENTS

1. The orthogonality of eigenvectors is a property of all symmetric matrices regardless of matrix size. Because the stress matrix is a symmetric matrix and the principal directions are the eigenvectors of the stress matrix, the result that principal directions are orthogonal is not surprising.

2. If $\sigma_m = \sigma_n$, then all directions in the plane containing the vectors $\{p_n\}$ and $\{p_m\}$ are principal directions.

EXAMPLE 1.7

Determine the principal stresses and the principal directions from the stress matrix at a point on a submarine hull given as follows:

$$\begin{bmatrix} 8 & 0 & 6 \\ 0 & 5 & 0 \\ 6 & 0 & -8 \end{bmatrix} \text{MPa}$$

PLAN

We can write the equivalent form of Equation (1.22) by subtracting σ_p from the diagonal term in each row and set the determinants to zero. Noting that the off-diagonal terms in the second column and row are zero, we can use the element of the diagonal as the pivot to simplify calculations. After finding the eigenvalues, we can find the eigenvectors.

SOLUTION

We can write Equation (1.22) for the given stress matrix as Equation (E1).

$$\begin{bmatrix} (8 - \sigma_p) & 0 & 6 \\ 0 & (5 - \sigma_p) & 0 \\ 6 & 0 & (-8 - \sigma_p) \end{bmatrix} \begin{Bmatrix} p_x \\ p_y \\ p_z \end{Bmatrix} = 0 \qquad \text{(E1)}$$

For a nontrivial solution, we can set the determinant of the matrix as zero by using the second row as the pivot row, to obtain Equation (E2).

$$(5 - \sigma_p)[(8 - \sigma_p)(-8 - \sigma_p) - 36] = 0 \quad \text{or} \quad (5 - \sigma_p)[-64 + \sigma_p^2 - 36] = 0 \qquad \text{(E2)}$$

From the first term in Equation (E2), we see that one of the roots of the equation is +5. By setting the term in the square brackets equal to zero, we find that the other two roots as are +10 and −10. Thus the three principal stresses are as follows:

ANS. $\sigma_1 = 10$ MPa (T) $\sigma_2 = 5$ MPa (T) $\sigma_3 = 10$ MPa (C)

Principal Direction 1 Substituting σ_1 in the first two rows of Equation (E1), we obtain Equations (E3) and (E4).

$$-2p_x + 6p_z = 0 \quad \text{or} \quad p_x = 3p_z \qquad \text{(E3)}$$

$$-5p_y = 0 \quad \text{or} \quad p_y = 0 \qquad \text{(E4)}$$

Substituting Equations (E3) and (E4) into Equation (1.23), we obtain $p_z = \pm(1/\sqrt{10})$. From Equation (E3) we obtain $p_x = \pm(3/\sqrt{10})$. The two possible solutions for principal direction 1 are

ANS. $\begin{Bmatrix} \theta_x = 18.4° \\ \theta_y = 90° \\ \theta_z = 71.6° \end{Bmatrix}$ or $\begin{Bmatrix} \theta_x = 161.6° \\ \theta_y = 90° \\ \theta_z = 108.4° \end{Bmatrix}$

Checking Results: If we find the normal stress in principal direction 1 by using Equation (1.11a), we should obtain principal stress 1 as follows:

$$\sigma_1 = \{p_1\}^T \begin{bmatrix} 8 & 0 & 6 \\ 0 & 5 & 0 \\ 6 & 0 & -8 \end{bmatrix} \begin{Bmatrix} \dfrac{3}{\sqrt{10}} \\ 0 \\ \dfrac{1}{\sqrt{10}} \end{Bmatrix} = \begin{Bmatrix} \dfrac{3}{\sqrt{10}} \\ 0 \\ \dfrac{1}{\sqrt{10}} \end{Bmatrix}^T \begin{Bmatrix} \dfrac{30}{\sqrt{10}} \\ 0 \\ \dfrac{10}{\sqrt{10}} \end{Bmatrix} = \frac{90}{10} + \frac{10}{10} = 10 \qquad \text{Checks.}$$

Principal direction 2   On substituting σ_2 in the first two rows of Equation (E1), we find that the equation corresponding to the second row yields no information because it results in $0 = 0$. So we use the first and the third rows to obtain Equations (E5) and (E6).

$$3p_x + 6p_z = 0 \tag{E5}$$

$$6p_x - 13p_z = 0 \tag{E6}$$

The solution of Equations (E5) and (E6) yields $p_x = 0$ and $p_z = 0$, which from Equation (1.23) implies that $p_y = \pm 1$ Thus the solution for principal direction 2 can be

$$\textbf{ANS.} \quad \begin{Bmatrix} \theta_x = 90° \\ \theta_y = 0° \\ \theta_z = 90° \end{Bmatrix} \quad \text{or} \quad \begin{Bmatrix} \theta_x = 90° \\ \theta_y = 180° \\ \theta_z = 90° \end{Bmatrix}$$

Checking Results: If we find the normal stress in principal direction 2 by using Equation (1.11a), we should obtain principal stress 2 as follows:

$$\sigma_2 = \{p_2\}^T \begin{bmatrix} 8 & 0 & 6 \\ 0 & 5 & 0 \\ 6 & 0 & -8 \end{bmatrix} \begin{Bmatrix} 0 \\ 1 \\ 0 \end{Bmatrix} = \begin{Bmatrix} 0 \\ 1 \\ 0 \end{Bmatrix}^T \begin{Bmatrix} 0 \\ 5 \\ 0 \end{Bmatrix} = 5 \quad \text{Checks.}$$

Principal direction 3   In a similar manner, by substituting σ_3 in the equations in the first two rows of Equation (E1), we obtain Equations (E7) and (E8).

$$18p_x + 6p_z = 0 \quad \text{or} \quad p_z = -3p_x \tag{E7}$$

$$15p_y = 0 \quad \text{or} \quad p_y = 0 \tag{E8}$$

Substituting Equations (E7) and (E8) into Equation (1.23), we obtain $p_x = \pm(1/\sqrt{10})$. From Equation (E7) we obtain $p_z = \mp(3/\sqrt{10})$. The two possible solutions for principal direction 3 are as follows:

$$\textbf{ANS.} \quad \begin{Bmatrix} \theta_x = 71.6° \\ \theta_y = 90° \\ \theta_z = 161.6° \end{Bmatrix} \quad \text{or} \quad \begin{Bmatrix} \theta_x = 108.4° \\ \theta_y = 90° \\ \theta_z = 18.4° \end{Bmatrix}$$

Checking Results: If we find the normal stress in principal direction 3 by using Equation (1.11a), we should obtain principal stress 3 as follows:

$$\sigma_3 = \{p_3\}^T \begin{bmatrix} 8 & 0 & 6 \\ 0 & 5 & 0 \\ 6 & 0 & -8 \end{bmatrix} \begin{Bmatrix} \dfrac{1}{\sqrt{10}} \\ 0 \\ \dfrac{-3}{\sqrt{10}} \end{Bmatrix} = \begin{Bmatrix} \dfrac{1}{\sqrt{10}} \\ 0 \\ \dfrac{-3}{\sqrt{10}} \end{Bmatrix}^T \begin{Bmatrix} \dfrac{-10}{\sqrt{10}} \\ 0 \\ \dfrac{30}{\sqrt{10}} \end{Bmatrix} = \dfrac{-10}{10} + \dfrac{-90}{10} = -10 \quad \text{Checks.}$$

COMMENTS

1. The calculations of principal directions show that the y direction is the principal direction and the other two directions lie in the x-z plane. We could have recognized this from looking at the stress matrix and noting that the shear stresses are zero on the plane with the outward normal in the y direction. Had we recognized that the y direction is the principal direction at the start of the problem, we could have solved it like a two-dimensional problem and reduced the algebraic effort significantly.

2. In all our calculations of principal directions, we will always obtain two solutions because we shall have to take a square root when we use Equation (1.23). The two solutions of direction cosines always have opposite sign and correspond to the two opposite directions of a line that represents the normal direction to the principal planes. This observation is further reflected in the sum of the two solutions for each angle, which is always $180°$.

1.4.3   Characteristic Equation and Stress Invariants

We set the determinant of the matrix in Equation (1.22) to zero to obtain the characteristic equation, which is written in the following form.

$$\sigma_p^3 - I_1\sigma_p^2 + I_2\sigma_p - I_3 = 0 \qquad (1.25)$$

The roots[6] of the characteristic Equation (1.25) yield the principal stresses, as observed in Section 1.3.2.

The coefficients I_1, I_2, and I_3 are called stress invariants because the value of these quantities is independent of coordinate transformation, as was shown for plane stress problem in Example 1.4. Notice that the sign in the terms containing odd stress invariants is negative. The three stress invariants can be shown (see Problem 1.34) to be given by Equations (1.26a) through (1.26c).

$$I_1 = \sigma_{xx} + \sigma_{yy} + \sigma_{zz} \qquad (1.26a)$$

$$I_2 = \begin{vmatrix} \sigma_{xx} & \tau_{xy} \\ \tau_{yx} & \sigma_{yy} \end{vmatrix} + \begin{vmatrix} \sigma_{yy} & \tau_{yz} \\ \tau_{zy} & \sigma_{zz} \end{vmatrix} + \begin{vmatrix} \sigma_{xx} & \tau_{xz} \\ \tau_{zx} & \sigma_{zz} \end{vmatrix} \qquad (1.26b)$$

$$I_3 = \begin{vmatrix} \sigma_{xx} & \tau_{xy} & \tau_{xz} \\ \tau_{yx} & \sigma_{yy} & \tau_{yz} \\ \tau_{zx} & \tau_{zy} & \sigma_{zz} \end{vmatrix} \qquad (1.26c)$$

---

[6] The roots of equation $x^3 - I_1 x^2 + I_2 x - I_3 = 0$ are $x_1 = 2A \cos \alpha + I_1/3$, and $x_{2,3} = -2A \cos (\alpha \pm 60°)$ $+ I_1/3$, where $A = \sqrt{(I_1/3)^2 - I_2/3}$ and $\cos 3\alpha = [2(I_1/3)^3 - (I_1/3)I_2 + I_3]/(2A^3)$.

where the determinant of the quantity shown is enclosed in verticals: | |.

- The first stress invariant I_1 is the sum of diagonal terms of the stress matrix.
- The second stress invariant I_2 is the sum of minor determinants of the stress matrix.
- The third stress invariant I_3 is the determinant of the entire stress matrix.

Since stress invariants are independent of the coordinate system, the value of these quantities will be the same in a principal coordinate system. The stress matrix in a principal coordinate system can be written as shown in Equation (1.27a).

$$\begin{bmatrix} \sigma_1 & 0 & 0 \\ 0 & \sigma_2 & 0 \\ 0 & 0 & \sigma_3 \end{bmatrix} \tag{1.27a}$$

The stress invariants can be written in terms of principal stress as shown in Equations (1.27b) through (1.27d).

$$\boxed{I_1 = \sigma_1 + \sigma_2 + \sigma_3} \tag{1.27b}$$

$$\boxed{I_2 = \begin{vmatrix} \sigma_1 & 0 \\ 0 & \sigma_2 \end{vmatrix} + \begin{bmatrix} \sigma_2 & 0 \\ 0 & \sigma_3 \end{bmatrix} + \begin{bmatrix} \sigma_1 & 0 \\ 0 & \sigma_3 \end{bmatrix} = \sigma_1\sigma_2 + \sigma_2\sigma_3 + \sigma_3\sigma_1} \tag{1.27c}$$

$$\boxed{I_3 = \begin{vmatrix} \sigma_1 & 0 & 0 \\ 0 & \sigma_2 & 0 \\ 0 & 0 & \sigma_3 \end{vmatrix} = \sigma_1\sigma_2\sigma_3} \tag{1.27d}$$

The stress invariants, particularly the first and the third, provide a quick and easy check on the laboriously calculated principal stresses. The maximum shear stress can be found from Equation (1.20) and will be on a plane that is 45° to the principal planes as discussed in Section 1.3.3.

1.4.4   Octahedral Stresses

In Section 2.3, we will see that one of the theories of failure is based on the stresses on a octahedral plane.

Definition 16 A plane that makes equal angles with the principal planes is called an octahedral plane.

Figure 1.20 shows eight (octal) planes that make equal angles with the principal planes. The stresses on octahedral planes are referred to as octahedral stresses. We use n_1, n_2, and n_3 to represent the direction cosines of the outward normal of a plane measured from the principal axis. The normal stress and shear stress on a plane with direction

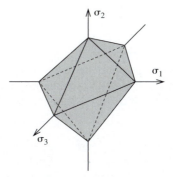

Figure 1.20 The eight octahedral planes.

cosines n_1, n_2, and n_3 can be shown (see Problem 1.29) to be given by Equations (1.28a) and (1.28b).

$$\sigma_{nn} = \sigma_1 n_1^2 + \sigma_2 n_2^2 + \sigma_3 n_3^2 \tag{1.28a}$$

$$\tau_{nt} = \sqrt{(\sigma_1^2 n_1^2 + \sigma_2^2 n_2^2 + \sigma_3^2 n_3^2) - \sigma_{nn}^2} \tag{1.28b}$$

Though the sign of the direction cosines changes with each of the eight octahedral planes, their magnitude is the same for all eight planes, that is, $|n_1| = |n_2| = |n_3| = 1/\sqrt{3}$. Substituting the direction cosines for the octahedral planes in Equation (1.28a) we obtain the normal stress given in Equation (1.29a).

$$\boxed{\sigma_{\text{oct}} = (\sigma_1 + \sigma_2 + \sigma_3)/3} \tag{1.29a}$$

Equation (1.29a) shows $\sigma_{\text{oct}} = I_1/3$, where I_1 is the first stress invariant. A more interesting interpretation of octahedral normal stress is obtained by considering a point in fluid. For a point in fluid, the shear stresses are zero and all normal stresses (principal stresses) are compressive and equal to the pressure at that point, that is, $\sigma_1 = \sigma_2 = \sigma_3 = -p$, where p is the hydrostatic pressure. From Equation (1.29a) we obtain $\sigma_{\text{oct}} = -p$; in other words, the octahedral normal stress is similar to the hydrostatic pressure.

Substituting the direction cosines for the octahedral planes in Equation (1.28b) we obtain the magnitude of shear stress, which can be simplified as shown in Equation (1.29b).

$$\tau_{\text{oct}} = \sqrt{\frac{(\sigma_1^2 + \sigma_2^2 + \sigma_3^2)}{3} - \frac{(\sigma_1 + \sigma_2 + \sigma_3)^2}{9}} \quad \text{or}$$

$$\tau_{\text{oct}} = \frac{1}{3}\sqrt{(2\sigma_1^2 + 2\sigma_2^2 + 2\sigma_3^2) - 2(\sigma_1\sigma_2 + \sigma_2\sigma_3 + \sigma_3\sigma_1)} \quad \text{or}$$

$$\boxed{\tau_{\text{oct}} = \frac{1}{3}\sqrt{(\sigma_1 - \sigma_2)^2 + (\sigma_2 - \sigma_3)^2 + (\sigma_3 - \sigma_1)^2}} \tag{1.29b}$$

In Section 2.3 it will be seen that ductile materials can withstand very large hydrostatic pressures and yielding is governed by τ_{oct} reaching a critical value.

EXAMPLE 1.8

The stresses at a point on the frame of a car are given by the stress matrix shown earlier, in Example 1.5. Determine (a) the principal stresses, (b) the third principal direction, (c) the maximum shear stress, and (d) the octahedral normal and shear stresses.

PLAN

(a) We can find the three stress invariants and find the roots of the characteristic equations from the formulas given in footnote 6 to obtain the principal stresses. (b) We can use Equation (1.23) along with the two equations from any two rows in the matrix and solve for the direction cosines. (c) The maximum shear stresses can be found from Equation (1.20). (d) The octahedral normal and shear stresses can be found from Equations (1.29a) and (1.29b).

SOLUTION

(a) We can find the three stress invariants from Equations (1.26a), (1.26b), and (1.26c). In calculating I_3, we use the first row as the pivots to evaluate the determinant of the matrix.

$$I_1 = 30 + 30 + 0 = 60 \tag{E1}$$

$$I_2 = \begin{vmatrix} 30 & 0 \\ 0 & 30 \end{vmatrix} + \begin{vmatrix} 30 & -10 \\ -10 & 0 \end{vmatrix} + \begin{vmatrix} 30 & 20 \\ 20 & 0 \end{vmatrix} = 900 - 100 - 400 = 400 \tag{E2}$$

$$I_3 = 30 \begin{vmatrix} 30 & -10 \\ -10 & 0 \end{vmatrix} + 20 \begin{vmatrix} 0 & 30 \\ 20 & -10 \end{vmatrix} + 0 = -3000 - 12{,}000 = -15{,}000 \tag{E3}$$

The characteristic equation is $\sigma_p^3 - 60\sigma_p^2 + 400\sigma_p + 15{,}000 = 0$. The roots of the equation can be found from the formulas given in footnote 6 as shown in Equations (E4) through (E9).

$$A = \sqrt{(60/3)^2 - 400/3} = 16.390 \tag{E4}$$

$$\cos 3\alpha = \frac{2(60/3)^3 - (60/3)400 + (-15{,}000)}{2(16.39)^3} = -0.8037 \tag{E5}$$

$$\alpha = 143.49/3 = 47.83 \tag{E6}$$

$$x_1 = 2(16.39)\cos 47.83 + 60/3 = 41.927 \tag{E7}$$

$$x_3 = (-2(16.39)\cos(-12.17) + (60)/3 = -11.925 \tag{E8}$$

$$x_2 = -2(16.39)\cos(107.83) + (60)/3 = 29.998 \tag{E9}$$

The principal stresses are

ANS. $\sigma_1 = 41.9$ MPa (T) $\sigma_2 = 30$ MPa (T) $\sigma_3 = 11.9$ MPa (C)

Checking Results: We can check our results by finding stress invariants 1 and 3 from Equations (1.27b) and (1.27d).

$$I_1 = 41.927 + 29.998 - 11.925 = 60 \qquad \text{Checks.}$$

$$I_3 = (41.927)(29.998)(-11.925) = -14,998 \qquad \text{Checks.}$$

(c) On substituting σ_3 into the equations corresponding to the first two rows of the matrix, we obtain Equations (E10) and (E11).

$$[30 - (-11.93)]p_x + (0)p_y + 20p_z = 0 \quad \text{or} \quad p_x = \frac{-20}{42.93}p_z = -0.477p_z \quad \text{(E10)}$$

$$(0)p_x + [30 - (-11.93)]p_y - 10p_z = 0 \quad \text{or} \quad p_y = \frac{10}{42.93}p_z = 0.239p_z \quad \text{(E11)}$$

Substituting Equations (E10) and (E11) in Equation (1.23), we obtain Equation (E12).

$$(-0.477p_z)^2 + (0.239p_z)^2 + p_z^2 = 1 \quad \text{or} \quad p_z = \pm 0.8823 \quad \text{(E12)}$$

From Equations (E10) and (E11) we obtain $p_x = \mp 0.421$ and $p_y = \pm 0.882$. The two possible solutions for principal direction 3 are as follows:

ANS. $\quad \begin{Bmatrix} \theta_x = 114.9° \\ \theta_y = 77.8° \\ \theta_z = 28.1° \end{Bmatrix} \quad \text{or} \quad \begin{Bmatrix} \theta_x = 65.1° \\ \theta_y = 102.2° \\ \theta_z = 151.9° \end{Bmatrix}$

Checking Results: If we use Equation (1.11a) to find the normal stress in principal direction 3, we should obtain principal stress 3 as follows:

$$\sigma_3 = \{p_3\}^T \begin{bmatrix} 30 & 0 & 20 \\ 0 & 30 & -10 \\ 20 & -10 & 0 \end{bmatrix} \begin{Bmatrix} -0.421 \\ 0.211 \\ 0.882 \end{Bmatrix} = \begin{Bmatrix} -0.421 \\ 0.211 \\ 0.882 \end{Bmatrix}^T \begin{Bmatrix} 5.01 \\ -2.49 \\ -10.53 \end{Bmatrix} = -11.92 \quad \text{Checks.}$$

(d) As per Equation (1.20), the maximum shear stress is the magnitude of the largest difference between the principal stresses. With principal stresses in an ordered state, this largest difference will be between principal stresses 1 and 3. Its value can be found as follows:

$$\tau_{max} = \left| \frac{\sigma_3 - \sigma_1}{2} \right| = \left| \frac{-11.925 - 41.927}{2} \right| = 26.926 \quad \textbf{ANS.} \; \tau_{max} = 26.9 \text{ MPa}$$

(e) Substituting principal stress values in Equations (1.29a) and (1.29b) yields the octahedral stresses.

$$\sigma_{oct} = \frac{41.927 + 29.998 + (-11.925)}{3} \quad \textbf{ANS.} \; \sigma_{oct} = 20 \text{ MPa (T)}$$

$$\tau_{oct} = \frac{1}{3}\sqrt{(41.927 - 29.998)^2 + [29.998 - (-11.925)]^2 + [(-11.925) - 41.927]^2}$$

ANS. $\tau_{oct} = 23.1$ MPa

COMMENTS

1. We could have found the octahedral normal stress from $\sigma_{oct} = I_1/3$ to obtain the same value.

2. The maximum shear stress is different from the octahedral shear stress. In Section 2.3 we will see how one failure theory uses maximum shear stress as a failure criterion and another uses octahedral shear stress. The results in this example highlight that two theories predict different failure values depending upon which failure criterion we use.

PROBLEM SET 1.2

1.23 Expand Equation (1.11a) and show that this equation is same as Equation (1.7a).

1.24 Expand Equation (1.11b) and show that this equation is same as Equation (1.7b).

1.25 The stresses at a point in plane stress on the body of a car were found to be $\sigma_{xx} = 30$ MPa (C), $\sigma_{yy} = 60$ MPa (T), and $\tau_{xy} = 40$ MPa. Use matrix methods to determine (a) the normal and shear stresses on a plane passing through the point that is 28° counterclockwise to the x axis, (b) the principal stresses and principal direction 1, and (c) the maximum shear stress at the point in plane stress.

1.26 The stresses at a point in plane stress on an aircraft fuselage were found to be $\sigma_{xx} = 45$ MPa (T), $\sigma_{yy} = 15$ MPa (T), and $\tau_{xy} = -20$ MPa. Use matrix methods to determine (a) the normal and shear stresses on a plane passing through the point that is 30° clockwise to the x axis, (b) the principal stresses and principal direction 1, and (c) the maximum shear stress at the point in plane stress.

1.27 The stresses at a point in plane stress on a sheet metal roof were found to be $\sigma_{xx} = 10$ ksi (C), $\sigma_{yy} = 20$ ksi (C), and $\tau_{xy} = 30$ ksi. Use matrix methods to determine (a) the normal and shear stresses on a plane passing through the point that is 42° counterclockwise to the x axis, (b) the principal stresses and principal direction 1, and (c) the maximum shear stress at the point in plane stress.

ANS. (a) $\sigma_{nn} = 45.4$ ksi (C), $\tau_{nt} = 1.84$ ksi (b) $\sigma_1 = 15.4$ ksi (T), $\sigma_2 = 45.4$ ksi (C) $\theta_x = 40.3°$, $\theta_y = 49.7°$ (c) $\tau_{max} = 30.4$ ksi

1.28 Starting with Equation (1.11b) and taking the transpose of both sides, show that $\tau_{tn} = \tau_{nt}$.

1.29 Starting with a stress matrix in a principal coordinate system, show that the normal and shear stresses on a plane with direction cosines n_1, n_2, and n_3 are given by Equations (1.28a) and (1.28b).

1.30 The stresses at a point on a bridge were determined to be $\sigma_{xx} = 8$ ksi (T), $\sigma_{yy} = 12$ ksi (T), and $\sigma_{zz} = 8$ ksi (C). Determine the normal stress on a plane that has an outward normal at 60°, –60°, and 45° to x, y, and z directions, respectively.

1.31 The stresses at a point on a crankshaft were determined to be $\tau_{xy} = 125$ MPa and $\tau_{xz} = -150$ MPa. Determine the normal stress on a plane that has an outward normal at 72.54°, 120°, and 35.67° to x, y, and z directions, respectively.

1.32 Show that the normal stress in three dimensions can be written as follows:

$$\sigma_{nn} = \sigma_{xx}n_x^2 + \sigma_{yy}n_y^2 + \sigma_{zz}n_z^2$$
$$+ 2\tau_{xy}n_xn_y + 2\tau_{yz}n_yn_z + 2\tau_{zx}n_zn_x \quad (1.30)$$

1.33 Starting with the wedge shown in Figure 1.19 (Section 1.4), derive the normal stress expression from equilibrium.
Note: $dA_x = n_x dA$, $dA_y = n_y dA$, and $dA_z = n_z dA$, where dA is the area of the inclined plane, and dA_x, dA_y, and dA_z are the areas of the planes with outward normal in the x, y, and z direction, respectively.

1.34 By expanding the determinant of the matrix in Equation (1.22), show that the invariants are as given by Equations (1.26a) through (1.26c).

1.35 The stresses at a point on the roof of a building are given by the following stress matrix:

$$\begin{bmatrix} 50 & 0 & -20 \\ 0 & 30 & 0 \\ -20 & 0 & 0 \end{bmatrix} \text{ksi}$$

Determine (a) the normal and shear stresses on a plane that has an outward normal at 45°, –60°, and –60° to x, y, and z directions, respectively, (b) the principal stresses, (c) the principal directions, (d) the maximum shear stress, and (e) the magnitude of the octahedral shear stress.

ANS.   (a) $\sigma_{nn} = 18.4$ ksi (T), $\tau_{nt} = 27.0$ ksi
(b) $\sigma_1 = 57.0$ ksi (T), $\sigma_2 = 30.0$ ksi (T),
$\sigma_3 = 7.0$ ksi (C)   (c) $\theta_{x1} = 160.7°$,
$\theta_{y1} = 90°$, $\theta_{z1} = 70.7°$, $\theta_{x2} = 90°$, $\theta_{y2} = 0°$,
$\theta_{z2} = 90°$, $\theta_{x3} = 70.7°$, $\theta_{y3} = 90°$, $\theta_{z3} = 19.3°$
(d) $\tau_{max} = 32$ ksi   (e) $\tau_{oct} = 26.25$ ksi

1.36 The stress at a point on a thick plate of a ship are given by the following stress matrix:

$$\begin{bmatrix} 8 & 0 & 0 \\ 0 & 10 & 5 \\ 0 & 5 & 10 \end{bmatrix} \text{ksi}$$

Determine (a) the normal and shear stresses on a plane that has an outward normal at 72.54°, 120°, and 35.67° to the x, y, and z directions, respectively, (b) the principal stresses, (c) the principal directions, (d) the maximum shear stress, and (e) the magnitude of the octahedral shear stress.

ANS.   (a) $\sigma_{nn} = 5.62$ ksi (T), $\tau_{nt} = 2.26$ ksi
(b) $\sigma_1 = 15.0$ ksi (T), $\sigma_2 = 8.0$ ksi (T),
$\sigma_3 = 5.0$ ksi (T)   (c) $\theta_{x1} = 90°$, $\theta_{y1} = 135°$,
$\theta_{z1} = 135°$, $\theta_{x2} = 0°$, $\theta_{y2} = 90°$, $\theta_{z2} = 90°$,
$\theta_{x3} = 90°$, $\theta_{y3} = 135°$, $\theta_{z3} = 45°$
(d) $\tau_{max} = 5$ ksi   (e) $\tau_{oct} = 4.19$ ksi

1.37 The stresses at a point on the frame of a car were determined by the finite element method and are as shown by the following stress matrix:

$$\begin{bmatrix} 18 & 12 & 9 \\ 12 & 12 & -6 \\ 9 & -6 & 6 \end{bmatrix} \text{ksi}$$

Determine (a) the normal and shear stresses on a plane that has an outward normal at 37°, 120°, and 70.43° to the x, y, and z directions, respectively, (b) the principal stresses, (c) the second principal direction, and (d) the magnitude of the octahedral shear stress.

1.38 The stresses at a point on a building frame were determined and are as shown by the following stress matrix:

$$\begin{bmatrix} 100 & 125 & -150 \\ 125 & 12 & -60 \\ -150 & -60 & 0 \end{bmatrix} \text{MPa}$$

Determine (a) the normal and shear stresses on a plane that has an outward normal at $60°$, $-60°$, and $45°$ to the x, y, and z directions, respectively, (b) the principal stresses, (c) the third principal direction, and (d) the magnitude of the octahedral shear stress.

ANS. (a) $\sigma_{nn} = 58.0$ MPa (C), $\tau_{nt} = 91.5$ MPa (b) $\sigma_1 = 278.4$ MPa (T), $\sigma_2 = 49.9$ MPa (C), $\sigma_3 = 116.5$ MPa (C) (c) $\theta_x = 48.9°$, $\theta_y = 108.8°$, $\theta_z = 47.1°$, (d) $\tau_{oct} = 172.6$ MPa

1.39 A *stress deviator* matrix is a stress matrix from which the effect of hydrostatic state of stress ($I_1/3$) is removed as follows:

$$\begin{bmatrix} \sigma_1 - I_1/3 & 0 & 0 \\ 0 & \sigma_2 - I_1/3 & 0 \\ 0 & 0 & \sigma_3 - I_1/3 \end{bmatrix}$$

Show that the stress invariants of the matrix are as follows:

$$J_1 = 0 \qquad (1.31a)$$

$$J_2 = I_2 + \frac{1}{3}I_1 = \frac{1}{6}[(\sigma_1 - \sigma_2)^2$$
$$+ (\sigma_2 - \sigma_3)^2 + (\sigma_3 - \sigma_1)^2] \qquad (1.31b)$$

$$J_3 = I_3 + \frac{1}{3}I_1 I_2 + \frac{2}{27}I_1^3 = \frac{1}{27}(2\sigma_1 - \sigma_2 - \sigma_3)$$

$$\times (2\sigma_2 - \sigma_3 - \sigma_1)(2\sigma_3 - \sigma_1 - \sigma_2) \qquad (1.31c)$$

The foregoing invariants of a stress deviator matrix are used in plastic flow theories for the calculation of plastic strains at a given load.

Note that $J_2 = \frac{3}{2}\tau_{oct}^2$.

1.40 Noting that the components of the stress vector S in terms of principal stresses can be written as $S_x = n_x \sigma_1$, $S_y = n_y \sigma_2$, and $S_x = n_z \sigma_3$, show that Equation (1.32) is valid.

$$\frac{S_x^2}{\sigma_1^2} + \frac{S_y^2}{\sigma_2^2} + \frac{S_z^2}{\sigma_3^2} = 1 \qquad (1.32)$$

Equation (1.32) is the equation of a stress ellipsoid, the surface of which is shown in the first quadrant in Figure P1.40. The stress ellipsoid graphically depicts that the principal stresses are the maximum and minimum stresses at a point. The surface of the ellipsoid describes all possible stress states at a point.

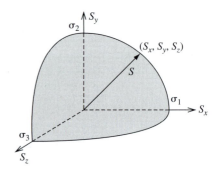

Figure P1.40 Stress ellipsoid.

1.5 AVERAGE NORMAL AND SHEAR STRAINS

Strain is a measure of the intensity of change in the shape of a body. To measure changes, we must consider the movement of points on the body, which requires a clear delineation of movement due to rigid body motion from movement associated with changes in the shape of a body. This delineation calls for the following definitions.

Definition 17 The total movement of a point with respect to a fixed reference coordinates is called *displacement*.

Definition 18 The relative movement of a point with respect to another point on the body is called *deformation*.

Thus displacement of a point represents changes due to deformation plus rigid body motion. In describing such changes, we must specify the reference value from which we are measuring each one. This leads to the following two definitions.

Definition 19 *Lagrangian strain* is computed from deformation by using the original undeformed geometry as the reference geometry.

Definition 20 *Eulerian strain* is computed from deformation by using the final deformed geometry as the reference geometry.

The Lagrangian description is usually used in solid mechanics. The Eulerian description is usually used in fluid mechanics. When a material undergoes very large deformations, such as in a soft rubber or in projectile penetration of metals, either description may be used, depending upon the analysis. In this book, we will use Lagrangian strain except for a few problems.

Recall that there are two kinds of strain—normal strains and shear strains. Normal strains are measures of change of length and shear strains are measures of change of angles.

Normal strain is usually designated by the Greek letter epsilon (ε). The change of length, also called deformation, is designated by the lowercase Greek letter delta (δ). The average normal Lagrangian strain of a line is the ratio of deformation δ of the line to the original length L_0 as follows:

$$\boxed{\varepsilon = \frac{\delta}{L_0}}$$

(1.33)

Definition 21 Elongations ($\delta > 0$) result in *positive* normal strains. Contractions ($\delta < 0$) result in *negative* normal strains.

In *small-strain* approximations ($\varepsilon < 0.01$), the deformation δ can be approximated by the deformation component in the direction of the undeformed line element. Figure 1.21 shows a bar AP that deforms to the position AP_1. For small-strain calculations, the deformed length of AP_1 is approximated by length AP_2, which is obtained by drawing a perpendicular from P_1 to the original direction AP. Another perspective on small-strain approximation is that the deformation δ is the component of deformation vector PP_1 in the original direction. The small-strain approximation results in linear analysis and greatly simplifies calculations.

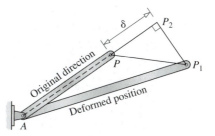

Figure 1.21 Approximation of a small normal strain.

Shear strain is usually designated by the Greek letter gamma (γ). The average Lagrangian shear strain is defined as a change of angle from the right angle as

$$\gamma = \frac{\pi}{2} - \alpha \qquad (1.34)$$

where the Greek letter alpha (α) is the final angle measured in radians (rad), and the Greek letter pi (π) = 3.14159 rad.

Definition 22 Decreases in the angle ($\alpha < \pi/2$) result in *positive* shear strain. Increases in the angle ($\alpha > \pi/2$) result in *negative* shear strain.

For small shear strain ($\gamma < 0.01$), the following approximation of the trigonometric functions may be used:

$$\tan\gamma \approx \gamma \qquad \sin\gamma \approx \gamma \qquad \cos\gamma \approx 1$$

1.5.1 Units of Average Strains

Equation (1.33) shows that normal strain is dimensionless, hence should have no units. However, to differentiate average strain and strain at a point (discussed in Section 1.6), units of length per unit length are used for average normal strains. Thus, average normal strains are reported with the units of inch/inch (or cm/cm, m/m, etc.). Average shear strains are reported in radians.

In reporting experimental results and for describing very large deformations, strains are given by means of a percentage change: that is, the right-hand side of Equations (1.33) and (1.34) is multiplied by 100 before the results are reported. Thus, a normal strain of 0.5% is equal to a strain of 0.005. The Greek letter mu (μ) is often used in reporting small strains because it stands for *micro* ($\mu = 10^{-6}$). Thus, a strain of 1000 μin/in represents a normal strain of 0.001 inch per inch.

EXAMPLE 1.9

Two bars are connected to a roller that slides in a slot as shown in Figure 1.22. Determine the strain in bar *AP* by finding the deformed length of *AP* (a) without small-strain approximation and (b) with small-strain approximation.

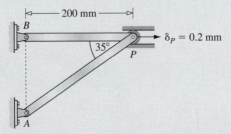

Figure 1.22 Illustration of Small-strain calculations for Example 1.9.

PLAN

(a) We can draw the two bars with exaggerated deformation and find the deformed length of bar AP if we use the cosine rule. (b) We can draw the exaggerated deformed shape and drop a perpendicular from the final position of point P onto the original direction of bar AP and use geometry to find the deformation of the bar AP.

SOLUTION

The length AP used in both methods can be found as $AP = 200/(\cos 35) = 244.155$ mm
(a) Let point P move to point P_1 as shown in Figure 1.23. The angle APP_1 is 145°. From triangle APP_1, we can find the length AP_1 by using the cosine formula as shown in Equation (E1).

$$AP_1 = \sqrt{AP^2 + PP_1^2 - 2(AP)(AP_1)\cos 145} = 244.3188 \text{ mm} \qquad \text{(E1)}$$

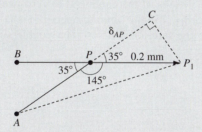

Figure 1.23 Exaggerated deformed shape for Example 1.9.

The average normal strain of member AP can be found as shown in Equation (E2).

$$\varepsilon_{AP} = \frac{AP_1 - AP}{AP} = \frac{0.1639}{244.15} \qquad \text{(E2)}$$

ANS. $\varepsilon_{AP} = 671.2 \ \mu\text{mm/mm}$

(b) We want the component of PP_1 in the direction of AP. We drop a perpendicular from P_1 onto the line in the direction AP and find the deformation of AP as shown in Equation (E3).

$$\delta_{AP} = 0.2 \cos 35 = 0.1638 \text{ mm} \qquad \text{(E3)}$$

The average normal strain of member AP is as shown in Equation (E4).

$$\varepsilon_{AP} = \frac{AP_1 - AP}{AP} = \frac{0.1638}{244.155} \qquad \text{(E4)}$$

ANS. $\varepsilon_{AP} = 671.01 \ \mu\text{mm/mm}$

COMMENTS

1. The strain value for part (a) differs from part (b) by 0.028%, which for most engineering calculations is insignificant.

2. Calculation of the deformation in part (a) is more intuitive but requires more algebraic effort than small-strain approximation in part (b). In analyzing a truss, this difference in algebraic effort can be significant.

1.6 STRAIN AT A POINT AND STRAIN–DISPLACEMENT EQUATIONS

Let u, v, and w be displacements in the x, y, and z directions, respectively. The *engineering strain at a point* can be visualized by considering the deformation of a cube of infinitesimal dimensions Δx, Δy, and Δz, shown in Figure 1.24.

As the dimensions of the cube tend to zero, we get the mathematical definition of strain at a point. Because the limiting operation is in a given direction, we obtain partial derivatives, not the total derivatives. The double subscript in the case of shear strain

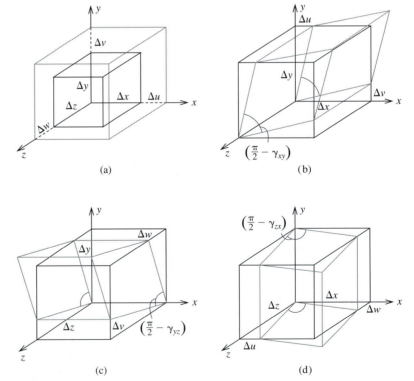

Figure 1.24 Strain components: (a) normal strains, (b) shear strain γ_{xy}, (c) shear strain γ_{yz}, and (d) shear strain γ_{zx}.

indicates the change of angle between two coordinate lines defined by the subscripts. We also record double subscripts for normal strain for sake of consistency, as well as for our subsequent matrix definition of strain. The mathematical definitions of engineering strains at a point are as given in Equations (1.35a) through (1.35f).

$$\varepsilon_{xx} = \lim_{\Delta x \to 0} \left(\frac{\Delta u}{\Delta x} \right) = \frac{\partial u}{\partial x} \tag{1.35a}$$

$$\varepsilon_{yy} = \lim_{\Delta y \to 0} \left(\frac{\Delta v}{\Delta y} \right) = \frac{\partial v}{\partial y} \tag{1.35b}$$

$$\varepsilon_{zz} = \lim_{\Delta z \to 0} \left(\frac{\Delta w}{\Delta z} \right) = \frac{\partial w}{\partial z} \tag{1.35c}$$

$$\gamma_{xy} = \gamma_{yx} = \lim_{\substack{\Delta x \to 0 \\ \Delta y \to 0}} \left(\frac{\Delta u}{\Delta y} + \frac{\Delta v}{\Delta x} \right) = \frac{\partial u}{\partial y} + \frac{\partial v}{\partial x} \tag{1.35d}$$

$$\gamma_{yz} = \gamma_{zy} = \lim_{\substack{\Delta y \to 0 \\ \Delta z \to 0}} \left(\frac{\Delta v}{\Delta z} + \frac{\Delta w}{\Delta y} \right) = \frac{\partial v}{\partial z} + \frac{\partial w}{\partial y} \tag{1.35e}$$

$$\gamma_{zx} = \gamma_{xz} = \lim_{\substack{\Delta x \to 0 \\ \Delta z \to 0}} \left(\frac{\Delta w}{\Delta x} + \frac{\Delta u}{\Delta z} \right) = \frac{\partial w}{\partial x} + \frac{\partial u}{\partial z} \tag{1.35f}$$

Equations (1.35d) through (1.35f) show that shear strain is symmetric and that the order of subscripts is immaterial. The symmetry of shear strain makes intuitive sense. The change of angle between the x and y directions is obviously the same as between the y and x directions. Note that the sign of shear strain depends only upon whether the angle increases or decreases, not upon the order in which we consider x and y.

Figure 1.25 shows the engineering strain components as an engineering strain matrix. The matrix is symmetric because of the symmetry of shear strain. Thus, in three-dimensional problems strain has nine components, but only six of the nine components are independent because of symmetry of shear strain.

We see that like stress, engineering strain has two subscripts, which would seem to suggest that engineering strain is also a second-order tensor like stress. However unlike stress, engineering strain does not satisfy certain coordinate transformation laws, which we will study in Section 1.8. Hence it is not a second-order tensor but is related to it, as shown in Equation (1.36).

$$\boxed{\begin{array}{l} \text{tensor normal strain} = \text{engineering normal strain} \\ \text{tensor shear strain} = \text{engineering shear strain/2} \end{array}} \tag{1.36}$$

$$\begin{bmatrix} \varepsilon_{xx} & \gamma_{xy} & \gamma_{xz} \\ \gamma_{yx} & \varepsilon_{yy} & \gamma_{yz} \\ \gamma_{zx} & \gamma_{zy} & \varepsilon_{zz} \end{bmatrix}$$

Figure 1.25 Engineering strain matrix.

In Section 1.8, we shall see that the factor of 1/2 that changes the engineering shear strain to tensor shear strain[7] permits the extension of stress transformation results to strain transformation.

If the displacement u is only a function of x [i.e., $u(x)$], then the partial derivative in Equation (1.35a) becomes an ordinary derivative resulting in the following equation:

$$\varepsilon_{xx} = \frac{du}{dx}(x) \tag{1.37}$$

1.6.1 Plane Strain

Plane strain is one of the two types of two-dimensional problems[8] encountered in the study of the mechanics of materials. By "two-dimensional," we imply that one of the coordinates does not play a role in the description of the problem. Choosing z to be that coordinate, we set all strains with subscript z to be zero, as shown in the strain matrix in Figure 1.26. Although four components of strain are needed in plane strain, because of the symmetry of shear strain, only three of the four components are independent.

The assumption of plane strain is often made in analyzing very thick bodies such as points around tunnels or mine shafts, or a point in the middle of a thick cylinder such as a submarine hull. In thick bodies, it is argued that to move, a point must push a lot of material in the thickness direction; hence the strains can be expected to be small in the thickness direction. Thus, the strain in the thickness direction is not zero but is small enough to neglect. Plane strain is a mathematical approximation made to simplify the analysis.

$$\begin{bmatrix} \varepsilon_{xx} & \gamma_{xy} & 0 \\ \gamma_{yx} & \varepsilon_{yy} & 0 \\ 0 & 0 & 0 \end{bmatrix}$$

Figure 1.26 Plane strain matrix.

EXAMPLE 1.10

A body under applied loads and body forces produce the displacement field

$$u = 0 \qquad v = Kx(y^2 - z^2) + Kaxz \qquad w = -2Kxyz - Kaxy$$

where u, v, and w are displacements in the x, y, and z directions, respectively. Determine all the strain components at $x = 2a$ in terms of K, a, y, and z.

---

[7] A convenient notation showing the relation of tensor strains and the derivatives is $\varepsilon_{ij} = \frac{1}{2}\left(\frac{\partial u_i}{\partial x_j} + \frac{\partial u_j}{\partial x_i}\right)$, where i and j can be 1, 2, or 3, x_1, x_2, and x_3 are the x, y, and z coordinates, and u_1, u_2, and u_3 are the displacements u, v, and w, respectively.

[8] The differences between plane stress and plane strain are discussed in Section 2.2.

PLAN

The strains can be found from the displacement functions given by means of Equations (1.35a) through (1.35f) and evaluated at $x = 2a$.

SOLUTION

From Equations (1.35a) through (1.35f) we obtain the strains as follows and evaluate them at $x = 2a$.

$$\varepsilon_{xx} = \frac{\partial u}{\partial x} \qquad\qquad \text{ANS. } \varepsilon_{xx} = 0$$

$$\varepsilon_{yy} = \frac{\partial v}{\partial y}\bigg|_{x=2a} = 2Kxy\big|_{x=2a} \qquad\qquad \text{ANS. } \varepsilon_{yy} = 4Kay$$

$$\varepsilon_{zz} = \frac{\partial w}{\partial z}\bigg|_{x=2a} = -2Kxy\big|_{x=2a} \qquad\qquad \text{ANS. } \varepsilon_{zz} = -4Kay$$

$$\gamma_{xy} = \left[\frac{\partial u}{\partial y} + \frac{\partial v}{\partial x}\right]\bigg|_{x=2a} = K(y^2 - z^2) + Kaz\big|_{x=2a} \qquad\qquad \text{ANS. } \gamma_{xy} = K(y^2 - z^2) + Kaz$$

$$\gamma_{yz} = \left[\frac{\partial v}{\partial z} + \frac{\partial w}{\partial y}\right]\bigg|_{x=2a} = [-2Kxz + Kax - 2Kxz - Kax]\big|_{x=2a} \qquad\qquad \text{ANS. } \gamma_{yz} = -8Kaz$$

$$\gamma_{xz} = \left[\frac{\partial w}{\partial x} + \frac{\partial u}{\partial z}\right]\bigg|_{x=2a} = [-2Kyz - Kay]\big|_{x=2a} \qquad\qquad \text{ANS. } \gamma_{xz} = -Ky[2z + a]$$

COMMENT

The example demonstrates the calculation of strains from given displacement functions. In Example 2.3 stresses will be calculated from these strains. In Chapter 3, the calculation of internal forces and moments from stresses will be discussed. In this manner it will be demonstrated that if the displacement field is known in an elastic body, then all the variables of the mechanics of materials can be found from the displacement functions.

1.7 FINITE DIFFERENCE APPROXIMATION OF STRAINS

When a derivative is approximated by using a finite length in place of infinitesimal length, we call it a finite difference approximation. From calculus we know that the derivative of a function is the slope of the tangent to the curve defined by the function. In calculating this slope of the tangent by means of finite lengths, do we

- Use the length in front of the point?

- Use the length behind the point?
- Use the length in front and behind to calculate an average value?

The choice made in answering these three questions leads to three methods: forward difference, backward difference, and central difference.

To distinguish among the three methods, consider a plot of u vs x, which we use to find the approximate value of strain ε_{xx} at point i in Figure 1.27. The strain at the point is the slope of the tangent to the actual function at point i.

- *Forward difference* approximates the slope of the tangent by using the point ahead of point i as

$$(\varepsilon_{xx})_i = \frac{u_{i+1} - u_i}{x_{i+1} - x_i} \qquad (1.38a)$$

- *Backward difference* approximates the slope of the tangent by using the point behind i as

$$(\varepsilon_{xx})_i = \frac{u_i - u_{i-1}}{x_i - x_{i-1}} \qquad (1.38b)$$

- *Central difference* takes the average value by using the point ahead and behind as

$$(\varepsilon_{xx})_i = \frac{1}{2}\left[\frac{u_{i+1} - u_i}{x_{i+1} - x_i} + \frac{u_i - u_{i-1}}{x_i - x_{i-1}}\right] \qquad (1.38c)$$

The central difference approximation bisects the slope of the lines that yield the forward and backward differences. As can be seen from Figure 1.27, the best approximation is from the central difference. But if the point is on the left-most boundary, there is no point before it and only a forward difference approximation can be used. Similarly, if the point is right-most boundary, only a backward difference approximation can be used.

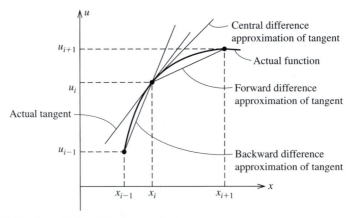

Figure 1.27 The three finite difference methods.

EXAMPLE 1.11

The displacements u and v in the x and y directions, respectively, were measured by Moiré interferometry.[9] Figure 1.28 gives the displacements of nine points on the body. Determine the strains ε_{xx}, ε_{yy}, and γ_{xy} at point 5 using (a) forward difference, (b) backward difference, and (c) central difference and the data in Table 1.3.

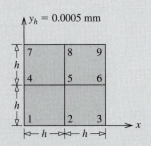

Figure 1.28 (a) Grid for Example 1.11.

TABLE 1.3 Displacements for Example 1.11

Point	u (μmm)	v (μmm)
1	0.000	0.0000
2	0.625	−0.3125
3	1.500	−0.5000
4	−0.500	−0.5625
5	0.250	−1.1250
6	1.250	−1.5625
7	−1.250	−1.250
8	−0.375	−2.0625
9	0.750	−2.7500

PLAN

(a) For forward difference we use points 5, 6, and 8. (b) For backward difference we use points 2, 4, and 5. (c) For central difference we use points 2, 4, 5, 6, and 8.

(a) <u>Forward difference</u>: Δx is the distance between points 5 and 6 and Δy is the difference between points 5 and 7. We use thus points 5, 6, and 8 to find strains.

$$\varepsilon_{xx} = \frac{u_6 - u_5}{x_6 - x_5} = \frac{1.0(10^{-6})}{0.0005} = 2000(10^{-6}) \quad \textbf{ANS.} \quad \varepsilon_{xx} = 2000 \text{ μmm/mm}$$

$$\varepsilon_{yy} = \frac{v_8 - v_5}{y_8 - y_5} = \frac{-0.9375(10^{-6})}{0.0005} = -1875(10^{-6}) \quad \textbf{ANS.} \quad \varepsilon_{yy} = -1875 \text{ μmm/mm}$$

$$\gamma_{xy} = \frac{u_8 - u_5}{y_8 - y_5} + \frac{v_6 - v_5}{x_6 - x_5} = \frac{(-0.625)(10^{-6})}{0.0005} + \frac{(-0.4375)(10^{-6})}{0.0005} = -2125(10^{-6})$$

$$\textbf{ANS.} \quad \gamma_{xy} = -2125 \text{ μrad}$$

(b) <u>Backward difference</u>: Δx is the distance between points 4 and 5 and Δy is the difference between points 2 and 5. We use the points 2, 4, and 5 to find the strains.

$$\varepsilon_{xx} = \frac{u_5 - u_4}{x_5 - x_4} = \frac{0.75(10^{-6})}{0.0005} = 1500(10^{-6}) \quad \textbf{ANS.} \quad \varepsilon_{xx} = 1500 \text{ μmm/mm}$$

---

[9] Moiré methods measure displacements of points on a grid. In moiré interferometry, displacements of grid points are measured by means of the interference of light. See Post [1993] in Appendix D for additional details.

$$\varepsilon_{yy} = \frac{v_5 - v_2}{y_5 - y_2} = \frac{-0.8125(10^{-6})}{0.0005} = -1625(10^{-6}) \quad \textbf{ANS.} \quad \varepsilon_{yy} = -1625 \ \mu mm/mm$$

$$\gamma_{xy} = \frac{u_5 - u_2}{y_5 - y_2} + \frac{v_5 - v_4}{x_5 - x_4} = \frac{(-0.375)(10^{-6})}{0.0005} + \frac{(-0.5625)(10^{-6})}{0.0005} = -1875(10^{-6})$$

$$\textbf{ANS.} \quad \gamma_{xy} = -1875 \ \mu rad$$

(c) <u>Central difference</u>: The central difference represents the mean value of the forward and backward difference values of strains.

$$\varepsilon_{xx} = \left(\frac{1}{2}\right)\left[\frac{u_6 - u_5}{x_6 - x_5} + \frac{u_5 - u_4}{x_5 - x_4}\right] \quad \textbf{ANS.} \quad \varepsilon_{xx} = 1750 \ \mu mm/mm$$

$$\varepsilon_{yy} = \left(\frac{1}{2}\right)\left[\frac{v_8 - v_5}{y_8 - y_5} + \frac{v_5 - v_2}{y_5 - y_2}\right] \quad \textbf{ANS.} \quad \varepsilon_{yy} = -1750 \ \mu mm/mm$$

$$\gamma_{xy} = \left(\frac{1}{2}\right)\left[\frac{u_8 - u_5}{y_8 - y_5} + \frac{v_6 - v_5}{x_6 - x_5} + \frac{u_5 - u_2}{y_5 - y_2} + \frac{v_5 - v_4}{x_5 - x_4}\right] \quad \textbf{ANS.} \quad \gamma_{xy} = -2000 \ \mu rad$$

COMMENTS

1. The three methods give three different results because the strains calculated are average strains. Average values depend upon the sample points used in averaging.

2. The results for central difference for all three strain components are between the results for forward and backward differences, which will always be the case, as elaborated by Figure 1.27.

1.8 STRAIN TRANSFORMATION

In the introductory course on the mechanics of materials, the strains (ε_{nn}, ε_{tt}, and γ_{nt}) in the coordinate system n-t-z shown Figure 1.11(c), were related to the strains (ε_{xx}, ε_{yy}, and γ_{xy}) in the Cartesian coordinate system as shown by Equations (1.39). The angle θ is measured from the x axis and is considered to be positive in the counterclockwise direction.

$$\varepsilon_{nn} = \varepsilon_{xx}\cos^2\theta + \varepsilon_{yy}\sin^2\theta + \gamma_{xy}\sin\theta\cos\theta \tag{1.39a}$$

$$\varepsilon_{tt} = \varepsilon_{xx}\sin^2\theta + \varepsilon_{yy}\cos^2\theta - \gamma_{xy}\sin\theta\cos\theta \tag{1.39b}$$

$$\gamma_{nt} = -2\varepsilon_{xx}\sin\theta\cos\theta + 2\varepsilon_{yy}\sin\theta\cos\theta + \gamma_{xy}(\cos^2\theta - \sin^2\theta) \tag{1.39c}$$

$$[\varepsilon] = \begin{bmatrix} \varepsilon_{xx} & \varepsilon_{xy} = \gamma_{xy}/2 & \varepsilon_{xz} = \gamma_{xz}/2 \\ \varepsilon_{yx} = \gamma_{yx}/2 & \varepsilon_{yy} & \varepsilon_{yz} = \gamma_{yz}/2 \\ \varepsilon_{zx} = \gamma_{zx}/2 & \varepsilon_{zy} = \gamma_{zy}/2 & \varepsilon_{zz} \end{bmatrix}$$

Figure 1.29 Tensor strain matrix obtained from an engineering strain matrix.

Equations (1.39) are similar to the stress transformation relations, Equations (1.7) with the difference that the coefficient of shear stress term is twice the coefficient of the shear strain term. This difference exists because we are using engineering strain instead of tensor strain. If we convert the engineering strains to tensor strains, by dividing all engineering shear strains by 2, then the results for the matrix method in stress transformation can be adapted for strain transformation. Figure 1.29 shows the tensor strain matrix that is obtained from the engineering strains.

Consider two perpendicular directions n and t with $\{n\}$ and $\{t\}$ representing the unit vectors in these directions, respectively. Analogous to Equations (1.11a) through (1.11c), we obtain the strain transformation equations for the normal and shear strain in the n-t directions given by Equations (1.40a) through (1.40c).

$$\boxed{\varepsilon_{nn} = \{n\}^T [\varepsilon]\{n\}} \qquad (1.40a)$$

$$\boxed{\varepsilon_{nt} = \{t\}^T [\varepsilon]\{n\}} \qquad (1.40b)$$

$$\boxed{\varepsilon_{tt} = \{t\}^T [\varepsilon]\{t\}} \qquad (1.40c)$$

It should be noted that the shear strain ε_{nt} in Equation (1.40b) is tensor shear strain, which must be multiplied by 2 to obtain the engineering shear strain γ_{nt}.

The strain vector in the direction of n is analogous to the stress vector and can be defined as shown in Equation (1.41).

$$\{E\} = [\varepsilon]\{n\} \qquad (1.41)$$

1.8.1 Principal Strains

We recall the following definitions from the introductory course on the mechanics of material.

Definition 23 The principal coordinate directions for strain are the axes in which the shear strain is zero.

Definition 24 Normal strains in the principal directions are called principal strains.

Definition 25 The greatest principal strain is called principal strain 1.

Definition 26 The angles the principal axis makes with the global coordinate system are called principal angles.

In plane strain, the shear strains with subscripts z are zero, which by Definition 23 implies that the principal direction lies along the z axis. In plane strain problems we will label the z direction as principal direction 3. There are three principal strains at a point, which are labeled as ε_1, ε_2, and ε_3.

The principal strains ε_p are the eigenvalues of the tensor strain matrix, and the principal directions $\{p\}$ are the eigenvectors of the strain matrix, which are found in exactly the same manner as the principal stresses and the associated principal directions. The characteristic equation analogous to Equation (1.25) is

$$\varepsilon_p^3 - I_1\varepsilon_p^2 + I_2\varepsilon_p - I_3 = 0 \tag{1.42}$$

and its roots of the characteristic yield the principal strains. The coefficients I_1, I_2, and I_3 are called the strain invariants because their value is independent of coordinate transformation. Notice that the sign in terms containing the odd stress invariants is negative.

The strain invariants I_1, I_2, and I_3, analogous to stress invariants in Equations (1.26a), (1.26b), and (1.26c) are as given in Equations (1.43a), (1.43b), and (1.43c).

$$\boxed{I_1 = \varepsilon_{xx} + \varepsilon_{yy} + \varepsilon_{zz}} \tag{1.43a}$$

$$\boxed{I_2 = \begin{vmatrix} \varepsilon_{xx} & \varepsilon_{xy} \\ \varepsilon_{yx} & \varepsilon_{yy} \end{vmatrix} + \begin{vmatrix} \varepsilon_{yy} & \varepsilon_{yz} \\ \varepsilon_{zy} & \varepsilon_{zz} \end{vmatrix} + \begin{vmatrix} \varepsilon_{xx} & \varepsilon_{xz} \\ \varepsilon_{zx} & \varepsilon_{zz} \end{vmatrix}} \tag{1.43b}$$

$$\boxed{I_3 = \begin{vmatrix} \varepsilon_{xx} & \varepsilon_{xy} & \varepsilon_{xz} \\ \varepsilon_{yx} & \varepsilon_{yy} & \varepsilon_{yz} \\ \varepsilon_{zx} & \varepsilon_{zy} & \varepsilon_{zz} \end{vmatrix}} \tag{1.43c}$$

The strain invariants in terms of principal strains are given in Equations (1.44a), (1.44b), and (1.44c).

$$\boxed{I_1 = \varepsilon_1 + \varepsilon_2 + \varepsilon_3} \tag{1.44a}$$

$$\boxed{I_2 = \begin{vmatrix} \varepsilon_1 & 0 \\ 0 & \varepsilon_2 \end{vmatrix} + \begin{bmatrix} \varepsilon_2 & 0 \\ 0 & \sigma_3 \end{bmatrix} + \begin{bmatrix} \varepsilon_1 & 0 \\ 0 & \varepsilon_3 \end{bmatrix} = \varepsilon_1\varepsilon_2 + \varepsilon_2\varepsilon_3 + \varepsilon_3\varepsilon_1} \tag{1.44b}$$

$$\boxed{I_3 = \begin{vmatrix} \varepsilon_1 & 0 & 0 \\ 0 & \varepsilon_2 & 0 \\ 0 & 0 & \varepsilon_3 \end{vmatrix} = \varepsilon_1\varepsilon_2\varepsilon_3} \tag{1.44c}$$

The strain invariants, particularly the first and the third, provide a quick and easy check on the laboriously calculated principal strains.

The maximum engineering shear strain is related to the difference in principal strains and can be expressed as shown in Equation (1.45).

$$\boxed{\frac{\gamma_{max}}{2} = \left| \max\left(\frac{\varepsilon_1 - \varepsilon_2}{2}, \frac{\varepsilon_2 - \varepsilon_3}{2}, \frac{\varepsilon_3 - \varepsilon_1}{2}\right) \right|} \tag{1.45}$$

EXAMPLE 1.12

The engineering strains at a point on a building frame were determined by means of the finite element method[10] as $\varepsilon_{xx} = 600~\mu$, $\varepsilon_{yy} = 400~\mu$, $\varepsilon_{zz} = 200~\mu$, $\gamma_{xy} = 800~\mu$, $\gamma_{yz} = 400~\mu$, $\gamma_{zx} = 600~\mu$. Determine (a) the normal strain along a line oriented at 40°, 60°, and 66.2° with the x, y, and z axes, respectively, (b) the principal strains and the maximum engineering shear strain, and (c) principal strain direction 1.

PLAN

(a) We can construct the tensor strain matrix from the given strains. We can then find the normal strain by first finding the strain vector from Equation (1.41) and then finding the normal strain from Equation (1.40a). (b) We can find the strain invariants and then find the roots of the characteristic equations by using the formulas given earlier in footnote 6 and thus obtain the principal strains. The maximum engineering shear strain can be found from Equation (1.45). (c) We can find the eigenvectors as we did in Example 1.8.

SOLUTION

The tensor strain matrix can be written as shown in Equation (E1).

$$[\varepsilon] = \begin{bmatrix} 600 & 400 & 300 \\ 400 & 400 & 200 \\ 300 & 200 & 200 \end{bmatrix} \mu \tag{E1}$$

(a) We can find the direction cosine from the given angles as shown in Equation (E2).

$$n_x = \cos 40 = 0.7660 \qquad n_y = \cos 60 = 0.5 \qquad n_z = \cos 66.2 = 0.4035 \tag{E2}$$

We can find the strain vector from Equation (1.41) and then the normal strain from Equation (1.40a) as shown in Equation (E3).

$$\{E\} = [\varepsilon]\{n\} = \begin{bmatrix} 600 & 400 & 300 \\ 400 & 400 & 200 \\ 300 & 200 & 200 \end{bmatrix} \begin{Bmatrix} 0.7660 \\ 0.5 \\ 0.4035 \end{Bmatrix} = \begin{Bmatrix} 780.65 \\ 587.1 \\ 410.5 \end{Bmatrix}$$

$$\varepsilon_{nn} = \{n\}^T [\varepsilon]\{n\} = \begin{Bmatrix} 0.7660 \\ 0.5 \\ 0.4035 \end{Bmatrix}^T \begin{Bmatrix} 780.65 \\ 587.1 \\ 410.5 \end{Bmatrix} = 597.86 + 293.55 + 165.6 \tag{E3}$$

ANS. $\varepsilon_{nn} = 1057~\mu$

[10] See Chapter 9 for additional details on the finite element method.

(b) We can find the three strain invariants from Equations (1.43a), (1.43b), and (1.43c). The determinant of the matrix is evaluated by using the first row as the pivots.

$$I_1 = 600 + 400 + 200 = 1200 \ \mu \tag{E4}$$

$$I_2 = \begin{vmatrix} 600 & 400 \\ 400 & 400 \end{vmatrix} + \begin{vmatrix} 400 & 200 \\ 200 & 200 \end{vmatrix} + \begin{vmatrix} 600 & 300 \\ 300 & 200 \end{vmatrix} = 150(10^3) \ \mu^2 \tag{E5}$$

$$I_3 = 600 \begin{vmatrix} 400 & 200 \\ 200 & 200 \end{vmatrix} - 400 \begin{vmatrix} 400 & 200 \\ 300 & 200 \end{vmatrix} + 300 \begin{vmatrix} 400 & 400 \\ 300 & 200 \end{vmatrix} = 4000(10^3) \ \mu^3 \tag{E6}$$

The characteristic equation thus is given by Equation (E7).

$$\varepsilon_p^3 - 1,200\varepsilon_p^2 + 150(10^3)\varepsilon_p - 4000(10^3) = 0 \tag{E7}$$

The roots of the equation can be found from the formulas given in footnote 6 and as shown by Equations (E8) through (E13).

$$A = \sqrt{(1200/3)^2 - 150(10^3)/3} = 331.67 \tag{E8}$$

$$\cos 3\alpha = 2(1200/3)^3 + -(1200/3)(150)(10^3)$$

$$+ [4000(10^3)]]\}/[2(331.67)^3] = 0.9867 \tag{E9}$$

$$\alpha = 9.356/3 = 3.119 \tag{E10}$$

$$x_1 = 2(331.67)\cos 3.119 + 1200/3 = 1062.36 \tag{E11}$$

$$x_2 = -2(331.67)\cos(63.119) + 1200/3 = 100.07 \tag{E12}$$

$$x_3 = -2(331.67)\cos(-56.88) + 1200/3 = 37.57 \tag{E13}$$

The principal strains are as follows:

ANS. $\varepsilon_1 = 1062 \ \mu \qquad \varepsilon_2 = 100 \ \mu \qquad \varepsilon_3 = 37.6 \ \mu$

The maximum engineering shear strain is

$$\frac{\gamma_{max}}{2} = \frac{\varepsilon_1 - \varepsilon_3}{2} \qquad \textbf{ANS.} \ \gamma_{max} = 1024.7 \ \mu$$

Checking Results: We can check our results by finding strain invariants 1 and 3 from Equations (1.44a) and (1.44c) as follows:

$$I_1 = 1062.36 + 100.07 + 37.57 = 1200 \qquad \text{Checks.}$$

$$I_3 = (1062.36)(100.07)(37.57) = 4.01(10^6) \qquad \text{Checks.}$$

(c) On substituting ε_1 in the equations corresponding to the first two rows of the matrix, we obtain Equations (E14) and (E15).

$$(600 - 1062.36)p_x + 400p_y + 300p_z = 0 \quad \text{or} \quad 400p_y + 300p_z = 462.4p_x \quad \text{(E14)}$$

$$400p_x + (400 - 1062.36)p_y + 200p_z = 0 \quad \text{or} \quad -662.4p_y + 200p_z = -400p_x \quad \text{(E15)}$$

Solving Equations (E14) and (E15) in terms of p_x we obtain Equations (E16) and (E17).

$$p_y = 0.7624 p_x \quad \text{(E16)}$$

$$p_z = 0.5247 p_x \quad \text{(E17)}$$

Noting that the sum of square of direction cosine is one, we obtain Equation (E18).

$$p_x^2 + (0.7624 p_x)^2 + (0.5247 p_x)^2 = 1 \quad \text{or} \quad p_x = \pm 0.7339 \quad \text{(E18)}$$

Substituting p_x into Equations (E16) and (E17), we obtain $p_y = \pm 0.5595$ and $p_z = \pm 0.3851$. Two solutions are possible for principal direction:

$$\textbf{ANS.} \quad \begin{Bmatrix} \theta_x = 42.8° \\ \theta_y = 56° \\ \theta_z = 67.4° \end{Bmatrix} \quad \text{or} \quad \begin{Bmatrix} \theta_x = 137.2° \\ \theta_y = 124° \\ \theta_z = 112.6° \end{Bmatrix}$$

Checking Results: If we use Equation (1.40a) to find the normal stress in principal direction 3, we should obtain principal stress 3 as follows:

$$\varepsilon_1 = \{p_1\}^T \begin{bmatrix} 600 & 400 & 300 \\ 400 & 400 & 200 \\ 300 & 200 & 200 \end{bmatrix} \begin{Bmatrix} 0.7339 \\ 0.5595 \\ 0.3851 \end{Bmatrix} = \begin{Bmatrix} 0.7339 \\ 0.5595 \\ 0.3851 \end{Bmatrix}^T \begin{Bmatrix} 779.7 \\ 594.4 \\ 409.1 \end{Bmatrix} = 1062.3 \quad \text{Checks.}$$

COMMENT

This example shows that once an engineering strain matrix has been converted to a tensor strain matrix, the calculations in strain transformation are similar to those in stress transformation.

PROBLEM SET 1.3

Problems on Small-Strain Approximation

1.41 Use a small-strain approximation to determine the deformation in bars AP and BP in Figure P1.41.

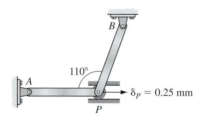

$\delta_P = 0.25$ mm

Figure P1.41

1.42 Use a small-strain approximation to determine the deformation in bars AP and BP in Figure P1.42.

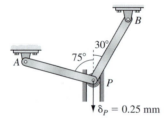

$\delta_P = 0.25$ mm

Figure P1.42

ANS. $\delta_{AP} = 0.0647$ mm;
$\delta_{BP} = 0.2165$ mm

1.43 Use a small-strain approximation to determine the deformation in bars AP and BP in Figure P1.43.

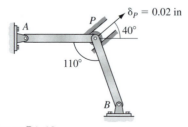

$\delta_P = 0.02$ in

Figure P1.43

1.44 The axial displacement due to the weight of a tapered circular bar that is hanging vertically was found as follows:

$$u(x) = \left[-19.44 + 1.44x - 0.01x^2 - \frac{933.12}{(72 - x)} \right]$$
$$\times (10^{-3}) \text{ in}$$

Determine the axial strain ε_{xx} at $x = 24$ ins.

ANS. $\varepsilon_{xx} = 555\ \mu\text{in/in}$

1.45 The axial displacement due to the weight of a tapered rectangular bar that is hanging vertically was found as follows:

$$u(x) = [-50x + 20x^2 + 2.5\ \ln(1 - 0.8x)]$$
$$\times (10^{-6}) \text{mm}$$

Determine the axial strain ε_{xx} at $x = 150$ mm.

1.46 The axial displacement in a quadratic one-dimensional finite element is as given in Figure P1.46 and in the following equation.

$$u(x) = \frac{u_1}{2a^2}(x - a)(x - 2a) - \frac{u_2}{a^2}(x)(x - 2a)$$
$$+ \frac{u_3}{2a^2}(x)(x - a)$$

Determine the strain at node 2.

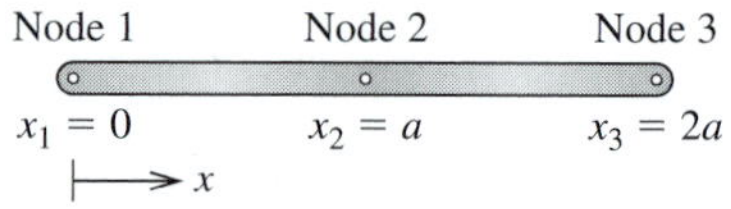

Node 1	Node 2	Node 3
$x_1 = 0$	$x_2 = a$	$x_3 = 2a$

$\longmapsto x$

Figure P1.46

1.47 The axial strain in a bar of length L was found to be $\varepsilon_{xx} = KL/(3L - 2x)$, $0 \le x \le L$, where K is a constant for a given material, loading, and cross-sectional dimensions. Determine the total extension in terms of K and L.

1.48 The axial strain in a bar of length L that is due to its own weight was found to be

$$\varepsilon_{xx} = K\left[4L - 2x - \frac{8L^3}{(4L - 2x)^2} \right] \qquad 0 \le x \le L$$

where K is a constant for a given material and cross-sectional dimensions. Determine the total extension in terms of K and L.

ANS. $u(L) - u(0) = 2KL^2$

1.49 Determine the total extension of the composite bar comprising N axial bars securely fastened together if the strain in the ith section is shown in Figure P1.49, with

$$\varepsilon_i = a_i, \qquad x_{i-1} \le x \le x_i.$$

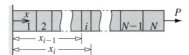

Figure P1.49

1.50 True strain (ε_T) is calculated from $d\varepsilon_T = (du)/(L_0 + u)$, where u is the deformation at any given instance and L_0 is the original undeformed length. Thus the increment in true strain is the ratio of change in length at any instant to the length at that given instant. If ε represents engineering strain, show that at any instant the relationship between true strain and engineering strain is given by the following:

$$\varepsilon_T = \ln(1 + \varepsilon) \qquad (1.46)$$

1.51 A differential element subjected to only normal strains is shown in Figure P1.51. The ratio of change in volume (ΔV) to the original volume (V) is called the volumetric strain (ε_V) or *dilatation*. Prove

$$\varepsilon_V = \Delta V / V = I_1 \qquad (1.47)$$

where I_1 is the first strain invariant.

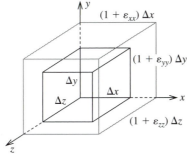

Figure P1.51

1.52 In the unsymmetric bending of beams, the displacements are as follows:

$$u = -y\frac{dv}{dx} - z\frac{dw}{dx} \qquad v = v(x) \qquad w = w(x)$$

Determine all the strain components at a point on the beam.

1.53 For the following displacements in the bending of a thin plate, determine all the strain components at a point on the plate.

$$u = -z\frac{\partial w}{\partial x} \qquad v = -z\frac{\partial w}{\partial y} \qquad w = w(x, y)$$

1.54 The displacements in a body are given by

$$u = 500x - 300y - 200 \ \mu\text{in}$$

$$v = -100x + 200y + 250 \ \mu\text{in}$$

Determine the strains ε_{xx}, ε_{yy}, and γ_{xy} at $x = 3$ in and $y = 2.5$ in.

ANS. $\varepsilon_{xx} = 500\,\mu$ $\varepsilon_{yy} = 200\,\mu$ $\gamma_{xy} = -400\,\mu$

1.55 The displacements in a body are given by

$$u = [0.5(x^2 - y^2) + 0.5xy](10^{-3}) \ \text{mm}$$

$$v = [0.25(x^2 - y^2) - xy](10^{-3}) \ \text{mm}$$

Determine the strains ε_{xx}, ε_{yy}, and γ_{xy} at $x = 5$ mm and $y = 7$ mm.

ANS. $\varepsilon_{xx} = 8500\,\mu\text{mm/mm}$
$\varepsilon_{yy} = -8500\,\mu\text{mm/mm}$ $\gamma_{xy} = -9000\,\mu\text{rad}$

1.56 In polar coordinates, let the displacement in the radial direction be given by u and displacement in the tangential direction be given by v. It can be shown that the strains in the r and θ coordinates are related to the displacements as follows:

$$\varepsilon_{rr} = \frac{\partial u}{\partial r} \qquad \varepsilon_{\theta\theta} = \frac{u}{r} + \frac{1}{r}\frac{\partial v}{\partial \theta}$$

$$\gamma_{x\theta} = \frac{1}{r}\frac{\partial u}{\partial \theta} + \frac{\partial v}{\partial r} - \frac{v}{r} \qquad (1.48)$$

The displacements at a point in polar coordinates are

$$u = K \cos\theta(\ln r)$$

$$v = -K \sin\theta(\ln r + 1)$$

where K is a constant. Use Equation 1.48 to determine the strains ε_{rr}, $\varepsilon_{\theta\theta}$, and $\gamma_{r\theta}$.

ANS. $\varepsilon_{rr} = (K \cos \theta)/r$
$\varepsilon_{\theta\theta} = -(K \cos \theta)/r \quad \gamma_{r\theta} = 0$

1.57 The displacements at a point in polar coordinates are

$$u = r^2(-C_1 \cos 3\theta + C_2 \sin 3\theta)$$

$$v = r^2(-C_1 \sin 3\theta + C_2 \cos 3\theta)$$

where C_1 and C_2 are constants. Use Equation (1.48) to determine the strains ε_{rr}, $\varepsilon_{\theta\theta}$, and $\gamma_{r\theta}$.

1.58 The displacements u and v in the x and y direction, respectively, were measured by Moiré interferometry. Displacements of 16 points on the body are as given in Table P1.58.

TABLE P1.58 Displacements for Problem 1.58

Point	u (μmm)	v (μmm)
1	0.000	0.000
2	−0.112	0.144
3	−0.128	0.256
4	−0.048	0.336
5	0.112	0.176
6	−0.032	0.224
7	−0.080	0.240
8	−0.032	0.224
9	0.128	0.384
10	−0.048	0.336
11	−0.128	0.256
12	−0.112	0.144
13	0.048	0.624
14	−0.160	0.480
15	−0.272	0.304
16	−0.288	0.096

Determine the strains ε_{xx}, ε_{yy}, and γ_{xy} at points 1 and 4 in Figure P1.58.

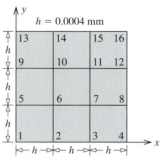

Figure P1.58

1.59 Determine the strains ε_{xx}, ε_{yy}, and γ_{xy} at points 13 and 16 in Figure P1.58.

ANS. point 13: $\varepsilon_{xx} = -520 \,\mu$mm/mm
$\varepsilon_{yy} = 600 \,\mu$mm/mm $\quad \gamma_{xy} = -560 \,\mu$mm/mm
point 16: $\varepsilon_{xx} = -40 \,\mu$mm/mm
$\varepsilon_{yy} = -120 \,\mu$mm/mm $\quad \gamma_{xy} = -960 \,\mu$mm/mm

---

1.60 Use central difference to determine the strains ε_{xx}, ε_{yy}, and γ_{xy} at point 6 in Figure P1.58.

ANS. $\varepsilon_{xx} = -240 \,\mu$mm/mm
$\varepsilon_{yy} = 240 \,\mu$mm/mm $\quad \gamma_{xy} = 160 \,\mu$mm/mm

---

1.61 Use central difference to determine the strains ε_{xx}, ε_{yy}, and γ_{xy} at point 7 in Figure P1.58.

1.62 Use central difference to determine the strains ε_{xx}, ε_{yy}, and γ_{xy} at point 10 in Figure P1.58.

1.63 Use central difference to determine the strains ε_{xx}, ε_{yy}, and γ_{xy} at point 11 in Figure P1.58.

1.64 At a point on the frame of an aircraft, the engineering strains were determined as $\varepsilon_{xx} = 1000 \,\mu$, $\varepsilon_{yy} = -400 \,\mu$, $\varepsilon_{zz} = 200 \,\mu$, $\gamma_{xy} = 300 \,\mu$, $\gamma_{yz} = 0$, and $\gamma_{zx} = 0$. Determine (a) the normal strain along a line oriented at 70.5°, 48.2°, and 48.2° with the x, y, and z axis, respectively, (b) the principal strains and maximum engineering shear strain, and (c) the principal strain direction 1.

ANS. $\varepsilon_{nn} = 89.3 \,\mu$, $\varepsilon_1 = 1015.9 \,\mu$,
$\varepsilon_2 = 200.0 \,\mu$, $\varepsilon_3 = -415.9 \,\mu$, $\gamma_{max} = 1431.8 \,\mu$,
$\theta_x = 6.0°$, $\theta_y = 83.9°$, $\theta_z = 90.0°$

1.65 At a point on a building frame, the engineering strains were determined as $\varepsilon_{xx} = 720\ \mu$, $\varepsilon_{yy} = 320\ \mu$, $\varepsilon_{zz} = 0$, $\gamma_{xy} = 210\ \mu$, $\gamma_{yz} = 0$, and $\gamma_{zx} = 210\ \mu$. Determine (a) the normal strain along a line oriented at 70.5°, 48.2°, and 48.2° with the x, y, and z axis, respectively, and (b) the principal strains and maximum engineering shear strain, and (c) the principal strain direction 1.

ANS. $\varepsilon_{nn} = 315.8\ \mu$, $\varepsilon_1 = 760\ \mu$, $\varepsilon_2 = 296.1\ \mu$, $\varepsilon_3 = -15.7\ \mu$, $\gamma_{max} = 775.7\ \mu$, $\theta_x = 15.4°$, $\theta_y = 76.7°$, $\theta_z = 82.4°$

1.9 CLOSURE

In this chapter we saw that stress on a surface is an internal distributed force system that is a vector quantity, while stress at a point is a second-order tensor whose sign is determined from the direction of the outward normal of an imaginary surface through the point and the direction of the internal force. We also saw that engineering strain, which is not a second-order tensor, may be transformed into a second-order tensor by dividing all shear strains at that point by a factor of 2. Principal stresses and strains are the eigenvalues, and principal directions are the eigenvectors, of the stress and tensor strain matrices. We saw that the principal stresses and strains at a point are unique, which implies that the coefficients of the characteristic equation are unique. The coefficients of the characteristic equations are the stress and strain invariants that do not depend upon the coordinate system—an observation we used to check our calculations of principal stresses and strains.

We saw that small-strain approximation is a linear relationship between displacements and strains that is valid for strains less than 1%. Small-strain approximations allow significant simplification, including calculation of normal strain, in which we approximate the actual deformation by the component of the deformation in the original direction of the line element, and shear strain, in which the sine and tangent functions of shear strain are approximated by their arguments. The derivatives of displacements in small strains can be approximated by the finite difference method. Central difference should be used when possible because this method produces more accurate results. However forward and backward differences may have to be used near boundaries.

2 Material Description

A material can be described qualitatively and quantitatively. Qualitative description by adjectives such as elastic, linear, ductile, and tough forms the engineering language of material description. Quantitative descriptions of a material are the equations relating stresses and strains that are established experimentally, usually from a tension or torsion test made on machines such as those shown in Figure 2.1. A material model, also called a constitutive model, is the qualitative and quantitative approximation of material behavior.

The number of parameters or material constants to be experimentally determined depends upon the material model chosen to relate stresses and strains. In this chapter we discuss several linear material models. Some inelastic and nonlinear material models are discussed in Chapter 5.

In a tension test, there is only one stress component. How do we apply the value of failure stress from the tension test to a state of stress in which all nine stress components are present? The answer to this question is provided by failure theories, some of which we will study in this chapter.

The elementary structural theories studied in an introductory course on the mechanics of materials give formulas obtained by making many simplifying assumptions. In regions of sudden changes in geometries and cracks, the basic assumptions of the elementary theories are no longer valid. Can we extrapolate results from our formulas into these regions? We will study two concepts by which we can extrapolate the results of elementary theory for use in these regions near sudden changes in geometry and cracks. These two concepts are *stress concentration factor* and *stress intensity factor.*

The learning objectives in this chapter are:

1. To understand the definitions in and differences between various linear material models.

2. To understand the statements and the applications of failure theories.

3. To understand the concepts and applications of stress concentration factor and stress intensity factor in analysis and design.

Figure 2.1 Machines for testing (a) tension and (b) torsion. (*Courtesy of Professor I. Miskioglu.*)

2.1 LINEAR MATERIAL MODEL

In the mechanics of materials, the experimental relationship between stresses and strains is incorporated into the logical framework of mechanics to produce formulas for the analysis and design of structural members. If a material model does not fit the experimental data well, there will be a high degree of error in the theoretical predictions. Yet a material model that fits the experimental data very accurately may be so complex that no analytical model (theory) can be built. Thus, the choice of material model is dictated both by the experimental data and by the accuracy needs of the analysis.

The simplest material model is a linear relationship between stresses and strains. With no additional assumptions, the linear relationship of the six strain components to six stress components can be written as Equation (2.1a).

$$
\begin{aligned}
\varepsilon_{xx} &= C_{11}\sigma_{xx} + C_{12}\sigma_{yy} + C_{13}\sigma_{zz} + C_{14}\tau_{yz} + C_{15}\tau_{zx} + C_{16}\tau_{xy} \\
\varepsilon_{yy} &= C_{21}\sigma_{xx} + C_{22}\sigma_{yy} + C_{23}\sigma_{zz} + C_{24}\tau_{yz} + C_{25}\tau_{zx} + C_{26}\tau_{xy} \\
\varepsilon_{zz} &= C_{31}\sigma_{xx} + C_{32}\sigma_{yy} + C_{33}\sigma_{zz} + C_{34}\tau_{yz} + C_{35}\tau_{zx} + C_{36}\tau_{xy} \\
\gamma_{yz} &= C_{41}\sigma_{xx} + C_{42}\sigma_{yy} + C_{43}\sigma_{zz} + C_{44}\tau_{yz} + C_{45}\tau_{zx} + C_{46}\tau_{xy} \\
\gamma_{zx} &= C_{51}\sigma_{xx} + C_{52}\sigma_{yy} + C_{53}\sigma_{zz} + C_{54}\tau_{yz} + C_{55}\tau_{zx} + C_{56}\tau_{xy} \\
\gamma_{xy} &= C_{61}\sigma_{xx} + C_{62}\sigma_{yy} + C_{63}\sigma_{zz} + C_{64}\tau_{yz} + C_{65}\tau_{zx} + C_{66}\tau_{xy}
\end{aligned}
\tag{2.1a}
$$

Equation (2.1a) can be written in matrix form as Equation (2.1b).

$$
\begin{Bmatrix}
\varepsilon_{xx} \\
\varepsilon_{yy} \\
\varepsilon_{zz} \\
\gamma_{yz} \\
\gamma_{zx} \\
\gamma_{xy}
\end{Bmatrix}
=
\begin{bmatrix}
C_{11} & C_{12} & C_{13} & C_{14} & C_{15} & C_{16} \\
C_{12} & C_{22} & C_{23} & C_{24} & C_{25} & C_{26} \\
C_{13} & C_{23} & C_{33} & C_{34} & C_{35} & C_{36} \\
C_{41} & C_{42} & C_{43} & C_{44} & C_{45} & C_{46} \\
C_{51} & C_{52} & C_{53} & C_{54} & C_{55} & C_{56} \\
C_{61} & C_{62} & C_{63} & C_{64} & C_{65} & C_{66}
\end{bmatrix}
\begin{Bmatrix}
\sigma_{xx} \\
\sigma_{yy} \\
\sigma_{zz} \\
\tau_{yz} \\
\tau_{zx} \\
\tau_{xy}
\end{Bmatrix}
\tag{2.1b}
$$

The matrix C_{ij} in Equation (2.1b) is called the compliance matrix. Equation (2.1b) implies that we need 36 material constants to describe the most general linear relationship between stresses and strains. However, it can be shown that the matrix formed by the constants C_{ij} is a symmetric matrix; that is, $C_{ij} = C_{ji}$, where i and j can be any number from 1 to 6. This symmetry can be proved by using the requirement that the strain energy[1] must always be positive, but the proof of symmetry is beyond the scope of this book. The symmetry requirement reduces the number of independent constants to 21 for the most general linear relationship between stress and strain.

Equation (2.1b) presupposes that the relation between the stress and the strain in the x direction is different from the relation of stress and strain in the y or z directions. Alternatively stated, Equation (2.1b) implies that if we apply a force (stress) in the x direction and observe the deformation (strain), this deformation will be different from the deformation that would be produced if we applied the same force in the y direction. This phenomenon is not observable by the naked eye for most metals, but for metals at the crystal-size level, the number of constants needed to describe the stress–strain relationship does depend upon the crystal structure. Thus we must ask whether we are conducting the analysis at eye level or at crystal-size level. For analysis at eye level, we average the impact of the crystal structure, to obtain the simplest material description, defined as follows.

Definition 1 An *isotropic material* has stress–strain relationships that are independent of the orientation of the coordinate system at a point.

An *anisotropic* material is a material that is not isotropic. The most general anisotropic material requires 21 independent material constants to describe its linear stress–strain relationships. An *isotropic* body requires only *two[2] independent* material constants to describe its linear stress–strain relationships. Some of the factors that influence whether we treat a material as isotropic or anisotropic are the degree of difference in material properties with orientation, the scale at which the analysis is being conducted, and the kind of information that is desired from the analysis.

Homogeneity is another approximation that is often used to describe a material behavior.

Definition 2 A material is said to be *homogeneous* if the material properties are the same at all points on the body. Alternatively, if the material constants C_{ij} are functions of the coordinates x, y, or z, the material is called nonhomogeneous.

Isotropic–homogeneous, anisotropic–homogeneous, isotropic–nonhomogeneous, and anisotropic–nonhomogeneous are all possible descriptions of material behavior.

Most materials at the atomic level, crystalline level, or grain-size level are nonhomogeneous. The treatment of a material as homogeneous or nonhomogeneous depends once more on the type of information that is required from the analysis. Homogenization of material properties is the process of averaging different material properties by an overall material property. Any body can be treated as a homogeneous body if the scale at which analysis is conducted is made sufficiently large.

---

[1] Strain energy is the energy stored in a material owing to deformation, as discussed in Section 7.1.

[2] The general proof is beyond the scope of the book.

There are 31 types of crystal. Crystalline bodies can be grouped into classes for the purpose of defining the independent material constants needed in the linear stress–strain relationship. In between the isotropic material and the most general anisotropic material, there are material groups of several types, some of which are discussed briefly in Sections 2.1.1 through 2.1.4.

2.1.1 Monoclinic Materials

The z plane is the plane of symmetry in monoclinic materials. This implies that the stress–strain relationship is the same in the positive z and negative z directions. In matrix form, the stress–strain relationship for monoclinic materials is given by Equation (2.2).

$$
\begin{Bmatrix} \varepsilon_{xx} \\ \varepsilon_{yy} \\ \varepsilon_{zz} \\ \gamma_{yz} \\ \gamma_{zx} \\ \gamma_{xy} \end{Bmatrix} = \begin{bmatrix} C_{11} & C_{12} & C_{13} & 0 & 0 & C_{16} \\ C_{12} & C_{22} & C_{23} & 0 & 0 & C_{26} \\ C_{13} & C_{23} & C_{33} & 0 & 0 & C_{36} \\ 0 & 0 & 0 & C_{44} & C_{45} & 0 \\ 0 & 0 & 0 & C_{45} & C_{55} & 0 \\ C_{16} & C_{26} & C_{36} & 0 & 0 & C_{66} \end{bmatrix} \begin{Bmatrix} \sigma_{xx} \\ \sigma_{yy} \\ \sigma_{zz} \\ \tau_{yz} \\ \tau_{zx} \\ \tau_{xy} \end{Bmatrix} \tag{2.2}
$$

Equation (2.2) shows that a *monoclinic material* requires 13 independent material constants in three dimensions.

2.1.2 Orthotropic Materials

Orthotropic materials have two orthogonal planes of symmetry. That is, if we rotate a sample by 90° about the x or the y axis, we will obtain the same stress–strain relation. In matrix form, the stress–strain relationship for orthotropic materials is given by Equation (2.3).

$$
\begin{Bmatrix} \varepsilon_{xx} \\ \varepsilon_{yy} \\ \varepsilon_{zz} \\ \gamma_{yz} \\ \gamma_{zx} \\ \gamma_{xy} \end{Bmatrix} = \begin{bmatrix} C_{11} & C_{12} & C_{13} & 0 & 0 & 0 \\ C_{12} & C_{22} & C_{23} & 0 & 0 & 0 \\ C_{13} & C_{23} & C_{33} & 0 & 0 & 0 \\ 0 & 0 & 0 & C_{44} & 0 & 0 \\ 0 & 0 & 0 & 0 & C_{55} & 0 \\ 0 & 0 & 0 & 0 & 0 & C_{66} \end{bmatrix} \begin{Bmatrix} \sigma_{xx} \\ \sigma_{yy} \\ \sigma_{zz} \\ \tau_{yz} \\ \tau_{zx} \\ \tau_{xy} \end{Bmatrix} \tag{2.3}
$$

Equation (2.3) shows that an *orthotropic material* requires nine independent constants in three dimensions. For orthotropic materials, the normal strains are not affected by the shear stresses, and the shear strains are not affected by the normal stresses, which is not the case for monoclinic and general anisotropic materials. If we transform the coordinate system from the x-y-z coordinate system, the form of Equation (2.3) will change. In the transformed coordinate system, many of the zeros in the compliance matrix may become nonzero, the normal strains may be affected by the shear stresses, and the shear strains may be affected by the normal stresses. Thus the form of equations in Equation (2.3) is valid for a specific coordinate system, the *material coordinate system*.

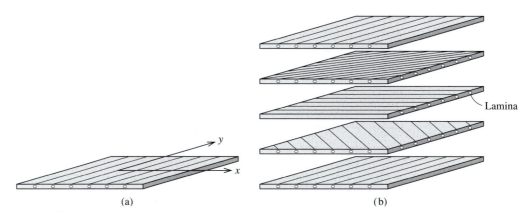

Figure 2.2 (a) A composite lamina. (b) Construction of a laminated composite.

To demonstrate the physical significance of material coordinate systems, consider the lamina of a composite material shown in Figure 2.2(a). Such a lamina would be produced by laying long fibers in a given direction and binding them together with epoxy. Fibers are inherently stiffer and stronger than the bulk material. The increase in strength and stiffness is due to the reduction of defects and alignment of crystals along the fiber axis. Clearly the mechanical properties will be different in the direction of the fiber and in the direction perpendicular to the fiber. If the properties of the fiber and the epoxies are averaged[3] (homogenized), each lamina can be regarded as an orthotropic material, and the directions parallel and perpendicular to the fiber are the material axis directions.

Laminae with different fiber orientations can be put together to create a composite laminate,[4] as shown in Figure 2.2(b). The overall properties of the laminate can be controlled by the orientation of the fibers and the stacking sequence of the laminae. For certain stacking sequences, the laminate will respond like an orthotropic material.

For plane stress problems, the orthotropic stress–strain relationships using engineering constants are as written in Equations (2.4a) through (2.4d),

$$\varepsilon_{xx} = \frac{\sigma_{xx}}{E_x} - \frac{\nu_{yx}}{E_y}\sigma_{yy} \qquad (2.4a)$$

$$\varepsilon_{yy} = \frac{\sigma_{yy}}{E_y} - \frac{\nu_{xy}}{E_x}\sigma_{xx} \qquad (2.4b)$$

$$\gamma_{xy} = \frac{\tau_{xy}}{G_{xy}} \qquad (2.4c)$$

$$\frac{\nu_{yx}}{E_y} = \frac{\nu_{xy}}{E_x} \qquad (2.4d)$$

---

[3] See Section 4.4 on micromechanical properties of lamina composites.

[4] Continuous-fiber composites technology is still expensive in comparison to that of metals, but the significant weight reduction justifies its use in the aerospace industry and in specialty sports equipment.

where E_x and E_y are the moduli of elasticity in the x and y (material axis) directions, ν_{xy} and ν_{yx} are the Poisson ratios obtained when the tension test is conducted in the x and y directions, respectively, and G_{xy} is the shear modulus of elasticity. Note that Equation (2.4d) implies that of the five material constants, only four are independent.

2.1.3 Transversely Isotropic Materials

Transversely isotropic materials are isotropic in a plane. That is, rotation by an arbitrary angle about the z axis does not change the stress–strain relations, and the material is isotropic in the x-y plane. In matrix form, the stress–strain relationship for transversely isotropic material is given by Equation (2.5).

$$
\begin{Bmatrix} \varepsilon_{xx} \\ \varepsilon_{yy} \\ \varepsilon_{zz} \\ \gamma_{yz} \\ \gamma_{zx} \\ \gamma_{xy} \end{Bmatrix} =
\begin{bmatrix}
C_{11} & C_{12} & C_{13} & 0 & 0 & 0 \\
C_{12} & C_{11} & C_{13} & 0 & 0 & 0 \\
C_{13} & C_{13} & C_{33} & 0 & 0 & 0 \\
0 & 0 & 0 & C_{44} & 0 & 0 \\
0 & 0 & 0 & 0 & C_{44} & 0 \\
0 & 0 & 0 & 0 & 0 & 2(C_{11}-C_{12})
\end{bmatrix}
\begin{Bmatrix} \sigma_{xx} \\ \sigma_{yy} \\ \sigma_{zz} \\ \tau_{yz} \\ \tau_{zx} \\ \tau_{xy} \end{Bmatrix}
\tag{2.5}
$$

Equation (2.5) implies that *transversely isotropic material* requires five independent material constants in three dimensions. Common transversely isotropic materials are fiberglass panels and other short-fiber composites.[5] One way of producing short-fiber composites is to spray fibers onto epoxy and cure the resulting mixture. The random orientation of the fibers results in overall (homogenized) isotropic material behavior within the plane of the panel. Perpendicular to the fibers, we expect a different material response, and hence the stress–strain relation perpendicular to the plane of the panel (fibers) is different, resulting in an overall transversely isotropic material response.

2.1.4 Isotropic Materials

An *isotropic material* requires only two independent material constants. Rotation about the x, y, or z axis by any arbitrary angle results in the same stress–strain relationship. In matrix notation, the stress–strain relationship for isotropic material is given by Equation (2.6).

$$
\begin{Bmatrix} \varepsilon_{xx} \\ \varepsilon_{yy} \\ \varepsilon_{zz} \\ \gamma_{yz} \\ \gamma_{zx} \\ \gamma_{xy} \end{Bmatrix} =
\begin{bmatrix}
C_{11} & C_{12} & C_{12} & 0 & 0 & 0 \\
C_{12} & C_{11} & C_{12} & 0 & 0 & 0 \\
C_{12} & C_{12} & C_{11} & 0 & 0 & 0 \\
0 & 0 & 0 & 2(C_{11}-C_{12}) & 0 & 0 \\
0 & 0 & 0 & 0 & 2(C_{11}-C_{12}) & 0 \\
0 & 0 & 0 & 0 & 0 & 2(C_{11}-C_{12})
\end{bmatrix}
\begin{Bmatrix} \sigma_{xx} \\ \sigma_{yy} \\ \sigma_{zz} \\ \tau_{yz} \\ \tau_{zx} \\ \tau_{xy} \end{Bmatrix}
\tag{2.6}
$$

---

[5] Chopped fibers, which are cheaper to produce than continuous-fiber composites, are finding increasing use in the automobile and marine industries for designing secondary structures, such as body panels.

Equation (2.6) can be written using three constants[6]: the modulus of elasticity E, the Poisson ratio ν, and the shear modulus of elasticity G as shown in Equations (2.7a) through (2.7f).

$$\varepsilon_{xx} = [\sigma_{xx} - \nu(\sigma_{yy} + \sigma_{zz})]/E \qquad (2.7a)$$

$$\varepsilon_{yy} = [\sigma_{yy} - \nu(\sigma_{zz} + \sigma_{xx})]/E \qquad (2.7b)$$

$$\varepsilon_{zz} = [\sigma_{zz} - \nu(\sigma_{xx} + \sigma_{yy})]/E \qquad (2.7c)$$

$$\gamma_{xy} = \tau_{xy}/G \qquad (2.7d)$$

$$\gamma_{yz} = \tau_{yz}/G \qquad (2.7e)$$

$$\gamma_{zx} = \tau_{zx}/G \qquad (2.7f)$$

Comparing Equations (2.7a) through (2.7f) with Equation (2.6), we obtain the following:

$$C_{11} = 1/E \qquad C_{12} = -\nu/E \qquad 2(C_{11} - C_{12}) = 1/G$$

Substituting C_{11} and C_{12} in the last relationship, we obtain

$$G = \frac{E}{2(1 + \nu)} \qquad (2.8)$$

The set of Equations (2.7a) through (2.7f), along with Equation (2.8), are called the *generalized* (version of) *Hooke's law.* It is valid only for linear, elastic, isotropic materials at any point, including a point in a nonhomogeneous material. An alternative form[7] for Equations (2.7a) through (2.7c) that may be easier to remember is the matrix form given by Equation (2.9).

$$\begin{Bmatrix} \varepsilon_{xx} \\ \varepsilon_{yy} \\ \varepsilon_{zz} \end{Bmatrix} = \frac{1}{E} \begin{bmatrix} 1 & -\nu & -\nu \\ -\nu & 1 & -\nu \\ -\nu & -\nu & 1 \end{bmatrix} \begin{Bmatrix} \sigma_{xx} \\ \sigma_{yy} \\ \sigma_{zz} \end{Bmatrix} \qquad (2.9)$$

The generalized Hooke's law for isotropic material is valid for *any orthogonal coordinate system.* Thus, we could write equivalent forms for Equations (2.7a) through (2.7f) for cylinderical (polar) coordinates (r, θ, z) and for spherical coordinates (r, θ, ϕ) (see Problems 2.1 and 2.2). The principal coordinate systems for stresses and strains are also orthogonal. For *only* isotropic materials the principal directions for stresses and strains are identical, hence we can write the generalized Hooke's law in principal coordinate form as shown in Equations (2.10a) through (2.10c).

$$\varepsilon_1 = [\sigma_1 - \nu(\sigma_2 + \sigma_3)]/E \qquad (2.10a)$$

---

[6] There are other constants used to describe material properties (see Problem 2.14 and footnote 8). For isotropic materials, however, only two are independent constants (i.e., all other constants can be found if any two constants are known).

[7] Another alternative is $\varepsilon_{ii} = [(1 + \nu)\sigma_{ii} - \nu I_1]/E$ where $I_1 = \sigma_{xx} + \sigma_{yy} + \sigma_{zz}$.

$$\boxed{\varepsilon_2 = [\sigma_2 - \nu(\sigma_3 + \sigma_1)]/E} \tag{2.10b}$$

$$\boxed{\varepsilon_3 = [\sigma_3 - \nu(\sigma_1 + \sigma_2)]/E} \tag{2.10c}$$

It should be emphasized that the Equations (2.10a) through (2.10c) relating principal stresses and principal strains are not valid for orthotropic and other anisotropic materials because the principal directions for stresses and strains are different.

The normal stresses can be solved in terms of normal strains by using the Equations (2.7a) through (2.7c) to obtain[8] Equation (2.11).

$$\sigma_{xx} = [(1-\nu)\varepsilon_{xx} + \nu\varepsilon_{yy} + \nu\varepsilon_{zz}]\frac{E}{(1-2\nu)(1+\nu)}$$

$$\sigma_{yy} = [(1-\nu)\varepsilon_{yy} + \nu\varepsilon_{zz} + \nu\varepsilon_{xx}]\frac{E}{(1-2\nu)(1+\nu)} \tag{2.11}$$

$$\sigma_{zz} = [(1-\nu)\varepsilon_{zz} + \nu\varepsilon_{xx} + \nu\varepsilon_{yy}]\frac{E}{(1-2\nu)(1+\nu)}$$

2.2 PLANE STRESS AND PLANE STRAIN

If we take the definitions of plane stress and plane strain and apply them to Equations (2.7a) through (2.7f), we will obtain the right-most matrices in Figure 2.3. The differences between the pair of two-dimensional idealizations of material behavior are in the zero and nonzero values of the normal strain and normal stress in the z direction. In plane stress, $\sigma_{zz} = 0$, which from Equation (2.7c) implies that the normal strain in the z direction is $\varepsilon_{zz} = -\nu/E(\sigma_{xx} + \sigma_{yy})$. In plane strain, $\varepsilon_{zz} = 0$, which from Equation (2.7c) implies that the normal stress in the z direction is $\sigma_{zz} = \nu(\sigma_{xx} + \sigma_{yy})$.

To better appreciate the difference between plane stress and plane strain, consider the two plates shown in Figure 2.4, on which only compressive normal stresses in the x and y

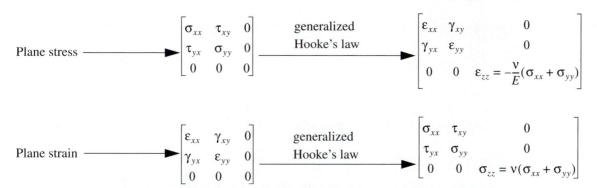

Figure 2.3 Stress and strain matrices in plane stress and plane strain.

[8]An alternative form that is easier to remember is $\sigma_{ii} = 2G\varepsilon_{ii} + \lambda(I_1)$, where i can be x, y, or z; $I_1 = \varepsilon_{xx} + \varepsilon_{yy} + \varepsilon_{zz}$; G is the shear modulus; and $\lambda = (2G\nu)/(1-2\nu)$ is called Lamé's constant, named after G. Lamé (1795–1870). See Timoshenko [1983] in Appendix D for additional details.

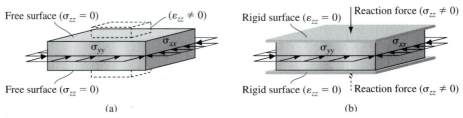

Figure 2.4 (a) Plane stress. (b) Plane strain.

direction are applied. The top and bottom surfaces on the plate in Figure 2.4(a) are free surfaces (plane stress); but because the plate is free to expand, the deformation (strain) in the z direction is not zero. The plate in Figure 2.4(b) is constrained from expanding in the z direction by the rigid plates. As the material pushes on the plate, a reaction force develops, and this reaction force results in a nonzero value of normal stress in the z direction.

Though the example in Figure 2.4 helps explain the difference, it should be emphasized that plane stress or plane strain often is an approximation made to simplify analysis. The approximation of plane stress is often made in thin plates and shells, such as the skin of an aircraft or the walls of pressure vessels. The approximation of plane strain is often made in thick plates and shells, such as submarine hulls or points beneath the earth's surface. A difference of an order of magnitude or more between the thickness and other dimensions is generally used to classify a body as "thin."

It should be recognized that in plane strain and plane stress there are only three independent quantities—even though the number of nonzero quantities is more than three. For example, if we know σ_{xx}, σ_{yy}, and τ_{xy}, we can calculate ε_{xx}, ε_{yy}, γ_{xy}, ε_{zz}, and σ_{zz} for plane stress and plane strain. Similarly, if we know ε_{xx}, ε_{yy}, and γ_{xy}, we can calculate σ_{xx}, σ_{yy}, τ_{xy}, σ_{zz}, and ε_{zz} for plane stress and plane strain. Thus, in both plane stress and plane strain, the number of independent stress or strain components is three, although the number of nonzero components is greater than three. Example 2.1 elaborates on the differences between plane stress and plane strain and the differences between nonzero and independent quantities.

EXAMPLE 2.1

The stresses at a point on steel were found to be $\sigma_{xx} = 15$ ksi (T), $\sigma_{yy} = 30$ ksi (C), and $\tau_{xy} = 25$ ksi. Using $E = 30{,}000$ ksi and $G = 12{,}000$ ksi, determine the strains ε_{xx}, ε_{yy}, γ_{xy}, ε_{zz}, and stress σ_{zz} under two assumptions: (a) the point is in plane stress and (b) the point is in plane strain.

PLAN

In both cases the shear strain is the same and can be calculated by using Equation (2.7d). (a) For plane stress $\sigma_{zz} = 0$, and the strains ε_{xx}, ε_{yy}, and ε_{zz} can be found from Equations (2.7a), (2.7b), and (2.7c), respectively. (b) For plane strain $\varepsilon_{zz} = 0$, and Equation (2.7c) can be used to find σ_{zz}. The stresses σ_{xx}, σ_{yy}, and σ_{zz} can be substituted in Equations (2.7a) and (2.7b) to calculate the normal strains ε_{xx} and ε_{yy}.

SOLUTION

From Equation (2.7d), $\gamma_{xy} = \tau_{xy}/G = 25/12(10^3)$, or   **ANS.** $\gamma_{xy} = 2083\,\mu$

The Poisson ratio can be found from Equation (2.8) as shown in Equation (E1).

$$G = \frac{E}{2(1+\nu)} \qquad 12,000 = \frac{30,000}{2(1+\nu)} \qquad \nu = 0.25 \qquad\qquad \text{(E1)}$$

(a) *Plane stress:* the normal strains in the x, y, and z directions are found from Equations (2.7a), (2.7b), and (2.7c), respectively.

$$\varepsilon_{xx} = \frac{[\sigma_{xx} - \nu(\sigma_{yy} + \sigma_{zz})]}{E} = \frac{[15 - 0.25(-30)]}{30,000} \qquad \text{or} \qquad \textbf{ANS.} \quad \varepsilon_{xx} = 750\,\mu$$

$$\varepsilon_{yy} = \frac{[\sigma_{yy} - \nu(\sigma_{zz} + \sigma_{xx})]}{E} = \frac{[(-30) - 0.25(15)]}{30,000} \qquad \text{or} \qquad \textbf{ANS.} \quad \varepsilon_{yy} = -1125\,\mu$$

$$\varepsilon_{zz} = \frac{[\sigma_{zz} - \nu(\sigma_{xx} + \sigma_{yy})]}{E} = \frac{[0 - 0.25(15 - 30)]}{30,000} \qquad \text{or} \qquad \textbf{ANS.} \quad \varepsilon_{zz} = 125\,\mu$$

(b) *Plane strain:* from Equation (2.7c), we have $\varepsilon_{zz} = [\sigma_{zz} - \nu(\sigma_{xx} + \sigma_{yy})]/E = 0$, or

$$\sigma_{zz} = \nu(\sigma_{xx} + \sigma_{yy}) = 0.25(15 - 30) \qquad \text{or} \qquad \textbf{ANS.} \quad \sigma_{zz} = 3.75 \text{ ksi (C)}$$

The normal strains in the x and y directions are found from Equations (2.7a) and (2.7b).

$$\varepsilon_{xx} = \frac{[\sigma_{xx} - \nu(\sigma_{yy} + \sigma_{zz})]}{E} = \frac{[15 - 0.25(-30 - 3.75)]}{30,000} \qquad \text{or } \textbf{ANS.} \quad \varepsilon_{xx} = 781.2\,\mu$$

$$\varepsilon_{yy} = \frac{[\sigma_{yy} - \nu(\sigma_{zz} + \sigma_{xx})]}{E} = \frac{[(-30) - 0.25(15 - 3.75)]}{30,000} \qquad \text{or } \textbf{ANS.} \quad \varepsilon_{yy} = -1094\,\mu$$

COMMENTS

1. The three independent quantities in this problem were σ_{xx}, σ_{yy}, and τ_{xy}. Knowing these, we were able to find all the strains in plane stress and plane strain.

2. The difference in the values of the strains came from the zero value of σ_{zz} for plane stress and the value of 3.75 ksi (C) in plane strain.

EXAMPLE 2.2

The stresses $\sigma_{xx} = 4$ ksi (T), $\sigma_{yy} = 10$ ksi (C), and $\tau_{xy} = 4$ ksi were calculated at a point on a free surface of an orthotropic composite material. Determine (a) principal stresses 1 and 2 and principal direction 1 for stresses and (b) principal strains 1 and 2 and principal direction 1 for strains. Use the following values for the material constants: $E_x = 7500$ ksi, $E_y = 2500$ ksi, $G_{xy} = 1250$ ksi, and $\nu_{xy} = 0.3$.

PLAN

(a) We can find the eigenvalues of the stress matrix to obtain the principal stresses and find the eigenvector associated with principal stress 1. (b) By substituting the stresses and material constants in Equations (2.4a) through (2.4d), the strains ε_{xx}, ε_{yy}, and γ_{xy} can be found. We can find the eigenvalues of the tensor strain matrix to find both the principal strains and the eigenvector associated with principal strain 1.

SOLUTION

(a) We can write the matrix form in unknowns of direction cosines as shown in Equation (E1).

$$\begin{bmatrix} 4 - \sigma_p & 4 \\ 4 & -10 - \sigma_p \end{bmatrix} \begin{Bmatrix} p_x \\ p_y \end{Bmatrix} = 0 \tag{E1}$$

The following characteristic equation can be obtained from Equation (E1):

$$[(4 - \sigma_p)(-10 - \sigma_p) - 16] = 0 \quad \text{or} \quad \sigma_p^2 + 6\sigma_p - 56 = 0 \tag{E2}$$

The two roots in Equation (E2) are 5.062 and 11.062. Following the convention that $\sigma_1 > \sigma_2$, we obtain the principal stresses: **ANS.** $\sigma_1 = 5.062$ ksi (T) $\sigma_2 = 11.062$ ksi (C)

Checking Results: We note that $\sigma_1 + \sigma_2 = -6$ and $\sigma_{xx} + \sigma_{yy} = -6$, which validates the results in accordance with Equation (1.8).

We can substitute σ_1 in the first row of Equation (E1) to obtain Equation (E3).

$$-1.062 p_x + 4 p_y = 0 \quad \text{or} \quad p_x = 3.7665 p_y \tag{E3}$$

We note that the square of direction cosines is 1. We thus obtain Equation (E4).

$$(3.7665 p_y)^2 + (p_y)^2 = 1 \quad \text{or} \quad p_y = \pm 0.2566 \tag{E4}$$

From Equation (E3) we can find two solutions for p_x (i.e., $p_x = \pm 0.9665$). The two possible solutions for principal direction 1 are as follows:

ANS. $\begin{Bmatrix} \theta_x = 14.87° \\ \theta_y = 75.13° \end{Bmatrix}$ or $\begin{Bmatrix} \theta_x = -165.1° \\ \theta_y = 104.87° \end{Bmatrix}$

Checking Results: If we use Equation (1.11a) to find the normal stress in principal direction 1, we should obtain principal stress 1 as follows:

$$\sigma_1 = \{p\}^T \begin{bmatrix} 4 & 4 \\ 4 & -10 \end{bmatrix} \begin{Bmatrix} 0.9665 \\ 0.2566 \end{Bmatrix} = \begin{Bmatrix} 0.9665 \\ 0.2566 \end{Bmatrix}^T \begin{Bmatrix} 4.8924 \\ 1.300 \end{Bmatrix} = 5.062 \qquad \text{Checks.}$$

(b) From Equation (2.4d), we obtain Equation (E5).

$$\nu_{yx} = \frac{E_y \nu_{xy}}{E_x} = \left(\frac{2500}{7500}\right)(0.3) = 0.1 \tag{E5}$$

Substituting the stresses and material constants in Equations (2.4a) through (2.4c), we obtain the strains given in Equations (E6) through (E8).

$$\varepsilon_{xx} = \frac{\sigma_{xx}}{E_x} - \frac{\nu_{yx}}{E_y}\sigma_{yy} = \frac{4}{7500} - \frac{0.1}{2500}(-10) = 0.933(10^{-3}) = 933\ \mu \tag{E6}$$

$$\varepsilon_{yy} = \frac{\sigma_{yy}}{E_y} - \frac{\nu_{xy}}{E_x}\sigma_{xx} = \frac{(-10)}{2500} - \frac{(0.3)}{7500}(4) = -4.160(10^{-3}) = -4160\ \mu \tag{E7}$$

$$\gamma_{xy} = \frac{\tau_{xy}}{G_{xy}} = \frac{4}{1250} = 3.200(10^{-3}) = 3200\ \mu \tag{E8}$$

We can write the strain tensor matrix form in unknowns of the direction cosines as Equation (E9).

$$\begin{bmatrix} 933 - \varepsilon_p & 1600 \\ 1600 & -4160 - \varepsilon_p \end{bmatrix} \begin{Bmatrix} p_x \\ p_y \end{Bmatrix} = 0 \tag{E9}$$

The characteristic equation can be obtained from the Equation (E9) as shown in Equation (E10).

$$[(933 - \varepsilon_p)(-4160 - \varepsilon_p) - 1600^2] = 0 \qquad \text{or}$$

$$\varepsilon_p^2 + 3227\sigma_p - 6441.3(10^3) = 0 \tag{E10}$$

The two roots in Equation (E10) are 1393.93 and −4620.93. Following the convention that $\varepsilon_1 > \varepsilon_2$, we obtain the principal strains as **ANS.** $\varepsilon_1 = 1393.93\ \mu$ $\quad \varepsilon_2 = -4620.93\ \mu$

Checking Results: We note that $\varepsilon_1 + \varepsilon_2 = -3227$ and $\varepsilon_{xx} + \varepsilon_{yy} = -3227$, which checks. We can substitute ε_1 in the first row of Equation (E9) to obtain Equation (E11).

$$-460.93p_x + 1600p_y = 0 \qquad \text{or} \qquad p_x = 3.4712p_y \tag{E11}$$

We note that the square of the direction cosines is one. We thus obtain Equation (E12).

$$(3.4712p_y)^2 + (p_y)^2 = 1 \qquad \text{or} \qquad p_y = \pm 0.2768 \tag{E12}$$

From Equation (E11) we can find two solutions for p_x as $p_x = \pm 0.9609$. The two possible solution for principal direction 1 are as follows:

$$\textbf{ANS.} \quad \begin{Bmatrix} \theta_x = 16.07° \\ \theta_y = 73.93° \end{Bmatrix} \quad \text{or} \quad \begin{Bmatrix} \theta_x = 163.93° \\ \theta_y = 106.07° \end{Bmatrix}$$

Checking Results: If we use Equation (1.40a) to find the normal strain in principal direction 1, we should obtain principal strain 1 as follows:

$$\varepsilon_1 = \{p\}^T \begin{bmatrix} 933 & 1600 \\ 1600 & -4160 \end{bmatrix} \begin{Bmatrix} 0.9609 \\ 0.2768 \end{Bmatrix} = \begin{Bmatrix} 0.9609 \\ 0.2768 \end{Bmatrix}^T \begin{Bmatrix} 1339.46 \\ 385.87 \end{Bmatrix} = 1393.9 \qquad \text{Checks.}$$

COMMENTS

1. The results in this example show that for orthotropic materials, the principal directions for stresses and strains are different.

2. For isotropic materials, however, the principal directions for stresses and strains are same. For a given value of stress, if we change the material constants for the isotropic material, the strain values will be different, but the result for the principal angle for strain will not change. However, if we change the material constants for orthotropic materials, not only do we change the strain values, but the principal angle for strain may also change because there may be a change in the degree of orthotropy (i.e., the degree of difference in material constants in the x and y directions).

3. The first two comments highlight some of the reasons why intuitive experience, usually based on work with isotropic materials, can be misleading when one is working with composite materials, which generally are not isotropic. In such cases, mathematical rigor can provide answers, which once confirmed by experiment, can form a new knowledge base on which to develop intuitive understanding.

EXAMPLE 2.3

A body under applied loads and body forces produces the following displacement field

$$u = 0 \qquad v = Kx(y^2 - z^2) + Kaxz \qquad w = -2Kxyz - Kaxy$$

where u, v, and w are displacements in the x, y, and z directions, respectively. Assume a linear, elastic, isotropic, homogeneous material with modulus of elasticity E and Poisson ratio ν. Determine all the stress components at $x = 2a$ in terms of K, a, E, ν, y, and z.

PLAN

The strains from the displacement field were found in Example 1.10. The stresses can be found by using Equations (2.11), (2.7d), (2.7e), and (2.7f).

SOLUTION

From Example 1.10 we have the strains shown in Equations (E1) and (E2).

$$\varepsilon_{xx} = 0 \qquad \varepsilon_{yy} = 4Kay \qquad \varepsilon_{zz} = -4Kay \qquad \text{(E1)}$$

$$\gamma_{xy} = K(y^2 - z^2) + Kaz \qquad \gamma_{yz} = -8Kaz \qquad \gamma_{xz} = -Ky[2z + a] \qquad \text{(E2)}$$

From Equations (2.11), (2.7d), (2.7e), and (2.7f) we obtain the stresses as calculated in Equations (E3) through (E8).

$$\sigma_{xx} = [\nu(4Kay) + \nu(-4Kay)]\frac{E}{(1 - 2\nu)(1 + \nu)} \qquad \text{or} \qquad \textbf{ANS.} \quad \sigma_{xx} = 0 \qquad \text{(E3)}$$

$$\sigma_{yy} = [(1 - \nu)(4Kay) + \nu(-4Kay)]\frac{E}{(1 - 2\nu)(1 + \nu)} \qquad \text{or} \qquad \text{(E4)}$$

$$\textbf{ANS.} \quad \sigma_{yy} = \frac{4EKay}{(1 + \nu)}$$

$$\sigma_{zz} = [(1 - \nu)(-4Kay) + \nu(4Kay)]\frac{E}{(1 - 2\nu)(1 + \nu)} \qquad \text{or} \qquad \text{(E5)}$$

$$\textbf{ANS.} \quad \sigma_{zz} = -\left[\frac{4EKay}{(1 + \nu)}\right]$$

$$\tau_{xy} = G\gamma_{xy} \qquad \text{or} \qquad \textbf{ANS.} \quad \tau_{xy} = \frac{EK}{2(1 + \nu)}[(y^2 - z^2) + az] \qquad \text{(E6)}$$

$$\tau_{yz} = G\gamma_{yz} = \frac{E}{2(1 + \nu)}[-8Kaz] \qquad \text{or} \qquad \textbf{ANS.} \quad \tau_{yz} = -\left[\frac{4KEaz}{(1 + \nu)}\right] \qquad \text{(E7)}$$

$$\tau_{xz} = G\gamma_{yz}\nu = \frac{E}{2(1 + \nu)}[-Ky(2z + a)] \qquad \text{or} \qquad \text{(E8)}$$

$$\textbf{ANS.} \quad \tau_{xz} = -\left[\frac{EKy}{2(1 + \nu)}(2z + a)\right]$$

COMMENT

Knowing the stresses at a cross section located at $x = 2a$, one could find principal stresses at any point on the cross section. Also, as will be discussed in Chapter 3, one could find the equivalent internal forces and moments.

PROBLEM SET 2.1

Problems on Material Description

2.1 Use modulus of elasticity E, Poisson's ratio ν, and shear modulus of elasticity G to write the generalized version of Hooke's law for an isotropic material in cylindrical coordinates (r, θ, z).

2.2 Use modulus of elasticity E, Poisson's ratio ν, and shear modulus of elasticity G to write the generalized version of Hooke's law for isotropic material in spherical coordinates (r, θ, ϕ).

2.3 Use $E = 200$ GPa and $\nu = 0.32$ to calculate ε_{xx}, ε_{yy}, γ_{xy}, ε_{zz}, and σ_{zz} (a) assuming plane stress and (b) assuming plane strain for the following stresses at a point:

$$\sigma_{xx} = 100 \text{ MPa (T)} \qquad \sigma_{yy} = 150 \text{ MPa (T)}$$
$$\tau_{xy} = -125 \text{ MPa}$$

ANS.   (a) $\varepsilon_{xx} = 260$ μ, $\varepsilon_{yy} = 590$ μ, $\gamma_{xy} = -1650$ μ, $\varepsilon_{zz} = -400$ μ   (b) $\varepsilon_{xx} = 132$ μ, $\varepsilon_{yy} = 462$ μ, $\sigma_{zz} = 80$ MPa (T)

2.4 Use $E = 70$ GPa and $G = 28$ GPa, to calculate ε_{xx}, ε_{yy}, γ_{xy}, ε_{zz}, and σ_{zz} (a) assuming plane stress and (b) assuming plane strain for the following stresses:

$$\sigma_{xx} = 225 \text{ MPa (C)} \qquad \sigma_{yy} = 125 \text{ MPa (T)}$$
$$\tau_{xy} = 150 \text{ MPa}$$

2.5 Use $E = 30,000$ ksi and $\nu = 0.3$ to calculate ε_{xx}, ε_{yy}, γ_{xy}, ε_{zz}, and σ_{zz} (a) assuming plane stress and (b) assuming plane strain for the following stresses at a point:

$$\sigma_{xx} = 22 \text{ ksi (C)} \qquad \sigma_{yy} = 25 \text{ ksi (C)}$$
$$\tau_{xy} = -15 \text{ ksi}$$

2.6 Calculate σ_{xx}, σ_{yy}, τ_{xy}, σ_{zz}, and ε_{zz}, assuming that the point is in plane stress, for the following:

$$\varepsilon_{xx} = -800 \text{ μ} \qquad \varepsilon_{yy} = -1000 \text{ μ} \qquad \gamma_{xy} = -500 \text{ μ}$$
$$E = 30,000 \text{ ksi} \qquad \nu = 0.3$$

2.7 Calculate σ_{xx}, σ_{yy}, τ_{xy}, σ_{zz}, and ε_{zz}, assuming that the point is in plane stress, for the following:

$$\varepsilon_{xx} = -3000 \text{ μ} \qquad \varepsilon_{yy} = 1500 \text{ μ} \qquad \gamma_{xy} = 2000 \text{ μ}$$
$$E = 70 \text{ GPa} \qquad G = 28 \text{ GPa}$$

2.8 Calculate σ_{xx}, σ_{yy}, τ_{xy}, σ_{zz}, and ε_{zz}, assuming that the point is in plane stress, for the following:

$$\varepsilon_{xx} = 50 \text{ μ} \qquad \varepsilon_{yy} = 75 \text{ μ} \qquad \gamma_{xy} = -25 \text{ μ}$$
$$E = 2000 \text{ psi} \qquad G = 800 \text{ psi}$$

2.9 Starting from the generalized Hooke's law, obtain Equation (2.11).

2.10 For a point in plane stress, show

$$\sigma_{xx} = [\varepsilon_{xx} + \nu\varepsilon_{yy}] \frac{E}{(1 - \nu^2)} \qquad \text{and}$$

$$\sigma_{yy} = [\varepsilon_{yy} + \nu\varepsilon_{xx}] \frac{E}{(1 - \nu^2)} \qquad (2.12)$$

2.11 For a point in plane stress, show

$$\varepsilon_{zz} = -\left(\frac{\nu}{1 - \nu}\right)(\varepsilon_{xx} + \varepsilon_{yy}) \qquad (2.13)$$

2.12 For a point in plane strain, show

$$\varepsilon_{xx} = [(1 - \nu)\sigma_{xx} - \nu\sigma_{yy}]\frac{(1 + \nu)}{E} \qquad \text{and}$$

$$\varepsilon_{yy} = [(1 - \nu)\sigma_{yy} - \nu\sigma_{xx}]\frac{(1 + \nu)}{E} \qquad (2.14)$$

2.13 For a point in plane strain, show

$$\sigma_{xx} = \frac{E[(1 - \nu)\varepsilon_{xx} + \nu\varepsilon_{yy}]}{(1 - 2\nu)(1 + \nu)} \qquad \text{and}$$

$$\sigma_{yy} = \frac{E[(1 - \nu)\varepsilon_{yy} + \nu\varepsilon_{xx}]}{(1 - 2\nu)(1 + \nu)} \qquad (2.15)$$

2.14 Prove

$$p = -K\varepsilon_v \qquad p = -\left(\frac{\sigma_{xx} + \sigma_{yy} + \sigma_{zz}}{3}\right)$$

$$K = \frac{E}{3(1 - 2\nu)} \qquad (2.16)$$

where K is called the *bulk modulus, p* is called the hydrostatic pressure (because at a point in fluid the normal stresses in all directions are equal to $-p$), and ε_V is the volumetric strain given by Equation (1.47). Note that at $\nu = 1/2$ there is no change in volume regardless of the value of stresses. Such materials are called *incompressible materials.*

2.15 An orthotropic material has the following properties: $E_x = 7500$ ksi, $E_y = 2500$ ksi, $G_{xy} = 1250$ ksi, and $\nu_{xy} = 0.25$. Determine the principal direction 1 for the stresses and strains at a point on a free surface where the following strains were measured: $\varepsilon_{xx} = -400$ μ, $\varepsilon_{yy} = 600$ μ, and $\gamma_{xy} = -500$ μ.

2.16 An orthotropic material has the following properties: $E_x = 7500$ ksi, $E_y = 2500$ ksi, $G_{xy} = 1250$ ksi, and $\nu_{xy} = 0.25$. Determine the principal direction 1 for the stresses and strain at a point on a free surface where the following stresses were computed: $\sigma_{xx} = 10$ ksi (T), $\sigma_{yy} = 7$ ksi (C), and $\tau_{xy} = 5$ ksi.

ANS. Stresses: $\bar{\theta}_1 = 15.23°$ or $-164.77°$
Strains: $\theta_1 = 20.2°$ or $-159.8°$

2.17 An orthotropic material has the following properties: $E_x = 50$ MPa, $E_y = 18$ MPa, $G_{xy} = 9$ MPa, and $\nu_{xy} = 0.25$. Determine the principal direction 1 for the stresses and strains at a point on a free surface where the following strains were measured: $\varepsilon_{xx} = 800$ μ, $\varepsilon_{yy} = 200$ μ, and $\gamma_{xy} = 300$ μ.

ANS. Stress: $\bar{\theta}_1 = 4.45°$ or $-175.55°$
Strains: $\theta_1 = 13.3°$ or $-166.7°$

2.18 An orthotropic material has the following properties: $E_x = 50$ MPa, $E_y = 18$ MPa, $G_{xy} = 9$ MPa, and $\nu_{xy} = 0.25$. Determine the principal direction 1 for the stresses and strain at a point on a free surface where the following stresses were computed: $\sigma_{xx} = 70$ MPa (C), $\sigma_{yy} = 49$ MPa (C), and $\tau_{xy} = -30$ MPa.

2.3 FAILURE THEORIES

The maximum strength of a material is its atomic strength. In bulk materials, however, the distribution of flaws (impurities, microholes, microcracks, etc.) creates large stress gradients. These large stress gradients cause the bulk strength of the material to be orders of magnitude lower than the atomic strength of the material. Failure theories assume a homogeneous material in which the effects of flaws have been averaged [9] in some manner. This assumption of homogeneity results in average material strength values that are adequate for most problems in engineering design and analysis.

For homogeneous–isotropic-materials, the characteristic failure stress is either the yield stress or the ultimate stress, usually obtained from a uniaxial tensile test, a test in which there is only one nonzero stress component. How do we relate the one stress component of a uniaxial stress state to the stress components in two- and three-dimensional states of stress? The theories that attempt to answer this question are called failure theories, but no one answer is applicable to all materials.

---

[9] Micromechanics tries to account for the some of the flaws and nonhomogeneity in predictions of the strength of a material; but extrapolation from the microlevel to the macrolevel requires some form of averaging or homogenization.

TABLE 2.1 Synopsis of Failure Theories

	Ductile Material	Brittle Material
Characteristic failure stress	Yield stress	Ultimate stress
Theories	1. Maximum shear stress	1. Maximum normal stress
	2. Maximum octahedral shear stress	2. Modified Mohr

Definition 3 A failure theory is a statement about the relationship of stress components to material failure characteristics values.

We shall consider four theories (see Table 2.1). Maximum shear stress theory and the maximum octahedral shear stress theory are generally used for ductile materials. In ductile materials, failure is characterized by yield stress. The maximum normal stress theory and modified Mohr's theory are generally used for brittle materials. In brittle materials, failure is characterized by ultimate stress.

2.3.1　Maximum Shear Stress Theory

For ductile materials, the theory of maximum shear stress is postulated as in Definition 4.

Definition 4 A material will fail when the maximum shear stress exceeds the shear stress at the yield obtained from a uniaxial tensile test.

The failure criterion described in Definition 4 is given by Equation (2.17),

$$\tau_{max} \leq \tau_{yield} \tag{2.17}$$

where τ_{yield} is the maximum shear stress at the yield point found in a uniaxial tension test. The maximum shear stress is given by the Equation (1.20). Substituting the stresses at the yield point in a uniaxial tension test as $\sigma_1 = \sigma_{yield}$, $\sigma_2 = 0$, and $\sigma_3 = 0$ in Equation (1.20); equating the result to τ_{yield}, we obtain the maximum shear stress at yield point as follows:

$$\tau_{yield} = \frac{\sigma_{yield}}{2} \tag{2.18}$$

Substituting Equations (2.18) and (1.20) into Equation (2.17), we obtain the following failure criterion:

$$\boxed{\left| max(\sigma_1 - \sigma_2, \sigma_2 - \sigma_3, \sigma_3 - \sigma_1) \right| \leq \sigma_{yield}} \tag{2.19}$$

Equation (2.19) is referred to as *Tresca's yield criterion*. If we plot each principal stress on an axis, Equation (2.19) will give us the failure envelope. For plane stress problems, the failure envelope is shown in Figure 2.5.

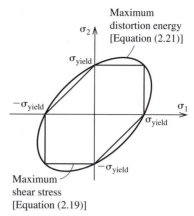

Figure 2.5 Failure envelopes for ductile materials in plane stress.

2.3.2 Maximum Octahedral Shear Stress Theory[10]

Definition 5 postulates the maximum octahedral shear stress theory for ductile materials.

> **Definition 5** A material will fail when the maximum octahedral shear stress exceeds the octahedral shear stress at the yield obtained from a uniaxial tensile test.

The failure criterion described in Definition 5 is given by

$$\tau_{oct} \leq \overline{\tau_{yield}} \tag{2.20}$$

where $\overline{\tau_{yield}}$ is the octahedral shear stress at yield point in a uniaxial tensile test. Substituting $\sigma_1 = \sigma_{yield}$, $\sigma_2 = 0$, and $\sigma_3 = 0$, the stresses at yield point in a uniaxial tension test, into the expression of octahedral shear stress in Equation (1.29b), we obtain $\overline{\tau_{yield}} = \sqrt{2}\sigma_{yield}/3$. Substituting this and Equation (1.29b) into Equation (2.20), we obtain Equation (2.21).

$$\frac{1}{\sqrt{2}}\sqrt{(\sigma_1 - \sigma_2)^2 + (\sigma_2 - \sigma_3)^2 + (\sigma_3 - \sigma_1)^2} \leq \sigma_{yield} \tag{2.21}$$

The left-hand side of Equation (2.21) is referred to as *von Mises stress*. Because von Mises stress σ_{von} is extensively used in the design of structures and machines, we formally define it as follows:

$$\boxed{\sigma_{von} = \frac{1}{\sqrt{2}}\sqrt{(\sigma_1 - \sigma_2)^2 + (\sigma_2 - \sigma_3)^2 + (\sigma_3 - \sigma_1)^2}} \tag{2.22}$$

The failure criterion represented by Equation (2.21) is sometimes referred to as the von Mises yield criterion and is stated as follows:

$$\boxed{\sigma_{von} \leq \sigma_{yield}} \tag{2.23}$$

---

[10] This theory is also called the maximum distortion strain energy theory. See Example 7.3 for additional details.

Note that in this theory we assumed that the normal stress on the octahedral plane has no influence on failure. In Section 1.4.4 it was shown that the octahedral normal stress corresponds to the hydrostatic state of stress; thus in this theory we are assuming that hydrostatic pressure has a negligible effect on the yielding of ductile material—a conclusion that is confirmed by experimental observation for very ductile materials like aluminum.

Equations (2.19) and (2.21) are failure envelopes[11] in a space in which the axes are principal stresses. For a plane stress ($\sigma_3 = 0$) problem, we can show these failure envelopes as in Figure 2.5. Notice that the maximum octahedral shear stress envelope encompasses the maximum shear stress envelope. Experiments show that for most ductile materials, the maximum octahedral shear stress theory gives better results than the maximum shear stress theory, but maximum shear stress theory is simpler to use.

2.3.3   Maximum Normal Stress Theory

The maximum normal stress theory for brittle materials is postulated in Definition 6.

Definition 6 A material will fail when the maximum normal stress at a point exceeds the ultimate normal stress (σ_{ult}) obtained from a uniaxial tension test.

The theory gives good results for brittle materials provided the first principal stress is tensile or the tensile yield stress has the same magnitude as the yield stress in compression. Thus the failure criterion is as follows:

$$\left| \max(\sigma_1, \sigma_2, \sigma_3) \right| \leq \sigma_{ult} \tag{2.24}$$

For concrete and many other brittle materials, the ultimate stress in tension is far less than the ultimate stress in compression because microcracks tend to grow in tension and to close in compression. But the simplicity of the failure criterion makes the theory attractive, and it can be used *if principal stress 1 is tensile* and is the *dominant* principal stress.

2.3.4   Modified Mohr's Theory

The Mohr's theory for brittle materials is postulated in Definition 7.

Definition 7 A material will fail if a stress state is on the envelope that is tangent to the three Mohr's circles corresponding to uniaxial ultimate stress in tension, uniaxial ultimate stress in compression, and pure shear.

We can conduct three experiments and determine the ultimate stress in tension σ_T, the ultimate stress in compression σ_C, and the ultimate shear stress in pure shear τ_S.

---

[11] Failure envelopes separate the design space from failure space. In drawing failure envelopes, the convention that $\sigma_1 > \sigma_2$ is ignored. If the convention were enforced, there would be no envelope in the second quadrant and only an envelope below the 45° line would be admissible in the third quadrant, resulting in a very strange-looking envelope rather than the symmetric envelope that is shown in Figure 2.5.

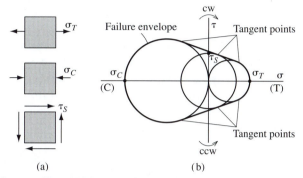

Figure 2.6 (a) Stress cubes. (b) Mohr's failure envelope.

The three stress states are shown on the stress cubes in Figure 2.6(a). A Mohr's circle for each of the three stress states is then drawn. Finally an envelope that is tangent to the three circles is drawn, which represents the failure envelope (Figure 2.6b). If a Mohr's circle corresponding to a stress state just touches the envelope at any point, the material is at incipient failure. If any part of the Mohr's circle for a stress state is outside the envelope, the material has failed at that point.

We can also plot the failure envelope of Figure 2.6 by using the principal stresses as the coordinate axis. In plane stress this envelope is represented by the solid line in Figure 2.7. For most brittle materials, the pure shear test is often ignored. In such a case the line tangent to the circles of uniaxial compression and tension would be a straight line in Figure 2.6. The resulting simplification for plane stress is shown as a dashed line in Figure 2.7 and describes what is called *Modified Mohr's theory.*

Figure 2.7 emphasizes the following points.

1. If both principal stresses are tensile, then the maximum normal stress must be less than the ultimate tensile strength.

2. If both principal stresses are negative, then the maximum normal stress must be less than the ultimate compressive strength.

3. If the principal stresses are of different signs, then for the modified Mohr's theory the failure is governed by Equation (2.25).

$$\left| \frac{\sigma_2}{\sigma_C} - \frac{\sigma_1}{\sigma_T} \right| \leq 1 \tag{2.25}$$

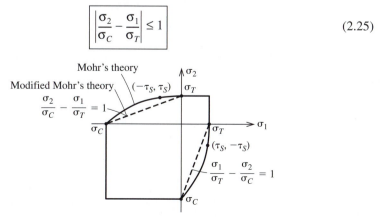

Figure 2.7 Failure envelope for plane stress according to Mohr's theory.

EXAMPLE 2.4

At a critical point on a machine part made of steel, the stress components are $\sigma_{xx} = 100$ MPa (T), $\sigma_{yy} = 50$ MPa (C), and $\tau_{xy} = 30$ MPa. Assuming that the point is in plane stress and the yield stress in tension is 220 MPa, determine the factor of safety[12] by using (a) the maximum shear stress theory and (b) the maximum octahedral shear stress theory.

PLAN

We can find the principal stresses by using Equation (1.19b) and maximum shear stress from Equation (1.20). (a) From Equation (2.18) we know that failure stress for the maximum shear stress theory is half the yield stress in tension, and we can use this fact to find the factor of safety. (b) We can find the von Mises stress from Equation (2.22); from the given yield stress in tension, we obtain the factor of safety.

SOLUTION

From Equations (1.19b) and (1.20) we can obtain the principal stresses and maximum shear stress as follows:

$$\sigma_{1,2} = \frac{100 - 50}{2} \pm \sqrt{\left(\frac{100 + 50}{2}\right)^2 + 30^2} = 25 \pm 80.8 \qquad \text{or}$$

$$\sigma_1 = 105.8 \text{ MPa} \qquad \sigma_2 = -55.8 \text{ MPa} \qquad \tau_{max} = \frac{\sigma_1 - \sigma_2}{2} = 80.8 \text{ MPa} \qquad \text{(E1)}$$

(a) The failure shear stress is half the yield stress in tension (i.e., 110 MPa). We divide this value by the maximum shear stress to obtain the factor of safety: $k_\tau = 110/80.8$, or

ANS. $k_\tau = 1.36$

(b) The von Mises stress can be found from Equation (2.22) as follows:

$$\sigma_{von} = \frac{1}{\sqrt{2}} \sqrt{[105.8 - (-55.8)]^2 + (-55.8)^2 + (105.8)^2} = 142.2 \text{ MPa} \qquad \text{(E2)}$$

We divide the failure stress of 220 MPa by the von Mises stress to get the following factor of safety:

$$k_\sigma = 220/142.2 \qquad \text{or} \qquad \textbf{ANS.} \quad k_\sigma = 1.55$$

---

[12] Factor of safety = failure-producing value divided by computed (allowable) value. In analysis, the computed value is determined, and the factor of safety is calculated from the definition. In design, the factor of safety is specified, and the allowable value is calculated from the definition.

COMMENTS

1. The failure envelopes corresponding to the yield stress of 220 MPa are shown in Figure 2.8. If we plot the coordinates $\sigma_1 = 105.8$ and $\sigma_2 = -55.8$, we will obtain point S. If we join the origin O to point S and draw the line, we get the load line for the given stress values. It may be verified by measuring (or calculating coordinates of T and V) that the following is true: $k_\tau = OS/OT = 1.36$ and $k_\sigma = OS/OV = 1.55$. Thus, failure envelopes provide a graphical interpretation of factors of safety. A factor of safety can be interpreted as a measure of how far the current design along a load line is from the failure envelope.

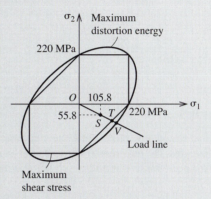

Figure 2.8 Failure envelope for Example 2.4.

2. Because the failure envelope for the maximum shear stress criterion is always inscribed inside the failure envelope of the maximum octahedral shear stress, the factor of safety based on the maximum octahedral shear stress will always be greater than the factor of safety based on the maximum shear stress.

EXAMPLE 2.5

Given a load P, the stresses at a point on a free surface were found to be $\sigma_{xx} = 3P$ ksi (C), $\sigma_{yy} = 5P$ ksi (T), and $\tau_{xy} = -2P$ ksi, where P is measured in kips. The brittle material has a tensile strength of 18 ksi and a compressive strength of 36 ksi. Determine the maximum value of the load P that can be applied on the structure by using the modified Mohr's theory.

PLAN

We can use Equation (1.19b) to determine the principal stresses in terms of P. From Equation (2.25) we can determine the maximum value of P.

SOLUTION

The principal stresses can be found by using Equation (1.19b) as shown in Equation (E1).

$$\sigma_{1,2} = \frac{(-3P + 5P)}{2} \pm \sqrt{\left(\frac{-3P - 5P}{2}\right)^2 + (2P)^2} = P \pm 4.47P$$

$$\sigma_1 = 5.57\,P \qquad \sigma_2 = -3.37\,P \tag{E1}$$

Substituting the principal stresses in Equation (2.25) and noting that $\sigma_T = 18$ ksi and $\sigma_C = -36$ ksi, we can obtain the maximum value of P as shown in Equation (E2)

$$\left|\frac{-3.37P}{-36} - \frac{5.57P}{18}\right| \le 1 \quad \text{or} \quad 0.2158P \le 1 \quad \text{or} \quad P \le 4.633 \tag{E2}$$

ANS. $P_{\text{max}} = 4.63$ kips

COMMENT

We could not have used the maximum normal stress theory for this material because the tensile and compressive strengths are significantly different, and it is the compressive strength that is the dominant strength, not the tensile strength.

2.4   SAINT-VENANT'S PRINCIPLE

Theories in the mechanics of materials are constructed by making assumptions regarding load, geometry, and material variations. These assumptions usually are not valid near concentrated forces or moments such as those that may be present near supports, near corners or/and holes, near interfaces of two materials, and at flaws such as cracks. Disturbances in the stress and displacement fields, however, dissipate rapidly as one moves away from the regions in which the assumptions of the theory are violated. The statement about of dissipation of a disturbance with distance is stated in Saint-Venant's principle as follows:

• Two statically equivalent load systems produce nearly the same stress in regions at a distance that is at least equal to the largest dimension in the loaded region.

Consider the two statically equivalent load systems shown in Figure 2.9. If we draw a stress cube under the concentrated force P in the Figure 2.9(a), we expect that the stresses will tend toward infinity because of the infinitesimal area of the stress cube. However, at a cross section that is at a distance W below the applied load, the stress distribution will be closer to uniform distribution, with a value equal to P divided by the cross-sectional area. In Figure 2.9(b) we have a distributed force that produces a total force of P. Though the distributed force is shown uniform, any distribution that does not produce a moment could be used. Thus, this distributed force is statically equivalent to the one on the left. By Saint-Venant's principle, the stress at a distance W in Figure 2.9(b) will be nearly uniform. In the region at a distance less than W, the stress distribution will be different and it is possible that shear stress components will be present as well. In a

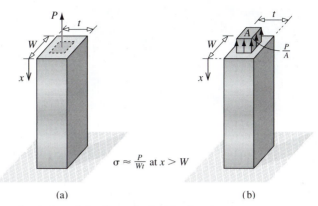

$$\sigma \approx \frac{P}{Wt} \text{ at } x > W$$

(a) (b)

Figure 2.9 Stress due to two statically equivalent load systems.

similar manner, changes in geometry and materials have local effects that can be ignored at distances. We will consider the effect of changes in geometry and an engineering solution to the problem in Section 2.5, on stress concentration.

The importance of Saint-Venant's principle is that it allows us to develop our theories with reasonable confidence away from the regions of stress concentration. These theories provide us with formulas for the calculation of nominal stresses. We can then use the stress concentration factor to obtain maximum stresses in regions of stress concentration where our theories are not valid.

2.5 STRESS CONCENTRATION FACTOR

Stress concentration is as described in Definition 8.

Definition 8 Large stress gradients in a small region is called stress concentration.

These large gradients could be due to sudden changes in geometry, material properties, or loading, as mentioned earlier. We know from Saint-Venant's principle, discussed in Section 2.4, that we can use our theoretical models to calculate stress away from the regions of large stress concentration. These stress values predicted by the theoretical models away from the regions of stress concentration are labeled "nominal stresses." Figure 2.10 shows photoelastic[13] pictures of two structural members under uniaxial tension. Large stress gradients near the circular cutout boundaries cause fringes to be formed. The fringe order can be used in calculating the stresses.

Definition 9 The stress predicted by theoretical models away from the regions of stress concentration is called *nominal stress*.

---

[13] Photoelastic materials are transparent materials that, affect the speed at which light passes through them when stressed and thus can used in the measurement of stress. See Burger [1993] in Appendix D for additional details.

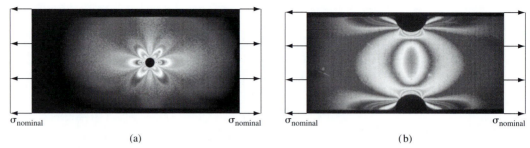

$\sigma_{nominal}$ $\sigma_{nominal}$ $\sigma_{nominal}$ $\sigma_{nominal}$

(a)   (b)

Figure 2.10 Photoelastic pictures showing stress concentration. (*Courtesy of Professor I. Miskioglu.*)

Stress concentration factor is an engineering concept that permits us to extrapolate the results of our elementary theory into the regions of large stress concentration, where the assumptions on which the theory is based are violated. The stress concentration factor K_{conc} is defined in Equation (2.26).

$$K_{conc} = \frac{\text{maximum stress}}{\text{nominal stress}} \qquad (2.26)$$

The stress concentration factor K_{conc} is found from charts, tables, or formulas that have been determined experimentally, numerically, analytically or by a combination of the three. Appendix C shows several graphs, which can be used in the calculation of stress concentration factors for problems in this book. Additional graphs can be found in handbooks[14] describing different situations. When we know the nominal stress and the stress concentration factor, we can estimate the maximum stress and use it in design or to estimate the factor of safety. Example 2.6 demonstrates the use of the stress concentration factor.

EXAMPLE 2.6

Finite element[15] analysis shows that a long structural component carries a uniform axial stress of 35 MPa (T). A hole in the center needs to be drilled for passing cables through the structural components. The yield stress of the material is 200 MPa. Determine the maximum diameter of the hole that can be drilled, using a factor of safety of 1.6, if failure due to yielding is to be avoided.

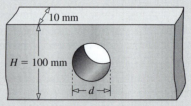

10 mm

$H = 100$ mm

$\leftarrow d \rightarrow$

Figure 2.11 Component geometry in Example 2.6.

---

[14] See Pilkey [1997] in Appendix D.

[15] See Chapter 9 for additional details on the finite element method.

PLAN

From the given factor of safety of 1.6 and the failure stress of 200 MPa, we can compute the allowable stress (maximum stress). Knowing the maximum stress, and the given gross nominal stress of 35 MPa, we can obtain the permissible stress concentration factor from Equation (2.26). From the plot of K_{gross} in Figure C.1 of Appendix C, we can estimate the ratio d/H. Knowing that the H is 100 mm, we can find the maximum diameter d of the hole.

SOLUTION

The failure stress of 200 MPa is divided by the factor of safety of 1.6 to obtain the allowable stress as shown in Equation (E1).

$$\sigma_{allow} = \frac{200}{1.6} = 125 \text{ MPa} \tag{E1}$$

The permissible stress concentration factor can be calculate by dividing σ_{allow} by the nominal stress of 35 MPa per Equation (2.26) to obtain Equation (E2).

$$K_{conc} \le \frac{125}{35} \qquad K_{conc} \le 3.57 \tag{E2}$$

From Figure C.1 of Appendix C we estimate the ratio d/H as 0.367. Substituting $H = 100$ mm in Equation (E2), we obtain Equation (E3).

$$\frac{d}{100} \le 0.367 \qquad d \le 36.7 \text{ mm} \tag{E3}$$

Thus the maximum permissible diameter to the nearest millimeter is

ANS. $d_{max} = 36$ mm

COMMENTS

1. The value of $d/H = 0.367$ was found from linear interpolation between value of $d/H = 0.34$, where the stress concentration factor is 0.35, and value of $d/H = 0.4$, where the stress concentration factor is 0.375. These points were used because they are easily read from the graph. Because we are rounding downward in Equation (E3), any value between 0.36 and 0.37 is acceptable; in other words, the third decimal place is immaterial.

2. Since we used the maximum diameter of 36 mm instead of 36.7 mm, the effective factor of safety will be slightly higher than the specified value of 1.6, which makes this design a conservative design.

3. Creating the hole will change the stress around it. In accordance with Saint-Venant's principle, the stress field far from the hole will not be significantly affected. This justifies the use of nominal stress without the hole in our calculation.

2.6 STRESS INTENSITY FACTOR

Stress intensity factor is another concept that permits us to extend our elementary theories and formulas to materials containing flaws such as small cracks. *Fracture mechanics,* the analysis of stresses and deformation of materials containing cracks, is beyond the scope of this book. But the concept of stress intensity factor and its application in design can be described by using the conclusions derived mathematically in fracture mechanics.

Consider a small elliptical hole in an infinite material as shown in Figure 2.12. The elasticity[16] solution shows that the maximum stress at the tip of the major axis (point *A*) can be found by using the stress concentration factor $K_{conc} = (1 + 2a/b)$, where *b* and *a* are half the diameter of the minor and the major axes, respectively. If we were to let the minor axis diameter *b* go to zero, the ellipse would become a crack; but our stress concentration factor would become infinite—the maximum stress at point *A* would become infinite. In other words, the use of the stress concentration concept for cracks implies that the moment a tiny crack is formed, the entire body should break because no material can sustain an infinite stress! We know this cannot be correct because we have often seen tiny cracks in materials and structures that continue to function well. The explanation is that the tip of the crack becomes slightly rounded,[17] but the blunting of the crack tip cannot be accurately predicted and accounted for in the theoretical elastic models. Furthermore, we also know that sometimes cracks grow and machine components (or structural members) break. Stress intensity factor is a concept that helps reconcile the infinite stress prediction of theoretical models to the reality of stable or unstable cracks in materials. This is accomplished by using results from elasticity that show that the stress components in the immediate vicinity of a crack tip can be written as in Equation (2.27),

$$\sigma_{ij} = K_{inten} f_{ij} \qquad (2.27)$$

where f_{ij} are functions[18] that depend upon the location of the point at which stress is found relative to the location of the crack tip. The functions f_{ij} become infinite at the crack tip, reflecting the behavior of the theoretical model, which is to ignore blunting of the crack tip. The factor K_{inten} is the stress intensity factor that depends upon the stress level and crack length for small cracks in very large bodies[19] (modeled as infinite bodies).

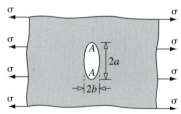

Figure 2.12 An elliptical hole in an infinite plate.

---

[16] See Timoshenko and Goodier [1951] in Appendix D.

[17] The rounding of crack tips is due to plasticity in ductile materials and to microfracturing in brittle materials.

[18] The functions are proportional to $1/\sqrt{r}$, where *r* is the radial distance from the crack tip. See Rolfe and Barsom [1977] in Appendix D for additional details.

[19] For bodies of finite size, the stress intensity factor also depends on the variables defining the geometry of the body.

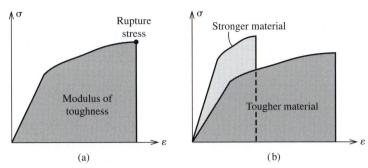

Figure 2.13 Two representations of modulus of toughness.

When the value of K_{inten} reaches a critical value K_{crit}, the crack starts to grow rapidly and the body breaks. The critical value of the stress intensity factor K_{crit} is a material property. The higher the value of K_{crit}, the greater the resistance of the material to crack growth and the tougher the material.

Recollect from the introductory course on the mechanics of materials that the "modulus of toughness" is equal to the strain energy density at rupture, as shown in Figure 2.13(a). Critical stress intensity factors are related to the modulus of toughness. We also note the distinction between strong materials, which are characterized by high ultimate stress, and tough materials, which are characterized by a large modulus of toughness, as shown in Figure 2.13(b).

We record the following observations.

- The stress intensity factor depends upon the stress level and the length of the crack.
- The critical stress intensity factor is a material property that is independent of stress level or crack length.
- A crack becomes unstable (the material breaks) when the stress intensity factor exceeds the critical stress intensity factor.

We need one more concept related to crack growth before we can use the stress intensity factor in design. Figure 2.14 shows three possible modes in which the two surfaces on either side of a crack can move relative to each other. Mode I is an opening apart in which tensile stresses cause the crack to grow. Mode II is a sliding mode due to in-plane shear stress. Mode III is a tearing mode, exemplified when we tear a piece of paper with both hands. Mode I is predominantly found in most failures and hence is

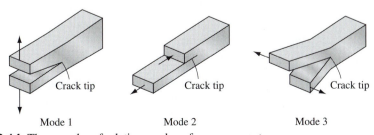

Figure 2.14 Three modes of relative crack surface movement.

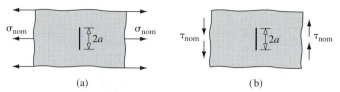

Figure 2.15 Stress intensity factors for mode I and mode II.

extensively studied. Mixed-mode failures are a subject of ongoing research. We will consider a simple model of mixed modes I and II in this book.

Figure 2.15 shows cracks for mode I and mode II in thin infinite planes. Fracture mechanics shows that the stress intensity factors for modes l and II for a crack in a thin infinite plane are as given in Equations (2.28a) and (2.28b),

$$K_{\mathrm{I}} = \sigma_{\mathrm{nom}}\sqrt{\pi a} \qquad (2.28a)$$

$$K_{\mathrm{II}} = \tau_{\mathrm{nom}}\sqrt{\pi a} \qquad (2.28b)$$

where σ_{nom} and τ_{nom} are the nominal normal and shear stress obtained from elementary theories. When both modes are present, we will use the following equivalent stress intensity factor:

$$K_{\mathrm{equiv}} = \sqrt{K_{\mathrm{I}}^2 + K_{\mathrm{II}}^2} \qquad (2.29)$$

When the equivalent stress intensity factor exceeds the critical stress intensity factor, the crack becomes unstable. Handbooks[20] supply formulas and graphs that can be used to obtain stress intensity factors in a variety of situations.

2.6.1   Analysis Procedure

Following are the steps for incorporating an existing crack into an analysis.

Step 1   Determine the state of stress at the point of the crack: use elementary theory that does not include a crack.

Step 2   To find the normal and shear stresses on the inclined crack surface. These normal and shear stresses on the crack surface are the nominal stresses.

Step 3   Use Equations (2.28a) and (2.28b) to find the stress intensity factors K_{I} and K_{II}.

Step 4   Use Equation (2.29) to find the equivalent stress intensity factor.

Step 5   Compare the equivalent stress intensity factor with the critical stress intensity factor for a material and decide whether the crack is or is not stable.

If no existing crack orientation is specified, we shall assume microcracks that will grow in mode I on a plane of maximum tensile normal stress (i.e., on principal plane 1). We record this observation for future use.

- Microcracks will be assumed to grow in mode I owing to principal stress 1 if the plane is in tension.

---

[20] See Murakami [1987] in Appendix D.

EXAMPLE 2.7

The propeller shaft of a submarine is subjected to a tensile axial stress and a torsional shear stress when the vessel reverses its direction. On display, the propeller shaft showed a crack at an angle of 27° to the axis of the shaft. At the point where crack was seen, the stresses are estimated as shown in Figure 2.16. The shaft material has a critical stress intensity factor of 140 $\text{ksi}\sqrt{\text{in}}$. The submarine is still in operation. At what crack length would you recommend that it be pulled out of water for repairs, assuming (a) that the detected crack could grow and (b) that there is no preexisting crack.

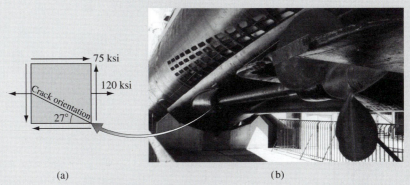

(a)   (b)

Figure 2.16 Crack in a propeller for Example 2.7.

PLAN

The submarine should be pulled out of water before the crack reaches critical length. (a) Following the procedure in Section 2.6.1, we can determine the critical crack length. (b) Upon determining principal stress 1 from Equation (1.19b) and the stress intensity factor K_I from Equation (2.28a), by equating K_I to the given critical value, we can find the critical crack length.

SOLUTION

Step 1 (a) Return to the analytical steps given at the beginning of Section 2.6.1. The stresses are given, so this step is accomplished.

Step 2 The angle of the normal to the plane containing the crack is 90° − 27° = 63°. Substituting this angle and $\sigma_{xx} = 120$, $\sigma_{yy} = 0$ and $\tau_{xy} = 75$ in Equations (1.7a) and (1.7b) we can obtain the normal and shear stress on the plane containing the crack as follows:

$$\sigma_\text{crack} = (120)\cos^2 63 + 2(75)\sin 63 \cos 63 = 85.4 \text{ ksi} \qquad (E1)$$

$$\tau_\text{crack} = -(120)\cos 63 \sin 63 + (75)(\cos^2 63 - \sin^2 63) = -92.6 \text{ ksi} \qquad (E2)$$

Step 3 Substituting the magnitudes of the stresses on the crack in Equations (2.28), we obtain the stress intensity factors for modes I and II as follows:

$$K_{\mathrm{I}} = (85.4)\sqrt{\pi a} = 151.4\sqrt{a} \text{ ksi}\sqrt{\text{in}} \qquad \text{(E3)}$$

$$K_{\mathrm{II}} = (92.6)\sqrt{\pi a} = 164.1\sqrt{a} \text{ ksi}\sqrt{\text{in}} \qquad \text{(E4)}$$

Step 4 and 5 Substituting Equations (E3) and (E4) in Equation (2.29), we obtain the equivalent stress intensity factor, which we can equate to the critical value of 140 ksi$\sqrt{\text{in}}$. We thus obtain the critical crack length as shown in Equation (E5).

$$K_{\mathrm{equiv}} = \sqrt{\left(151.4\sqrt{a_{\mathrm{cr}}}\right)^2 + \left(164.1\sqrt{a_{\mathrm{cr}}}\right)^2} = 223.3\sqrt{a_{\mathrm{cr}}} = 140 \qquad \text{or} \qquad \text{(E5)}$$

ANS. $2a_{\mathrm{cr}} = 0.786 \text{ in}$

(b) We can find principal stress 1 by using Equation (1.19b) as follows:

$$\sigma_1 = (120/2) + \sqrt{(120/2)^2 + 75^2} = 156 \text{ ksi} \qquad \text{(E6)}$$

From Equation (2.28a) we can find the mode I stress intensity factor and equate it to the critical value of 140 ksi$\sqrt{\text{in}}$. We thus obtain the critical crack length as follows:

$$K_1 = (156)\sqrt{\pi a_1} = 276.5\sqrt{a_1} = 140 \qquad \text{or} \qquad \textbf{ANS} \quad 2a_1 = 0.512 \text{ in} \qquad \text{(E7)}$$

COMMENTS

1. If there are no visible cracks, the crack growth is due to a mode I fracture. But if a crack has been formed from other causes, such as the shaft hitting something, then crack growth may be dictated by a mixed-mode fracture. But since the critical crack length for mode I is smaller than the critical crack length in mixed mode, it is possible that a new crack may start and cause failure before the existing crack becomes critical.

2. It is not surprising that the critical crack length in part (b) is smaller than that in part (a). For this problem, the magnitude of principal stress 1 is greater than the maximum shear stress on any plane. Hence, mode I is the dominant fracture mode.

2.7 FATIGUE

To appreciate the difference between static ultimate strength and fatigue strength, try to break a piece of wire (e.g., a paper clip) by pulling on it with both hands—you will not be able to break it because you cannot exceed the ultimate static stress of the material. Next take the same piece of wire and bend it first one way, then the other a few times. You will find that it breaks easily. The explanation of this phenomenon is as follows.

All materials are *assumed* to have microcracks. These cracks are not critical because their length is very small, and hence in static problems, the bulk strength of the material corresponds to the ultimate stress determined by a tension test. However, if the material

Striation
marks

Figure 2.17 Failure due to fatigue. (*Courtesy Professor I. Miskioglu*).

is subjected to cyclic loading, microcracks can grow till one or more reaches a critical length, at which time the remaining material breaks. The stress value at rupture in cyclic loading is significantly lower then the ultimate stress of the material.

Definition 10 Failure due to cyclic loading at stress levels significantly lower than the static ultimate stress is called *fatigue*.

Failure due to fatigue is like a brittle failure irrespective of whether the material is brittle or ductile.[21] We see that there are two phases of failure. In the first phase the microcracks grow, and regions of crack growth can be identified by the striation marks, also called beach marks, as shown in Figure 2.17. On examination of a fractured surface, the region of microcrack growth shows only small deformation. In the second phase, which begins once the critical crack length has been reached, the failure surface of the region shows significant deformation, as shown in Figure 2.17.

The following strategy is used in design to account for fatigue failure. Experiments are conducted at different magnitude levels of cyclic stress, and the number of cycles at which the material fails is recorded. A plot is made of stress (*S*) vs number of cycles (*N*) to failure. There is always significant scatter in the data. A curve through the data is constructed. This curve is called the *S-N* curve. At low levels of stress the failure may occur only after millions or even billions of cycles. To accommodate this large scale, a log scale is used for the number of cycles. Figure 2.18 shows some typical *S-N* curves. Notice that the curves approach a stress level asymptotically, implying that if stresses are kept below this level, the material will not fail under cyclic loading. This asymptotic level is called the endurance limit or fatigue strength.

Definition 11 The highest stress level for which a material will not fail under cyclic loading is called the *endurance limit* or *fatigue strength*.

It should be emphasized that a particular *S-N* curve for a material depends upon many factors such as manufacturing process, machining process, surface preparation, and operating environment. Thus two specimens made from same steel alloy but with different histories will yield different *S-N* curves. Care must be taken to use an *S-N* curve that corresponds as closely as possible to the actual situation.

---

[21] See Rolfe and Barsom [1977] in Appendix D for additional details.

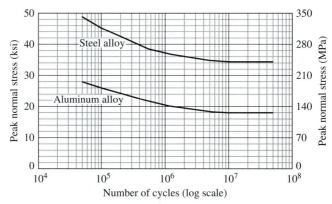

Figure 2.18 *S-N* curves.

In a typical preliminary design, static stress analysis would be conducted by using the peak load of the cyclic loading. In other words, we would determine the highest stress level that might occur in a cyclic loading. Using an appropriate *S-N* curve, the number of cycles to failure for the computed stress value is then calculated. This number of cycles to failure is the predicted life of the structural component. If the predicted life is unacceptable, then the component would be redesigned to lower the stress level and hence increase the number of cycles to failure.

EXAMPLE 2.8

The steel plate represented in Figure 2.19 has an *S-N* like the top curve shown in Figure 2.18. (a) Determine the maximum diameter of the hole to the nearest millimeter if the predicted life of 0.5 million cycles is desired for a uniform far field stress $\sigma = 75$ MPa. (b) For the hole radius in part (a), what percentage reduction in far field stress must occur if the predicted life is to be increased to 1 million cycles?

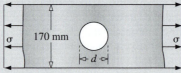

Figure 2.19 Steel plate for Example 2.8.

PLAN

(a) From Figure 2.18 we can find the maximum stress that the material can carry for 0.5 million cycles. We can use Equation (2.26) to obtain the gross stress concentration factor K_{gross}. From Figure C.1 of Appendix C, we can estimate the ratio d/H. Knowing that the value of H is 170 mm, we can find the diameter d of the hole. (b) The percentage reduction in the gross nominal stress is the same percentage reduction as in the maximum stress values in Figure 2.18 from 1 million cycles to 0.5 million cycles.

SOLUTION

(a) From Figure 2.18 the maximum allowable stress for 0.5 million cycles is estimated as 273 MPa. From Equation (2.26) the gross stress concentration factor is $K_{\text{gross}} = 273/75 = 3.64$. From Figure C.1 of Appendix C the value of the ratio d/H corresponding to $K_{\text{gross}} = 3.64$ is 0.374. Knowing that $H = 170$ mm, we find that d should be less than 63.58 mm. Thus the maximum permissible diameter to the nearest millimeter is 63 mm.

ANS. $d = 63$ mm

(b) From Figure 2.18 the maximum allowable stress for 1 million cycles is estimated as 259 MPa. Thus the percentage reduction in maximum allowable stress is $((273 - 259)/273)100 = 5.13\%$. The geometry is the same as in part (a); therefore the percentage reduction in far field stress should be the same as for the maximum allowable stress.

ANS.   The required percentage reduction is 5.13%.

COMMENT

A 5.13% reduction in stress value causes the predicted life cycle to double. There are many factors that can cause small changes in stress values resulting in very wide predictive life cycles. Our estimates of allowable stress in Figure 2.18, our estimate of the ratio d/H from Figure C.1 of Appendix C, the estimate of far field stress σ, and the tolerances of drilling the hole are some factors that can significantly affect our life prediction of the component. This emphasizes that the data used in predicting life cycles and failure due to fatigue must be of much higher accuracy than usually used in traditional engineering analysis.

PROBLEM SET 2.2

Failure Theory Problems

2.19   The stress components at a critical point that is in plane stress due to a force P are

$$\sigma_{xx} = 10P \text{ MPa (T)} \qquad \sigma_{yy} = 20P \text{ MPa (C)} \qquad \tau_{xy} = 5P \text{ MPa}$$

where P is in kilonewtons (kN). The material has a yield stress of 160 MPa as determined in a tension test. If yielding must be avoided, predict the maximum value of the force P by using (a) the maximum shear stress theory and (b) the maximum octahedral shear stress theory.

ANS.   (a) $P_{\text{max}} = 5.0$ kN   (b) $P_{\text{max}} = 5.75$ kN

2.20   A material has a tensile rupture strength of 18 ksi and a compressive rupture strength of 32 ksi. During use, a component made from this material showed the following stresses on a free surface at a critical point:

$$\sigma_{xx} = 9 \text{ ksi (T)} \qquad \sigma_{yy} = 6 \text{ ksi (C)} \qquad \tau_{xy} = -4 \text{ ksi}$$

Use the modified Mohr's theory to determine the factor of safety.

ANS. $k = 2.97$

2.21 For *plane stress*, show that the von Mises stress representing the left-hand side of Equation (2.21) can be written as

$$\sigma_{\text{von}} = \sqrt{\sigma_{xx}^2 + \sigma_{yy}^2 - \sigma_{xx}\sigma_{yy} + 3\tau_{xy}^2} \qquad (2.30a)$$

2.22 In Cartesian coordinates, the von Mises stress in three dimensions is given by:

$$\sigma_{\text{von}} = \sqrt{\sigma_{xx}^2 + \sigma_{yy}^2 + \sigma_{zz}^2 - \sigma_{xx}\sigma_{yy} - \sigma_{yy}\sigma_{zz} - \sigma_{zz}\sigma_{xx} + 3\tau_{xy}^2 + 3\tau_{yz}^2 + 3\tau_{zx}^2} \qquad (2.30b)$$

Show that for plane strain, Equation (2.30b) reduces to:

$$\sigma_{\text{von}} = \sqrt{(\sigma_{xx}^2 + \sigma_{yy}^2)(1 + v^2 - v) - \sigma_{xx}\sigma_{yy}(1 + 2v - 2v^2) + 3\tau_{xy}^2} \qquad (2.30c)$$

where v is the Poisson ratio of the material.

2.23 On a free surface of aluminum ($E = 10{,}000$ ksi, $v = 0.25$, $\sigma_{\text{yield}} = 24$ ksi), the strains recorded by the three strain gages shown in Figure P2.23 are $\varepsilon_a = -600\ \mu$ in/in, $\varepsilon_b = 500\ \mu$ in/in, and $\varepsilon_c = 400\ \mu$ in/in. By how much can the loads be scaled without exceeding the yield stress of aluminum at the point? Use the maximum shear stress theory.

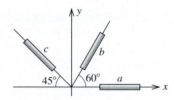

Figure P2.23

ANS. $k = 1.76$

2.24 A steel bar is axially loaded as shown in Figure P2.24. Determine the factor of safety for the bar if yielding is to be avoided. The yield stress for steel is 30 ksi. Use the stress concentration factor chart in Appendix C.

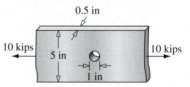

Figure P2.24

ANS. $k = 2.4$

2.25 The stress concentration factor for a stepped flat tension bar with shoulder fillets shown in Figure P2.25 was determined as

$$K_{\text{conc}} = 1.970 - 0.384\left(\frac{2r}{H}\right) - 1.018\left(\frac{2r}{H}\right)^2 + 0.430\left(\frac{2r}{H}\right)^3$$

The equation is valid only if $(H/d) > (1 + 2r/d)$ and $(L/H) > (5.784 - 1.89r/d)$. The nominal stress is P/dt. Make a chart for the stress concentration factor vs H/d for the following values of r/d: 0.2, 0.4, 0.6, 0.8, and 1.0. Use of a spreadsheet is recommended.

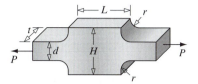

Figure P2.25

2.26 Determine the maximum normal stress in the stepped flat tension bar shown in Figure P2.25 for the following data: $P = 9$ kips, $H = 8$ in, $d = 3$ in, $t = 0.125$ in, and $r = 0.625$ in.

2.27 The aluminum stepped tension bar shown in Figure P2.25 is to carry a load $P = 56$ kN. The yield stress of aluminum is 160 MPa. The bar has $H = 300$ mm, $d = 100$ mm, $t = 10$ mm. For a factor of safety of 1.6, determine the minimum value r of the fillet radius if yielding is to be avoided.

ANS. $r = 40$ mm

2.28 The stress concentration factor for the flat tension bar with U-shaped notches shown in Figure P2.28 was determined as

$$K_{conc} = 3.857 - 5.066\left(\frac{4r}{H}\right) + 2.469\left(\frac{4r}{H}\right)^2 - 0.258\left(\frac{4r}{H}\right)^3$$

The nominal stress is P/Ht. Make a chart for the stress concentration factor vs r/d for the following values of H/d: 1.25, 1.50, 1.75, and 2.0. Use of a spreadsheet is recommended.

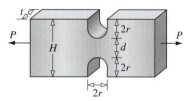

Figure P2.28

2.29 Determine the maximum normal stress in the flat tension bar shown in Figure P2.28 for the following data: $P = 150$ kN, $H = 300$ mm, $r = 15$ mm, and $t = 5$ mm.

2.30 A steel tension bar with U-shaped notches, of the type shown in Figure P2.28, is to carry a load $P = 18$ kips. The yield stress of steel is 30 ksi. The bar has $H = 9$ in, $d = 6$ in, $t = 0.25$ in. For a factor of safety of 1.4, determine the value of r if yielding is to be avoided.

Stress Intensity Factor Problems

2.31 The stresses at a point in plane stress are as follows:

$$\sigma_{xx} = 250 \text{ MPa (T)} \qquad \sigma_{yy} = 150 \text{ MPa (C)} \qquad \tau_{xy} = 200 \text{ MPa}$$

The critical stress intensity factor of the material is 33 MPa$\sqrt{\text{m}}$. Determine the critical crack length assuming that (a) a small crack exists on a plane $-40°$ from the x axis and (b) there is no preexisting crack.

ANS. (a) 7.02 mm (b) 6.25 mm

2.32 The stresses at a point in plane stress were found to be

$$\sigma_{xx} = 27 \text{ ksi (T)} \qquad \sigma_{yy} = 10 \text{ ksi (C)} \qquad \tau_{xy} = 15 \text{ ksi}$$

The critical stress intensity factor of the material is 20 ksi$\sqrt{\text{in}}$. Determine the critical crack length assuming that (a) a small crack exists on a plane $-25°$ from the x axis and (b) there is no preexisting crack.

2.33 The stress components at a critical point that is in plane stress due to a force P are

$$\sigma_{xx} = 10P \text{ MPa (T)} \qquad \sigma_{yy} = 20P \text{ MPa (C)} \qquad \tau_{xy} = 5P \text{ MPa}$$

where P is in kilonewtons. The critical stress intensity factor of the material is 33 MPa$\sqrt{\text{m}}$. For inspection purposes, the smallest crack length that is acceptable is 1 mm. What is the maximum force P that can be applied?

ANS. $P_{\text{max}} = 77$ kN

Fatigue Problems

2.34 A machine component is made from a steel alloy that has an *S-N* curve like that of the steel alloy shown in Figure 2.18. Estimate the service life of the component if the peak stress is reversed at the following rate.

(a) 40 ksi at 200 cycles per minute,

(b) 36 ksi at 250 cycles per minute, and

(c) 32 ksi at 300 cycles per minute,

2.35 A machine component is made from a aluminum alloy that has an *S-N* curve like that of the aluminum alloy shown in Figure 2.18. What should be the maximum permissible peak stress, in mega pascals,

(a) 17 hours of service at 100 cycles per minute.

(b) 40 hours of service at 50 cycles per minute.

(c) 80 hours of service at 20 cycles per minute.

2.36 A uniaxial stress acts on an aluminum plate with a hole as shown in Figure P2.36. The aluminum has the *S-N* curve shown for aluminum in Figure 2.18. Predict the number of cycles the plate could be used if $d = 3.2$ in and the far field stress $\sigma = 6$ ksi.

Figure P2.36

2.37 A uniaxial stress acts on an aluminum plate with a hole as shown in Figure P2.36. The aluminum has the *S-N* curve for aluminum shown in Figure 2.18. Determine the maximum diameter of the hole to the nearest 1/8 inch, if the predicted service life of 0.5 million cycles is desired for a uniform far field stress $\sigma = 6$ ksi.

2.38 A uniaxial stress acts on an aluminum plate with a hole as shown in Figure P2.36. The aluminum has the *S-N* curve for aluminum shown in Figure 2.18. Determine the maximum far field stress σ if the diameter of the hole is 2.4 in and a predicted service life of 0.75 million cycles is desired.

2.8 CLOSURE

In this chapter we saw that a linear model requires 21 independent material constants for materials that are fully anisotropic; 13 for monoclinic; 9 and 4 for orthotropic in three and two dimensions, respectively; and only 2 for isotropic materials. The stress–strain relationship for isotropic materials called the generalized Hooke's law is valid for any orthogonal system and is independent of the orientation of coordinate system. For all models other than isotropic, the stress–strain relationship is described in a material coordinate system and the principal coordinates for stresses and strains have different orientations.

We also saw that a material can fail when the material characteristic strength is exceeded as described by failure theories; when a crack grows in mode I or mixed mode; and when cyclic loading causes fatigue. We saw that sudden changes in geometry, loading, or material properties cause a stress concentration, which quickly dissipates with distance according to Saint-Venant's principle. In the next chapter we will study the stress analysis of one-dimensional structural elements in a way that accounts for these various forms of material failure.

3 Basic Structural Members

3.0 OVERVIEW

The frame of a building, the skeleton of a human being, and the frame of a car (Figure 3.1) all contain many one-dimensional structural members. Chair legs, utility poles, drills, bookshelves, and diving boards are among the countless applications of one-dimensional structural members. The theories and formulas for one-dimensional structural members vary in complexity depending upon the complexities in loading, material properties, and cross-sectional geometries. In this chapter we briefly review the simplest theories in a framework that permits ready incorporation of complexities, some of which are studied in this chapter and others in subsequent chapters.

In the introductory course on the mechanics of materials, the basic theories of axial members, torsion of circular shafts, and symmetric bending of beams were developed and studied. In this chapter a synopsis of all three theories is presented in tabular form to emphasize the *commonalities and differences* in the derivation of stress and deformation formulas. The limitations and assumptions that are made to get the simplest possible formulas are identified. In this and later chapters it is shown that complexities of interest can be added by dropping a limitation or an assumption. It will be seen that most complexities in all three structural members (axial rods, circular shafts, and symmetric beams) enter the derivation at the same point and are accounted for in nearly the same manner.

Discontinuous loading on a member is fairly common. In the introductory course these discontinuities in loading were accounted for by segmenting the structural member such that in each segment the loading was continuous. Relative deformations of segment ends were found, and the overall deformation was obtained by using the observation that deformation was continuous at the point of the loading discontinuity. This approach is algebraically tedious, particularly for beam-bending problems and statically indeterminate structural members. In this chapter the concept of discontinuity functions is introduced and effectively used for analyzing the deformation of statically determinate and indeterminate structural members that are subjected to discontinuous loads.

Figure 3.1 Assemblies of one-dimensional structural members. [*Photographs courtesy of Ulster County, Area Transit (a) and Stephen C. Burnett, copyright 2003 (b).*]

The learning objectives in this chapter are as follows:

1. To understand the limitations of basic theory and how complexities may be added to the basic theories of axial members, torsion of circular shafts, and symmetric bending of beams.

2. To understand the concept and use of discontinuity functions in the analysis of structural members subjected to discontinuous loads.

3.1 LOGIC IN STRUCTURAL ANALYSIS

The logic symbolically shown in Figure 3.2 is used for developing both the simple theories discussed in the introductory course on the mechanics of materials and the advanced theories encountered in graduate courses. It is possible to start at any point in the logic and move either clockwise (solid arrows ➝) or counterclockwise (open arrows ⇾). It is not possible to relate displacement directly to external forces without imposing limitations and making assumptions regarding the geometry of the body, material behavior, and external loading. This is emphasized by the absence of arrows directly linking displacements and external forces in Figure 3.2.

The logic shown in Figure 3.2 is intrinsically very modular; that is, *the equations* (kinematics, material models, static equivalency, and equilibrium[1]) *linking two variables are mutually independent*. Thus, if complexity is added in one part, then the theory can be rederived with minimum change. This will be demonstrated many times in this book through examples, in posttext problems, and in Chapters 4, 5, and 6. Understanding the modular nature of this logic and learning how complexities can be incorporated to refine basic theories is an important learning objective in this book. In this section three examples demonstrate the application of this logic.

---

[1] In Chapter 7 we will see that energy methods are an alternative to Newtonian methods of equilibrium equations.

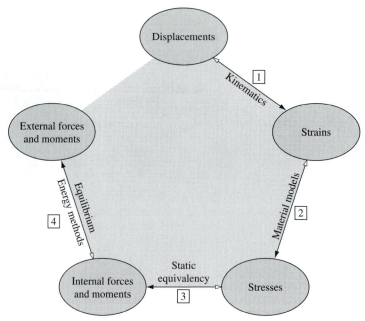

Figure 3.2 Logic in structural analysis.

EXAMPLE 3.1

Two thin bars are securely attached to a rigid plate, as shown in Figure 3.3. The cross-sectional area of each bar is 20 mm^2. The force F_{ext} is to be placed such that the rigid plate moves only horizontally by 0.05 mm without rotating. Determine the force F_{ext} and its location h for the following cases.

Case 1: Both bars are made from steel with modulus of elasticity of $E = 200$ GPa.

Case 2: Bar 1 is made of steel ($E = 200$ GPa), and bar 2 is made of aluminum ($E = 70$ GPa).

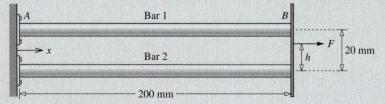

Figure 3.3 Axial bars in Example 3.1.

PLAN

The relative displacement of point B with respect to A is 0.05 mm, from which we can find the axial strain. By multiplying the axial strain by the modulus of elasticity, we can obtain the axial stress. By multiplying the axial stress by the cross-sectional area, we can obtain the internal axial force in each bar. We can draw the free body diagram of the rigid plate and by equilibrium obtain the force F_{ext} and its location h.

SOLUTION

1. **Strain calculations:** The displacement of B is $u_B = 0.05$ mm. Point A is built into the wall, hence has a zero displacement. The normal strain in both rods is the same and can be found as shown in Equation (E1).

$$\varepsilon_1 = \varepsilon_2 = \frac{u_B - u_A}{x_B - x_A} = \frac{0.05}{200} = 250 \ \mu\text{mm/mm} \qquad \text{(E1)}$$

2. **Stress calculations:** From Hooke's law, $\sigma = E\varepsilon$, we can find the normal stress in each bar for the two cases as shown in Equations (E2) through (E5).

   Case 1: Since E and ε_1 are the same for both bars, the stress is the same in both bars. We obtain the following:

$$\sigma_1 = E_1\varepsilon_1 = 200(10^9)(250)(10^{-6}) \qquad \text{or} \qquad \sigma_1 = 50 \text{ MPa (T)} \qquad \text{(E2)}$$

$$\sigma_2 = 50 \text{ MPa (T)} \qquad \text{(E3)}$$

   Case 2: Because E is different for the two bars, the stress in the two bars is different and can be found as follows:

$$\sigma_1 = E_1\varepsilon_1 = 200(10^9)(250)(10^{-6}) \qquad \text{or} \qquad \sigma_1 = 50 \text{ MPa (T)} \qquad \text{(E4)}$$

$$\sigma_2 = E_2\varepsilon_2 = 70(10^9)(250)(10^{-6}) \qquad \text{or} \qquad \sigma_2 = 17.5 \text{ MPa (T)} \qquad \text{(E5)}$$

3. **Internal forces:** Assuming that the normal stress is uniform in each bar, the equivalent internal normal force is $N = \sigma A$, where $A = 20 \text{ mm}^2 = 20(10^{-6}) \text{ m}^2$. The calculations for the two cases are shown in Equations (E6) through (E9).

   Case 1: Both bars have the same internal force as because stress and cross-sectional area are same:

$$N_1 = \sigma_1 A_1 = 50(10^6)(20)(10^{-6}) \qquad \text{or} \qquad N_1 = 1000 \text{ N (T)} \qquad \text{(E6)}$$

$$N_2 = 1000 \text{ N (T)} \qquad \text{(E7)}$$

   Case 2: Because stresses are different in each bar, the equivalent internal force is different for each bar:

$$N_1 = \sigma_1 A_1 = 50(10^6)(20)(10^{-6}) \qquad \text{or} \qquad N_1 = 1000 \text{ N (T)} \qquad \text{(E8)}$$

$$N_2 = \sigma_2 A_2 = 17.5(10^6)(20)(10^{-6}) \qquad \text{or} \qquad N_2 = 350 \text{ N (T)} \qquad \text{(E9)}$$

4. **External force:** We make an imaginary cut through the bars and draw the free body diagram of the rigid plate (Figure 3.4). We note that the internal axial force is tensile in all cases and that is recognized in drawing the free body diagram.

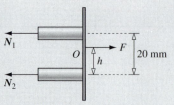

Figure 3.4 Free body diagram for Example 3.1.

By equilibrium of forces in the x direction, we obtain Equations (E10).

$$F_{ext} = N_1 + N_2 \tag{E10}$$

By equilibrium of moment about point O in Figure 3.4 we obtain:

$$N_1(20 - h) - N_2(h) = 0 \qquad \text{or} \tag{E11}$$

$$h = \frac{20(N_1)}{N_1 + N_2} \tag{E12}$$

We can substitute the value of internal forces for each case in Equations (E10) and (E12) and obtain F_{ext} and h as shown for the next two cases.

Case 1: Substituting Equations (E6) and (E7) into Equations (E10) and (E12), we obtain the desired results.

$$F_{ext} = 1000 + 1000 \qquad \text{or} \qquad \textbf{ANS.} \quad F_{ext} = 2000 \text{ N}$$

$$h = \frac{20(1000)}{1000 + 1000} \qquad \text{or} \qquad \textbf{ANS.} \quad h = 10 \text{ mm}$$

Case 2: Substituting Equations (E8) and (E9) into Equations (E10) and (E12) we obtain the following results:

$$F_{ext} = 1000 + 350 \qquad \text{or} \qquad \textbf{ANS.} \quad F_{ext} = 1350 \text{ N}$$

$$h = \frac{20(1000)}{1000 + 350} \qquad \text{or} \qquad \textbf{ANS.} \quad h = 14.81 \text{ mm}$$

COMMENTS

1. Note that the kinematic equation (E1) and the equilibrium equations (E10) and (E12) are independent of the material properties of the two bars. Both bars, irrespective of the material, were subjected to the same axial strain, which is the fundamental kinematic assumption in the development of the theory for axial members. As shall be seen, case 1 corresponds to a homogeneous cross section, while case 2 is analogous to a laminated bar in which the nonhomogeneity affects the stress distribution across the cross section.

2. The summation on the right-hand side of Equation (E10) can be rewritten as $\sum_{i=1}^{n=2} \sigma_i \Delta A_i$, where σ_i is the normal stress in the ith bar, ΔA_i is the cross-sectional area of the ith bar, and $n = 2$ reflects the two bars in this problem. If we had n bars attached to the rigid plate, the total axial force would be given by summation over n bars. As we increase the number of bars n to infinity, the cross-sectional area ΔA_i will tend to zero (infinitesimal area written as dA) as we try to fit an infinite number of bars on the same plate—resulting in a continuous body with the summation replaced by an integral over the cross-sectional area. The integrals that convert stresses into forces are the equations of static equivalency.

3. If the external force is located at any point other than that given by the value of h, the plate will rotate. This emphasizes that for pure axial problems with no bending, a point on the cross section must be found such that the internal moment from the axial stress distribution is zero.

EXAMPLE 3.2

Two thin bars of hard rubber have a cross-sectional area of 20 mm². The bars are attached to a rigid disk of radius 20 mm, as shown in Figure 3.5. The applied torque T_{ext}, causes the rigid disk to rotate by an angle of 0.04 rad about the axis of the disk. Determine the applied torque for the following cases:

Case 1: Assume that the hard rubber is governed by the linear version of Hooke's law $\tau = G_1 \gamma$, where $G_1 = 280$ MPa.

Case 2: Assume that the hard rubber is governed by the nonlinear stress–strain relation $\tau = G_2 \gamma^{0.8}$, where $G_2 = 300$ MPa.

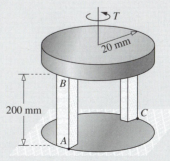

Figure 3.5 Geometry for Example 3.2.

PLAN

We can relate the rotation ($\Delta\phi = 0.04$) of the disk, the radius ($\rho = 0.02$ m) of the disk, and the length (0.2 m) of the bars to the shear strain in the bars. We can find the shear stress in each bar for the two cases from the given stress–strain relationship. By assuming uniform shear stress in each bar, we can find the shear force. By drawing the free body diagram of the rigid disk, we can find the applied torque T_{ext}.

SOLUTION

1. **Strain calculations:** Figure 3.6(a) shows an approximate deformed shape of the two bars. The two bars have the same length and are located at the same radial distance; hence the shear strain in bar C is the same as in bar A. The shear strain can be calculated as shown in Equations (E1) through (E3).

$$BB_1 = (0.02)(\Delta\phi) = 0.0008 \text{ m} \tag{E1}$$

$$\tan \gamma_A \approx \gamma_A = \frac{BB_1}{AB} = 0.004 \tag{E2}$$

$$\gamma_C = 0.004 \tag{E3}$$

2. **Stress calculations:** We can find shear stresses for the two cases by using Hooke's law.

Case 1: We use $\tau = G_1\gamma$ to obtain the shear stresses as shown in Equation (E4).

$$\tau_A = \tau_C = 280(10^6)(0.004) = 1.12(10^6) \text{ N/m}^2 \tag{E4}$$

Case 2: We use $\tau = G_2\gamma^{0.8}$ to obtain the shear stresses as shown in Equation (E5)

$$\tau_A = \tau_C = 300(10^6)(0.004)^{0.8} = 3.621(10^6) \text{ N/m}^2 \tag{E5}$$

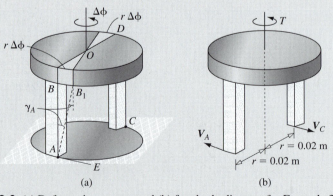

Figure 3.6 (a) Deformed geometry and (b) free body diagram for Example 3.2.

3. **Internal forces:** Assuming uniform shear stresses across the cross section, we obtain the shear forces by $V = A\tau$, where $A = 20(10^{-6})$ m^2 as shown Equations (E6) and (E7).

Case 1: We use Equation (E4) to obtain Equation (E6).

$$V_A = V_C = 1.12(10^6)(20)(10^{-6}) = 22.4 \text{ N} \tag{E6}$$

Case 2: We use Equation (E5) to obtain Equation (E7).

$$V_A = V_C = 3.621(10^6)(20)(10^{-6}) = 72.410 \text{ N} \tag{E7}$$

4. **External torque:** We make imaginary cuts through the bars and draw the free body diagram of the top part as shown in Figure 3.6(b). By equilibrium of the moment about the axis of the disk, we can obtain the external torque as follows:

$$T_{ext} = (r)(V_A) + (r)(V_C) \tag{E8}$$

Case 1: Substituting Equation (E6) into Equation (E8), we obtain the external torque.

$$T_{ext} = (0.02)(22.4) + (0.02)(22.4) \quad \textbf{ANS.} \quad T_{ext} = 0.896 \text{ N} \cdot \text{m}$$

Case 2: Substituting Equation (E7) into Equation (E8), we obtain the external torque.

$$T_{ext} = (0.02)(72.410) + (0.02)(72.410) \quad \textbf{ANS.} \quad T_{ext} = 2.9 \text{ N} \cdot \text{m}$$

COMMENTS

1. The kinematic equations of small strain given by Equations (E2) and (E3) and the equilibrium equation (E8) are independent of the stress–strain relationship.

2. The shear stress acts on a surface with outward normal in the direction of the length of the bar, which is also the axis of the disk. The shear force acts along in the tangent direction to the circle of radius r. If we label the direction of the axis x and the tangent direction θ, then in accordance with the notation described in Section 1.2, the shear stress is represented by $\tau_{x\theta}$.

3. The summation in Equation (E8) can be rewritten as $\sum_{i=1}^{2} r\tau\Delta A_i$, where τ is the shear stress acting at the radius r, and ΔA_i is the cross-sectional area of the ith bar. If we had n bars attached to the disk at the same radius, the total torque would be given by $\sum_{i=1}^{n} r\tau\Delta A_i$. As we increase the number of bars from n to infinity, the cross-sectional area ΔA_i will tend to zero (infinitesimal area written as dA) as we try to fit an infinite number of bars on the same circle—resulting in a continuous body with the summation replaced by an integral. The integral that converts shear stress into torque represents equations of static equivalency.

EXAMPLE 3.3

A canoe is tied down on top of a car by means of stretch cords as shown in Figure 3.7. The undeformed length of the stretch cord is 40 in, the initial diameter is $d = 1/2$ in, and the modulus of elasticity E is 510 psi. Marks were made on the cord every 2 inches to produce a total of 20 segments. The stretch cord is symmetric with respect to the top of the canoe. The starting point of the first segment is on the carrier rail of the car, and the end point of the tenth segment is on the top of the canoe. The measured length of each segment is shown in Table 3.1. We have two cases.

Case 1: Assume small strain: that is, $\varepsilon_s = \dfrac{\delta}{L_0}$.

Case 2: Assume large strain, given by $\varepsilon_L = \dfrac{\delta}{L_0} + \dfrac{1}{2}\left(\dfrac{\delta}{L_0}\right)^2$.

where δ is the deformation and L_0 is the original length of a segment. Determine (a) the tension in the cord of each segment and (b) the force exerted by the cord on the carrier of the car.

Figure 3.7 Canoe attached to car for Example 3.3. (*Courtesy of Dana Fransworth*, www.outdoortravels.com.)

TABLE 3.1 Cord Division for Example 3.3

Segment Number	Deformed Length (in)	Segment Number	Deformed Length (in)
1	3.4	6	3.4
2	3.4	7	3.1
3	3.4	8	2.7
4	3.4	9	2.3
5	3.4	10	2.2

PLAN

The undeformed length of each segment is $L_0 = 2$ in, and the final length is as given in Table 3.1. We can use a spreadsheet to find the average normal strain for each segment from the equations for the two cases. The modulus of elasticity will give us the average normal stress in each segment for the two cases. Knowing the diameter of the cord, we can find the cross-sectional area of the cord and multiply it by the normal stress to obtain the tension in each segment. If we make an imaginary cut in the cord just above A, we see that the tension in the first segment of the cord is the force exerted on the carrier of the car.

SOLUTION

1. **Deformation calculations:** The deformation of each segment can be found as $\delta = L_f - 2$, where, L_f is the deformed length. The deformation for each segment is shown in column 3 of Table 3.2.

TABLE 3.2 Solution for Example 3.3

Segment Number	Deformed Length (in)	Deformation	Case 1: Small Strain			Case 2: Large Strain		
			Strain	Stress (psi)	Tension (lb)	Strain	Stress (psi)	Tension (lb)
1	3.4	1.4	0.70	357	70.10	0.95	481.95	94.63
2	3.4	1.4	0.70	357	70.10	0.95	481.95	94.63
3	3.4	1.4	0.70	357	70.10	0.95	481.95	94.63
4	3.4	1.4	0.70	357	70.10	0.95	481.95	94.63
5	3.4	1.4	0.70	357	70.10	0.95	481.95	94.63
6	3.4	1.4	0.70	357	70.10	0.95	481.95	94.63
7	3.1	1.1	0.55	280.5	55.08	0.70	357.64	70.22
8	2.7	1.7	0.35	178.5	35.05	0.41	209.74	41.18
9	2.3	1.3	0.15	76.5	15.02	0.16	82.24	16.15
10	2.2	1.2	0.10	51	10.01	0.11	53.55	10.51

2. **Strain calculations:** The average normal strain can be calculated for the two cases from the given equations, and data are shown for each segment in columns 4 and 7 of Table 3.2.

3. **Stress calculation:** From Hooke's law we can find the stress as follows:

$$\sigma = E\varepsilon = 510\varepsilon \text{ psi} \tag{E1}$$

The stress calculation are shown for each segment in columns 5 and 8 of Table 3.2

4. **Internal force calculations:** The cross-sectional area can be found as shown in Equation (E2).

$$A = \frac{\pi d^2}{4} = \frac{\pi(0.5)^2}{4} = 0.1963 \text{ in}^2 \tag{E2}$$

The internal tension can be found as shown in Equation (E3).

$$T = \sigma A = 0.1963\sigma \text{ lb} \tag{E3}$$

The data for calculating, internal tension for the two cases are shown in columns 6 and 9 of Table 3.2.

5. **Reaction force calculation:** We can make a cut just above A and draw the free body diagram as shown in Figure 3.8, where T_1 is the internal tension of the cord in the first segment and R is the force exerted on the carrier. By force equilibrium, we obtain Equation (E4).

$$R = T_1 \tag{E4}$$

Figure 3.8 Free body diagram for Example 3.3.

The reaction force can be written as follows:

ANS.   Case 1: $R = 70.1$ lb

ANS.   Case 2: $R = 94.6$ lb

COMMENTS

1. Unlike Examples 3.1 and 3.2, in this example the kinematic equations changed. However, Hooke's law given by Equation (E1), the static equivalency given by Equation (E2), and the equilibrium equation given by Equation (E4) remain unchanged.

2. The additional complexities of a nonlinear stress–strain curve for the rubber cord could be incorporated into the analysis. The stress values would change, but Equations (E2) and (E4) would remain unchanged.

3. Examples 3.1 through 3.3 demonstrate the application and modular structure of the logic shown in Figure 3.2. By identifying equations that do not change, we can add complexities one at a time to improve the accuracy of the solution. In a similar manner, complexities can be added to the basic theories discussed in the next section. Which complexity to include depends upon the individual case and our need for accuracy.

3.2   BASIC THEORIES OF ONE-DIMENSIONAL STRUCTURAL MEMBERS

This section presents the basic theory on axial members, symmetric bending of beams, and torsion of circular shafts in tabular form to emphasize the *commonalities and differences* in the derivation of various formulas. We will start by reviewing the simplest possible theories, discussed in the introductory course on the mechanics of materials. The assumptions that lead to simplifications will be identified. We will drop some of these assumptions and incorporate complexities that are of practical interest in later sections and chapters. It will be seen that these complexities enter the theoretical derivation at the same level and are accounted for in nearly the same manner for all three structural members: axial rods, symmetric beams, and circular shafts.

Table 3.3 is a synopsis of the derivation the elementary theories. Additional details of the derivation can be found in Section A.12 of Appendix A. Assumptions and equations with suffixes -A, -B, and -T refer to axial, bending, and torsion, respectively. An assumption without any suffix is applicable to all three theories. This notation is used to facilitate the demonstration, in subsequent sections and chapters, that complexities enter the derivation at the same level and are accounted for in nearly a similar manner for all three structural members.

3.2.1   Limitations

The following limitations are common to theories about axial members, torsion of circular shafts, and symmetric bending of beams.

1. The length of the member is significantly greater (approximately 10 times) than the greatest dimension in the cross section. Approximation across the cross section is now possible because the region of approximation is small.

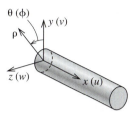

Figure 3.9 Coordinate systems and deformation variables.

2. We are away from regions of stress concentration, where displacements and stresses can be three-dimensional. The results from the simplified theories can be extrapolated into regions of stress concentration as described in Section 2.5. In a similar manner, results can be extrapolated into regions near cracks by means of stress intensity factors, described in Section 2.6.

3. The variation of external loads or changes in the cross-sectional area is gradual except in regions of stress concentration. The theory of elasticity shows that this limitation is necessary; otherwise approximations across the cross section would be untenable.

4. The external loads are such that the axial, torsion, and bending problems can be studied individually. This requires not only that the applied loads be in a given direction, but also that the loads pass through specific points associated with the cross section. The location of these points is determined within the development of the theories.

Figure 3.9 shows a slender member with the x axis along the axial direction and y and z in the plane of the cross section. We will use this coordinate system for axial and bending problems. For torsion problems we will use a polar coordinate system (ρ, θ, x).

For torsion and bending, we impose the following additional limitations.

- **Torsion:** We limit ourselves to circular cross sections.[2] This permits us to use arguments of axisymmetry in deducing deformation.

- **Bending:** We limit ourselves to beam cross sections that have a plane of symmetry to which the loading is restricted. This limitation ensures symmetric bending. It is assumed that the x-y plane is the plane of symmetry and that bending is occurring about the z axis (i.e., cross sections rotate about the z axis). This limitation will be dropped in Section 6.1 on the unsymmetric bending of beams.

3.2.2 Elementary Theories of One-Dimensional Structural Members

The variables used in Table 3.3, a synopsis of the derivation of the elementary theories studied in the introductory course on the mechanics of materials, are defined as follows.

---

[2] See Sections 6.4 and 8.8 for theory on noncircular shafts, and Problems 3.32 and 3.33.

TABLE 3.3 Synopsis of Elementary Theories of One-Dimensional Structural Members

	Axial	Bending	Torsion
Deformations			
Assumption 1	Deformations are not a functions of time.		
Assumption 2	2-A: Plane sections remain plane and parallel.	2a-B: Squashing deformation is significantly smaller than deformation due to bending. 2b-B: Plane sections before deformation remain plane after deformation. 2c-B: A plane perpendicular to the beam axis remains *nearly* perpendicular after deformation.	2a-T: Plane sections perpendicular to the axis remain plane during deformation. 2b-T: All radial lines rotate by equal angle during deformation on a cross section. 2c-T: Radial lines remain straight during deformation.
	$u = u_0(x)$ (3.1-A)	$v = v(x)$ (3.1a-B) $u = -y\dfrac{dv}{dx}$ (3.1b-B)	$\phi = \phi(x)$ (3.1-T)
Strains			
Assumption 3	The strains are small.		
	$\varepsilon_{xx} = \dfrac{du_0}{dx}(x)$ (3.2-A)	$\varepsilon_{xx} = -y\dfrac{d^2v}{dx^2}(x)$ (3.2-B)	$\gamma_{x\theta} = \rho\dfrac{d\phi}{dx}(x)$ (3.2-T)
Stresses			
Assumption 4 The material is isotropic. Assumption 5 There are no inelastic strains. Assumption 6 The material is elastic. Assumption 7 Stress and strain are linearly related.			
Using Hooke's law	$\sigma_{xx} = E\dfrac{du_0}{dx}(x)$ (3.3-A)	$\sigma_{xx} = -Ey\dfrac{d^2v}{dx^2}(x)$ (3.3-B)	$\tau_{x\theta} = G\rho\dfrac{d\phi}{dx}(x)$ (3.3-T)

Internal Forces and Moments

	-A	-B	-T
Static equivalency	$N = \int_A \sigma_{xx}\,dA$ (3.4a-A) $M_z = -\int_A y\sigma_{xx}\,dA = 0$ (3.4b-A) $M_y = -\int_A z\sigma_{xx}\,dA = 0$ (3.4c-A)	$N = \int_A \sigma_{xx}\,dA = 0$ (3.4a-B) $M_z = -\int_A y\sigma_{xx}\,dA$ (3.4b-B) $V_y = \int_A \tau_{xy}\,dA$ (3.4c-B)	$T = \int_A \rho\,\tau_{x\theta}\,dA$ (3.4-T)
Sign convention	$+\sigma_{xx}$ $=$ $+N$	$+\tau_{xy}$ $=$ $+V_y$ σ_{xx} Distribution — Compressive positive y face $=$ $+M_z$	$+\tau_{x\theta}$ $=$ $+T$ — Outward normal

Substituting stresses into internal forces and moments and noting $\dfrac{du_0}{dx}$, $\dfrac{d^2v}{dx^2}$, and $\dfrac{d\phi}{dx}$ are functions of x only, while the integration is with respect to y and z

	-A	-B	-T
Origin location	$\int_A yE\,dA = 0$ (3.5-A) $N = \dfrac{du_0}{dx}\int_A E\,dA$ (3.6-A)	$\int_A yE\,dA = 0$ (3.5-B) $M_z = \dfrac{d^2v}{dx^2}\int_A Ey^2\,dA$ (3.6-B)	$T = \dfrac{d\phi}{dx}\int_A G\rho^2\,dA$ (3.6-T)

Assumption 8 The material is homogeneous across the cross section.

	-A	-B	-T
The origin is at the centroid of the cross section.	$\int_A y\,dA = 0$ (3.7-A) $\dfrac{du_0}{dx} = \dfrac{N}{EA}$ (3.8-A)	$\int_A y\,dA = 0$ (3.7-B) $\dfrac{d^2v}{dx^2} = \dfrac{M_z}{EI_{zz}}$ (3.8-B)	$\dfrac{d\phi}{dx} = \dfrac{T}{GJ}$ (3.8-T)
	A = cross sectional area EA = axial rigidity	I_{zz} = second area moment of inertia about z EI_{zz} = bending rigidity	J = polar moment of the area GJ = torsional rigidity

(continues)

TABLE 3.3 *(continued)*

Axial	Bending	Torsion

Stress Formulas

Substituting Equations (3.8-A), (3.8-B), and (3.8-T) into Equations (3.3-A), (3.3-B), and (3.3-T)

Axial	Bending	Torsion
$\sigma_{xx} = \dfrac{N}{A}$ (3.9-A)	$\sigma_{xx} = -\left(\dfrac{M_z y}{I_{zz}}\right)$ (3.9-B)	$\tau_{x\theta} = \dfrac{T\rho}{J}$ (3.9-T)
	See Section 3.2.3 for shear stresses in bending.	

Deformation Formulas

Assumption 9   The material is homogeneous between x_1 and x_2.
Assumption 10   The structural member is not tapered between x_1 and x_2.
Assumption 11   The external loads do not change with x between x_1 and x_2.
Integrating Equations (3.8-A) and (3.8-T)

Axial	Bending	Torsion
$u_2 - u_1 = \dfrac{N(x_2 - x_1)}{EA}$ (3.10-A)	See Section 3.4.3 for beam deflection. (3.10-B)	$\phi_2 - \phi_1 = \dfrac{T(x_2 - x_1)}{GJ}$ (3.10-T)

Equilibrium Equations

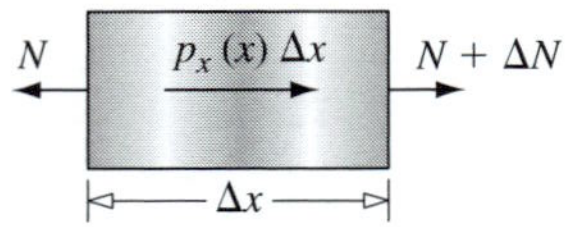

$$\dfrac{dN}{dx} = -p_x(x) \qquad (3.11\text{-A})$$

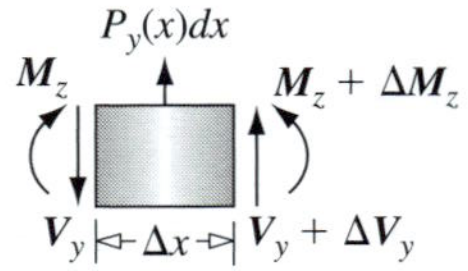

$$\dfrac{dV_y}{dx} = -p_y(x) \qquad (3.11\text{a-B})$$

$$\dfrac{dM_z}{dx} = -V_y \qquad (3.11\text{b-B})$$

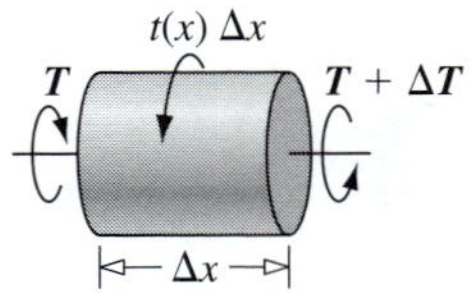

$$\dfrac{dT}{dx} = -t(x) \qquad (3.11\text{-T})$$

Differential Equations

Substituting Equations (3.8-A), (3.8-B), and (3.8-T) into Equations (3.11-A), (3.11a-B), (3.11b-B), and (3.11-T)

Axial	Bending	Torsion
$\dfrac{d}{dx}\left(EA\dfrac{du_0}{dx}\right) = -p_x(x)$ (3.12-A)	$\dfrac{d^2}{dx^2}\left(EI_{zz}\dfrac{d^2 v}{dx^2}\right) = p_y(x)$ (3.12-B)	$\dfrac{d}{dx}\left(GJ\dfrac{d\phi}{dx}\right) = -t(x)$ (3.12-T)

> **Definition 1** The displacements u, v, and w will be considered to be positive in the positive x, y, and z direction, respectively.
>
> **Definition 2** The rotation ϕ of the cross section will be considered to be positive counter clockwise with respect to the x axis.
>
> **Definition 3** The external distributed torque per unit length $t(x)$ is positive counterclockwise with respect to the x axis.
>
> **Definition 4** The external distributed forces per unit length $p_x(x)$ and $p_y(x)$ are considered to be positive in the positive x and y direction, respectively.
>
> **Definition 5** On a free body diagram, the positive internal axial force N is shown as tensile.
>
> **Definition 6** On a free body diagram, the positive internal shear force V_y is shown in the same direction as positive τ_{xy}.
>
> **Definition 7** On a free body diagram, the positive internal bending moment M_z is shown to put the positive y face in compression.
>
> **Definition 8** On a free body diagram, positive internal torque T is shown counterclockwise with respect to the outward normal.

If the sign conventions for internal forces and moments are followed in drawing free body diagrams, the formulas in Table 3.3 will give the right sign for displacement and stress components. The displacement direction can then be interpreted in accordance with the foregoing definitions, and the direction of the stress components can be determined by using subscripts as discussed in Chapter 1.

If the internal forces and moments are drawn in a manner that serves to equilibrate the external forces and moments, the formulas in Table 3.3 should be used only to determine the magnitude of the displacement and stress components; the direction of these components must be done by inspection.

The derivation of shear stress in bending is a beautiful demonstration of engineering approximation to overcome the limitations of consistent logic. This is discussed separately in Section 3.2.3. The calculation of beam deflection is also discussed separately, since external loads generally vary with x and require boundary and continuity conditions to determine the beam deflection.

EXAMPLE 3.4

A stepped circular beam in a turbomachine is modeled as shown in Figure 3.10. If the allowable bending normal stress is 200 MPa, determine the smallest fillet radius that can be used at section B for a maximum bending load estimated to be $P = 200$ N.

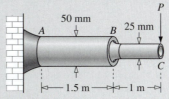

Figure 3.10 Stepped beam for Example 3.4.

PLAN

The nominal bending stress just to the right of B can be found by using Equation (3.9-B). The maximum stress concentration factor can be found by using the given allowable bending normal stress of 200 MPa. From the graph in Figure C.4, we can find the smallest fillet radius.

SOLUTION

An imaginary cut just to the right of B is made and the free body diagram drawn as shown in Figure 3.11. By moment equilibrium about point O, we obtain Equation (E1).

$$M_z = -P(1) = -200 \text{ N} \cdot \text{m} \qquad (E1)$$

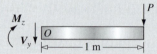

Figure 3.11 Free body diagram for Example 3.4.

The area moment of inertia is given by Equation (E2).

$$I_{zz} = \frac{\pi}{64}(0.025)^4 = 19.17(10^{-9}) \text{ m}^4 \qquad (E2)$$

The maximum nominal bending normal stress is at top or bottom, where $y_{max} = (12.5)(10^{-3})$ m. From Equation (3.9-B) we obtain Equation (E3).

$$\left|\sigma_{nominal}\right| = \left|\frac{-(-200)(\pm 12.5)(10^{-3})}{19.17(10^{-9})}\right| = 130.4(10^6) \text{ N/m}^2 \qquad \text{or}$$

$$\left|\sigma_{nominal}\right| = 130.4 \text{ MPa} \qquad (E3)$$

We note that $\left|\sigma_{max}\right| \leq 200 \text{ MPa}$. From our definition of stress concentration factor, we obtain Equation (E4):

$$\left|\sigma_{max}\right| = K_{conc}\left|\sigma_{nominal}\right| \qquad \text{or} \qquad 130.4 K_{conc} \leq 200 \qquad \text{or} \qquad K_{conc} \leq 1.53 \qquad (E4)$$

From Figure C.4, we obtain the approximate value of r/d corresponding to $D/d = 2$ and $K_{conc} = 1.53$ as follows:

$$r/d = 0.15 \qquad \text{or} \qquad r = (0.15)(25) \qquad \text{or} \qquad \textbf{ANS.} \quad r = 3.75 \text{ mm}$$

COMMENT

The stresses would be significantly higher without the fillet. The theoretical solution of elasticity for stresses at the reentrant corner with no fillet is infinite.

EXAMPLE 3.5

The hollow shaft of a ship's propeller is modeled as shown in Figure 3.12. The shaft has an outside diameter of 100 mm and an inside diameter of 75 mm. Determine the maximum value of P for the following cases.

Case I: Use the maximum octahedral shear stress theory and a yield strength of $\sigma_{yield} = 280$ MPa.

Case II: Use a critical stress intensity factor of $K_{crit} = 33$ MPa$\sqrt{m}$, and note that the small crack shown at point A in Figure 3.12 has a length of 4 mm.

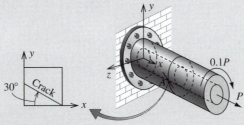

Figure 3.12 Hollow cylinder for Example 3.5.

PLAN

The axial normal stress and torsional shear stress in terms of P can be found from Equations (3.9-A) and (3.9-T). Case I: The principal stresses and the von Mises stress in terms of P can be found from Equations (1.19b) and (2.22) and compared with the given σ_{yield} to obtain the maximum value of P. Case II: The normal and shear stresses on the plane containing the crack can be found in terms of P from Equations (1.7a) and (1.7b); K_I, K_{II}, and K_{equiv} can be found in terms of P from Equations (2.28a), (2.28b), and (2.29). One limit on P can be determined by comparing K_{equiv} and K_{crit}. The second limit on P can be found assuming that microcracks grow as a result of principal stress 1. From the two limits, the maximum value of P can be found.

SOLUTION

The area A and the polar area moment J of a cross section can be found as shown in Equations (E1).

$$A = \frac{\pi}{4}(0.1^2 - 0.075^2) = 3.436(10^{-3}) \text{ m}^2$$
$$J = \frac{\pi}{32}(0.1^4 - 0.075^4) = 6.711(10^{-6}) \text{ m}^4 \tag{E1}$$

Figure 3.13 shows the free body diagram after an imaginary cut has been made through point A. The internal torque is drawn in accordance with the sign convention. Equilibrium of forces and moment yields Equations (E2).

$$N = P \quad \text{and} \quad T = -0.1P \tag{E2}$$

Figure 3.13 Free body diagram for Example 3.5.

From Equations (3.9-A) and (3.9-T), we obtain the axial stress and the torsional shear stress as shown in Equations (E3) and (E4). Note that point A is on the surface, where $\rho_A = 0.05$ m.

$$\sigma_{xx} = \frac{N}{A} = \frac{P}{3.436(10^{-3})} = 0.291P(10^3) \ \text{N/m}^2 \qquad \text{(E3)}$$

$$\tau_{x\theta} = \frac{T\rho_A}{J} = \frac{(-0.1P)(0.05)}{6.711(10^{-6})} = -0.745P(10^3) \ \text{N/m}^2 \qquad \text{(E4)}$$

The direction of torsional shear stress can be determined either by subscript or by inspection as in Figure 3.14(a) and (b). Figure 3.14(c) shows the stresses on the stress cube. We obtain the stresses in the Cartesian coordinates as shown in Equation (E5).

$$\sigma_{xx} = 0.291P(10^{-3}) \ \text{MPa (T)} \qquad \sigma_{yy} = 0 \qquad \tau_{xy} = 0.745P(10^{-3}) \ \text{MPa} \qquad \text{(E5)}$$

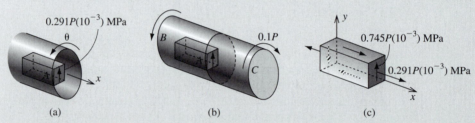

(a)   (b)   (c)

Figure 3.14 Direction of shear stress: (a) by subscript and (b) by inspection. (c) Stress cube.

Case I: Substituting the stresses from Equation (E5) into Equation (1.19b) we obtain principal stresses 1 and 2 as shown in Equation (E6).

$$\sigma_{1,2} = \left[\frac{0.291P(10^{-3}) + 0}{2}\right] \pm \sqrt{\left[\frac{0.291P(10^{-3}) - 0}{2}\right]^2 + [0.745P(10^{-3})]^2} \qquad \text{or}$$

$$\sigma_1 = 0.9046P(10^{-3}) \ \text{MPa} \qquad \sigma_2 = -0.6136P(10^{-3}) \ \text{MPa} \qquad \text{(E6)}$$

We note that the third principal stress (σ_3) is zero, as all stress component with subscript z are zero. The equivalent von Mises stress can be found from Equation (2.22) as shown in Equation (E7).

$$\sigma_{\text{von}} = \frac{P(10^{-3})}{\sqrt{2}} \sqrt{(0.9046 + 0.6136)^2 + (-0.6136 - 0)^2 + (0 - 0.9046)^2}$$

$$= 1.322P(10^{-3}) \ \text{MPa} \qquad \text{(E7)}$$

Noting that von Mises stress should be less than or equal to the yield stress, we obtain the maximum value of P:

$$1.322P(10^{-3}) \le 280 \qquad \text{or} \qquad P \le 211.66(10^3) \ \text{N} \qquad \textbf{ANS.} \quad P_{\text{max}} = 211.6 \ \text{kN}$$

Case II: The normal to the plane containing the crack is at $\theta = 90° - 30° = 60°$ from the x axis. Substituting this value of θ and stresses in Equation (E5) into Equations (1.7a) and (1.7b), we obtain the normal and shear stress on the plane as in Equations (E8) and (E9).

$$\sigma_{nn} = [(0.291)\cos^2 60 + (0)\sin^2 60 + 2(0.745)\sin 60 \,\cos 60]P(10^{-3}) \qquad \text{or}$$

$$\sigma_{nn} = 0.718P(10^{-3}) \text{ MPa} \qquad\qquad\qquad\qquad\qquad (E8)$$

$$\tau_{nt} = [-(0.291)\cos 60 \,\sin 60 + (0)\sin 60 \,\cos 60 + (0.745)(\cos^2 60 - \sin^2 60)]P(10^{-3})$$

$$\tau_{nt} = -0.4985P(10^{-3}) \qquad\qquad\qquad\qquad\qquad (E9)$$

The crack length is $2a = 4\,(10^{-3})$ m or $a = 2(10^{-3})$ m. The stress intensity factor for modes I and II can be found from Equations (2.28a) and (2.28b) as shown in Equations (E10) and (E11).

$$K_{\text{I}} = \sigma_{\text{nom}}\sqrt{\pi a} = 0.718P(10^{-3})\sqrt{\pi(2)(10)^{-3}} = 0.0569(10^{-3})P \qquad (E10)$$

$$K_{\text{II}} = \tau_{\text{nom}}\sqrt{\pi a} = 0.4985P(10^{-3})\sqrt{\pi(2)(10^{-3})} = 0.0395(10^{-3})P \qquad (E11)$$

We can find the equivalent stress intensity factor from Equation (2.29) and compare it to the critical stress intensity factor to obtain one limit on the maximum value of P as shown in Equation (E12).

$$K_{\text{equiv}} = \sqrt{K_{\text{I}}^2 + K_{\text{II}}^2} = (10^{-3})P\sqrt{0.0569^2 + 0.0395^2} = 0.0693(10^{-3})P \le 33 \qquad \text{or}$$

$$P \le 476.4(10^3) \text{ N} \qquad\qquad\qquad\qquad\qquad (E12)$$

To check whether a microcrack may grow as a result of principal stress 1, we calculate the mode I stress intensity factor with principal stress 1 as the nominal normal stress and compare the critical stress intensity factor to determine another limit on P as shown in Equation (E13).

$$K_{\text{I}} = \sigma_1\sqrt{\pi a} = 0.9046P(10^{-3})\sqrt{\pi(2)(10^{-3})} = 0.0717(10^{-3})P \le 33 \qquad \text{or}$$

$$P \le 460(10^3) \text{ N} \qquad\qquad\qquad\qquad\qquad (E13)$$

The maximum value of P is that which satisfies both Equations (E11) and (E12).

ANS. $P_{\text{max}} = 460$ kN

COMMENT

The problem demonstrates that in designs, the critical stress intensity factor and other limiting constraints such as yield stress are used in similar ways.

EXAMPLE 3.6

Show that the simplest differential equation for beam vibrations under a dynamic distributed force $p_y(x, t)$ is

$$\frac{\partial^2 v}{\partial^2 t} + c^2 \frac{\partial^4 v}{\partial^2 x} = \frac{p_y(x, t)}{\rho_m A} \tag{3.13-B}$$

where ρ_m is the material mass density and $c = \sqrt{EI_{zz}/\rho_m A}$. Neglect rotational inertia.

PLAN

The deflection v is now a function of x and time t; that is, Assumption 1 in Table 3.3 is now not valid. This makes ordinary derivatives of x on v into partial derivatives of x. With this difference accounted for, all equations up to equilibrium equations in Table 3.3 are valid. In equilibrium equations, the inertial forces must be included to obtain Equation (3.13-B).

SOLUTION

Equation (3.8-B) can be written as Equation (E1).

$$M_z = EI_{zz}\frac{\partial^2 v}{\partial x^2} \tag{E1}$$

Figure 3.15 shows a differential element of the beam. The mass of the differential element is $\rho_m A \Delta x$ and the linear acceleration is $\partial^2 v / \partial t^2$. The inertial force acting on the differential element is shown on the right in Figure 3.15.

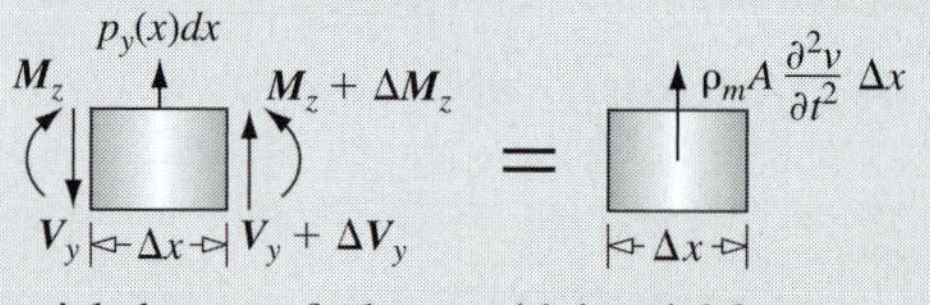

Figure 3.15 Differential element of a beam with inertial force.

By the force equilibrium in Figure 3.15, we obtain Equation (E2).

$$V_y + \Delta V_y - V_y + p_y(x, t)\Delta x = \rho_m A \frac{\partial^2 v}{\partial t^2}\Delta x \tag{E2}$$

Dividing by Δx and taking the limit $\Delta x \to 0$, we obtain Equation (E3).

$$\lim_{\Delta x \to 0}\left(\frac{\Delta V_y}{\Delta x}\right) + p_y(x, t) = \rho_m A \frac{\partial^2 v}{\partial t^2} \quad \text{or} \quad \frac{\partial V_y}{\partial x} + p_y(x, t) = \rho_m A \frac{\partial^2 v}{\partial t^2} \tag{E3}$$

By the moment equilibrium about the center, we obtain Equation (E4).

$$M_z + \Delta M_z + (V_y + \Delta V_y)\frac{\Delta x}{2} + V_y\frac{\Delta x}{2} - M_z = 0 \qquad \text{(E4)}$$

Dividing by Δx and taking the limit $\Delta x \to 0$, we obtain Equation (E5):

$$\lim_{\Delta x \to 0}\left(\frac{\Delta M_z}{\Delta x} + \Delta V_y\frac{\Delta x}{2}\right) = -V_y \qquad \text{or} \qquad \frac{\partial M_z}{\partial x} = -V_y \qquad \text{(E5)}$$

Substituting Equation (E1) into Equation (E5), we obtain Equation (E6).

$$V_y = -\frac{\partial}{\partial x}\left(EI_{zz}\frac{\partial^2 v}{\partial x^2}\right) \qquad \text{(E6)}$$

Substituting Equation (E6) into Equation (E3), we obtain Equation (E7).

$$-\frac{\partial^2}{\partial x^2}\left(EI_{zz}\frac{\partial^2 v}{\partial x^2}\right) + p_y(x, t) = \rho_m A\frac{\partial^2 v}{\partial t^2} \qquad \text{(E7)}$$

Assuming that EI_{zz} is constant, we can rewrite Equation (E7) as Equation (E8).

$$-\frac{EI_{zz}}{\rho_m A}\frac{\partial^2}{\partial x^2}\left(\frac{\partial^2 v}{\partial x^2}\right) + \frac{p_y(x, t)}{\rho_m A} = \frac{\partial^2 v}{\partial t^2} \qquad \text{or} \qquad \textbf{ANS.} \quad \frac{\partial^2 v}{\partial t^2} + c^2\frac{\partial^4 v}{\partial x^4} = \frac{p_y(x, t)}{\rho_m A} \qquad \text{(E8)}$$

COMMENTS

1. In this example only the equilibrium equation has changed. The kinematic equations, the constitutive equation (Hooke's law), and the equivalency equations did not change. The differential equations for axial members and circular shafts can be changed in a similar manner to include dynamic terms. See Problems 3.25 and 3.28.

2. The primary change is the additional inertial force term in Equation (E2). If we set $\partial^2 v/\partial t^2 = 0$ in Equation (3.13-B) and multiply by $\rho_m A$, we obtain the differential equation (3.12-B), as expected.

3. We neglected the rotational inertia term $\rho_m I_{zz}(\partial^2/\partial t^2)(\partial v/\partial x)\Delta x$ in deriving the beam vibration equation because this term is usually much smaller than the translational term we included in Figure 3.15. Inclusion of this term in differential equations is not difficult, but the increase in complexity of finding the solution of the resulting differential equation does not justify the small correction.

EXAMPLE 3.7

In Timoshenko beams,[3] the assumption of planes remaining perpendicular to the axis of the beam is dropped to account for shear. The derivation starts with the displacement field

$$u = -y\psi(x) \qquad v = v(x) \tag{3.14a-B}$$

where ψ is the angle of rotation of the plane of cross section from the vertical. Show that the simplest differential equations for Timoshenko beams is given by the following equations:

$$\frac{d}{dx}\left[GA\left(\frac{dv}{dx} - \psi\right)\right] = -p_y \tag{3.14b-B}$$

$$\frac{d}{dx}\left(EI_{zz}\frac{d\psi}{dx}\right) = -GA\left(\frac{dv}{dx} - \psi\right) \tag{3.14c-B}$$

PLAN

We follow the logic shown in Figure 3.2. The strain–displacement equations (1.35a) and (1.35d) are as before—by differentiating the given displacements, we can find normal strain ε_{xx} and shear strain γ_{xy}. Hooke's law is same as before—the normal and shear stress can be found. The static equivalency equations (3.4b-B) and (3.4c-B) are the same as before—by substituting the normal and shear stress, we can obtain new expressions for M_z and V_y. The equilibrium equations (3.11a-B) and (3.11b-B) are the same as before—by substituting M_z and V_y we can obtain Equations (3.14b-B) and (3.14c-B).

SOLUTION

1. **Strains:** The normal strain and shear strain can be obtained from Equations (1.35a) and (1.35d) as shown in Equations (E1).

$$\varepsilon_{xx} = \frac{\partial u}{\partial x} = -y\frac{d\psi(x)}{dx} \qquad \text{and} \qquad \gamma_{xy} = \frac{\partial u}{\partial y} + \frac{\partial v}{\partial x} = \frac{dv}{dx} - \psi(x) \tag{E1}$$

2. **Stresses:** By Hooke's law we obtain the stress as Equations (E2).

$$\sigma_{xx} = E\varepsilon_{xx} = -Ey\frac{d\psi(x)}{dx} \qquad \text{and} \qquad \tau_{xy} = G\gamma_{xy} = G\left(\frac{dv}{dx} - \psi(x)\right) \tag{E2}$$

---

[3] See Fung [1965] or Reddy [1993] in Appendix D.

3. **Internal forces and moments:** The foregoing stresses can be substituted in Equations (3.4b-B) and (3.4c-B). Noting that ψ and v are functions of x only and do not vary across the cross section, and assuming that the material is homogeneous across the cross section (E and G do not vary across the cross section), we obtain Equations (E3) and (E4).

$$M_z = -\int_A y\sigma_{xx} = \int_A Ey^2\frac{d\psi}{dx}dA = E\frac{d\psi}{dx}\int_A y^2 dA = EI_{zz}\frac{d\psi}{dx} \qquad \text{(E3)}$$

$$V_y = \int_A \tau_{xy}dA = \int_A G\left(\frac{dv}{dx} - \psi(x)\right)dA = GA\left[\frac{dv}{dx} - \psi(x)\right] \qquad \text{(E4)}$$

4. **Equilibrium:** We can substitute Equations (E3) and (E4) into the equilibrium equations (3.11a-B) and (3.11b-B) to obtain the desired results.

$$\frac{dV_y}{dx} = -p_y \qquad \text{or} \qquad \textbf{ANS.} \quad \frac{d}{dx}\left[GA\left(\frac{dv}{dx} - \psi(x)\right)\right] = -p_y$$

$$\frac{dM_z}{dx} = -V_y \qquad \text{or} \qquad \textbf{ANS.} \quad \frac{d}{dx}\left[EI_{zz}\frac{d\psi}{dx}\right] = -GA\left[\frac{dv}{dx} - \psi(x)\right]$$

COMMENTS

1. In this example the kinematic equations have changed. The constitutive equation (Hooke's law), the equivalency equations, and the equilibrium equations did not change. In a similar manner, equations for torsion of an elliptical shaft can be developed starting with the given displacement equations in Problem 3.32.

2. In classical beam theory, where Assumption 2c-B is applicable, the rotation of the cross section is related to deflection of the beam as $\psi = dv/dx$. Thus, Assumption 2c-B is a simplifying assumption that reduces the number of unknown variables from two to one.

3. Substituting Equation (3.14b-B) into Equation (3.14c-B), we obtain

$$\frac{d}{dx}\left[\frac{d}{dx}\left(EI_{zz}\frac{d\psi}{dx}\right)\right] = p_y$$

Substituting $\psi = dv/dx$, we obtain the differential equation (3.12-B).

4. The differential equations for Timoshenko beam vibrations can be developed by including inertial forces in the equilibrium equations as we did in Example 3.6.

3.2.3   Shear Stress in Thin Symmetric Beams

Assumption 2c-B implies that the shear strain γ_{xy} must be so small that we can neglect it in kinematic equations. By Hooke's law this implies that τ_{xy} should be small. In beam bending, a check on the validity of the analysis is to compare the maximum shear stress

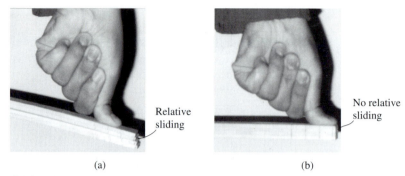

Figure 3.16 Effects of shear stress in bending: (a) separate beams and (b) glued beams.

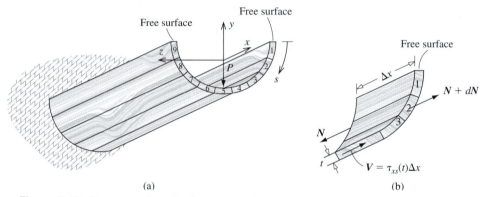

Figure 3.17 Shear stress in a circular cross section.

τ_{xy} with the maximum normal stress σ_{xx} for the *entire* beam. If the two stress components are comparable, then the shear strain cannot be neglected in kinematic considerations and our theory is not valid.[4] The maximum bending normal stress σ_{xx} in the beam should be nearly an order of magnitude greater than the maximum bending shear stress τ_{xy}.

Figure 3.16(a) shows the bending of four separate wooden strips, while Figure 3.16(b) shows the bending of the same four strips after they had been glued together. In the Figure 3.16(a) each wooden strip slides relative to the others in the axial direction, but in the Figure 3.16(b) the relative sliding is prevented by the shear resistance of the glue; that is, shear stress in the glue is developed to prevent the relative sliding between the strips. One may thus hypothesize that in any beam there will be shear stress on a imaginary surface parallel to the axis of the beam.

Consider a circular cross section that is glued together from nine wooden strips as shown in Figure 3.17. The bending load P, causes the moment M_z to vary along the length of the beam. Thus from Equation (3.9-B) we know that the magnitude of the normal stress σ_{xx} will change along the length of the beam. From the evidence of the photographs in Figure 3.16, we know that shear stress will exist at each glued surface to resist

---

[4] Timoshenko beam theory as presented in Example 3.7 should be used.

relative sliding by the wood strips. The shear stress in bending must balance the variation of normal stress σ_{xx} along the length of the beam.[5]

The outward normal of the surface will be in a different direction for each glue surface on which we consider the shear stress. If we define a tangential coordinate "s" that is in the direction of the tangent to the centerline of the cross section, the outward normal to the glued surface will be in the s direction and the shear stress will be τ_{sx}. Once more, by symmetry of shear stresses, $\tau_{xs} = \tau_{sx}$. At a point, if the s direction and the y direction are the same, then τ_{xs} will equal plus or minus τ_{xy}. If the s direction and the z direction are the same at a point, then τ_{xs} will equal plus or minus τ_{xz}.

In Figure 3.17(b), which shows a differential element of the beam, the shear force that balances the change in axial force N_s is shown on only one surface. The surface on the other end of the free body diagram is assumed to be a free surface; that is, the shear stress is zero on these other surfaces.

Definition 9 The direction of the s coordinate is from the free surface toward the point at which the shear stress is being calculated.

In a beam cross section, the top, bottom, and side surfaces are always assumed to be surfaces on which the shear stress is zero. Furthermore we state another assumption:

Assumption 12-B The beam is thin perpendicular to the centerline of the cross section (i.e., t is small).

Based on Assumption 12-B we can neglect the shear stress normal to the centerline because the surfaces on either side of the centerline are free surfaces. Furthermore, we can assume that the shear stress τ_{sx} is uniform in the direction normal to the centerline. By equilibrium of forces in the x direction in Figure 3.17b, we obtain $(N_s + dN_s) - N_s + \tau_{sx} t \, dx = 0$, or

$$\tau_{sx} t = -\frac{dN_s}{dx} = -\frac{d}{dx} \int_{A_s} \sigma_{xx} dA \tag{3.15-B}$$

In Equation (3.15-B), the identification of the area A_s is critical, so we formally define it as follows:

Definition 10 Area A_s is the area between the free surface and the point at which the shear stress is being evaluated.

Substituting Equation (3.9-B) into Equation (3.15-B), and noting that the moment M_z and the area moment of inertia I_{zz} do not vary over the cross section, we obtain

$$\tau_{sx} t = \frac{d}{dx} \left[\frac{M_z}{I_{zz}} \int_{A_s} y \, dA \right] = \frac{d}{dx} \left[\frac{M_z Q_z}{I_{zz}} \right] \tag{3.16-B}$$

---

[5] From the field of "elasticity," it is known that in the absence of body forces, equilibrium at a point requires $\partial \sigma_{xx}/\partial x + \partial \tau_{yx}/\partial y + \partial \tau_{zx}/\partial z = 0$ (see Section 8.1.3 and Problem 1.22). Thus, if σ_{xx} varies with x, then τ_{yx} (or τ_{xy}) must vary with y, and τ_{zx} (or τ_{xz}) must vary with z.

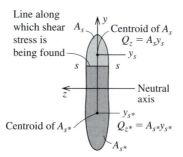

Figure 3.18 Calculation of Q_z.

where Q_z is referred to as the first moment of the area A_s and is defined as

$$Q_z = \int_{A_s} y\, dA \tag{3.17-B}$$

We make the following assumption.

Assumption 13-B The beam is not tapered.

Assumption 13-B implies that I_{zz} and Q_z are not functions of x, and these quantities can be taken outside the derivative sign, we obtain $\tau_{sx}t = (Q_z/I_{zz})(dM_z/dx)$. Substituting Equation (3.11b-B), the relationship between shear force V_y and moment M_z, we obtain

$$\tau_{sx}t = -\left(\frac{Q_z V_y}{I_{zz}}\right) \qquad \text{or} \qquad \boxed{\tau_{sx} = \tau_{xs} = -\left(\frac{V_y Q_z}{I_{zz}t}\right)} \tag{3.18-B}$$

From Equation (3.17-B), we note that Q_z is the first moment of the area A_s about the z axis. Figure 3.18 shows the area A_s between the top free surface and the point at which shear stress is being found (the s–s line). The integral in Equation (3.17-B) is the numerator in the definition of the centroid of the area A_s. Analogous to the moment due to a force, the first moment of an area can be found by placing the area A_s at its centroid and finding the moment about the neutral axis. That is, Q_z *is the product of area A_s and the distance of the centroid of the area A_s from the neutral axis*, as shown in Figure 3.18. Alternatively, Q_z can be found by using the bottom surface as the free surface, shown as Q_{z*} in Figure 3.18. In the introductory course on the mechanics of materials it was shown that Q_z is *maximum at the neutral axis*; hence bending shear stress is maximum at the neutral axis of a cross section.

We note that since y is measured from the centroid, the first moment of the entire cross-sectional area is zero; hence $Q_z + Q_{z*} = 0$. The equation implies that Q_z and Q_{z*} will have the same magnitude but opposite signs. Thus, we will get the same magnitude of the shear stress whether we use Q_z or Q_{z*} in Equation (3.18-B). But which gives the correct sign (direction)? The answer is that both will give the correct sign, provided the following point is remembered. The s direction in Equation (3.18-B) is measured from the free surface used in the calculation of Q_z.

Definition 11 The product of the shear stress and the thickness is called the *shear flow*.

Shear flow is designated by the symbol q, defined as follows:

$$q = \tau_{xs}t \tag{3.19-B}$$

The units of shear flow q are *force per unit length*. The terminology is from fluid flow in channels, but the term is used extensively for discussing shear stresses in thin cross sections, probably because of the visual image of a flow it conveys in the discussion of shear stress direction. The direction of shear flow (shear stress) can be determined by inspection as elaborated next and in Example 3.8.

Draw the shear flow along the centerline of the cross section in a direction that will satisfy the following rules:

1. The resultant force in the y direction is in the same direction as V_y.

2. The resultant force in the z direction is zero.

3. It is symmetric about the y axis. This requires shear flow to change direction as one crosses the y axis on the centerline. Sometimes this will imply that shear stress is zero at the point(s) where centerline intersects the y axis.

We can find the magnitude and the direction of the bending shear stress in two ways:

1. Use Equation (3.18-B) to find the magnitude of the shear stress. Use the rules just listed to determine the direction of shear stress.

2. Follow the sign convention for determining the shear force V_y. The shear stress is found from Equation (3.18-B), and the direction of shear stress is determined from the subscripts.

3.2.4 Stresses and Strains

Table 3.4 shows the stresses and strains in the three structural elements discussed in Section 3.2.3.

TABLE 3.4 Stresses and Strains

Axial		Symmetric Bending			Torsion	
Stresses	**Strains**	**Stresses**		**Strains**	**Stresses**	**Strains**
		About the z Axis	**About the y Axis**			
$\sigma_{xx} = \dfrac{N}{A}$	$\varepsilon_{xx} = \dfrac{\sigma_{xx}}{E}$	$\sigma_{xx} = -\left(\dfrac{M_z y}{I_{zz}}\right)$	$\sigma_{xx} = -\left(\dfrac{M_y z}{I_{yy}}\right)$	$\varepsilon_{xx} = \dfrac{\sigma_{xx}}{E}$	$\sigma_{xx} = 0$	$\varepsilon_{xx} = 0$
$\sigma_{yy} = 0$	$\varepsilon_{yy} = -\left(\dfrac{\nu\sigma_{xx}}{E}\right)$	$\sigma_{yy} = 0$	$\sigma_{yy} = 0$	$\varepsilon_{yy} = -\left(\dfrac{\nu\sigma_{xx}}{E}\right)$	$\sigma_{yy} = 0$	$\varepsilon_{yy} = 0$
$\sigma_{zz} = 0$	$\varepsilon_{zz} = -\left(\dfrac{\nu\sigma_{xx}}{E}\right)$	$\sigma_{zz} = 0$	$\sigma_{zz} = 0$	$\varepsilon_{zz} = -\left(\dfrac{\nu\sigma_{xx}}{E}\right)$	$\sigma_{zz} = 0$	$\varepsilon_{zz} = 0$
$\tau_{xy} = 0$ $\tau_{xz} = 0$ $\tau_{yz} = 0$	$\gamma_{xy} = 0$ $\gamma_{xz} = 0$ $\gamma_{yz} = 0$	$\tau_{xs} = -\left(\dfrac{V_y Q_z}{I_{zz} t}\right)$ $\tau_{yz} = 0$	$\tau_{xs} = -\left(\dfrac{V_z Q_y}{I_{yy} t}\right)$ $\tau_{yz} = 0$	$\gamma_{xs} = \dfrac{\tau_{xs}}{G}$ $\gamma_{yz} = 0$	$\tau_{x\theta} = \dfrac{T\rho}{J}$ $\tau_{yz} = 0$	$\gamma_{x\theta} = \dfrac{\tau_{x\theta}}{G}$ $\gamma_{yz} = 0$

Stress formulas for bending about the y axis can be obtained from the formulas for bending about the z axis by interchanging y and z. The generalized Hooke's law can be used to obtain the strains from the formulas derived for stresses. The bending shear stress τ_{xs} and the torsional shear stress $\tau_{x\theta}$ could be τ_{xy} or τ_{xz} or neither, depending upon the location of the point. However, our stress and strain transformation formulas are in Cartesian coordinate systems. Thus, we need to transform the shear stresses into stresses in Cartesian coordinates. This can be done by first showing the stresses from the formula on a stress cube and then examining the stresses on the stress cube in a Cartesian coordinate system.

EXAMPLE 3.8

Assuming a positive shear force V_y, sketch the direction of the shear flow along the centerline on the thin cross sections shown in Figure 3.19.

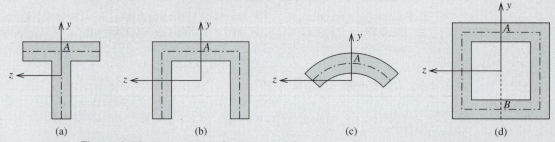

(a) (b) (c) (d)

Figure 3.19 Cross sections for Example 3.8.

PLAN

The positive shear force V_y will be upward in the positive y direction. At point A, the centerline intersects the y axis. Thus on either side of A, the shear flow will be in the opposite direction. In cross section of Figure 3.19(d), the shear flow will also change direction at point B. We can determine the direction of the flow in each cross section to satisfy the rules listed at the end of Section 3.2.3.

SOLUTION

(a) On the cross section shown in Figure 3.20(a), the shear flow (shear stress) from C to A will be in the positive y direction because V_y on the cross section is in the positive y direction. At point A in the flange, the flow will break in two and go in opposite directions, as shown in Figure 3.20(a). The resultant force due to shear flow from A to D will cancel the force due to shear flow from A to E, satisfying the condition of zero resultant force in the z direction and the condition of symmetric flow about the y axis.

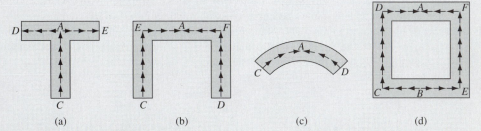

Figure 3.20 Shear flow for Example 3.8.

(b) On the cross section shown in Figure 3.20(b), the shear flow from C to E and from D to F will be in the positive y direction to satisfy the condition of symmetry about the y axis and to have the same direction as V_y. In the flange, the two flows will approach point A from opposite direction. The resultant force due to the shear flow from A to E will cancel the force due to the shear flow from A to F, satisfying the condition of zero resultant force in the z direction and the condition of symmetric flow about the y axis.

(c) On the cross section shown in Figure 3.20(c), the shear flows from points C and D will approach point A in the opposite direction. This ensures the condition of symmetry, and the condition of zero force in the z direction is met. The y components of the two shear flows are in the positive y direction, satisfying the condition that shear flow should result in positive V_y.

(d) The shear flow from C to D and the shear flow from E to F must be in the positive y direction to satisfy the condition of symmetry about the y axis and to have the same direction as V_y. At points A and B, the shear flows must change direction to ensure symmetric shear flows about the y axis. The force from the shear flow in BC and DA will cancel the force from the shear flow in BE and FA, ensuring the condition of zero force in the z direction.

COMMENTS

1. It should be noted that the shear flow (shear stress) is zero at the following points because they are on the free surface: points C, D, and E in Figure 3.20(a); points C and D in Figure 3.20(b) and (c).

2. At point A in Figure 3.20(b), (c), and (d), the shear flow will be zero, but not in Figure 3.20(a). This can be verified by calculating Q_z at point A for the two cases. The conclusion leads to the following analogy for fluid flow. In Figure 3.20(a), the shear flows at point A in branches AD and AE add up to the value of the shear flow at point A in branch CA. With no other branch at point A in Figure 3.20 (b), (c), and (d), the values of the shear flow are equal and opposite, which is possible only if the value of shear flow is zero.

3. The word "flow" invokes an image that makes visualizing the direction of shear stress a little easier.

4. By examining the direction of stress components in a Cartesian system, we can determine whether a stress component is positive or negative τ_{xy} or τ_{xz}, as shown in Figure 3.21. Note that τ_{xy} in all cases is positive, which means that it must be in the direction of the shear force V_y. But τ_{xz} can be positive or negative, depending upon the location of the point.

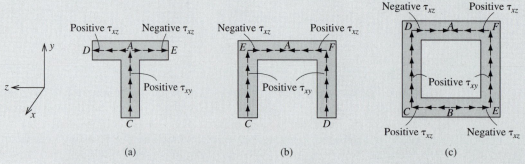

Figure 3.21 Direction and sign of stress components for Example 3.8.

EXAMPLE 3.9

A positive shear force of $V_y = 30$ N acts on the thin cross sections shown in Figure 3.22. Determine the shear flow along the centerlines and sketch it.

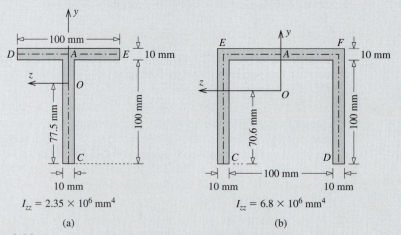

$$I_{zz} = 2.35 \times 10^6 \text{ mm}^4$$

(a)

$$I_{zz} = 6.8 \times 10^6 \text{ mm}^4$$

(b)

Figure 3.22 Cross sections for Example 3.9 (not drawn to scale).

PLAN

Since V_y and I_{zz} are known in Equation (3.18-B), the shear flow along the centerline can be determined if Q_z is determined along the centerline. Noting that the cross section is symmetric about the y axis, we see that the shear flow needs to be found on only one side of the y axis.

SOLUTION

(a) Figure 3.23 shows the areas A_s that can be used for finding shear flows in DA and CA of the cross section shown in Figure 3.22(a). The parameters s_1 and s_2 are defined from the free surface to the point at which the shear flow is to be found. The distance from the centroid of the areas A_s to the z axis can be found, and Q_z for each case calculated as shown in Equations (E1) and (E2).

$$Q_1 = (s_1)(0.01)(0.105 - 0.0775) \quad \text{or} \quad Q_1 = 0.275 s_1 (10^{-3}) \text{ m}^3 \quad \text{(E1)}$$

$$Q_2 = (s_2)(0.01)[-(0.0775 - s_2/2)] \quad \text{or}$$

$$Q_2 = -(0.775 s_2 - 5 s_2^2)(10^{-3}) \text{ m}^3 \quad \text{(E2)}$$

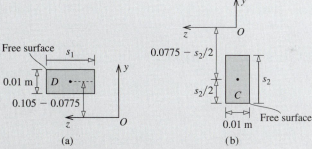

Figure 3.23 Calculation of Q_z in part (a) of Example 3.9.

Substituting V_y, I_{zz}, and Equations (E1) and (E2) into Equation (3.18-B), allows us to find the shear flow, in DA and CA of the cross section shown in Figure 3.22(a), as indicated in Equations (E3) and (E4).

$$q_1 = -\left[\frac{(30)0.275 s_1 (10^{-3})}{2.35(10^{-6})}\right] \text{ N/m} \quad \text{or} \quad q_1 = -3.51 s_1 \text{ kN/m} \quad \text{(E3)}$$

$$q_2 = -\left[\frac{(30)[-(0.775 s_2 - 5 s_2^2)](10^{-3})}{2.35(10^{-6})}\right] \text{ N/m} \quad \text{or}$$

$$q_2 = [9.89 s_2 - 63.83 s_2^2] \text{ kN/m} \quad \text{(E4)}$$

The shear flow q_1 is negative, implying that the direction of the flow is opposite to the direction of s_1. The values of q_1 can be calculated from Equation (E3) and plotted as shown in Figure 3.24(a). By symmetry, the flow in AE can also be plotted. The values of q_2 are positive between C and A, implying that the flow is in the direction of s_2. The values of q_2 can be calculated from Equation (E4) and plotted as shown in Figure 3.24(a).

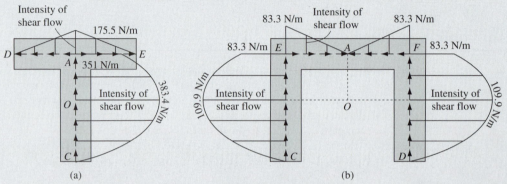

(a)   (b)

Figure 3.24 Shear flow on the cross sections of Example 3.9.

(b) Figure 3.25 shows the areas A_s that can be used for finding the shear flows in CE and EA of the cross section shown in Figure 3.22(b). In CE the parameter s can vary between points C and E. The same parameter s can be used for EA, but now its value is restricted between points E and A. The distance from the centroid of the areas A_s to the z axis can be found, and Q_z for each case can be calculated as shown in Equations (E5) and (E6).

For $0 \le s \le 0.10$ (i.e., between C and E),

$$Q_1 = (s)(0.01)[-(0.0706 - s/2)] \qquad \text{or}$$

$$Q_1 = -(0.706s - 5s^2)(10^{-3}) \text{ m}^3 \qquad\qquad \text{(E5)}$$

For $0.10 \le s \le 0.16$ (i.e., between E and F),

$$Q_2 = (0.01)(0.1)[-(0.0706 - 0.05)] + (s - 0.10)(0.01)[(0.105 - 0.0706)] \qquad \text{or}$$

$$Q_2 = (0.344s - 0.055)(10^{-3}) \text{ m}^3 \qquad\qquad \text{(E6)}$$

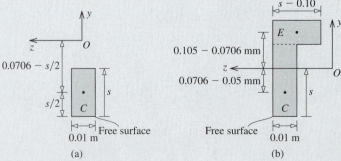

(a)   (b)

Figure 3.25 Calculation of Q_z in part (b) of Example 3.9.

Substituting V_y, I_{zz}, and Equations (E5) and (E6) into Equation (3.18-B), allows us to find the shear flow in CE and EA of the cross section shown in Figure 3.22(b), as indicated in Equations (E7) and (E8).

$$q_1 = -\left[\frac{(30)[(-(0.706s - 5s^2))(10^{-3})]}{6.8(10^{-6})}\right] \text{N/m} \quad \text{or}$$

$$q_1 = [3.115s - 22.06s^2] \text{ kN/m} \tag{E7}$$

$$q_2 = -\left[\frac{(30)(0.344s - 0.055)(10^{-3})}{6.8(10^{-6})}\right] \text{N/m} \quad \text{or}$$

$$q_2 = [0.243 - 1.157s] \text{ kN/m} \tag{E8}$$

In Equations (E7) and (E8), the shear flows q_1 and q_2 are positive, implying that the flows are in the positive s direction. The values of q_1 and q_2 can be calculated from Equation (E7) and plotted as shown in Figure 3.24(a). By using symmetry, the flows in DF and FA can also be plotted as shown in Figure 3.24(a).

COMMENTS

1. Figure 3.24 shows that the shear flow in the flanges varies linearly, while the shear flow in the web varies quadratically and is maximum at the neutral axis.

2. In Example 3.8, the direction of flow was determined by inspection, whereas in this example it was determined by using the formulas. Comparison of Figure 3.24 with Figure 3.20 yields the same results. Thus, inspection could be used as a check on the results of the formulas. Alternatively, the formulas could be used for determining the magnitude of the shear flow (stress) and inspection to determine the direction of the shear flow.

3. In Figure 3.24(a), the flow value at point A in CA is 351 N/m, which is the sum of the flows in AD and AE. In Figure 3.24(b), the flows in EA and FA approach A from opposite directions, and hence the flow at A is zero. Also notice that the flow at E in CE is the same as in EA in Figure 3.24(b). Thus the behavior of shear flow is similar to that of fluid flow in a channel.

4. Point E is not at the junction of web and flange in Figure 3.24(b). Should the material between E and the web–flange junction be considered to be part of the flange or the web? Slight differences in numerical results will occur, depending on how the material near the corner at E is considered in calculating Q_2, but these differences are within the accuracies of our approximations and have little impact on engineering design.

PROBLEM SET 3.1

3.1 The roller in Figure P3.1 slides in a slot by the amount $\delta_P = 0.25$ mm in the direction of the force F. Both bars have a cross sectional area of $A = 100$ mm² and a modulus of elasticity $E = 200$ GPa. Bar AP and bar BP have lengths of $L_{AP} = 200$ mm and $L_{BP} = 250$ mm, respectively. Use the logic in Figure 3.2, to determine the applied force F.

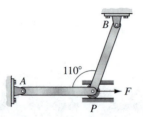

Figure P3.1

3.2 The roller in Figure P3.2 slides in a slot by the amount $\delta_P = 0.25$ mm in the direction of the force F. Both bars have a cross-sectional area of $A = 100$ mm² and a modulus of elasticity $E = 200$ GPa. Bar AP and bar BP have lengths of $L_{AP} = 200$ mm and $L_{BP} = 250$ mm, respectively. Use the logic in Figure 3.2, to determine the applied force F.

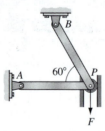

Figure P3.2

3.3 The roller in Figure P3.3 slides in a slot by the amount $\delta_P = 0.25$ mm in the direction of the force F. Both bars have a cross-sectional area of $A = 100$ mm² and a modulus of elasticity $E = 200$ GPa. Bar AP and bar BP have lengths of $L_{AP} = 200$ mm and $L_{BP} = 250$ mm, respectively. Use the logic in Figure 3.2, to determine the applied force F.

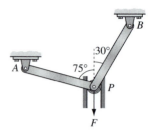

Figure P3.3

ANS. $F = 16.7$ kN

3.4 Draw the shear stress due to torsion on the stress cubes at points A and B in Figure P3.4. Is the shear stress positive or negative τ_{xy}?

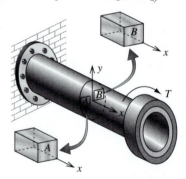

Figure P3.4

3.5 Draw the shear stress due to torsion on the stress cubes at points A and B in Figure P3.5. Is the shear stress positive or negative τ_{xy}?

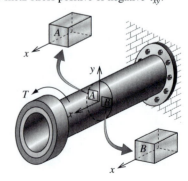

Figure P3.5

3.6 The allowable axial stress in the stepped axial rod shown in Figure P3.6 is 20 ksi. (a) Determine the smallest fillet radius that can be used at section B. (b) What is the percentage increase in the factor of safety if the fillet radius in part (a) is doubled? Use the stress concentration graphs given in Appendix C (Figure C.1–C.4).

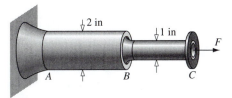

Figure P3.6

3.7 The tapered bar shown in Figure P3.8 has a cross-sectional area that varies with x as $A = K(2L - 0.25x)^2$. Determine the elongation of the bar in terms of P, L, E, and K.

3.8 The tapered bar shown in Figure P3.8 has a cross-sectional area that varies with x as $A = K(4L - 3x)$. Determine the elongation of the bar in terms of P, L, E, and K.

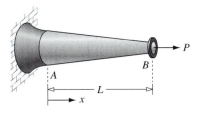

Figure P3.8

ANS. $u = 0.4621(P/EK)$

3.9 The radius of the tapered shaft shown in Figure P3.9 varies as $R = (r/L)(2L - 0.25\,x)$. In terms of T, L, G, and r, determine (a) the rotation of wheel B and (b) the maximum shear stress in the shaft.

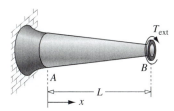

Figure P3.9

ANS. $\phi_B - \phi_A = 0.523\,T_{ext}L/(Gr^4)$
$\tau_{max} = 0.1188\,T_{ext}/r^3$

3.10 Determine the contraction due to its own weight of the column shown in Figure P3.10. The specific weight is $\gamma = 0.28$ lb/in^3, the modulus of elasticity is $E = 3600$ ksi, the length is $L = 120$ in, and the radius varies as $R = \sqrt{240 - x}$, where R and x are in inches.

Figure P3.10

3.11 Determine the contraction due to its own weight of the column shown in Figure P3.10. The specific weight is $\gamma = 24$ kN/m^3, the modulus of elasticity is $E = 25$ GPa, the length is $L = 10$ m, and the radius varies as $R = 0.5\,e^{-0.07x}$, where R and x are in meters.

3.12 A thin tube of has an outer diameter of 310 mm and a thickness of 10 mm. The tube material has a critical stress intensity factor of 33 MPa$\sqrt{m}$. The smallest crack length of 2 mm represents the limit of the flaw detection system. What is the maximum torque that should be permitted to act on the tube?

ANS. $T_{max} = 806$ kN $\cdot$ m

3.13   A thin cylindrical tube with an outer diameter of 5 inches is fabricated by butt-welding 1/16-inch-thick plate along a spiral seam as shown in Figure P3.13. The critical stress intensity factor for the material is 22 ksi $\sqrt{\text{in}}$. If $T = 25$ in · kips and $P = 10$ kips, determine the critical crack length of a crack along the seam.

Figure P3.13

3.14   A thin cylindrical tube with an outer diameter of 5 inches is fabricated by butt-welding 1/16-inch-thick plate along a spiral seam as shown Figure P3.13. A through crack of 0.07 inch was observed in the seam. The critical stress intensity factor for the material is 22 ksi $\sqrt{\text{in}}$. If $P = 4$ kips, determine the maximum torque T that the tube can transmit.

3.15   Assuming a positive shear force V_y, (a) sketch the direction of the shear flow along the centerline on the thin cross section shown in Figure P3.15. (b) At points A, B, C, and D, determine whether the stress component is τ_{xy} or τ_{xz} and whether it is positive or negative.

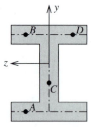

Figure P3.15

3.16   Assuming a positive shear force V_y, (a) sketch the direction of the shear flow along the centerline on the thin cross section shown Figure P3.16. (b) At points A, B, C, and D, determine whether the stress component is τ_{xy} or τ_{xz} and whether it is positive or negative.

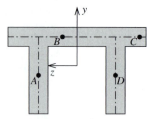

Figure P3.16

3.17   Assuming a positive shear force V_y, (a) sketch the direction of the shear flow along the centerline on the thin cross section shown Figure P3.17. (b) At points A, B, C, and D, determine whether the stress component is τ_{xy} or τ_{xz} and whether it is positive or negative.

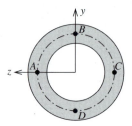

Figure P3.17

3.18   A positive shear force $V_y = 10$ kN acts on the thin cross section shown in Figure P3.18. The cross section has a uniform thickness of 10 mm. Determine the equation of shear flow along the centerlines and sketch it.

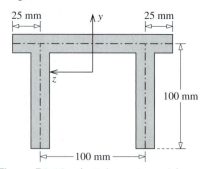

Figure P3.18   (not drawn to scale)

3.19 A shear force $V_y = -20$ kN acts on the thin cross section shown in Figure P3.19. The cross section has a uniform thickness of 10 mm. Determine the equation of shear flow along the centerlines and sketch it.

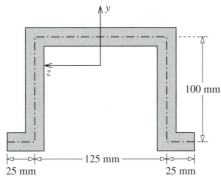

Figure P3.19 (not drawn to scale)

3.20 The hoist shown in Figure P3.20 was used to lift heavy loads in a mining operation. Member *EF* supported loads that only if unsymmetric with respect to the pulley. Otherwise it carried no load and can be neglected in the stress analysis. The yield stress of steel is 30 ksi. For a factor of safety of 1.5, determine the maximum load *W*. Use the maximum shear stress theory.[6]

3.21 The highway sign shown in Figure P3.21 uses a 16-inch hollow pipe as a vertical post and 12-inch hollow pipe for the horizontal arms. The pipes are one inch thick. Assume that a uniform wind pressure of *p* acts on the sign boards and on the pipes. Note that the pressure on the pipes acts on the projected area of *Ld*, where *L* is the length of pipe and *d* is the diameter of the pipe. The yield stress of the pipes is 40 ksi. For a factor of safety of 2, determine the maximum wind pressure. Use the maximum octahedral shear stress theory.

Figure P3.21

ANS. $p_{max} = 140.5$ lb/ft^2

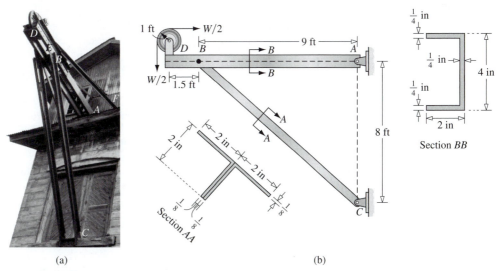

(a)   (b)

Figure P3.20

---

[6] Though the load on section *BB* is not passing through the plane of symmetry, the theory of symmetric bending can still be used because of the structural symmetry.

3.22 The displacement in the x direction in a beam cross section is given by $u = u_0(x) - y(dv/dx)(x)$. Assuming small strains and linear, elastic, isotropic, homogeneous material with no inelastic strains, show

$$N = EA\frac{du_0}{dx} - EAy_c\frac{d^2v}{dx^2} \quad \text{and}$$

$$M_z = EAy_c\frac{du_0}{dx} + EI_{zz}\frac{d^2v}{dx^2}$$

where y_c is the y coordinate of the centroid of the cross section measured from some arbitrary origin, A is the cross-sectional area, I_{zz} is the area moment of inertia about the z axis, and N and M_z are the internal axial force and internal bending moment, respectively. Note that if y is measured from the centroid of the cross section (i.e., $y_c = 0$), the axial and bending problems decouple. In such a case, show

$$\sigma_{xx} = \frac{N}{A} - \frac{M_z y}{I_{zz}}$$

3.23 A beam resting on an elastic foundation has a distributed spring force that depends upon the deflections at a point acting as shown in Figure P3.23. Show that the differential equation governing the deflection of the beam is:

$$\frac{d^2}{dx^2}\left(EI_{zz}\frac{d^2v}{dx^2}\right) + kv = p_y$$

where k is the foundation modulus (i.e., the spring constant per unit length).

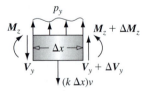

Figure P3.23 Elastic foundation effect.

3.24 Determine the elongation of the rotating bar shown in Figure P3.24 in terms of the rotating speed ω, density ρ, length L, and cross-sectional area A. (*Hint:* The body force per unit volume is $\rho\omega^2x$.)

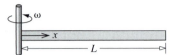

Figure P3.24

3.25 Consider the dynamic equilibrium of the differential element shown in Figure P3.25, where N is the internal force, ρ_m is the mass density, A is the cross-sectional area, and $\partial^2 u/\partial t^2$ is the acceleration.

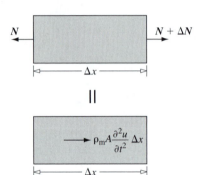

Figure P3.25 Dynamic equilibrium of an axial member.

Derive the *wave equation*

$$\frac{\partial^2 u}{\partial t^2} = c^2\frac{\partial^2}{\partial x^2}(u) \qquad (3.20\text{-A})$$

The material constant c, the velocity of the propagation of sound in the material, is $\sqrt{E/\rho_m}$.

3.26 Show that the functions $f(x - ct)$ and $g(x + ct)$ satisfy Equation (3.20-A), the wave equation.

3.27 Show that the following solution satisfies Equation (3.13-B).

$$v(x, t) = G(x)H(t)$$

$$G(x) = A \cos \omega x + B \sin \omega x + C \cosh \omega x + D \sinh \omega x \qquad (3.21)$$

$$H(t) = E \cos(c\omega^2)t + D \sin(c\omega^2)t$$

3.28 Consider the dynamic equilibrium of the differential element shown in Figure P3.28, where T is the internal torque, ρ_m is the mass density, J is the polar area moment of inertia, and $\partial^2\phi/\partial t^2$ is the angular acceleration.

For $c = \sqrt{G/\rho_m}$, derive the following differential equation for torsional vibration of shafts:

$$\frac{\partial^2\phi}{\partial t^2} = c^2\frac{\partial^2\phi}{\partial x^2} \qquad (3.22\text{-T})$$

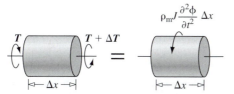

Figure P3.28 Dynamic equilibrium in torsion.

3.29 Show that Equation (3.22-T) is satisfied by the solution

$$\phi = \left(A\ \cos\frac{\omega x}{c} + B\ \sin\frac{\omega x}{c}\right)$$
$$\times (C\ \cos\ \omega t + D\ \sin\ \omega t) \qquad (3.23\text{-T})$$

where A, B, C, and D are constants that are determined from the boundary conditions and the initial conditions, and ω is the frequency of vibration.

3.30 The displacements for a shaft with an elliptical cross section under torsion[7] are

$$u = \phi Kyz \qquad v = -\phi xz \qquad w = \phi xy$$

where the constant K is related to the semiaxis of the ellipse a-b as $K = (b^2 - a^2)/(b^2 + a^2)$. The

angle of twist per unit length ϕ may be treated as constant for this problem. The axis of the shaft is in the x direction. Determine the relationship between the internal torque T and ϕ on a *cross section* in terms of the shear modulus G, a, and b. Also find the stress components in terms of the internal torque T.

ANS. $\phi = \dfrac{T(a^2+b^2)}{\pi a^3 b^3 G}$

$$\tau_{xy} = -\left(\frac{2Tz}{\pi ab^3}\right) \qquad \tau_{xz} = \left(\frac{Ty}{\pi a^3 b}\right)$$

3.31 A body under applied loads and body forces produces the displacement field given by

$$u = 0 \quad v = Kx(y^2 - z^2) + Kaxz$$
$$w = -2Kxyz - Kaxy$$

where u, v, and w are displacements in the x, y, and z directions, respectively. Figure P3.31 shows the internal forces and moments (stress resultants) on a cross section. Assuming a linear, elastic, isotropic, homogeneous material, determine all the internal forces and moments in terms of modulus of elasticity, Poisson's ratio ν, and the constants K and a. (See Examples 1.10 and 2.3, earlier.)

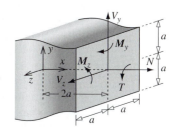

Figure P3.31

---

[7] The solution for noncircular shafts under torsion was found by Saint-Venant (1797–1886).

3.32 The displacements of a point on a cross section of a noncircular shaft are

$$u = \psi(y, z)\frac{d\phi}{dx} \qquad v = -xz\frac{d\phi}{dx}$$
$$w = xy\frac{d\phi}{dx}$$

(3.24)

where u, v, and w are the displacements in the x, y, and z directions, respectively, $d\phi/dx$ is the rate of twist and is considered to be *constant*, and $\psi(x, y)$, called the warping function,[8] describes the movement of points out of the plane of the cross section.

Show that the shear stresses for the noncircular shaft are given by

$$\tau_{xy} = G\left(\frac{\partial\psi}{\partial y} - z\right)\frac{d\phi}{dx} \qquad \text{and}$$

(3.25)

$$\tau_{xz} = G\left(\frac{\partial\psi}{\partial z} + y\right)\frac{d\phi}{dx}$$

3.33 Show that for circular shafts [$\psi(x, y) = 0$] the equations in Problem 3.32 reduce to Equations (3.8-T) and (3.9-T).

3.3 DISCONTINUITY FUNCTIONS

A structural member usually has distributed loads and point loads applied to it. Point forces cause jumps in the internal forces, and point moments cause jumps in internal moments as one crosses the point loads from one side to the other. The discontinuity functions permit us to model these jumps such that a single expression for distributed load can be used over the length of the entire structural member. This single expression, when substituted into the equilibrium equations (3.11-A), (3.11a-B), (3.11b-B), and (3.11-T) and integrated, yields a single expression for the internal forces and moments that incorporates the jumps in these forces and moments. The use of just one expression has two advantages: (1) At any point on the member, the values of the internal forces and moments can be obtained by substituting the appropriate value of x. These internal forces and moments can be used for finding stress at any point on the cross section. (2) Unlike the internal forces and moments, which are discontinuous at points of concentrated forces and moments, the displacements and rotations are continuous. The displacements and rotations are obtained by integrating Equations (3.8-A), (3.8-B), and (3.8-T).[9] Without discontinuity functions, we would have to find the displacements and rotations before and after the concentrated load points and then impose the continuity conditions, *explicitly* to get the integration constants. With the discontinuity functions, the single expression of the internal forces and moments can be integrated directly, and the continuity conditions are *implicitly* satisfied. The reduces the algebraic effort significantly.

We shall now define the discontinuity functions. In Section 3.4, we will use the discontinuity functions for calculating axial displacements and the torsional rotation and deflection of beams.

3.3.1 Definition of Discontinuity Functions

Consider a distributed load p and an equivalent load $P = p\varepsilon$ as shown in Figure 3.26. Suppose we now let the intensity of the distributed load continuously increase to infinity,

---

[8] Equations of elasticity show that the warping function satisfies the Laplace equation: $\partial^2\psi/\partial y^2 + \partial^2\psi/\partial z^2 = 0$.

[9] Alternatively, we can start from Equations (3.12-A), (3.12-B), and (3.12-T) and integrate.

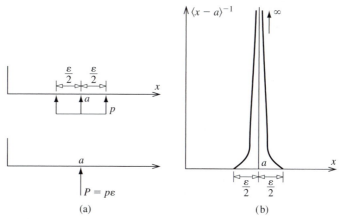

Figure 3.26 The delta function.

while we decrease the length over which the distributed force is applied to zero in such a manner that the area $p\varepsilon$ remains a finite quantity; then we obtain a concentrated force P applied at $x = a$. Mathematically, this is stated as follows:

$$P = \lim_{p \to \infty} \lim_{\varepsilon \to 0} (p\varepsilon) \tag{3.26}$$

Rather than write the limit operations to represent a concentrated force, the following notation is used for representing the concentrated force: $P\langle x - a \rangle^{-1}$.

The function $\langle x - a \rangle^{-1}$ is called the *Dirac delta function*, or just the delta function. The delta function is zero except in an infinitesimal region near a. As x tends toward a, the delta function tends to infinity, but the area under the function is equal to 1. Mathematically, the delta function is defined as follows:

$$\langle x - a \rangle^{-1} = \begin{cases} 0 & x \neq a \\ \infty & x \to a \end{cases} \qquad \int_{(a-\varepsilon)}^{(a+\varepsilon)} \langle x - a \rangle^{-1} dx = 1 \tag{3.27}$$

Now consider the following integral of the delta function: $\int_{-\infty}^{x} \langle x - a \rangle^{-1} dx$. The lower limit of minus infinity emphasizes that the point is before a. If $x < a$, then in the interval of integration, the delta function is zero at all points, and hence the integral value is zero. If $x > a$, then we can write the integral as the sum of three integrals:

$$\int_{-\infty}^{(a-\varepsilon)} \langle x - a \rangle^{-1} dx + \int_{(a-\varepsilon)}^{(a+\varepsilon)} \langle x - a \rangle^{-1} dx + \int_{(a+\varepsilon)}^{x} \langle x - a \rangle^{-1} dx$$

The first and the third integrals are zero because the delta function is zero at all points in the interval of integration, while the second integral is equal to 1, per Equation (3.27). Thus, the integral $\int_{-\infty}^{x} \langle x - a \rangle^{-1} dx$ is zero before a and one after a. The integral, which is called the *step function*, is represented by the notation $\langle x - a \rangle^{0}$ and is defined in Equation (3.28) and shown in Figure 3.27(a).

$$\langle x - a \rangle^{0} = \int_{-\infty}^{x} \langle x - a \rangle^{-1} dx = \begin{cases} 0 & x < a \\ 1 & x > a \end{cases} \tag{3.28}$$

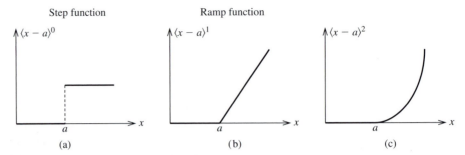

Figure 3.27 Discontinuity (step) functions.

Now consider the integral of the step function,

$$\int_{-\infty}^{x} \langle x - a \rangle^0 dx$$

If $x < a$, then in the interval of integration the step function is zero at all points, and hence the integral value is zero. If $x > a$, then we can write the integral as sum of two integrals:

$$\int_{-\infty}^{a} \langle x - a \rangle^0 dx + \int_{a}^{x} \langle x - a \rangle^0 dx$$

The first integral is zero because the step function is zero at all points in the interval of integration, while the second integral value is $(x - a)$. The integral $\int_{-\infty}^{x} \langle x - a \rangle^0 dx$, called the *ramp function,* is represented by the notation $\langle x - a \rangle^1$ and shown in Figure 3.27(b). Proceeding in this manner, we can define an entire class of functions that are mathematically represented as follows:

$$\langle x - a \rangle^n = \begin{Bmatrix} 0 & x \le a \\ (x - a)^n & x > a \end{Bmatrix} \tag{3.29}$$

We can also generate the integral formula of Equation (3.30) from Equation (3.29):

$$\int_{-\infty}^{x} \langle x - a \rangle^n dx = \frac{\langle x - a \rangle^{n+1}}{n + 1} \qquad n \ge 0 \tag{3.30}$$

We define one more function, called the *doublet function.* It is represented by the notation $\langle x - a \rangle^{-2}$, mathematically defined in Equation (3.31),

$$\langle x - a \rangle^{-2} = \begin{Bmatrix} 0 & x \ne a \\ \infty & x \to a \end{Bmatrix} \qquad \int_{-\infty}^{x} \langle x - a \rangle^{-2} dx = \langle x - a \rangle^{-1} \tag{3.31}$$

The delta function $\langle x - a \rangle^{-1}$ and the doublet function $\langle x - a \rangle^{-2}$ become infinite at $x = a$. Alternatively stated, these functions are singular at $x = a$ and are referred to as *singularity functions.*

Definition 12 The *discontinuity functions* consist of the entire class of functions $\langle x - a \rangle^n$ for positive and negative n.

Note that the discontinuity function is zero if the argument is zero or negative. By differentiating Equations (3.28), (3.30), and (3.31), we can obtain the formulas in Equation (3.32).

$$\frac{d}{dx}\langle x-a\rangle^{0} = \langle x-a\rangle^{-1}$$

$$\frac{d}{dx}\langle x-a\rangle^{-1} = \langle x-a\rangle^{-2} \quad\quad (3.32)$$

$$\frac{d}{dx}\langle x-a\rangle^{n} = n\langle x-a\rangle^{n-1} \quad\quad n \geq 1$$

3.4 BOUNDARY VALUE PROBLEMS

Definition 13 A boundary value problem is a mathematical listing of the differential equations, the boundary conditions, and any other conditions necessary to solve the differential equations.

The differential equations for the axial, bending, and torsion cases are given by Equations (3.12-A), (3.12-B), and (3.12-T), respectively (see Table 3.3). The distributed loads $p_x(x)$, $p_y(x)$, and $t(x)$ in the differential equations are known functions. If there is a jump in loading, then it may be possible to represent the loading by using the discontinuity functions as shown in the sections that follow.

The boundary conditions, which are conditions on the internal forces/moments or displacements/rotations at the two ends of the member, can be written either by inspection or by drawing a free body diagram of an infinitesimal element at the end, as discussed in the sections that follow.

The "other conditions" referred to in Definition 13 are generally the continuity conditions on the displacements and rotations. However, the use of the discontinuity functions eliminates the need for these continuity conditions, resulting in a significant decrease in algebraic effort, as will be seen in a number of examples.

3.4.1 Axial Displacement

Figure 3.28 shows a template of an axial bar in which a concentrated force is applied at $x = a$. If we make an imaginary cut in the region $x < a$, the internal axial force $N = 0$. If we make an imaginary cut in the region $x > a$, the internal axial force $N = -F$. We can use the discontinuity function to write the internal force as $N = -F\langle x-a\rangle^{0}$. Substituting this internal force into Equation (3.11-A) and using Equation (3.32), we obtain $p_x = F\langle x-a\rangle^{-1}$. These two equations are shown as template equations in Figure 3.28. If the point force on the actual member is in the direction shown in Figure 3.28, the internal force and distributed force are as given. If the point force is opposite to that shown in Figure 3.28, the signs are reversed in the template equations.

The axial displacement $u_0(x)$ can be found from integrating Equation (3.8-A), provided we can find N as a function of x. For statically indeterminate problems and problems in which the distributed load p_x is a complex function, it may be preferable to find

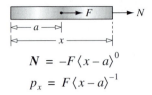

$$N = -F\langle x - a\rangle^0$$
$$p_x = F\langle x - a\rangle^{-1}$$

Figure 3.28 Template and template equations for use of discontinuity functions in axial members.

$u_0(x)$ by starting with Equation (3.12-A) and integrating two times. For convenience, the differential equation (3.12-A), given in Table 3.3, is repeated as Equation (3.33a-A).

- Differential equation

$$\boxed{\frac{d}{dx}\left(EA\frac{du_0}{dx}\right) = -p_x(x)}$$

(3.33a-A)

The integration of the differential equation will generate two constants that must be determined from boundary conditions. The first integration will generate $EA(du_0/dx)$, which is the internal axial force N. The second integration will generate the axial displacement u_0. At each end of the axial member, either the axial displacement u_0 or the internal axial force N must be specified.

- Boundary condition: At each end specify:

$$\boxed{u_0 \quad \text{or} \quad N}$$

(3.33b-A)

To determine the boundary condition on the internal axial force N, we make an imaginary cut an infinitesimal distance (ε) from the end (say point A) and draw a free body diagram of the infinitesimal element as in Figure 3.29. The internal axial force is drawn in tension in accordance with the sign convention given in Definition 5.

By equilibrium of forces and by taking the limit $\varepsilon \to 0$, we obtain the following boundary condition:

$$\lim_{\varepsilon \to 0}(F_{ext} - N_A - p_x(x_A)\varepsilon) = 0 \quad \text{or} \quad N_A = F_{ext}$$

This equation shows that the distributed axial force does not affect the boundary condition on the internal axial force. The value of the internal axial force N at the end of an axial bar is equal to the concentrated external axial force applied at the end. If either the free body diagram or the internal force N is not drawn in accordance with the sign convention given in Definition 5, the likelihood of getting a sign error increases. If F_{ext} is zero, we can write the boundary condition of zero value of N by inspection.

Figure 3.29 Boundary condition on the internal axial force.

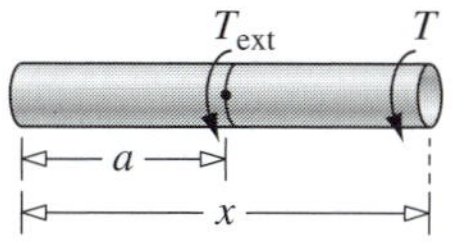

$$T = -T_{ext}\langle x - a \rangle^0$$
$$t = T_{ext}\langle x - a \rangle^{-1}$$

Figure 3.30 Template and template equations for use of discontinuity functions in torsions.

3.4.2 Torsional Rotation

Figure 3.30 shows a template of a shaft on which a concentrated torque is applied at $x = a$. If we make an imaginary cut in the region $x < a$, the internal torque $T = 0$. If we make an imaginary cut in the region $x > a$, the internal torque is $T = -T_{ext}$. We can use the discontinuity function to write the internal torque as $T = -T_{ext}\langle x - a \rangle^0$. Substituting this internal torque into Equation (3.11-T) and using Equation (3.32), we obtain $t = T_{ext}\langle x - a \rangle^{-1}$. These two equations are shown as template equations. If the concentrated torque T on the actual shaft is in the direction shown in Figure 3.30, the internal torque and the distributed torque are as given. If the concentrated torque is opposite to that shown in Figure 3.30, the signs are reversed in the template equations.

The rotation $\phi(x)$ can be found from integrating Equation (3.8-T), provided we can find T as a function of x. For statically indeterminate problems and problems in which the distributed load t is a complex function, it may be preferable to find $\phi(x)$ by starting with Equation (3.12-T) and integrating two times. For convenience, the differential equation (3.12-T) is repeated as (3.34a-T).

- Differential equation

$$\boxed{\frac{d}{dx}\left(GJ\frac{d\phi}{dx} \right) = -t(x)}$$

(3.34a-T)

The integration of the differential equation will generate two constants that must be determined from boundary conditions. The first integration will generate $GJ(d\phi/dx)$, which is the internal torque T. The second integration will generate the rotation ϕ. At each end of the axial member, either the rotation ϕ or the internal torque T must be specified.

- Boundary condition: At each end specify

$$\boxed{\phi \quad \text{or} \quad T}$$

(3.34b-T)

To determine the boundary condition on internal torque T, we make an imaginary cut an infinitesimal distance (ε) from the end (say point A) and draw a free body diagram of the infinitesimal element as shown in Figure 3.31. The internal torque T is drawn in accordance with the sign convention given in Definition 8.

By equilibrium of moments and by taking the limit $\varepsilon \to 0$, we obtain the boundary condition

$$\lim_{\varepsilon \to 0}(T_{ext} - T_A - t(x_A)\varepsilon) = 0 \quad \text{or} \quad T_A = T_{ext}$$

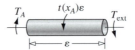

Figure 3.31 Boundary condition on the internal torque.

Which shows that the distributed torque does not affect the boundary condition on the internal torque. The value of the internal torque T at the end of a shaft is equal to the concentrated external torque applied at the end. If either the free body diagram or the internal torque T is not drawn in accordance with the sign convention given in Definition 8, the likelihood of getting a sign error increases. If T_{ext} is zero, we can write the boundary condition of zero value of T by inspection.

EXAMPLE 3.10

The column shown in Figure 3.32 has a specific weight of $\gamma = 0.1$ lb/in³, modulus of elasticity of $E = 4000$ ksi, and cross-sectional area $A = 100$ in². Determine (a) the movement of the rigid plate at C and (b) the reaction force at A.

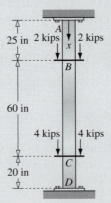

Figure 3.32 Column for Example 3.10.

PLAN

By using the discontinuity functions and the fact that the distributed force per unit length is $p = \gamma A$, we can use Equation (3.12-A) to write the differential equation. Zero displacements at A and D are the boundary conditions. The boundary value problem can be solved to obtain the displacement as a function of x. (a) The movement of the plate at $C u_c$ can be found by substituting $x = 85$ in the displacement expression. (b) The reaction force R_A is equal to the internal axial force at $x = 0$.

SOLUTION

The weight per unit length is $p = \gamma A = 10$ lb/in. The concentrated forces at B and C are in the increasing direction of x as in Figure 3.28. We may thus write the total distributed force as follows:

$$p_x = [10 + 4000\langle x - 25\rangle^{-1} + 8000\langle x - 85\rangle^{-1}] \text{ lb/in} \qquad (E1)$$

Upon substituting Equation (E1) in Equation (3.12-A), we will obtain the differential equation. Noting that the displacement at $x = 0$ and $x = 105$ inches is zero, we obtain the following boundary value problem.

- Differential equation

$$\frac{d}{dx}\left(EA\frac{du}{dx}\right) = -[10 + 4000\langle x - 25\rangle^{-1} + 8000\langle x - 85\rangle^{-1}] \qquad (E2)$$

- Boundary conditions

$$u(0) = 0 \qquad (E3)$$

$$u(105) = 0 \qquad (E4)$$

Integrating Equation (E2) and using Equation (3.28), we obtain

$$EA\frac{du}{dx} = -[10x + 4000\langle x - 25\rangle^{0} + 8000\langle x - 85\rangle^{0}] + c_1 \qquad (E5)$$

Integrating Equation (E5) and using Equation (3.30), we obtain

$$EAu = -[5x^2 + 4000\langle x - 25\rangle^{1} + 8000\langle x - 85\rangle^{1}] + c_1 x + c_2 \qquad (E6)$$

Substituting Equation (E6) into Equation (E3) and noting that discontinuity functions with negative arguments are zero, we obtain

$$EAu(0) = -[0 + 4000\langle -25\rangle^{1} + 8000\langle -85\rangle^{1}] + c_1(0) + c_2 \text{ or } c_2 = 0 \qquad (E7)$$

Substituting Equation (E6) into Equation (E4), we obtain

$$EAu(105) = -[5(105)^2 + 4000\langle 105 - 25\rangle^{1} + 8000\langle 105 - 85\rangle^{1}] + c_1(105) = 0 \quad \text{or}$$

$$-[5(105)^2 + 4000(80) + 8000(20)] + c_1(105) = 0 \qquad \text{or} \qquad c_1 = 5096.4 \quad (E8)$$

Substituting Equations (E7) and (E8) and the value of $EA = 4000(10^3)(100) = 400(10^6)$ into Equation (E6), we obtain

$$400(10^6)u = -[5x^2 + 4000\langle x - 25\rangle^{1} + 8000\langle x - 85\rangle^{1}] + (5096.4)x \qquad (E9)$$

(a) Substituting $x = 85$ in Equation (E9), we obtain:

$$400(10^6)u_c = -[5(85)^2 + 4000\langle 60\rangle^{1} + 8000\langle 0\rangle^{1}] + (5096.4)(85)$$

ANS. $u_c = 0.393(10^{-3})$ in

(b) Substituting $x = 0$ in Equation (E6) and using equation (E8), we obtain the internal force at A as follows:

$$N_A = EA\frac{du}{dx}\bigg|_{x=0} = -[10(0) + 4000\langle -25\rangle^{0} + 8000\langle -85\rangle^{0}] + c_1 = c_1 = 5096.4 \text{ lb}$$

By inspection, the reaction force will be upward, and its magnitude will be equal to N_A. Thus we obtain the following answer.

ANS. $R_A = 5096$ lb upward

COMMENTS

1. We did not draw any free body diagram to find the internal axial force or displacement. This is because Equation (3.12-A) is an equilibrium equation.

2. In this problem we could by inspection determine the direction of the reaction force at A. For problems in which such is not the case, we must get the direction by making an infinitesimal cut away from A, writing equilibrium equation, and drawing a free body diagram.

3. Without the use of the discontinuity functions this work would be very tedious, for we would have to solve a statically indeterminate problem that required us to find the relative displacement for each segment by integration.

EXAMPLE 3.11

The external torque on a drill bit varies linearly to a maximum intensity of q in·lb/in as shown in Figure 3.33. If the drill bit diameter is d, its length L, and the modulus of rigidity G, determine the relative rotation of the end of the drill bit with respect to the chuck.

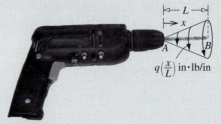

Figure 3.33 Distributed torque on the drill bit of Example 3.11.

PLAN

We can substitute the given distributed torque in Equation (3.12-T) to obtain the differential equation. The boundary conditions are as follows: at point B the internal torque will be zero, and at point A the rotation is ϕ_A. Solving the boundary value problem, we obtain the relative rotation of point B with respect to point A.

SOLUTION

The distributed torque on the drill bit is counterclockwise with respect to the x axis; thus we can substitute $t(x) = q(x/L)$ in Equation (3.12-T) to obtain the differential equation. To complete the statement of boundary value problem, we list the following boundary conditions as discussed in the plan.

- Differential equation

$$\frac{d}{dx}\left(GJ\frac{d\phi}{dx}\right) = -q\left(\frac{x}{L}\right) \tag{E1}$$

- Boundary conditions

$$T(L) = GJ\frac{d\phi}{dx}\bigg|_{x=L} = 0 \tag{E2}$$

$$\phi(0) = \phi_A \tag{E3}$$

Integrating Equation (E1), we obtain Equation (E4).

$$GJ\frac{d\phi}{dx} = -q\left(\frac{x^2}{2L}\right) + c_1 \tag{E4}$$

Substituting Equation (E4) into Equation (E2), we obtain Equation (E5).

$$-q\left(\frac{L^2}{2L}\right) + c_1 = 0 \quad \text{or} \quad c_1 = \frac{qL}{2} \tag{E5}$$

Substituting Equation (E5) into Equation (E4) and integrating, we obtain Equation (E6).

$$GJ\phi = -q\left(\frac{x^3}{6L}\right) + \frac{qL}{2}x + c_2 \tag{E6}$$

Substituting Equation (E6) into Equation (E3), we obtain Equation (E7).

$$GJ\phi_A = 0 + 0 + c_2 \quad \text{or} \quad c_2 = GJ\phi_A \tag{E7}$$

Substituting Equation (E7) into equation (E6) and simplifying, we obtain Equation (E8).

$$GJ\phi = -q\left(\frac{x^3}{6L}\right) + \frac{qL}{2}x + GJ\phi_A \quad \text{or} \quad \phi - \phi_A = \frac{q}{6GJL}(-x^3 + 3L^2x) \tag{E8}$$

Substituting $x = L$ and $\phi = \phi_B$ and $J = (\pi d^4)/32$ in Equation (E8), we obtain the relative rotation as follows:

$$\phi_B - \phi_A = \frac{q}{6G(\pi d^4/32)L}(-L^3 + 3L^3) \quad \textbf{ANS.} \quad \phi_B - \phi_A = \left(\frac{32qL^2}{3\pi Gd^4}\right) \text{ccw}$$

Dimension check:

$$q \to O(FL/L) \to O(F) \qquad d \to O(L) \qquad L \to O(L) \qquad G \to O(F/L^2)$$

$$\phi \to O(\ ) \qquad \frac{qL^2}{Gd^4} \to O\left(\frac{FL^2}{(F/L^2)L^4}\right) \to O(\ ) \qquad \text{Checks.}$$

COMMENTS

1. We did not need to draw a free body diagram to find the internal torque because Equation (3.12-T), from which we started, is an equilibrium equation.

2. After substituting for c_1, Equation (E4) can be written as

$$\frac{d\phi}{dx} = \frac{(q/2L)(L^2 - x^2)}{G(\pi d^4/32)}.$$

Integration from A to B can be done as well:

$$\int_{\phi_A}^{\phi_B} d\phi = \left(\frac{16q}{\pi GLd^4}\right) \int_{x_A=0}^{x_B=L} (L^2 - x^2)dx \qquad \text{or} \qquad \phi_B - \phi_A = \left(\frac{16q}{\pi GLd^4}\right)\left(L^2 x - \frac{x^3}{3}\right)\Bigg|_0^L$$

which gives the same answer as before.

3. If at the end of the drill bit there is a concentrated torque in the model, $T(x = L)$ in Equation (E1) will not be zero; rather, it will be equal in magnitude to the concentrated torque. To get the right sign for the internal torque, we will need to make an imaginary cut at an infinitesimal distance from $x = L$ and draw a free body diagram, as was done in Figure 3.31.

4. The discontinuity functions are not needed in this problem because there are no concentrated torques between A and B.

3.4.3   Beam Deflection

Figure 3.34 shows a free body diagram of a beam in which a concentrated force, moment, or the start of distributed force occurs at $x = a$. If we make an imaginary cut in the region $x < a$, then the internal bending moment $M_z = 0$. If we make an imaginary cut in region $x > a$, then the internal bending moment M_z can be found by means of the moment equilibrium, as shown in Figure 3.34. The moment can then be written by using the definition of discontinuity functions. By substituting Equation (3.11a-B) into Equation (3.11b-B), we obtain Equation (3.35-B).

$$\frac{d^2 M_z}{dx^2} = p_y(x) \qquad\qquad\qquad (3.35\text{-B})$$

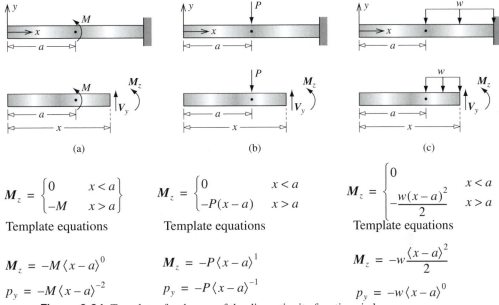

$$M_z = \begin{cases} 0 & x < a \\ -M & x > a \end{cases}$$

Template equations

$$M_z = \begin{cases} 0 & x < a \\ -P(x-a) & x > a \end{cases}$$

Template equations

$$M_z = \begin{cases} 0 & x < a \\ -\dfrac{w(x-a)^2}{2} & x > a \end{cases}$$

Template equations

$$M_z = -M\langle x-a \rangle^0$$

$$p_y = -M\langle x-a \rangle^{-2}$$

$$M_z = -P\langle x-a \rangle^1$$

$$p_y = -P\langle x-a \rangle^{-1}$$

$$M_z = -w\frac{\langle x-a \rangle^2}{2}$$

$$p_y = -w\langle x-a \rangle^0$$

Figure 3.34 Templates for the use of the discontinuity functions in beams.

By substituting the moment expressions for each case obtained in Figure 3.34 into Equation (3.35-B), we obtain an expression for the distributed force for each case as shown in Figure 3.34.

The template equations are associated with the coordinate system shown in Figure 3.34. If the coordinate system is changed, corresponding changes must be made in the template equations. If the external forces and moments are in the directions shown on the template free body diagram, then the expressions for M_z and p_y will be in accordance with the template equations. If the external forces and moments are opposite, then the signs will have to be reversed in the template equations.

Second-order differential equation The deflection of the beam as a function of x [i.e., $v(x)$] represents a curve often referred to as the *elastic curve*. Provided we can find the internal moment M_z as a function of x, we can find $v(x)$ by integrating Equation (3.8-B) twice. The differential equation (3.8-B) is rewritten for convenience in the following form.

- Second-order differential equation

$$EI_{zz}\frac{d^2v}{dx^2} = M_z(x)$$

(3.36a-B)

Integration of the differential equation will generate two constants that must be determined from the boundary conditions. The first integration will generate dv/dx, which is the slope of the elastic curve or, alternatively, the rotation of the cross section about the z axis in accordance with Assumption 2c-B of Table 3.3. The second integration will generate the deflection v. At each end of the axial member, either the slope dv/dx or the deflection v must be specified.

• Boundary conditions for second order: At each end specify:

$$\boxed{\dfrac{dv}{dx} \qquad \text{or} \qquad v}$$

(3.36b-B)

Fourth-order differential equation: For statically indeterminate problems and problems in which the distributed load p_y is a complex function, it *may* be preferable to find $v(x)$ by starting with Equation (3.12-B) and integrating four times. For convenience, the differential equation (3.12-B) is repeated.

• Fourth-order differential equation

$$\boxed{\dfrac{d^2}{dx^2}\!\left(EI_{zz}\dfrac{d^2v}{dx^2}\right) = p_y(x)}$$

(3.37a-B)

Integration of this differential equation will generate four constants that will have to be determined from the boundary conditions. Integrating once will result in the shear force given by Equation (3.37b-B), which is obtained by substituting the moment expression [Equation (3.36a-B)] into the equilibrium condition [Equation (3.12-B)] to obtain

$$V_y = -\dfrac{d}{dx}\!\left(EI_{zz}\dfrac{d^2v}{dx^2}\right)$$

(3.37b-B)

The second integration will result in the moment, and the third and the fourth integrations will result in the slope and the deflection, respectively. Thus conditions could be imposed on any of the four quantities v, dv/dx, M_z, or V_y. To understand how these conditions are determined, we generalize a principle discussed in statics for determining the reaction force and/or moments. Recall that in drawing free body diagrams, we used the following principles for determining reaction forces and moment at the supports.

1. If a point cannot move in a given direction, then a reaction force opposite to the direction acts at that support point.

2. If a line cannot rotate about an axis in a given direction, then a reaction moment opposite to the direction acts at that support.

If we were to make an imaginary cut very close to the support and then draw a free body diagram, we would find that the internal shear force V_y is equal to the reaction force, and the internal moment M_z is equal to the reaction moment. Thus, the first observation implies that if a point cannot move, (i.e., the deflection v is zero at the point), then the shear force is not known because the reaction force is not known. Similarly, the second observation implies that if a line cannot rotate around an axis passing through a point (i.e., dv/dx is zero), then the internal moment is not known because the reaction moment is not known. The reverse is equally true. Consider the free end of a cantilever beam. Clearly at the free end the internal moment and the internal shear force are zero; but the free end can deflect and rotate by any amount that is dictated by the loading. Thus, when we specify a value of shear force, we cannot specify displacement; and when we specify the value of an internal moment at a point, we cannot specify rotation. Thus, there are two sets in which the four quantities v, dv/dx, M_z, and V_y are grouped for purpose of determining the boundary conditions.

- *Group 1*: At a boundary point either the deflection v or the internal shear force V_y can be specified, but not both.
- *Group 2*: At a boundary point either the slope dv/dx or the internal bending moment M_z can be specified, but not both.

To generate four boundary conditions, two conditions are specified at each end of the beam. We choose one condition from group 1 and one from group 2; alternatively we can make the following statement.

- Boundary conditions for fourth-order differential equations: specify at each end:

$$
\begin{array}{ccc}
v & \text{or} & V_y \\
& \text{and} & \\
\dfrac{dv}{dx} & \text{or} & M_z
\end{array}
\tag{3.37c-B}
$$

To determine the boundary condition on the internal moment M_z, and the shear force V_y, we make an imaginary cut an infinitesimal distance (ε) from the end (say point A) and draw a free body diagram of the infinitesimal element as shown in Figure 3.29. The internal shear force and internal bending moment are drawn in accordance with the sign convention given in Definitions 6 and 7.

By equilibrium of forces and moment at point O and by taking the limit $\varepsilon \to 0$, we obtain the boundary conditions in Equations (3.38a-B) and (3.38b-B).

$$
\lim_{\varepsilon \to 0}(F_{ext} + V_A - p_y(x_A)\varepsilon) = 0 \qquad \text{or} \qquad V_A = F_{ext}
\tag{3.38a-B}
$$

$$
\lim_{\varepsilon \to 0}\left(M_A - M_{ext} + \varepsilon F_{ext} + p_y(x_A)\frac{\varepsilon^2}{2}\right) = 0 \qquad \text{or} \qquad M_A = M_{ext}
\tag{3.38b-B}
$$

These equations show that the distributed force does not affect the boundary condition on either the internal shear force or the bending moment. On a boundary point, the value of the internal shear force V_y and the bending moment M_z is equal to the concentrated external force and the concentrated external moment, respectively. If the free body diagram is not drawn, or if the internal shear force V_y and the bending moment M_z are not drawn in accordance with the sign convention, the likelihood of getting a sign error increases. If F_{ext} and M_{ext} are zero, we can write the boundary condition of zero value for shear force and moment by inspection.

A synopsis of the boundary value problems for axial, torsion, and symmetric bending is presented in Table 3.5.

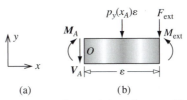

(a)   (b)

Figure 3.35 Boundary condition on the internal shear force and bending moment.

TABLE 3.5 Synopsis of Boundary Value Problems

	Axial	Torsion	Symmetric Bending	
			Second Order	**Fourth Order**
Differential equation	$\dfrac{d}{dx}\left(EA\dfrac{du_0}{dx}\right) = -p_x(x)$	$\dfrac{d}{dx}\left(GJ\dfrac{d\phi}{dx}\right) = -t(x)$	$EI_{zz}\dfrac{d^2v}{dx^2} = M_z(x)$	$\dfrac{d^2}{dx^2}\left(EI_{zz}\dfrac{d^2v}{dx^2}\right) = p_y(x)$
Boundary conditions at each end	u_0 or N	ϕ or T	$\dfrac{dv}{dx}$ or v	v or V_y and $\dfrac{dv}{dx}$ or M_z
Internal forces/ moments	$N = EA\dfrac{du_0}{dx}$	$T = GJ\dfrac{d\phi}{dx}$	$M_z = EI_{zz}\dfrac{d^2v}{dx^2}$	$V_y = -\dfrac{d}{dx}\left(EI_{zz}\dfrac{d^2y}{dx^2}\right)$

EXAMPLE 3.12

Use the discontinuity functions for the three templates shown in Figure 3.36 to write the moment and distributed force expressions.

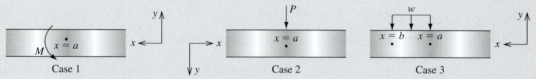

Figure 3.36 The three cases of Example 3.12.

PLAN

For cases I and II, we can make an imaginary cut after $x = a$ and draw the shear force and bending moment in accordance with the sign convention. By equilibrium we can obtain the moment expression and rewrite it by using the discontinuity functions. By differentiating twice, we can obtain the distributed force expression. For case III, we can write the expression for the distributed force by using the discontinuity functions and integrate twice to obtain the moment expression.

SOLUTION

Case I: We make an imaginary cut at $x > a$ and draw the free body diagram as shown in Figure 3.37(a). The moment and shear force are drawn in accordance with our sign convention.

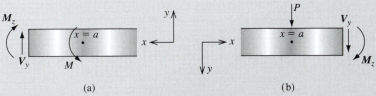

Figure 3.37 Free body diagrams for Example 3.12: (a) case 1 and (b) case II.

By moment equilibrium in Figure 3.37(a) we obtain Equation (E1).

$$M_z = M \qquad \text{(E1)}$$

Equation (E1) is valid only after $x > a$. We can use the step function to write the moment expression. By differentiating twice as indicated in Equation (3.35-B), we obtain the distributed force expression.

ANS. $\quad M_z = M \langle x - a \rangle^0 \qquad p_y = M \langle x - a \rangle^{-2}$

Case II: We make an imaginary cut at $x > a$ and draw the free body diagram as shown in Figure 3.37(b). The moment and shear force are drawn in accordance with our sign convention. By moment equilibrium in Figure 3.37(b) we obtain Equation (E2).

$$M_z = P(x - a) \qquad \text{(E2)}$$

Equation (E2) is valid only after $x > a$. We can use the ramp function to write the moment expression. By differentiating twice as indicated in Equation (3.35-B), we obtain the distributed force expression.

ANS. $\quad M_z = P \langle x - a \rangle^1 \qquad p_y = P \langle x - a \rangle^{-1}$

Case III: The distributed force is in the negative y direction; its start can be represented by the step function at $x = a$. The end of the distributed force can also be represented by a step function, with an opposite sign used at the start to obtain Equation (E3).

ANS. $\quad p_y = -w \langle x - a \rangle^0 + w \langle x - b \rangle^0 \qquad \text{(E3)}$

Substituting Equation (E3) into Equation (3.35-B) and integrating twice, we can obtain the moment expression as shown in Equation (E4).

ANS. $\quad M_z = -\dfrac{w}{2} \langle x - a \rangle^2 + \dfrac{w}{2} \langle x - b \rangle^2 \qquad \text{(E4)}$

COMMENTS

1. The three cases presented could be part of a beam with more complex loading. But the contribution for each load would be calculated as shown in this example.

2. In obtaining Equation (E4) we did not write integration constants, since at this stage we were seeking only expressions for use in the calculation of displacements. However, when we integrate for displacements, we will write integration constants that will be determined from the boundary conditions.

3. In case III we did not have to draw the free body diagram. This is an advantage in problems where the distributed load changes character over the length of the beam. Thus, even for statically determinate beams, it may be advantageous to start with the fourth-order differential equation, rather than the second-order differential equation.

EXAMPLE 3.13

Determine the equation of the elastic curve in Figure 3.38 in terms of E, I, L, P, and x.

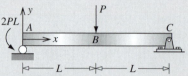

Figure 3.38 Beam and loading for Example 3.13.

PLAN

We can create two templates, one for an applied moment and one for the applied force. With the templates to guide us, we can write the moment expression in terms of the discontinuity functions. The second-order differential Equation (3.8-B) can be written and solved by using the zero deflection boundary conditions at A and C to obtain the elastic curve.

SOLUTION

We can create the two templates and obtain the moment expression for each template as shown in Figure 3.39.

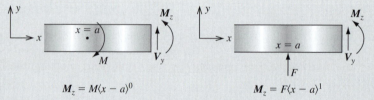

$$M_z = M\langle x - a \rangle^0 \qquad\qquad M_z = F\langle x - a \rangle^1$$

Figure 3.39 Templates and template equations for Example 3.13.

We can draw the free body diagram for the entire beam and find the reaction at A as $R_A = 3P/2$ upward. The templates in Figure 3.39 will guide us in writing the moment expressions. The reaction force is in the same direction as the force in the template; hence in the moment expression it will have the same sign shown in the template equation. The applied moment at point A is in the direction opposite to that shown in the template in Figure 3.39; hence the moment expression will have a negative sign with respect to that shown in the template equation. The force P has the sign opposite to that shown on the template, and hence the moment expression will have a negative sign. The moment expression can be written as in Equation (E1).

$$M_z = \frac{3P}{2}\langle x \rangle^1 - 2PL\langle x \rangle^0 - P\langle x - L \rangle^1 \qquad\qquad \text{(E1)}$$

Substituting Equation (E1) in Equation (3.8-B) and writing the zero deflection conditions at A and C, we obtain the boundary value problem described by Equations (E2) through (E4).

- Differential equation

$$EI_{zz}\frac{d^2v}{dx^2} = \frac{3P}{2}\langle x\rangle^1 - 2PL\langle x\rangle^0 - P\langle x-L\rangle^1 \qquad 0 \le x < 2L \qquad \text{(E2)}$$

- Boundary conditions

$$v(0) = 0 \qquad \text{(E3)}$$

$$v(2L) = 0 \qquad \text{(E4)}$$

Integrating Equation (E2) twice by means of Equation (3.30), we obtain Equations (E5) and (E6).

$$EI_{zz}\frac{dv}{dx} = \frac{3P}{4}\langle x\rangle^2 - 2PL\langle x\rangle^1 - \frac{P}{2}\langle x-L\rangle^2 + c_1 \qquad \text{(E5)}$$

$$EI_{zz}v = \frac{P}{4}\langle x\rangle^3 - PL\langle x\rangle^2 - \frac{P}{6}\langle x-L\rangle^3 + c_1 x + c_2 \qquad \text{(E6)}$$

Substituting Equation (E6) into Equation (E3) and noting that the discontinuity function is zero when the argument is zero or negative, we obtain Equation (E7).

$$\frac{P}{4}\langle 0\rangle^3 - PL\langle 0\rangle^2 - \frac{P}{6}\langle -L\rangle^3 + c_2 = 0 \qquad \text{or} \qquad c_2 = 0 \qquad \text{(E7)}$$

Substituting Equation (E6) into Equation (E4), we obtain Equation (E8).

$$\frac{P}{4}\langle 2L\rangle^3 - PL\langle 2L\rangle^2 - \frac{P}{6}\langle L\rangle^3 + c_1(2L) = 0 \qquad \text{or} \qquad c_1 = \frac{13}{12}PL^2 \qquad \text{(E8)}$$

Substituting Equations (E7) and (E8) into Equation (E6), we obtain the elastic curve shown in Equation (E9).

ANS. $$v = \frac{P}{12EI_{zz}}[3P\langle x\rangle^3 - 12PL\langle x\rangle^2 - 2P\langle x-L\rangle^3 + 13L^2 x] \qquad \text{(E9)}$$

COMMENTS

1. Without the discontinuity functions, we would have had to write two differential equations, one applicable in *AB* and another in *BC*. We would then have had to use the continuity of the displacements and slope at *B* to evaluate the two extra constants. The use of the discontinuity functions implicitly accounts for the continuity conditions by using a single moment expression for the entire beam.

2. We can easily obtain the deflection expressions for regions *AB* and *BC* from Equation (E9) as follows:

$$v = \frac{P}{12EI_{zz}}[3Px^3 - 12PLx^2 + 13L^2 x] \qquad 0 \le x \le L$$

$$v = \frac{P}{12EI_{zz}}[3Px^3 - 12PLx^2 - 2P(x-L)^3 + 13L^2 x] \qquad L \le x \le 2L$$

EXAMPLE 3.14

A beam with a bending rigidity of $EI = 42,000$ kN · m² is shown in Figure 3.40. Determine the deflection at point B.

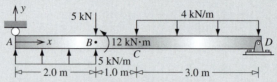

Figure 3.40 Beam and loading for Example 3.14.

PLAN

The coordinate system in this example is the same as in Example 3.13, and hence we can use the templates in Figure 3.39. Differentiating the template equations twice, we obtain the template equation for the distributed forces and write the distributed force expression in terms of the discontinuity functions. We can use Equation (3.12-B) and the boundary conditions at A and D to write the boundary value problem, which we then solve to obtain the elastic curve. Substituting $x = 2.0$ m in the elastic curve, we can obtain the deflection at B.

SOLUTION

The templates of Example 3.13 are shown in Figure 3.41. The moment expression is differentiated twice to obtain the template equations for the distributed force, as shown in Figure 3.41.

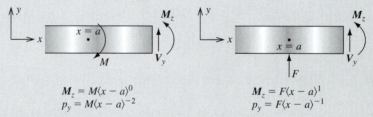

Figure 3.41 Templates and template equations for Example 3.14.

We note that the distributed force in section AB is positive, starts at zero, and ends at $x = 2$. The distributed force in section CD is negative, starts at $x = 3$, and is over the rest of the beam. We can use the template equations in Figure 3.41 to write the distributed force expression as shown in Equation (E1).

$$p_y = 5\langle x \rangle^0 - 5\langle x - 2 \rangle^0 - 4\langle x - 3 \rangle^0 - 5\langle x - 2 \rangle^{-1} - 12\langle x - 2 \rangle^{-2} \tag{E1}$$

Substituting Equation (E1) in Equation (3.12-B) and noting that displacement and moment are zero at the roller support, we obtain the boundary value problem described in Equations (E2) through (E6).

- Differential equation

$$\frac{d^2}{dx^2}\left(EI_{zz}\frac{d^2v}{dx^2}\right) = 5\langle x\rangle^0 - 5\langle x-2\rangle^0 - 4\langle x-3\rangle^0 - 5\langle x-2\rangle^{-1} - 12\langle x-2\rangle^{-2} \quad \text{(E2)}$$

- Boundary conditions

$$v(0) = 0 \quad \text{(E3)}$$

$$EI_{zz}\frac{d^2v}{dx^2}(0) = 0 \quad \text{(E4)}$$

$$v(6) = 0 \quad \text{(E5)}$$

$$EI_{zz}\frac{d^2v}{dx^2}(6) = 0 \quad \text{(E6)}$$

Integrating Equation (E2) twice, we obtain Equations (E7) and (E8).

$$\frac{d}{dx}\left(EI_{zz}\frac{d^2v}{dx^2}\right) = 5\langle x\rangle^1 - 5\langle x-2\rangle^1 - 4\langle x-3\rangle^1 - 5\langle x-2\rangle^0 - 12\langle x-2\rangle^{-1} + c_1 \quad \text{(E7)}$$

$$EI_{zz}\frac{d^2v}{dx^2} = \frac{5}{2}\langle x\rangle^2 - \frac{5}{2}\langle x-2\rangle^2 - 2\langle x-3\rangle^2 - 5\langle x-2\rangle^1 - 12\langle x-2\rangle^0 + c_1x + c_2 \quad \text{(E8)}$$

Substituting Equation (E8) into (E4), we obtain Equation (E9).

$$c_2 = 0 \quad \text{(E9)}$$

Substituting Equation (E8) into Equation (E6), we obtain Equation (E10).

$$\frac{5}{2}\langle 6\rangle^2 - \frac{5}{2}\langle 4\rangle^2 - 2\langle 3\rangle^2 - 5\langle 4\rangle^1 - 12\langle 4\rangle^0 + c_1(6) = 0 \quad \text{or} \quad c_1 = 0 \quad \text{(E10)}$$

Substituting Equations (E9) and (E10) into Equation (E8) and integrating twice, we obtain Equations (E11) and (E12).

$$EI_{zz}\frac{dv}{dx} = \frac{5}{6}\langle x\rangle^3 - \frac{5}{6}\langle x-2\rangle^3 - \frac{2}{3}\langle x-3\rangle^3 - \frac{5}{2}\langle x-2\rangle^2 - 12\langle x-2\rangle^1 + c_3 \quad \text{(E11)}$$

$$EI_{zz}v = \frac{5}{24}\langle x\rangle^4 - \frac{5}{24}\langle x-2\rangle^4 - \frac{2}{12}\langle x-3\rangle^4 - \frac{5}{6}\langle x-2\rangle^3 - 6\langle x-2\rangle^2 + c_3x + c_4 \quad \text{(E12)}$$

Substituting Equation (E12) into Equation (E3), we obtain Equation (E13).

$$c_4 = 0 \quad \text{(E13)}$$

Substituting Equation (E11) into Equation (E5), we obtain Equation (E14).

$$\frac{5}{24}\langle 6\rangle^4 - \frac{5}{24}\langle 4\rangle^4 - \frac{2}{12}\langle 3\rangle^4 - \frac{5}{6}\langle 4\rangle^3 - 6\langle 4\rangle^2 + c_3(6) = 0 \quad \text{or}$$

$$c_3 = -\frac{323}{36} = -8.97 \quad \text{(E14)}$$

Substituting Equations (E13) and (E14) into Equation (E12) and simplifying, we obtain the elastic curve shown in Equation (E15).

$$v = \frac{1}{72EI_{zz}}[15\langle x\rangle^4 - 15\langle x-2\rangle^4 - 12\langle x-3\rangle^4$$

$$- 60\langle x-2\rangle^3 - 432\langle x-2\rangle^2 - 646x] \quad \text{(E15)}$$

Substituting $x = 2$ into Equation (E15), we obtain the deflection at point B as follows:

$$v(2) = \frac{1}{72(42)(10^3)}[15\langle 2\rangle^4 - 15\langle 0\rangle^4 - 12\langle -1\rangle^4 - 60\langle 0\rangle^3 - 432\langle 0\rangle^2 - 646(6)] \quad \text{(E16)}$$

ANS. $v(2) = -1.2$ mm

COMMENTS

1. This is a statically determinate problem, and the distributed loads are uniform. Thus we could have used second-order differential equations, writing the moment expression in terms of the discontinuity functions. But this would have required creating a template with template equation for the distributed load, as was done in case III of Example 3.12.

2. This would be very tedious algebraically without the discontinuity functions.

EXAMPLE 3.15

In terms of E, I, w, L, and x, determine (a) the elastic curve and (b) the reaction force at A in Figure 3.42.

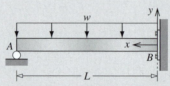

Figure 3.42 Beam and loading for Example 3.15.

METHOD 1 Fourth-order differential equation approach

PLAN

(a) We can use Equation (3.12-B) to write the fourth-order differential equation. The two boundary conditions at A are zero deflection and zero moment, and the two boundary conditions at B are zero deflection and zero slope. We can solve the boundary value problem and obtain the elastic curve. (b) After making an imaginary cut just to the right of A, we can draw a free body diagram and relate the reaction force to the shear force. The shear force at point A can be found by substituting $x = L$ into the solution obtained in part (a).

SOLUTION

(a) Noting that the distributed force is in the negative y direction, we can substitute $p_y = -w$ in Equation (3.12-B) and write the differential equation. The deflection at A is zero, but the slope depends upon the loading; hence the moment must be specified. Since there is no applied moment, we conclude that the internal moment at A is zero. The deflection and slope at point B are zero. We can write the boundary value problem statement as described in Equations (E1) through (E5).

- Differential equation

$$\frac{d^2}{dx^2}\left(EI_{zz}\frac{d^2v}{dx^2}\right) = -w \tag{E1}$$

- Boundary conditions

$$v(0) = 0 \tag{E2}$$

$$\frac{dv}{dx}(0) = 0 \tag{E3}$$

$$v(L) = 0 \tag{E4}$$

$$EI_{zz}\frac{d^2v}{dx^2}(L) = 0 \tag{E5}$$

Integrating Equation (E1) twice, we obtain Equations (E6) and (E7).

$$\frac{d}{dx}\left(EI_{zz}\frac{d^2v}{dx^2}\right) = -wx + c_1 \tag{E6}$$

$$EI_{zz}\frac{d^2v}{dx^2} = -\frac{wx^2}{2} + c_1x + c_2 \tag{E7}$$

Substituting Equation (E7) into Equation (E5), we obtain Equation (E8).

$$c_1L + c_2 = \frac{wL^2}{2} \tag{E8}$$

Integrating Equation (E7), we obtain Equation (E9).

$$EI_{zz}\frac{dv}{dx} = -\frac{wx^3}{6} + c_1\frac{x^2}{2} + c_2x + c_3 \tag{E9}$$

Substituting Equation (E9) into Equation (E3), we obtain Equation (E10).

$$c_3 = 0 \tag{E10}$$

Integrating Equation (E9) after substituting Equation (E10), we obtain Equation (E11).

$$EI_{zz}v = -\frac{wx^4}{24} + c_1\frac{x^3}{6} + c_2\frac{x^2}{2} + c_4 \tag{E11}$$

Substituting Equation (E11) into Equation (E2), we obtain Equation (E12).

$$c_4 = 0 \tag{E12}$$

Substituting Equations (E11) and (E12) into Equation (E4), we obtain Equation (E13).

$$c_1 \frac{L^2}{2} + c_2 L = \frac{wL^3}{6} \tag{E13}$$

Solving Equations (E8) and (E13) simultaneously, we obtain Equations (E14) and (E15).

$$c_1 = \frac{5wL}{8} \tag{E14}$$

$$c_2 = -\frac{wL^2}{8} \tag{E15}$$

Substituting Equations (E12), (E14), and (E15) into Equation (E11) and simplifying, we obtain the elastic curve as follows:

ANS. $\quad v(x) = -\dfrac{w}{48EI_{zz}}(2x^4 - 5Lx^3 + 3L^2x^2) \tag{E16}$

Dimension Check: We note that all terms in the parentheses on the right-hand side of Equation (E16) have the dimension of length to the fourth power [i.e., $O(L^4)$]. Thus Equation (E16) is dimensionally homogeneous. We can also check whether the left-hand side and any one term on the right have the same dimension as follows:

$$w \to O(F/L) \quad x \to O(L) \quad E \to O(F/L^2) \quad I_{zz} \to O(L^4)$$

$$v \to O(L) \quad \frac{wx^4}{EI_{zz}} \to O\left(\frac{(F/L)L^4}{(F/L^2)O(L^4)}\right) \to O(L) \quad \text{Checks.}$$

(b) We make an imaginary cut just to the right of point A (an infinitesimal distance) and draw the free body diagram of the left part, using the sign convention as shown in Figure 3.43. By force equilibrium in the y direction, we can relate the shear force at A to the reaction force at A.

$$R_A = V_A = V_y(L) \tag{E17}$$

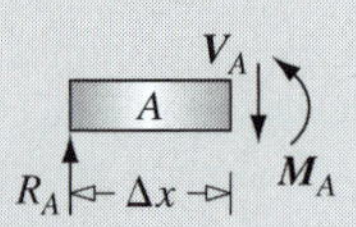

Figure 3.43 The infinitesimal equilibrium element at A in Example 3.15.

Substituting Equations (E6) and (E14) into Equation (3.37b-B), we obtain Equation (E18).

$$V_y(x) = -\frac{d}{dx}\left(EI_{zz}\frac{d^2v}{dx^2}\right) = wx - \frac{5wL}{8} \tag{E18}$$

Substituting Equation (E18) into Equation (E17), we obtain the reaction at A as

ANS. $R_A = \dfrac{3wL}{8}$

METHOD 2 Second-order differential equation approach

PLAN

We can make an imaginary cut at some arbitrary location x and use the left-hand part to draw the free body diagram. The moment expression will contain the reaction force at A as an unknown. The second-order differential equation (3.8-B) would generate two integration constants, leading to a total of three unknowns. We need three conditions: the displacement at A must be zero, and both the displacement and the slope at B must be zero. Solving the boundary value problem, we can obtain the elastic curve and the unknown reaction force at A.

SOLUTION

We make an imaginary cut at a distance x from the right wall and take the left part of length $(L - x)$ to draw a free body diagram by means of the sign convention for internal quantities as shown in Figure 3.44.

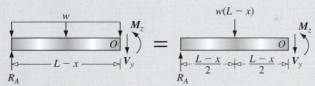

Figure 3.44 Free body diagram for Example 3.15.

Balancing the moment at point O, we obtain the moment expression as shown in Equation (E19).

$$M_z - R_A(L - x) + w\frac{(L - x)^2}{2} = 0 \quad \text{or} \quad M_z = R_A(L - x) - \frac{w}{2}(L^2 + x^2 - 2Lx) \quad \text{(E19)}$$

Substituting Equation (E19) into Equation (3.8-B) and writing the boundary conditions, we obtain the boundary value problem described by Equations (E20) through (E23).

- Differential equation

$$EI_{zz}\frac{d^2v}{dx^2} = R_A(L - x) - \frac{w}{2}(L^2 + x^2 - 2Lx) \quad \text{(E20)}$$

- Boundary conditions

$$v(0) = 0 \quad \text{(E21)}$$

$$\frac{dv}{dx}(0) = 0 \qquad \text{(E22)}$$

$$v(L) = 0 \qquad \text{(E23)}$$

Integrating Equation (E20), we obtain Equation (E24).

$$EI_{zz}\frac{dv}{dx} = R_A\left(Lx - \frac{x^2}{2}\right) - \frac{w}{2}\left(L^2x + \frac{x^3}{3} - Lx^2\right) + c_1 \qquad \text{(E24)}$$

Substituting Equation (E24) into Equation (E22) we obtain Equation (E25).

$$c_1 = 0 \qquad \text{(E25)}$$

Integrating Equation (E24) after substituting Equation (E25), we obtain Equation (E26).

$$EI_{zz}v = R_A\left(\frac{Lx^2}{2} - \frac{x^3}{6}\right) - \frac{w}{2}\left(\frac{L^2x^2}{2} + \frac{x^4}{12} - \frac{Lx^3}{3}\right) + c_2 \qquad \text{(E26)}$$

Substituting Equation (E26) into Equation (E21), we obtain Equation (E27).

$$c_2 = 0 \qquad \text{(E27)}$$

Substituting Equations (E26) and (E27) into Equation (E23), we obtain Equation (E28).

$$R_A\left(\frac{L^3}{2} - \frac{L^3}{6}\right) - \frac{w}{2}\left(\frac{L^4}{2} + \frac{L^4}{12} - \frac{L^4}{3}\right) = 0 \quad \text{or} \quad \textbf{ANS.} \quad R_A = \frac{3wL}{8} \qquad \text{(E28)}$$

Substituting Equations (E27) and (E28) into Equation (E26) and simplifying, we obtain the elastic curve.

$$\textbf{ANS.} \quad v(x) = -\frac{w}{48EI_{zz}}(2x^4 - 5Lx^3 + 3L^2x^2)$$

COMMENTS

1. Method 2 has less algebra than method 1 and should be used whenever possible.
2. The discontinuity functions are not needed because there is no change of loading inside the beam.
3. Suppose that in drawing the free body diagram for calculating the internal moment, we had taken the right part. Then we would have had the wall reaction force and the wall reaction moment in the moment expression; in other words, there would have been two unknown rather than one in the moment expression. In such cases we must use the static equilibrium equation for the entire beam to eliminate one of the unknowns. In other words, the moment expression can contain only as many unknowns as are represented by the degree of static redundancy.
4. The moment boundary condition given by Equation (E5) in method 1 is implicitly satisfied in Equation (E20). We can confirm this by substituting $x = L$ in Equation (E20).

EXAMPLE 3.16

A light pole is subjected to wind pressure that varies quadratically as shown in Figure 3.45. In terms of E, I, w, L, and x, determine (a) the deflection at the free end and (b) the ground reactions.

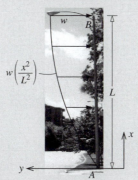

Figure 3.45 Beam and loading for Example 3.16.

PLAN

(a) Though the problem is statically determinate, finding the moment as a function of x by static equilibrium is difficult. We can use the fourth-order differential equation (3.12-B). The four boundary conditions are thus as follows: both the deflection and the slope at A are zero, and the moment and the shear force at B are zero. We can solve the boundary value problem and determine the elastic curve. By substituting $x = L$ in the elastic curve equation, we can obtain the deflection at the free end. (b) By making an imaginary cut just to the right of point A, we can relate the internal shear force and the internal moment at point A to the reactions at A. By substituting $x = 0$ in the expressions for moment and shear force, we can obtain the shear force and moment values at point A.

SOLUTION

If we note that the distributed force is in the negative y direction, we can use Equation (3.12-B) to write the differential equation. The four boundary conditions discussed in the plan can also be written to complete the boundary value problem statement as described in Equations (E1) through (E5).

- Differential equation

$$\frac{d^2}{dx^2}\left(EI_{zz}\frac{d^2v}{dx^2}\right) = -w\left(\frac{x^2}{L^2}\right) \tag{E1}$$

- Boundary conditions

$$v(0) = 0 \tag{E2}$$

$$\frac{dv}{dx}(0) = 0 \tag{E3}$$

$$EI_{zz}\frac{d^2v}{dx^2}\bigg|_{x=L} = 0 \tag{E4}$$

$$\frac{d}{dx}\left(EI_{zz}\frac{d^2v}{dx^2}\right)\Bigg|_{x=L} = 0 \tag{E5}$$

Integrating Equation (E1), we obtain Equation (E6).

$$\frac{d}{dx}\left(EI_{zz}\frac{d^2v}{dx^2}\right) = -\frac{wx^3}{3L^2} + c_1 \tag{E6}$$

Substituting Equation (E6) into Equation (E5), we obtain Equation (E7).

$$c_1 = \frac{wL}{3} \tag{E7}$$

Substituting Equation (E7) into Equation (E6) and integrating, we obtain Equation (E8).

$$EI_{zz}\frac{d^2v}{dx^2} = -\frac{wx^4}{12L^2} + \frac{wL}{3}x + c_2 \tag{E8}$$

Substituting Equation (E8) into Equation (E4), we obtain Equation (E9).

$$c_2 = -\frac{(wL^2)}{4} \tag{E9}$$

Substituting Equation (E9) into Equation (E8) and integrating, we obtain Equation (E10).

$$EI_{zz}\frac{dv}{dx} = -\frac{wx^5}{60L^2} + \frac{wLx^2}{6} - \frac{wL^2x}{4} + c_3 \tag{E10}$$

Substituting Equation (E10) into Equation (E3), we obtain Equation (E11).

$$c_3 = 0 \tag{E11}$$

Substituting Equation (E11) into Equation (E10) and integrating, we obtain Equation (E12).

$$EI_{zz}\frac{dv}{dx} = -\frac{wx^6}{360L^2} + \frac{wLx^3}{18} - \frac{wL^2x^2}{8} + c_4 \tag{E12}$$

Substituting Equation (E12) into Equation (E2), we obtain Equation (E13).

$$c_4 = 0 \tag{E13}$$

Substituting Equation (E13) into Equation (E12) and simplifying, we obtain Equation (E14).

ANS. $v(x) = -\left[\dfrac{w}{360EI_{zz}L^2}\right](x^6 - 20L^3x^3 + 45L^4x^2)$ \hfill (E14)

Dimension Check: We note that all terms in the second term on the right-hand side of Equation (E14) have the dimension of length to the sixth power. Thus the expression in Equation (E14) is dimensionally homogeneous. We can also check, whether the left-hand side and any one term on the right have the same dimension as follows:

$$w \rightarrow O(F/L) \qquad x \rightarrow O(L) \qquad E \rightarrow O(F/L^2) \qquad I_{zz} \rightarrow O(L^4)$$

$$v \rightarrow O(L) \qquad \frac{wx^6}{EI_{zz}L^2} \rightarrow O\left(\frac{(F/L)L^6}{(F/L^2)(L^4)(L^2)}\right) \rightarrow O(L) \qquad \text{Checks.}$$

(a) Substituting $x = L$ into Equation (E14), we obtain the deflection at the free end as

ANS. $$v(L) = -\left(\frac{13wL^4}{180EI_{zz}}\right)$$

(b) We make an imaginary cut just above $(\varepsilon \to 0)$ point A and take the bottom part to draw the free body diagram, as shown in Figure 3.46. By the equilibrium of forces and moments, we obtain Equations (E15) and (E16).

$$M_z(0) = -M_A \tag{E15}$$

$$V_y(0) = -R_A \tag{E16}$$

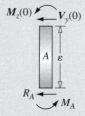

Figure 3.46 The infinitesimal equilibrium element at A for Example 3.16.

Substituting Equations (E8) and (E9) into Equation (3.36a-B) and Equations (E6) and (E7) into Equation (3.37b-B), we can obtain the bending moment and shear force expressions as shown in Equations (E17) and (E18).

$$M_z(x) = -\frac{wx^4}{12L^2} + \frac{wL}{3}x - \frac{wL^2}{4} \tag{E17}$$

$$V_y(x) = \frac{wx^3}{3L^2} - \frac{wL}{3} \tag{E18}$$

Substituting Equations (E17) and (E18) into Equations (E15) and (E16), we obtain the reaction force and reaction moment.

ANS. $R_A = \frac{wL}{3}$ $\qquad M_A = \frac{wL^2}{4}$

COMMENTS

1. We can check the directions of R_A and M_A by inspection. They are correct because they are the directions necessary for equilibrium of the external distributed force.
2. In drawing the free body diagram in Figure 3.46, the reaction force R_A and the reaction moment M_A can be drawn in any direction, but the internal quantities V_y and M_z must be drawn in accordance with the sign convention. Irrespective of the directions in which R_A and M_A are drawn, the final answer will be as just given. The sign in the equilibrium equations (E15) and (E16) will account for the assumed directions of the reactions.

PROBLEM SET 3.2

3.34 The frictional force per unit length on a cast iron pipe being pulled from the ground varies as a quadratic function as shown in Figure P3.34. Determine the force F needed to pull the pipe out of ground and the elongation of the pipe before the pipe slips in terms of the modulus of elasticity E, cross-sectional area A, length L, and maximum value of frictional force f_{max}.

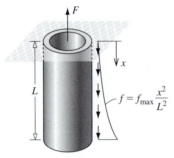

Figure P3.34

ANS. $F = f_{max} L/3;$

$u_L - u_0 = (f_{max} L^2)/(4EA)$

3.35 Figure P3.35 shows a rectangular steel ($E = 30,000$ ksi, $\nu = 0.25$) bar of 0.5 in thickness, there is a gap of 0.01 in between the section at D and a rigid wall before the forces are applied. Assuming that the applied forces are sufficient to close the gap, determine (a) the movement of section C with respect to the left wall and (b) the change in the depth d of segment CD. Use the discontinuity function to solve the problem.

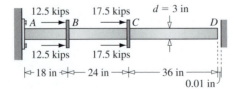

Figure P3.35

ANS. (a) 0.025 in   (b) -0.00031 in

3.36 The external torque on the drill bit shown in Figure P3.36 varies as a quadratic function to a maximum intensity of $q(x^2/L^2)$ in-lb/in, as shown. If the drill bit diameter is d, its length L, and modulus of rigidity G, determine (a) the maximum shear stress on the drill bit and (b) the relative rotation of the end of the drill bit with respect to the chuck.

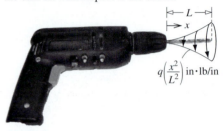

Figure P3.36

3.37 A solid circular steel ($G = 12,000$ ksi, $E = 30,000$ ksi) shaft of 4-inch diameter is loaded as shown in Figure P3.37. Determine the maximum shear stress in the shaft. Use the discontinuity functions to solve the problem.

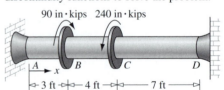

Figure P3.37

ANS. 11.1 ksi

3.38 An aluminum alloy ($G = 28$ GPa) hollow shaft has a critical stress intensity factor of 22 ksi $\sqrt{in}$. The shaft has a thickness of $1/4$ in and an outer diameter of 2 in and is loaded as shown in Figure P3.38. What is the critical crack length at which the shaft should be taken out of service?

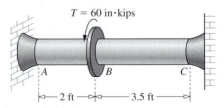

Figure P3.38

ANS. $2a_{crit} = 0.244$ in

3.39 A circular solid shaft is acted upon by torques as shown in Figure P3.39. Determine the rotation of the rigid wheel A with respect to the fixed end C in terms of q, L, G, and J.

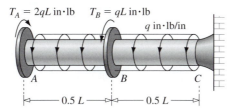

Figure P3.39

ANS. $\quad \phi_A = \left(\dfrac{qL^2}{GJ}\right) cw$

3.40 A uniform distributed torque of q in · lb/in is applied to the entire shaft as shown in Figure P3.40. In addition to the distributed torque, a concentrated torque of $T = 3\ qL$ in · lb is applied at section B. Let the shear modulus be G and radius of the shaft be r. Determine, in terms of q, L, G, and r, (a) the rotation of section at B and (b) the magnitude of maximum torsional shear stress in the shaft.

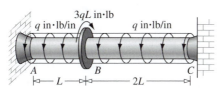

Figure P3.40

3.41 The displacement in the y direction in section AB of the beam shown in Figure P3.41 is $v(x) = (20x^3 - 40x^2)(10^{-6})$ in. If the bending rigidity EI is $135\ (10^6)$ lb · in², determine the reaction force and reaction moment at point A at the wall.

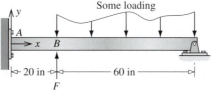

Figure P3.41

ANS. $\quad R_A = 16.2$ kips $\qquad M_A = 10.8$ in · kips

3.42 The displacement in the y direction in section AB of the beam shown in Figure P3.42 is given, in inches, by $v_1(x) = -3(x^4 - 20x^3)(10^{-6})$ and in BC by $v_2(x) = -8(x^2 - 100x + 1600)(10^{-3})$. If the bending rigidity EI is $135\ (10^6)$ lb · in², determine the reaction force at B.

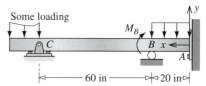

Figure P3.42

ANS. $\quad R_B = 145.8$ kips

3.43 The displacement in the y direction in section AB of the beam shown in Figure P3.43 is given, in inches, by $v_1 = 5(x^3 - 20x^2)(10^{-6})$ and in section BC by $v_2 = 5(x^3 - 800x + 8000)(10^{-6})$. If the bending rigidity EI is $135\ (10^6)$ lb · in², determine the moment M_B and the reaction force at B.

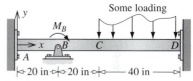

Figure P3.43

3.44 In terms of w, L, E, and I, determine the deflection and slope at $x = L$ of the beam shown in Figure P3.44.

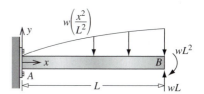

Figure P3.44

ANS. $\quad v = -\left(\dfrac{43wL^4}{180EI}\right)$

$\qquad v' = -\left(\dfrac{3wL^3}{5EI}\right)$

3.45 Determine the reaction and moment at the left wall on the beam shown in Figure P3.45, and also the slope at $x = L$.

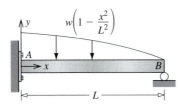

Figure P3.45

ANS. $R = \dfrac{61wL}{120}$ $M = \dfrac{11wL^2}{120}$

$v' = \dfrac{wL^3}{80EI}$

3.46 Determine the deflection and moment reaction at $x = L$ in terms of E, I, w, and L for the beam shown in Figure P3.46.

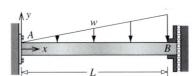

Figure P3.46

3.47 Determine the elastic curve and the reaction(s) at A for the beam shown in Figure P3.47.

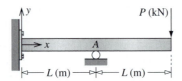

Figure P3.47

ANS. $R_A = \dfrac{5P}{2}$

3.48 Determine the elastic curve and the reaction(s) at A for the beam shown in Figure P3.48.

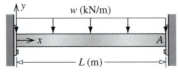

Figure P3.48

3.49 A linear spring that has a spring constant K is attached to a beam at one end as shown in Figure P3.49. In terms of w, E, I, L, and K. Use the discontinuity functions to write the boundary value problem, but do not integrate or solve.

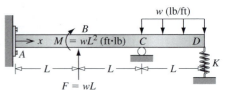

Figure P3.49

3.50 For the beam and loading shown in Figure P3.50, in terms of w, L, E, and I, determine (a) the equation of the elastic curve and (b) the deflection at $x = L$.

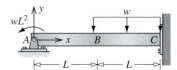

Figure P3.50

ANS. $v(L) = -\left(\dfrac{wL^4}{6EI}\right)$

3.51 For the beam and loading shown in Figure P3.51, in terms of w, L, E, and I, determine (a) the equation of the elastic curve and (b) the deflection at $x = L$.

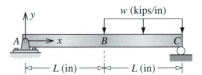

Figure P3.51

ANS. $v(L) = \dfrac{-5wL^4}{48EI}$

3.52 For the beam and loading shown in Figure P3.52, in terms of w, L, E, and I, determine (a) the equation of the elastic curve and (b) the deflection at $x = L$.

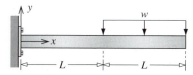

Figure P3.52

3.53 Determine the deflection of the beam at point C in terms of E, I, w, and L for the beam shown in Figure P3.53.

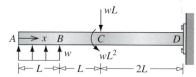

Figure P3.53

3.54 Determine the deflection of the beam at point C in terms of E, I, w, and L for the beam shown in Figure P3.54.

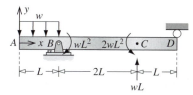

Figure P3.54

ANS. $v(3L) = \dfrac{-2wL^4}{9EI}$

3.5 SYMMETRIC BENDING OF CURVED BEAMS

The flexure formula of Equation (3.9-B) is applicable to the symmetric bending of straight beams. It can also be applied to beams with slight curvatures, that is, beams with $r_c/h > 5$, where r_c is the radius of curvature of the line joining the centroids of the cross sections as shown in Figure 3.47, and h is the depth of the beam cross section. In this section we develop stress formulas for sharply curved ($r_c/h < 5$) beams that have a plane of symmetry, with the loading in the plane of symmetry. The deflection of a point on a curved beam is described in Section 7.7.

We will use cylindrical coordinates (r, θ, z) as shown in Figure 3.47. We will assume that bending is about the z axis (i.e., the cross section rotates about the z axis). As in the symmetric bending of straight beams, our objective is to find the location of the neutral axis (the value of R) and the bending normal stress ($\sigma_{\theta\theta}$) at any point with the radial coordinate ρ on the cross section. In the initial development, we shall assume that the only load is that of a moment about the z axis. Later we will relax this condition and include transverse forces.

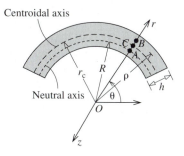

Figure 3.47 Curved beam geometry.

The theoretical development of the stress formula follows lines very similar to those we used in discussing the symmetric bending of straight beams. Assumptions of similar kinds are made in the development of the theory, with the primary difference coming from a new kinematic expression for the normal strains.

3.5.1   Kinematics

We make the following assumptions, repeated here from Table 3.3, to obtain the simplest theory for symmetric bending of curved beams.

Assumption 1 Deformations are not a function of time.

Assumption 2a-B Squashing deformation is significantly smaller than deformation due to bending.

Assumption 2b-B Plane sections before deformation remain plane after deformation.

Assumption 2c-B A plane perpendicular to the beam axis remains *nearly* perpendicular after deformation.

Assumption 3 The strains are small.

Assumption 1 implies that we have a static problem and that the applied loads are not functions of time. Assumption 2a-B implies that the changes in the radial dimensions are negligible. Assumption 2b-B implies that the deformation in the θ direction is a linear function of ρ. As we shall see, however, unlike the case of straight beams, this will not result in a linear function of ρ for the normal strain $\varepsilon_{\theta\theta}$. Assumption 2c-B implies that the shear strain $(\gamma_{r\theta})$ is nearly zero. As in the case of straight beams, this assumption is strictly valid if there are no shear forces and the applied load is only a bending moment about the z axis. Assumption 3 implies that we can use small-strain approximations, resulting in a linear theory.

Consider a small segment of a curved beam element that subtends an angle $\Delta\theta$ at the center of curvature O as shown in Figure 3.48(a). Line AA represents the neutral axis, and line BB is any line for which we are trying to find the strain. We show the deformed shape relative to the left radial line AB, which is assumed to remain perpendicular to lines AA and BB in accordance with Assumptions 2b and 2c. Line AA, being the neutral axis, does not see any strain, while deformation causes point B to move to point B_1, and the center of curvature then moves to point O_1—both movements are significantly exaggerated in Figure 3.48.

The arc length obtained by using O and O_1 as the centers results in $R\Delta\theta = R_1\Delta\theta_1$. The original length is $BB = \rho\Delta\theta$, and the final length is $BB_1 = (R_1 + \rho - R)\Delta\theta_1$. The normal tangential strain of line BB can be written as follows:

$$\varepsilon_{\theta\theta} = \frac{BB_1 - BB}{BB} = \frac{(R_1 + \rho - R)\Delta\theta_1 - \rho\Delta\theta}{\rho\Delta\theta}$$

$$= \frac{R\Delta\theta + (\rho - R)\Delta\theta_1 - \rho\Delta\theta}{\rho\Delta\theta} = \left(\frac{R - \rho}{\rho}\right)\left(\frac{\Delta\theta - \Delta\theta_1}{\Delta\theta}\right)$$

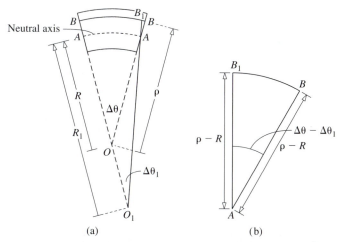

Figure 3.48 (a) Exaggerated original and deformed curved beam element. (b) Enlarged portion of (a).

We define $\psi = (\Delta\theta - \Delta\theta_1/\Delta\theta)$ as a relative rotation of a cross section that does not change in the radial direction. The tangential strain can be written as in Equation (3.39).

$$\varepsilon_{\theta\theta} = -\left(1 - \frac{R}{\rho}\right)\psi \qquad (3.39)$$

An alternative derivation of the tangential strain can be obtained by considering Figure 3.48(b). The radial length AB_1 is nearly equal to AB as a result of Assumption 2a-B. From the geometry in Figure 3.48(a), we note that angle O_1OA is $180 - \Delta\theta$; hence angle OAO_1 is $\Delta\theta - \Delta\theta_1$, which implies that angle BAB_1 is $\Delta\theta - \Delta\theta_1$. The arc length BB_1 for small strains can be approximated as $BB_1 = (\rho - R)(\Delta\theta - \Delta\theta_1)$. Noting that BB_1 is a contraction, we can write the tangential strain as

$$\varepsilon_{\theta\theta} = -\frac{BB_1}{BB} = -\left[\frac{(\rho - R)(\Delta\theta - \Delta\theta_1)}{\rho\Delta\theta}\right] = -\left(1 - \frac{R}{\rho}\right)\psi \qquad (3.40)$$

Equation (3.40) is same as Equation (3.39) but was obtained from a different perspective. In Equation (3.39) the only variable to change across the cross section is ρ. Equation (3.39) shows that the tangential strain for curved beams does not vary linearly across the cross section; rather, it varies hyperbolically, reaching zero when $\rho = R$.

3.5.2 Material Model

With the motivation of developing the simplest theory, we use the simplest material model, namely Hooke's law. To use Hooke's law, we make the following assumptions, repeated from Table 3.3.

> **Assumption 4** The material is isotropic.
>
> **Assumption 5** There are no inelastic strains.
>
> **Assumption 6** The material is elastic.
>
> **Assumption 7** Stress and strain are linearly related.

The dominant normal stress[10] is $\sigma_{\theta\theta}$, which can be written as Equation (3.41).

$$\sigma_{\theta\theta} = E\varepsilon_{\theta\theta} = -E\left(1 - \frac{R}{\rho}\right)\psi \tag{3.41}$$

3.5.3 Static Equivalency

The normal stress $\sigma_{\theta\theta}$ can be replaced by an equivalent internal moment $\boldsymbol{M}_z$ and a zero axial force as shown in Figure 3.49.

By considering the forces in the axial direction and the moment about the z axis through O in Figure 3.49, we obtain Equations (3.42a) and (3.42b),

$$N = \int_A \sigma_{\theta\theta} \, dA = 0 \tag{3.42a}$$

$$\boldsymbol{M}_z = -\int_A \rho\sigma_{\theta\theta} \, dA \tag{3.42b}$$

where A is the cross-sectional area, and $\boldsymbol{M}_z$ is an internal moment. The associated sign convention for drawing on a free body diagram is given in Definition 14.

> **Definition 14** The direction of positive internal moment $\boldsymbol{M}_z$ on a free body diagram must be such that it puts a point on the outer surface into compression.

Definition 14 implies that the positive direction of $\boldsymbol{M}_z$ on a free body diagram tends to straighten out the curved beam.

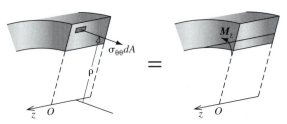

Figure 3.49 Static equivalency in curved beams.

---

[10] We assume that σ_{zz} is zero because there are no loads in the z direction and that σ_{rr} is zero because the outside and inside surfaces are free surfaces. Elasticity, however, shows that σ_{rr} does vary across the cross section, although its maximum value is significantly smaller than the maximum value of $\sigma_{\theta\theta}$.

3.5.4 Location of the Neutral Axis

Substituting Equation (3.41) into Equation (3.42a) and noting that ψ and R do not change across the cross section, we obtain Equation (3.43).

$$\int_A E\left(1 - \frac{R}{\rho}\right)\psi \, dA = \psi\left[\int_A E\left(1 - \frac{R}{\rho}\right) dA\right] = \psi\left[\int_A E \, dA - R\int_A \frac{E \, dA}{\rho}\right] = 0 \qquad \text{or}$$

$$R = \frac{\int_A E \, dA}{\int_A \frac{E \, dA}{\rho}} \tag{3.43}$$

Assumption 8 The material is homogeneous across the cross section.

Assumption 8 implies that the modulus of elasticity E does not change across the cross section and can be taken outside the integral in Equation (3.43) to yield Equation (3.44).

$$\boxed{R = \frac{A}{\int_A \frac{dA}{\rho}}} \tag{3.44}$$

The value of R locates the neutral axis. The integral in the denominator is a geometric quantity that depends only upon the shape of the cross section. Table 3.6 shows the values of the integral in Equation (3.44) for several common basic shapes that can be used for finding integrals of more complex cross sections that are composites of made from these shapes.

3.5.5 Stress Formula

Substituting Equation (3.41) into Equation (3.42b) and noting that ψ and R do not change across the cross section, we obtain Equation (3.45).

$$M_z = \psi\int_A E\rho\left(1 - \frac{R}{\rho}\right) dA = \psi\left[\int_A E\rho \, dA - R\int_A E \, dA\right] \tag{3.45}$$

TABLE 3.6 Values of Integrals in Curved Beams

Shape					
A	$a(r_2 - r_1)$	$\frac{1}{2}a(r_2 - r_1)$	$\frac{1}{2}(a_1 + a_2)(r_2 - r_1)$	πa^2	πab
$\int_A \frac{dA}{\rho}$	$a\ln\frac{r_2}{r_1}$	$\frac{ar_2}{r_2 - r_1}\ln\frac{r_2}{r_1} - a$	$a_2 - a_1 + \frac{r_2 a_1 - r_1 a_2}{r_2 - r_1}\ln\frac{r_2}{r_1}$	$2\pi\left(r_c - \sqrt{r_c^2 - a^2}\right)$	$\frac{2\pi b}{a}\left(r_c - \sqrt{r_c^2 - a^2}\right)$

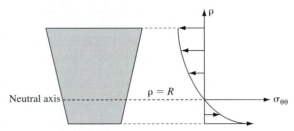

Figure 3.50 Hyperbolic variation of bending normal stress across the cross section of a curved beam.

If Assumption 8 is applicable, then E can be taken outside the integral. Noting the definition of centroid, we obtain Equation (3.46),

$$\int_A \rho \, dA = r_c A \tag{3.46}$$

where r_c is the radius of the line passing through the centroid as shown in Figure 3.47. Substituting Equation (3.46) into Equation (3.45) and noting that E is a constant across the cross section for a homogeneous material, we obtain Equation (3.47).

$$M_z = E\psi[r_c A - RA] \qquad \text{or} \qquad E\psi = \frac{M_z}{A(r_c - R)} \tag{3.47}$$

Substituting Equation (3.47) into Equation (3.41), we obtain the stress formula in Equation (3.48),

$$\sigma_{\theta\theta} = -\left[\frac{M_z}{A(r_c - R)}\left(1 - \frac{R}{\rho}\right)\right] \tag{3.48}$$

where A is the cross-sectional area, r_c is the radius of the line joining the centroids, and R is the radius of the neutral axis as determined from Equation (3.44); are all geometric properties of the cross section. The internal moment M_z does not change across the cross section. The only variable that changes across the cross section is the radial coordinate ρ. Thus, the stress distribution across the cross section of a curved beam is *hyperbolic*, as shown symbolically for a cross section in Figure 3.50.

There are *two possible ways* in which the internal bending moment M_z may be found:

1. On a free body diagram M_z is always drawn in accordance with the sign convention in Definition 14. An equilibrium equation then is used to get a positive or negative value for M_z. Now positive values of stress $\sigma_{\theta\theta}$ from Equation (3.48) are tensile, and negative values of $\sigma_{\theta\theta}$ are compressive.

2. The internal bending moment M_z is drawn at the imaginary cut in a direction selected to equilibrate the external loads. Since inspection is being used in determining the direction of M_z, Equation (3.48) should be used only for determining the magnitude. The tensile and compressive properties of $\sigma_{\theta\theta}$ must be determined by inspection.

3.5.6 Combined Loading

Forces applied to hooks and rings not only will subject curved beam members to bending moment but will also subject a cross section of the beam to axial and shear forces.

Assumption 2b-B (that the plane perpendicular to the beam axis remains nearly perpendicular after deformation) implies that the shear strain, hence the shear stress, is small. Thus, we can use our theory, provided the maximum shear stress ($\tau_{r\theta}$) in bending is an order of magnitude smaller than the maximum bending normal stress ($\sigma_{\theta\theta}$) as obtained from Equation (3.48). An estimate of the maximum shear stress ($\tau_{r\theta}$) can be obtained by dividing the maximum shear force by the cross-sectional area.

A uniform axial stress would be replaced by an internal axial force acting at the centroid of the cross section. Thus, on our free body diagram we shall assume that the *internal axial force is acting at the centroid of the cross section* (see Problem 3.55). Our theory for curved beams results in a linear theory; hence by the principle of superposition, we obtain Equation (3.49).

$$\boxed{\sigma_{\theta\theta} = \frac{N}{A} - \left[\frac{M_z}{A(r_c - R)} \left(1 - \frac{R}{\rho}\right) \right]} \tag{3.49}$$

EXAMPLE 3.17

A beam with a 2 in × 2 in square cross section is subjected to moment $M_{ext} = 30$ in·kips. Determine the maximum bending normal stress for the following three cases.

Case 1: The beam is straight.
Case 2: The beam is curved, and the radius of the inside surface is 5 in.
Case 3: The beam is curved, and the radius of the inside surface is 20 in.

PLAN

For case 1, we can find the area moment of inertia for a rectangular cross section. Noting that the maximum stress occurs at $y_{max} = \pm1$ in the maximum bending normal, the stress can be found from Equation (3.9-B). For cases 2 and 3, the neutral axis (value of R) can be found from Equation (3.44) and Table 3.6. The radius of the line joining the centroids is the inner radius plus one. The stress at the inside and outside radius can be found in terms from Equation (3.48), and thus the maximum value determined.

SOLUTION

Case 1: The area moment of inertia is $I_{zz} = \frac{1}{12}(2)(2^3) = 1.333$ in^4.

Substituting $y_{max} = \pm1$ in, the maximum bending normal stress can be found from Equation (3.9-B) as shown in Equation (E1).

$$\sigma_{max} = -\left(\frac{M_z y_{max}}{I_{zz}}\right) = -\left(\frac{(30)(\pm1)}{1.333}\right) = \mp22.5 \tag{E1}$$

ANS. $\sigma_{max} = 22.5$ ksi (C,T)

Case 2: The inner radius is $r_1 = 5$ in, the outer radius is $r_2 = 7$ in, and $a = 2$ in. From Table 3.6 for a rectangular cross section, we obtain Equation (E2).

$$\int_A \frac{dA}{\rho} = a\ln\frac{r_2}{r_1} = 2\ln\frac{7}{5} = 0.6729 \text{ in} \tag{E2}$$

The cross-sectional area is $A = 4 \text{ in}^2$. From Equation (3.44) we obtain Equation (E3).

$$R = \frac{A}{\int_A \frac{dA}{\rho}} = \frac{4}{0.6729} = 5.944 \text{ in} \tag{E3}$$

The radius of the line joining the centroids is $r_c = 5 + 1 = 6$ in. On inner surface $\rho_i = r_1 = 5$ in. From Equation (3.48) we obtain Equation (E4).

$$(\sigma_{\theta\theta})_i = -\left[\frac{M_z}{A(r_c - R)}\left(1 - \frac{R}{\rho_i}\right)\right] = -\left[\frac{(30)}{(4)(6 - 5.944)}\left(1 - \frac{5.944}{5}\right)\right] = 25.298 \text{ ksi} \tag{E4}$$

On the outer surface $\rho_o = r_2 = 7$ in. From Equation (3.48) we obtain Equation (E5).

$$(\sigma_{\theta\theta})_o = -\left[\frac{M_z}{A(r_c - R)}\left(1 - \frac{R}{\rho_o}\right)\right] = -\left[\frac{(30)}{(4)(6 - 5.944)}\left(1 - \frac{5.944}{7}\right)\right] = -20.214 \text{ ksi} \tag{E5}$$

Comparing Equations (E3) and (E4), we obtain the maximum bending stress as

ANS. $\sigma_{max} = 25.3 \text{ ksi (C)}$

Case 3: The inner radius is $r_1 = 20$ in, the outer radius is $r_2 = 22$ in, and $a = 2$ in. From Table 3.6 for a rectangular cross section, we obtain Equation (E6).

$$\int_A \frac{dA}{\rho} = a\ln\frac{r_2}{r_1} = 2\ln\frac{22}{20} = 0.1906 \text{ in} \tag{E6}$$

The cross-sectional area is $A = 4 \text{ in}^2$. From Equation (3.44) we obtain Equation (E7).

$$R = \frac{A}{\int_A \frac{dA}{\rho}} = \frac{4}{0.1906} = 20.984 \text{ in} \tag{E7}$$

The radius of the line joining the centroids is $r_c = 20 + 1 = 21$ in. On the inner surface $\rho_i = r_1 = 20$ in. From Equation (3.48) we obtain Equation (E8).

$$(\sigma_{\theta\theta})_i = -\left[\frac{M_z}{A(r_c - R)}\left(1 - \frac{R}{\rho_i}\right)\right] = -\left[\frac{(30)}{(4)(21 - 20.984)}\left(1 - \frac{20.984}{20}\right)\right] \tag{E8}$$

$$= 23.236 \text{ ksi}$$

On the outer surface $\rho_o = r_2 = 22$ in. From Equation (3.48) we obtain Equation (E9).

$$(\sigma_{\theta\theta})_o = -\left[\frac{M_z}{A(r_c - R)}\left(1 - \frac{R}{\rho_o}\right)\right] = -\left[\frac{(30)}{(4)(21 - 20.984)}\left(1 - \frac{20.984}{22}\right)\right] \tag{E9}$$

$$= -21.805 \text{ ksi}$$

Comparing Equations (E7) and (E8), we obtain the maximum bending stress.

ANS. $\sigma_{max} = 21.8$ ksi (C)

COMMENTS

1. Figure 3.51 shows the stress distribution on the cross section as one moves from the bottom (inner) to the top (outer) surface. For the straight beam of case 1, the stress distribution is linear and symmetric about the centroidal axis, which is also the neutral axis. For curved beams, the stress distribution is hyperbolic; the neutral axis and the centroidal axis are different, and the maximum stress on the inner surface is greater than the stress on the outer surface.

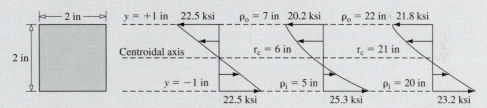

Figure 3.51 Stress distribution across the cross section (far left) for the three cases in Example 3.17.

2. For case 3 the ratio of r_c/h is 10.5, and the difference in maximum stress from the straight beam is 3.1%. For case 2 the ratio of r_c/h is 3, and the difference in maximum stress from the straight beam is 12.4%. The results indicate that as the radius of curvature increases, the results of the curved beam tends toward those of the straight beam. Also notice that the neutral axis moves toward the centroidal axis as the radius of curvature increases.

3. For case 3: $(1 - R/\rho_i) = -0.049206$ and $(r_c - R) = 0.015883$, which implies that we divide one small number by another small number. This can lead to large errors unless calculations are carried out to many significant figures.

EXAMPLE 3.18

An I-section is formed into a ring as shown in Figure 3.52(a). The yield stress of the material is 200 MPa. Determine the maximum force P that can be applied if yielding is to be avoided.

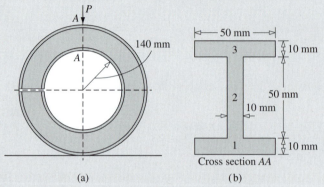

(a)   (b)

Figure 3.52 (a) Ring geometry and (b) I-section for Example 3.18.

PLAN

The integral in Equation (3.44) can be written as the sum of the integrals over each of the three rectangular areas of the I-section shown in Figure 3.52(b). The values of the three integrals can be found from Table 3.6. The radius of the neutral axis can be calculated from Equation (3.44). The maximum moment will occur at a cross section on the horizontal centerline opposite the gap. By drawing a free body diagram, we can find the axial force and bending moment in terms of P. The total normal stress on the inner and outer surfaces in terms of P can be found from Equation (3.49). Knowing that this stress should be less than or equal to the yield stress, we can find the maximum value of P.

SOLUTION

The integral in Equation (3.44) can be written as $\int_A (dA/\rho) = \int_{A_1} (dA/\rho) + \int_{A_2} (dA/\rho) +$

$\int_{A_3} (dA/\rho)$. The calculation of the integral is shown in Table 3.7. From Equation (3.44) and Table 3.7, we obtain the radius of the neutral axis as shown in Equation (E1).

$$R = \frac{A}{\int_A \frac{dA}{\rho}} = \frac{1500}{8.76597} = 171.1 \text{ mm} = 0.1711 \text{ m} \qquad (E1)$$

TABLE 3.7 Calculation of the Radius of the Neutral Axis in Example 3.18

Section	A (mm²)	r_1 (mm)	r_2 (mm)	a (mm)	$\int_A \dfrac{dA}{\rho}$ (mm)
1	500	140	150	50	$50 \ln\left(\dfrac{150}{140}\right) = 3.44964$
2	500	150	200	10	$10 \ln\left(\dfrac{200}{150}\right) = 2.87682$
3	500	200	210	50	$50 \ln\left(\dfrac{210}{200}\right) = 2.43951$
Total	1500				8.76597

Figure 3.53 shows the free body diagram that results after an imaginary cut has been made through the horizontal centerline. We expect the internal moment to be maximum as the moment arm is maximum. The internal axial force N acts at the centroid of the cross section. From geometry, the radius of the centroidal axis is $r_c = 140 + 35 = 175$ mm $= 0.175$ m. The internal forces and moment are drawn on the free body diagram in the positive direction, in accordance with the sign convention.

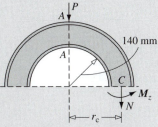

Figure 3.53 Free body diagram for Example 3.18.

By the equilibrium of forces and moments about C, we obtain Equations (E2).

$$N = -P \qquad \text{and} \qquad M_z = -Pr_c = -0.175P \qquad (E2)$$

On the inner surface $\rho_i = 0.14$ m, and on the outer surface $\rho_o = 0.21$ m. The normal stress, which can be found on these surfaces from Equation (3.49), should be less than or equal to 200 MPa, yielding limits on P as shown in Equations (E3) and (E4).

$$\left|(\sigma_{\theta\theta})_i\right| = \left| \frac{N}{A} - \left[\frac{M_z}{A(r_c - R)}\left(1 - \frac{R}{\rho_i}\right) \right] \right|$$

$$= \left| -\frac{P}{1.5(10^{-3})} - \frac{(-0.175P)}{1.5(10^{-3})(0.175 - 0.1711)}\left(1 - \frac{0.1711}{0.14}\right) \right|$$

$$\left|(\sigma_{\theta\theta})_i\right| = \left| -666.67P - 6645.3P \right| = \left| -7311.97P \right| \le 200(10^6) \qquad \text{or}$$

$$P \le 27.35(10^3) \text{ N} \qquad (E3)$$

$$\left|(\sigma_{\theta\theta})_o\right| = \left|\frac{N}{A} - \left[\frac{M_z}{A(r_c - R)}\left(1 - \frac{R}{\rho_o}\right)\right]\right|$$

$$= \left|-\frac{P}{1.5(10^{-3})} - \frac{(-0.175P)}{1.5(10^{-3})(0.175 - 0.1711)}\left(1 - \frac{0.1711}{0.21}\right)\right|$$

$$\left|(\sigma_{\theta\theta})_o\right| = \left|-666.67P + 5541.3P\right| = \left|4874.64P\right| \leq 200(10^6) \quad \text{or}$$

$$P \leq 41.03(10^3) \text{ N} \qquad\qquad (E4)$$

The maximum value of P must satisfy the inequalities in Equations (E3) and (E4). Thus the maximum value of P is **ANS.** $P_{max} = 27.3$ kN

COMMENTS

1. Table 3.7 shows how the radius of the neutral axis for complex cross sections can be found by using the formulas given in Table 3.6.

2. By inspection we can see that the normal stress due to bending alone will be compressive on the inside surface and tensile on the outer surface. With axial stresses also in compression, we expect the normal stress to be maximum on the inside.

PROBLEM SET 3.3

3.55 The total strain from axial and bending forces on a curved beam can be written as

$$\varepsilon_{\theta\theta} = \varepsilon_a - \left(1 - \frac{R}{\rho}\right)\psi \qquad (3.50)$$

where ε_a is a uniform axial strain across a cross section, R is the radius of the neutral axis, ρ is the radial coordinate of the point at which the tangential strain is evaluated, and ψ is a constant. Assuming that Assumptions 4 through 8 are valid, show that the tangential stress is given by Equation (3.49), provided N is placed at the centroid of the cross section and R is given by Equation (3.44).

3.56 A curved beam having an inside radius of 5 inches and a T cross section is shown in Figure P3.56. For a moment of $M = 40$ in·kips, determine the maximum tensile and compressive tangential normal stresses.

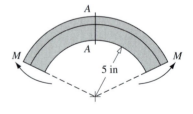

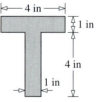

Cross section AA

Figure P3.56

ANS. $\sigma_{max} = 5.1$ ksi (T); $\sigma_{max} = 1.8$ ksi (C)

3.57 A curved beam having an inside radius of 6 inches has the cross section shown in Figure P3.57. Point C locates the centroid of the cross section. If the magnitude of the maximum tangential normal stress that is permitted is 20 ksi, determine the maximum moment M that can be applied.

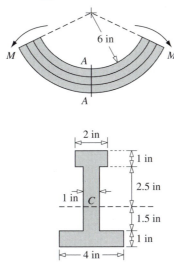

Cross section *AA*

Figure P3.57

3.58 A curved beam having an inside radius of 250 mm has the cross section shown in Figure P3.58. Point C locates the centroid of the cross section. If the maximum permitted tangential normal stresses are 140 MPa in tension and 120 MPa in compression, determine the maximum moment M that can be applied.

ANS. $M_{max} = 6.87$ kN · m

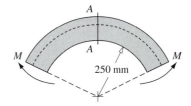

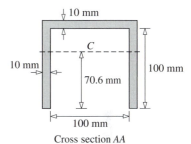

Cross section *AA*

Figure P3.58

3.59 A square tube with an outer dimension of $a = 4$ inches and a wall thickness of 0.5 inch is formed into a half-ring with an inner radius $\rho_i = 10$ inches as shown in Figure P3.59. Determine the maximum tangential normal stress if the applied force is $P = 8$ kips.

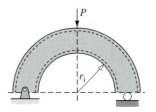

Figure P3.59

3.60 A square tube with an outer dimension of 4 inches and a wall thickness of 0.5 inch is formed into a half-ring with an inner radius $r_i = 10$ inches as shown in Figure P3.59. Determine the maximum load P that can be applied if the allowable bending normal stress is 24 ksi.

3.61 A circular bar (diameter $d = 3$ in) is formed into a hook as shown in Figure P3.61. Determine the maximum tangential normal stress if the hook lifts a load of $P = 2$ kips.

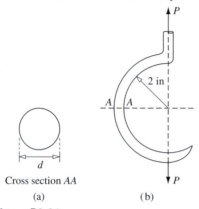

Cross section AA

(a)   (b)

Figure P3.61

ANS. $\sigma_{max} = 4.2$ ksi (T)

3.62 A circular bar (diameter $d = 3$ in) is formed into a hook as shown in Figure P3.61. Determine the maximum load P that can be lifted by the hook if the allowable tangential normal stress is 20 ksi.

3.63 A circular tube is formed into a ring with an inner radius $r_i = 150$ mm as shown in Figure P3.63. The tube has an outer diameter of $d = 100$ mm and a thickness of 20 mm. Determine the maximum tangential normal stress if the applied load $P = 20$ kN. Assume that the gap does not close.

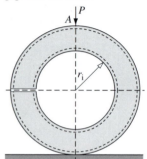

Cross section AA

Figure P3.63

ANS. $\sigma_{max} = 56.3$ MPa (T)

3.64 A circular tube is formed into a ring with an inner radius $r_i = 200$ mm as shown in Figure P3.63. The tube has an outer diameter of $d = 75$ mm and a thickness of 10 mm. If the allowable tangential normal stress is 120 MPa, determine the maximum force P that can be applied.

3.6   CLOSURE

The logic shown in Figure 3.2 emphasizes that the *kinematic equations* between displacements and strains, the *constitutive equations* between strains and stresses, the *static equivalency* equations between stresses and internal forces/moments, and the *equilibrium equations* between internal forces/moments and external forces/moments can be treat as independent modules, permitting one to add complexities to one module without changing the fundamental equations in the other modules. Thus, dynamics effects and elastic foundation effects change the equilibrium equations without changing the kinematics equations, constitutive equations, and equivalency equations. In the Timoshenko beam, we changed the kinematic equations but not the equations in other modules. In Chapter 4, Composite Structural Members, we drop the assumption of material homogeneity across the cross section that we make in the module of static equivalency, but there are no other changes. In Chapter 5, Inelastic Structural Behavior, we drop the assumptions related to Hooke's law, thus changing the constitutive equations but nothing else. In Chapter 6, Thin-Walled Structural Members, we drop the restrictions on the geometry of the cross sections of symmetric beams and circular shafts, thus changing the kinematic equations without changing the fundamental equations in other modules. In Chapter 7, Energy Methods, we replace the entire module of equilibrium equations with energy method equations but do not change the equations of the other modules. An understanding of the modular character of the logic, and how complexities are incorporated not only will help you with the remaining chapters in this book but equally significantly, will give you the capability of adding complexities to one-dimensional structures not covered here.

In this chapter we also saw the discontinuity functions and their applications in solving boundary value problems for statically determinate and indeterminate axial members, the torsion of circular shafts, and the symmetric bending of beams. We will use discontinuity functions whenever possible in the remaining chapters.

4 Composite Structural Members

4.0 OVERVIEW

Fishing rods, bicycle frames, the wings and control surfaces of aircraft, tennis racquets, boat hulls, storage tanks, reinforced concrete bars, wooden beams stiffened with steel strips, laminated shafts, and fiberglass automobile bodies are examples from the vast variety of applications of composite structural members (Figure 4.1). In this chapter we develop the formulas for deformation and stresses for composite axial bars, shafts, and beams, as well as formulas for obtaining macroscopic material properties from the material constants of the constituents of the composites.

In Section 3.1, while developing simplified theories on structural members, we made the assumption of material homogeneity across the cross section. This assumption is not valid for composite members. If we assume that different materials are perfectly bonded together, we can still make our kinematic assumptions (Assumptions 1–3 in Table 3.3). If our Assumptions 4 through 7 about material behavior are still valid, so that we can still use Hooke's law, then all the equations before Assumption 8 on material homogeneity in Table 3.3 are still applicable. The static equivalency equations relating stresses to the internal forces and moments are still valid. The integrals in static equivalency can be written as the sum of the integrals over each material in the cross section, and new formulas for deformations and stresses in terms of the internal forces and moments can be derived. The equilibrium equations relating the internal forces and moments are independent of the assumption of material homogeneity; thus all the techniques we have developed for finding the internal forces and moments for determinate and statically indeterminate structures can be used for composite structures.

The mechanical properties of a composite from a tension test are macroscopic (bulk) properties. The theories of micromechanics attempt to derive the macroscopic properties from the individual constituent properties, which may be microscopic in size. One such theory is based on the equations for the mechanics of materials. This theory results in equations that are called the "rule of mixtures" and the "inverse rule of mixtures," by which the macroscopic mechanical properties of a composite can be obtained from the constituent properties, as shall be seen in Section 4.4.

Figure 4.1 Examples of composite structural members. (*Photographs of bicycle and aircraft courtesy of Giant Bicycles, U.S.A., and U.S. Air Force, photograph by Staff Sgt Mark Borosch.*)

The learning objectives of this chapter are as follows:

1. To understand the incorporation and implications of material nonhomogeneity across the cross section in the theories for axial members, circular shafts in torsion, and the symmetric bending of beams.

2. To understand the use of the rule of mixture, and the inverse rule of mixtures in obtaining the macroscopic properties of a composite from its constituent properties.

4.1 COMPOSITE AXIAL MEMBERS

We assume that all materials, like the laminated cross section in Figure 4.2, are securely bonded to each other so that our kinematic assumption of plane sections remaining plane and parallel is still valid. We further assume that we still have small strains, that all materials are linear-elastic-isotropic materials, and that there are no inelastic strains. In other words, all the assumptions before Assumption 8 in Table 3.3 are still valid. Thus, Equation (3.6-A), which precedes Assumption 8, is still valid and forms our starting point for deriving the axial formulas for composite bars.

Consider the laminated structure shown in Figure 4.2. Each material has a modulus of elasticity E_i, which is constant over the material cross-sectional area A_i. Suppose there are n materials in the cross section. We can write the integral over the cross section in Equation (3.6-A) as the sum of the integrals over each material as shown in Equation (4.1a).

$$N = \frac{du}{dx}\int_A E\ dA = \frac{du}{dx}\left[\int_{A_1} E_1 dA + \int_{A_2} E_2 dA + \cdots + \cdots + \int_{A_n} E_n dA\right] \qquad (4.1a)$$

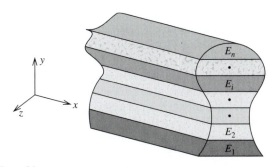

Figure 4.2 A laminated bar.

In Equation (4.1a), E_i is a constant in each integral and can be taken outside the integral. The remaining integral is the area A_i. We thus obtain Equation (4.1b).

$$N = \frac{du}{dx}[E_1 A_1 + E_2 A_2 + \cdots + \cdots + E_n A_n] \tag{4.1b}$$

Equation (4.1b) can be written more compactly as Equation (4.1c).

$$N = \frac{du}{dx}\left[\sum_{j=1}^{n} E_j A_j\right] \tag{4.1c}$$

Equation (4.1c) shows that the axial rigidity of the composite bar is the sum of the axial rigidities of all the materials. The strain at any cross section can be written as Equation (4.2).

$$\varepsilon_{xx} = \frac{du}{dx} = \frac{N}{\sum_{j=1}^{n} E_j A_j} \tag{4.2}$$

The axial stress in the ith material is $(\sigma_{xx})_i = E_i \varepsilon_{xx}$. Substituting Equation (4.2), we obtain Equation (4.3).

$$(\sigma_{xx})_i = \frac{N E_i}{\sum_{j=1}^{n} E_j A_j} \tag{4.3}$$

We assume that the axial rigidity and the internal axial force do not change with x between x_1 and x_2. In other words, Assumptions 9 through 11 in Table 3.3 are still applicable to composite axial bars. With the validity of Assumptions 9 through 11, du/dx is constant between x_1 and x_2, and we can write $du/dx = (u_2 - u_1)/(x_2 - x_1)$ in Equation (4.2) to obtain Equation (4.4).

$$u_2 - u_1 = \frac{N(x_2 - x_1)}{\sum_{j=1}^{n} E_j A_j} \tag{4.4}$$

Equations (4.3) and (4.4), which are applicable to composite bars, now replace Equations (3.9-A) and (3.10-A) for the homogeneous cross section. If we consider a homogeneous materials, then $E_1 = E_2 = \cdots = E_i \cdots = E_n = E$. Substituting this in Equations (4.3) and (4.4), we obtain Equations (3.9-A) and (3.10-A) for homogeneous materials, as expected.

The internal axial force N in Equations (4.3) and (4.4) represents the statically equivalent axial force *over the entire cross section,* as it did for the homogeneous cross section. The analysis techniques for finding the internal axial force N at a cross section remains the same as before. However, the location at which the internal axial force acts on the cross section depends upon the distribution of the material across the cross section, as emphasized in Example 4.1.

Equation (4.2) emphasizes that the normal axial strain ε_{xx} is uniform across the cross section. The normal axial stress $(\sigma_{xx})_i$ is uniform only in each material; it changes with each material point because the value of E_i changes with each material, as shown in Equation (4.3). Thus, the normal axial stress is piecewise uniform across the cross section. This is elaborated further in Example 4.2.

EXAMPLE 4.1

For the laminated, symmetric, linear elastic bar in Figure 4.2, show that the location of the origin (η_c) can be found from the formula

$$\eta_c = \frac{\sum_{i=1}^{n} \eta_i E_i A_i}{\sum_{i=1}^{n} E_i A_i} \tag{4.5}$$

where η_i is the location of the centroid of the ith material as measured from a common datum line.

PLAN

The location of the origin for a linear elastic material is given by Equation (3.5-A). We write the integral over the entire cross section as the sum of the integrations over all the materials. In each material the modulus of elasticity is a constant and can be taken outside the integral, whereupon the origin can be determined.

SOLUTION

The location of the origin of y is determined from Equation (3.5-A), repeated for convenience as Equation (E1).

$$\int_A yE \, dA = 0 \tag{E1}$$

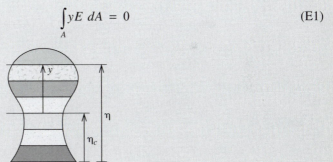

Figure 4.3 Coordinate description for Example 4.1.

Figure 4.3 shows that the relationship between y and η is given by Equation (E2).

$$y = \eta - \eta_c \tag{E2}$$

Substituting Equation (E2) into Equation (E1), we obtain Equation (E3).

$$\int_A \eta E \, dA - \int_A \eta_c E \, dA = 0 \tag{E3}$$

Noting that η_c is a constant, we obtain Equation (E4).

$$\eta_c = \frac{\int_A \eta E \, dA}{\int_A E \, dA} \tag{E4}$$

Writing the integration over the area as the sum of the integrations over all the materials and noting that E_i is a constant within each A_i, we can write Equation (E4) as Equation (E5).

$$\eta_c = \frac{\sum_{i=1}^{n}\int_{A_i}\eta E_i dA}{\sum_{i=1}^{n}\int_{A_i}E_i dA} = \frac{\sum_{i=1}^{n}E_i\int_{A_i}\eta \, dA}{\sum_{i=1}^{n}E_i\int_{A_i}dA} \tag{E5}$$

From definition of centroid for η_i, we note that $\int_{A_i}\eta \, dA = \eta_i A_i$ and $\int_{A_i}dA = A_i$. Substituting these two identities into Equation (E5), we obtain Equation (4.5).

COMMENTS

1. If the external axial forces do not pass through the point defined by Equation (4.5), then the laminated rod will undergo axial deformation and will also bend.

2. If $E_1 = E_2 = \cdots = E_i \cdots = E_n = E$, then the modulus of elasticity in the numerator and denominator of Equation (4.5) can be taken outside the summation, and we obtain Equation (E6)

$$\eta_c = \frac{\sum_{i=1}^{n}\eta_i A_i}{\sum_{i=1}^{n}A_i} \tag{E6}$$

where η_c defines the centroid for a homogeneous body.

EXAMPLE 4.2

A homogeneous wooden cross section and a cross section in which the wood is reinforced with steel are shown in Figure 4.4. The normal strain for both cross sections is uniform: $\varepsilon_{xx} = -200 \ \mu$. The modulus of elasticity for steel is $E_{st} = 30,000$ ksi, and for wood it is $E_w = 8000$ ksi. (a) Plot the σ_{xx} distribution for each of the two cross sections shown. (b) Use Equation (3.4a-A) from Table 3.3 to calculate the equivalent internal axial force N for each cross section.

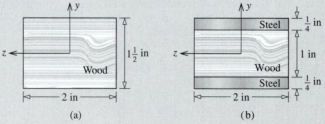

Figure 4.4 Cross sections for Example 4.2: (a) homogeneous and (b) laminated.

PLAN

(a) We can use Hooke's law to find the stress values for each material. After noting that the stress is uniform in each material, we can plot it across the cross section. (b) For a homogeneous cross section, we can perform the integration in Equation (3.4a-A) directly. For a nonhomogeneous cross section we can write the integral in Equation (3.4a-A) as the sum of the integrals over steel and wood and then perform the integration to get N.

SOLUTION

From Hooke's law we can write Equations (E1) and (E2).

$$(\sigma_{xx})_{\text{wood}} = (8)(10^3)(-200)(10^{-6}) = -1.6 \text{ ksi} \tag{E1}$$

$$(\sigma_{xx})_{\text{st}} = (30)(10^3)(-200)(10^{-6}) = -6 \text{ ksi} \tag{E2}$$

For the homogeneous cross section, the stress distribution is as given in Equation (E1), but for the laminated case it switches from Equation (E1) to (E2), depending upon the value of the location of the point at which stress is being evaluated, as shown in Figure 4.5.

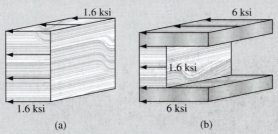

Figure 4.5 Stress distribution for Example 4.2: (a) homogeneous and (b) laminated cross sections.

(a) **Homogeneous cross section**: Substituting the stress distribution for the homogeneous cross section in Equation (3.4a-A) and integrating, we obtain the equivalent internal axial force as shown in Equation (E3).

$$N = \int_A (-1.6)dA = -1.6A = -(1.6)(2)(1.5) = -4.8 \qquad \text{or} \tag{E3}$$

ANS. $N = 4.8$ kips (C)

(b) **Laminated cross section:** The stress value changes as we move across the cross section. Let $A_{\text{St,b}}$ and $A_{\text{St,t}}$ represent the cross sectional area of the steel at bottom and top. Let A_{wood} represent the cross-sectional area of the wood. We can write the integral in Equation (3.4a-A) as a sum of three integrals, substitute the stress values of Equations (E1) and (E2), and perform the integral as follows:

$$N = \left[\int_{A_{\text{St,b}}} \sigma_{xx}dA + \int_{A_{\text{wood}}} \sigma_{xx}dA + \int_{A_{\text{St,t}}} \sigma_{xx}dA \right]$$

$$= \left[\int_{A_{\text{St,b}}} (-6)dA + \int_{A_{\text{wood}}} (-1.6)dA + \int_{A_{\text{St,t}}} (-6)dA \right] \tag{E4}$$

$$N = [(-6)A_{St,b} + (-1.6)A_{wood} + (-6)A_{St,t}]$$
$$= [(-6)(2)(1/4) + (-1.6)(2)(1) + (-6)(2)(1/4)]$$

ANS. $N = 9.2$ kips (C)

COMMENTS

1. Writing the integral in the internal axial force as the sum of the integrals over each material, as is done in Equation (E4), is equivalent to calculating the internal force carried by each material and then summing it, as shown in Figure 4.6.

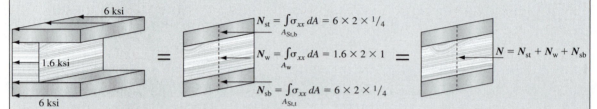

Figure 4.6 Statically equivalent internal force for the laminated cross section in Example 4.2.

2. An alternative calculations for internal force is as follows. We can find the axial rigidity of the bar as

$$\sum E_j A_j = [(30{,}000)(2)(1/4)](2) + (8000)(2)(1) = 46{,}000 \text{ kips}$$

Substituting the strain and axial rigidity in Equation (4.1c), we obtain the internal force

$$N = (46{,}000)(-200)(10^{-6}) = -9.2 \text{ kips}$$

as before.

3. The cross section is symmetric geometrically as well as materially. Thus, we can determine the location of the origin to be on the line of symmetry. Suppose a lower steel strip is not present. Then we must determine the vertical distance from bottom (or top), where the equivalent force will have to be placed as given by Equation (4.5).

4. The example demonstrates that although strain is uniform across a cross section, stress is not. We considered material nonhomogeneity in this example. In a similar manner, we will consider other material models in Chapter 5, including elastic–perfectly plastic and material models that have nonlinear stress–strain curves.

EXAMPLE 4.3

A cast iron pipe ($E_{Fe} = 100$ GPa) and a copper pipe ($E_{Cu} = 130$ GPa) are adhesively bonded together as shown in Figure 4.7. The outer diameters of the two pipes are 50 and 70 mm, respectively, and the wall thickness of each pipe is 10 mm. Determine (a) the displacement of end D with respect to end A and (b) the axial stresses in the iron and copper in the bonded region.

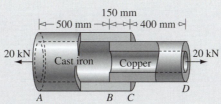

Figure 4.7 Pipes for Example 4.3.

PLAN

(a) Pipe segments *AB* and *CD* have homogeneous cross sections, and we can use Equation (3.10-A) to find the relative displacements of the segment ends. However, segment *BC* is a composite pipe, and we can use Equation (4.4) to find the relative displacement of the section at *C* with respect to the section at *B*. (b) We can use Equation (4.3) to find the axial stress in each material.

SOLUTION

(a) The cross-sectional areas of cast iron (A_{Fe}) and copper (A_{Cu}) can be calculated as follows:

$$A_{Fe} = \frac{\pi}{4}(0.07^2 - 0.05^2) = 1.885(10^{-3})m^2 \tag{E1}$$

$$A_{Cu} = \frac{\pi}{4}(0.05^2 - 0.03^2) = 1.257(10^{-3})m^2 \tag{E2}$$

The axial rigidities for each segment of pipe can be found as shown in Equations (E3), (E4), and (E5).

$$E_{Fe}A_{Fe} = 1.885(10^{-3})(100)(10^9) = 188.5(10^6) \tag{E3}$$

$$E_{Cu}A_{Cu} = 1.257(10^{-3})(130)(10^9) = 163.4(10^6) \tag{E4}$$

$$E_{BC}A_{BC} = \sum E_iA_i = E_{Fe}A_{Fe} + E_{Cu}A_{Cu} = 351.9(10^6) \tag{E5}$$

We can see that regardless of where we make our imaginary cut in the pipe, the internal force value is the same—$N = 20$ kN. By substituting Equations (E3), (E4), and (E5) into Equations (3.10-A) and (4.4), we obtain the following relative displacements of the various segments of pipe:

$$u_B - u_A = \frac{N(x_B - x_A)}{E_{Fe}A_{Fe}} = \frac{(20)(10^3)(0.5)}{188.5(10^6)} = 53.05(10^{-6}) \tag{E6}$$

$$u_C - u_B = \frac{N(x_C - x_B)}{\sum E_iA_i} = \frac{(20)(10^3)(0.15)}{351.9(10^6)} = 8.53(10^{-6}) \tag{E7}$$

$$u_D - u_C = \frac{N(x_D - x_C)}{E_{Cu}A_{Cu}} = \frac{(20)(10^3)(0.4)}{163.4(10^6)} = 48.96(10^{-6}) \qquad \text{(E8)}$$

We can add Equations (E6), (E7), and (E8) to obtain the relative deformation of the section at D with respect to the section at A as shown in Equation (E9).

$$u_D - u_A = [53.05 + 8.53 + 48.96](10^{-6}) = 110.54(10^{-6})\text{m} \qquad \text{(E9)}$$

ANS. $u_D - u_A = 0.1105$ mm

(b) The stress in each material in the bonded region can be calculated by using Equation (4.3), as shown Equations (E10) and (E11).

$$(\sigma_{BC})_{Fe} = \frac{N E_{Fe}}{\sum E_i A_i} = \frac{(20)(10^3)(100)(10^9)}{351.9(10^6)} = 5.68(10^6) \qquad \text{(E10)}$$

ANS. $(\sigma_{BC})_{Fe} = 5.68$ MPa (T)

$$(\sigma_{BC})_{Cu} = \frac{N E_{Cu}}{\sum E_i A_i} = \frac{(20)(10^3)(130)(10^9)}{351.9(10^6)} = 7.39(10^6) \qquad \text{(E11)}$$

ANS. $(\sigma_{BC})_{Cu} = 7.39$ MPa (T)

COMMENTS

1. The analytic procedure is the same for homogeneous and nonhomogeneous cross sections. The difference is in the formulas used in the calculation of the stress and the relative displacement.

2. The axial force carried by each material in the bonded region can be found by multiplying the respective axial stress by the material cross-sectional area. This yields $N_{Fe} = 10.71$ kN and $N_{Cu} = 9.23$ kN. Notice that the total axial force on the cross section is $N = N_{Fe} + N_{Cu}$.

EXAMPLE 4.4

We must design a 200 mm × 200 mm reinforced concrete bar of length 2.5 m (see Figure 4.8) to carry a compressive axial force of 1000 kN. The allowable contraction of the bar is 3 mm. Determine, to the nearest kilogram, the minimum amounts of concrete and steel that can be used in constructing the reinforced concrete bar. Table 4.1 gives the material properties.

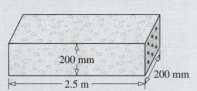

Figure 4.8 Reinforced concrete bar for Example 4.4.

TABLE 4.1 Material Properties for Example 4.4

	Concrete	Steel
Modulus of elasticity	$E_{conc} = 20$ GPa	$E_{st} = 200$ GPa
Density	2400 kg/m³	7870 kg/m³
Allowable stress	12 MPa	150 MPa

PLAN

From Equation (4.3), we can determine the axial rigidity that is needed to meet the limitation on allowable stress in concrete and steel. From Equation (4.4), we can determine the axial rigidity that is needed to meet the limitation on allowable deformation. The value of the axial rigidity that satisfies all the limitations on stress and deformation can thus be determined, whereupon the cross-sectional areas of the steel and concrete can be found. By multiplying the areas by the given length, we can find the volume, which we can multiply by density to obtain the amount of each constituent in kilograms.

SOLUTION

In the cross section of the reinforced bar, let A_{conc} and A_{st} represent the cross-sectional areas of concrete and steel, respectively. The axial rigidity of the reinforced bar can be written as Equation (E1).

$$\sum E_j A_j = (20 A_{conc} + 200 A_{st})(10^9) \tag{E1}$$

Noting that the axial force $N = 1000$ kN, we can use Equation (4.3) to write the stress in concrete and in steel. These stresses must be less than the allowable value, and we obtain two conditions on axial rigidity, as shown in Equations (E2) and (E3).

$$(\sigma_{xx})_{conc} = \frac{(N E_{conc})}{\sum E_j A_j} = \frac{1000(10^3)(20)(10^9)}{(20 A_{conc} + 200 A_{st})(10^9)} = \frac{20(10^6)}{(20 A_{conc} + 200 A_{st})} \le 12(10^6) \quad \text{or}$$

$$20 A_{conc} + 200 A_{st} \ge 1.667 \tag{E2}$$

$$(\sigma_{xx})_{st} = \frac{(N E_{st})}{\sum E_j A_j} = \frac{1000(10^3)(200)(10^9)}{(20 A_{conc} + 200 A_{st})(10^9)} = \frac{200(10^6)}{(20 A_{conc} + 200 A_{st})} \le 150(10^6) \quad \text{or}$$

$$20 A_{conc} + 200 A_{st} \ge 1.333 \tag{E3}$$

The contraction of the bar $u_2 - u_1$ should be less than 3 mm, or $3(10^{-3})$ m, over a length $x_2 - x_1 = 2.5$ m. From Equation (4.4) we can obtain another limitation on axial rigidity, as shown in Equation (E4).

$$u_2 - u_1 = \frac{N(x_2 - x_1)}{\sum E_j A_j} = \frac{1000(10^3)(2.5)}{(20 A_{conc} + 200 A_{st})(10^9)} \leq 3(10^{-3}) \qquad \text{or}$$

$$20 A_{conc} + 200 A_{st} \geq 0.833 \qquad \qquad \text{(E4)}$$

If axial rigidity is 1.667, then we meet the three limitation given by Equations (E2), (E3), and (E4). For the minimum amount, we use the equality sign in Equation (E2). Noting that the total cross-sectional area is $A = (200)(200)$ mm^2 or $A = 0.04$ m^2, we can write Equations (E5) and (E6).

$$20 A_{conc} + 200 A_{st} = 1.667 \qquad \qquad \text{(E5)}$$

$$A_{conc} + A_{st} = 0.04 \qquad \qquad \text{(E6)}$$

Solving Equations (E5) and (E6), we obtain Equations (E7) and (E8).

$$A_{conc} = 35.183(10^{-3}) \text{m}^2 \qquad \qquad \text{(E7)}$$

$$A_{st} = 4.817(10^{-3}) \text{m}^2 \qquad \qquad \text{(E8)}$$

Multiplying the areas in Equations (E7) and (E8) by the length, which is 2.5 m, we obtain the volume of each material. Multiplying the volumes by the density, we obtain the mass of concrete m_{conc} and mass of steel m_{st}, as shown in Equations (E9) and (E10).

$$m_{conc} = 35.183(10^{-3})(2.5)(2400) = 211.1 \qquad \qquad \text{(E9)}$$

ANS. $m_{conc} = 211$ kg

$$m_{st} = 4.817(10^{-3})(2.5)(7870) = 94.77 \qquad \qquad \text{(E10)}$$

ANS. $m_{st} = 95$ kg

COMMENTS

1. Since the amount of concrete is rounded down and the amount of steel up, the reinforced concrete bar will be stronger and stiffer than is required by the design constraints.
2. Suppose we use circular steel bars of diameter of 25 mm. Then each bar will have a cross-sectional area of $0.491(10^{-3})$ m^2. We can divide Equation (E8) by the area of each bar to obtain 9.8, which we round upward to get 10 steel bars. Now we have an estimate of the number of steel bars and the amount of concrete we need to construct the reinforced concrete bar. But since we rounded up (i.e., used more steel), we will have to reduce the amount of concrete used to ensure that the total volume is equal to that shown in Figure 4.8.

PROBLEM SET 4.1

4.1 The strain at a cross section of the axial rod shown in Figure P4.1 is assumed to have a uniform value of $\varepsilon_{xx} = 200\ \mu$. (a) Plot the stress distribution across the laminated cross section. (b) Determine the equivalent internal axial force N and its distance from the bottom of the cross section. Use $E_{Al} = 100$ GPa, $E_{wood} = 10$ GPa, and $E_{st} = 200$ GPa.

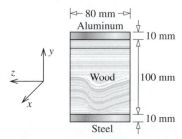

Figure P4.1

ANS. $N = 64$ kN (T); $y_N = 46.3$ mm from bottom

4.2 A reinforced concrete bar is constructed by embedding 2 in $\times$ 2 in square iron rods in concrete as shown in Figure P4.2. Assuming a uniform strain of $\varepsilon_{xx} = -1500\ \mu$ in the cross section, (a) plot the stress distribution across the cross section and (b) determine the equivalent internal axial force N. Use $E_{Fe} = 25{,}000$ ksi and $E_{conc} = 3000$ ksi.

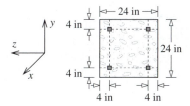

Figure P4.2

ANS. $N = 3120$ kips (C)

4.3 A wooden rod $(E_{wood} = 2000$ ksi$)$ and a steel strip $(E_{st} = 30{,}000$ ksi$)$ are fastened securely to each other and to the rigid plates as

shown in Figure P4.3. Determine (a) the location h of the line along which the external forces must act to produce no bending and (b) the maximum axial stress in the steel and wood components.

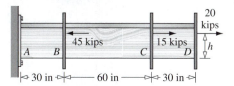

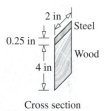

Cross section

Figure P4.3

4.4 A solid steel circular bar $(E_{st} = 200$ GPa$)$ that is 4 m long has a shaft of diameter 80 mm that extends through and is attached to a bar of hollow brass $(E_{Cu/Zn} = 100$ GPa$)$ with an outside diameter of 120 mm, as shown in Figure P4.4. Determine (a) the displacement of point C with respect to the wall and (b) the maximum axial stress in both metals.

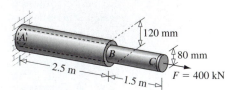

Figure P4.4

ANS. $u_C = 1.21$ mm $(\sigma_{Cu/Zn})_{max} = 24.5$ MPa (T) $(\sigma_{st})_{max} = 79.6$ MPa (T)

4.5 A concrete column for use in a building is modeled as shown in Figure P4.5. The axial forces represent the weights of the floors that the column transmits to the ground. The column is created by reinforcing the concrete with nine steel circular bars of diameter 1 inch. The moduli of elasticity for concrete and steel are $E_{conc} = 4500$ ksi and $E_{st} = 30,000$ ksi. Determine (a) the maximum axial stress in concrete and steel and (b) the contraction of the column.

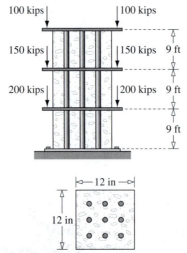

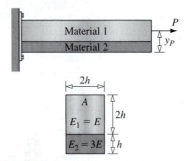

Figure P4.5

4.6 A cross section of a bar is made from two materials as shown in Figure P4.6. Assume that the parallel sections remain parallel [i.e., $\varepsilon_{xx} = \dfrac{du}{dx}(x)$]. In terms of the variables P, E, and h, determine (a) the location (y_P) of force P on the cross section so that there is only axial deformation and no bending and (b) the axial stress at point A.

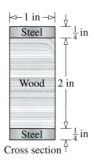

Figure P4.6

ANS. $\eta_C = 1.1\,h$ $\sigma_{xx} = 0.1P/h^2$

4.7 A 2 in × 1 in wooden bar is reinforced with 1/4 in steel bars to create laminated bars as shown in Figure P4.7. Two pieces of these laminated bars are securely fastened to a rigid plate loaded as shown. The modulus of elasticity is 30,000 ksi for steel and 8000 ksi for wood. Determine the movement of the rigid plate and the maximum stress in both the steel and the wood.

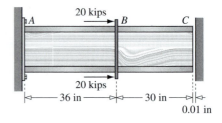

Figure P4.7

ANS. $\delta_B = 0.0266$ in $\sigma_{st} = 22.1$ ksi (T)
$\sigma_A = 5.9$ ksi (T)

4.8 A column for use in a building is modeled as shown in Figure P4.8. The column is constructed by reinforcing concrete with nine steel circular bars of diameter 1 inch. The modulus of elasticity for concrete is $E_{conc} = 4500$ ksi and $E_{st} = 30,000$ ksi for steel. Determine the maximum axial stress in both the concrete and the steel.

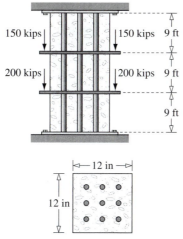

Figure P4.8

4.9 Two 1 in thick wooden boards (E_{wood} = 1800 ksi) are joined together by means of two 1/8 in thick aluminum (E_{Al} = 10,000 ksi) sheets as shown in Figure P4.9. The allowable stress in wood is 1.5 ksi, the allowable stress in aluminum is 12 ksi, and the total elongation of the joint length AD is to be limited to 0.05 inch. Determine the maximum axial force F that can be applied.

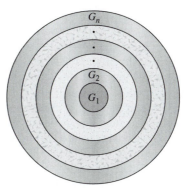

Figure P4.9

ANS. $F_{max} = 4.86$ kips

4.2 COMPOSITE SHAFTS

We assume that all material cross sections are axisymmetric as shown in Figure 4.9. We assume that all materials are securely bonded to each other; that is, there is no relative rotation at any interface of the material. This assumption of secure bonding is necessary to preserve our kinematic assumptions: that there is no warping and that all radial lines in a cross section rotate by equal amounts. We further assume that we still have small strains and that all materials are linear-elastic-isotropic. In other words, all assumptions prior to Assumption 8 in Table 3.3 are still valid. Thus, Equation (3.6-T), which precedes Assumption 8, is still valid and forms our starting point for deriving the torsional formulas for composite shafts.

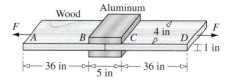

Figure 4.9 Cross section of a composite shaft.

Consider the laminated cross section shown in Figure 4.9. Each material has a shear modulus of elasticity G_i that is constant over the material cross-sectional area A_i. Suppose there are n materials in the cross section. The integral in Equation (3.6-T) can be written as the sum of the integrals over each material as follows:

$$T = \frac{d\phi}{dx}\left[\int_{A_1} G_1\rho^2 dA + \int_{A_2} G_2\rho^2 dA + \cdots + \cdots + \int_{A_n} G_n\rho^2 dA\right] \tag{4.6a}$$

Here G_i is a constant in each integral and can be taken outside the integral. The remaining integral is the polar moment J_i of the ith material. We thus obtain

$$T = \frac{d\phi}{dx}[G_1 J_1 + G_2 J_2 + \cdots + \cdots + G_N J_N] \tag{4.6b}$$

Equation (4.6b) can be written more compactly as follows:

$$\boxed{T = \frac{d\phi}{dx}\left[\sum_{j=1}^{n} G_j J_j\right]} \tag{4.7}$$

Equation (4.7) shows that the torsional rigidity of the composite shaft is the sum of the torsional rigidities of all the materials. We can write Hooke's law for the ith material as $(\tau_{x\theta})_i = G_i\rho(d\phi/dx)$, where $(\tau_{x\theta})_i$ is the torsional shear stress in the ith material. Substituting Equation (4.7) into Hooke's law, we obtain Equation (4.8).

$$\boxed{(\tau_{x\theta})_i = \frac{G_i\rho T}{\left[\sum_{j=1}^{n} G_j J_j\right]}} \tag{4.8}$$

We assume that torsional rigidity and the internal torque are not functions of x. That is, between x_1 and x_2, Assumptions 9 through 11 in Section 3.1 are applicable to composite shafts. Then $d\phi/dx$ is constant between x_1 and x_2, and we can write $d\phi/dx = (\phi_2 - \phi_1)/(x_2 - x_1)$ in Equation (4.7) to obtain Equation (4.9).

$$\boxed{\phi_2 - \phi_1 = \frac{T(x_2 - x_1)}{\left[\sum_{j=1}^{n} G_j J_j\right]}} \tag{4.9}$$

Equations 4.8 and 4.9, which are applicable to composite shafts, now replace Equations (3.9-T) and (3.10-T) for the homogeneous cross section. If we consider a homogeneous material, then $G_1 = G_2 = \cdots = G_i \cdots = G_n = G$. Substituting this in Equations 4.8 and 4.9, we obtain Equations (3.9-T) and (3.10-T) for homogeneous materials, as expected.

The internal torque T in Equations (4.8) and (4.9), represents the statically equivalent torque *over the entire cross section,* as it did for the homogeneous cross section. The analytic techniques for finding the internal torque T at a cross section remain the same as before.

The torsional shear *strain* still varies linearly across the cross section. The shear stress in each material, $(\tau_{x\theta})_i$, given by Equation (4.7), is a piecewise linear function that changes its slope with each material, as demonstrated in Example 4.5.

EXAMPLE 4.5

A homogeneous cross section made of brass and a composite cross section of brass and steel are shown in Figure 4.10. The shear modulus of elasticity for brass is $G_{\text{Cu/Zn}} = 40$ GPa, and for steel it is $G_{\text{st}} = 80$ GPa. The shear strain in polar coordinates at the cross section is $\gamma_{x\theta} = 0.08\rho$, where ρ is in meters. (a) Write expressions for $\tau_{x\theta}$ as a function of ρ and plot the shear strain and shear stress distribution across both cross sections. (b) For each cross section, determine the statically equivalent internal torques.

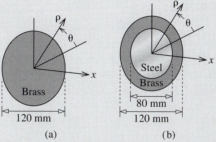

(a)   (b)

Figure 4.10 (a) Homogeneous and (b) composite cross sections for Example 4.5.

PLAN

(a) We can use Hooke's law to find the shear stress distribution as a function of ρ in each material. (b) Each shear stress distribution can be substituted into Equation (3.4-T) and the equivalent internal torque obtained by integration.

SOLUTION

(a) From Hooke's law, we can write Equations (E1) and (E2).

$$(\tau_{x\theta})_{\text{Cu/Zn}} = (40)(10^9)(0.08\rho) = 3200\rho \text{ MPa} \qquad \text{(E1)}$$

$$(\tau_{x\theta})_{\text{st}} = (80)(10^9)(0.08\rho) = 6400\rho \text{ MPa} \qquad \text{(E2)}$$

For the homogeneous cross section, the stress distribution is as given in Equation (E1), but for the composite section it switches from Equation (E2) to (E1) depending upon the value of ρ. We can write the shear stress distribution for both cross sections as a function of ρ as follows:

Homogeneous cross section

$$\text{ANS.} \quad \tau_{x\theta} = 3200\rho \text{ MPa} \qquad (0.00 \leq \rho < 0.06) \qquad \text{(E3)}$$

Composite cross section

$$\text{ANS.} \quad \tau_{x\theta} = \begin{cases} 6400\rho \text{ MPa} & 0.00 \leq \rho < 0.04 \\ 3200\rho \text{ MPa} & 0.04 < \rho \leq 0.06 \end{cases} \qquad \text{(E4)}$$

The shear strain and shear stress can now be plotted as a function of ρ as shown in Figure 4.11.

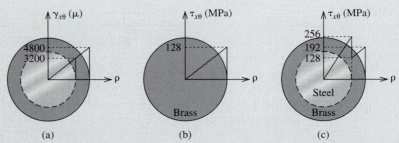

Figure 4.11 Properties as functions of ρ for Example 4.5. (a) Shear strain distribution. (b) Shear stress distribution in a homogeneous cross section. (c) Shear stress distribution in a composite cross section.

(b) The differential area dA is the area of a ring of radius ρ and thickness $d\rho$ [i.e., $dA = (2\pi\rho)d\rho$]. Equation (3.4-T) can be written as Equation (E5).

$$T = \int_0^{0.06} \rho\tau_{x\theta}(2\pi\rho)d\rho \qquad \text{(E5)}$$

Homogeneous cross section: Substituting Equation (E3) into Equation (E5) and integrating, we obtain the equivalent internal torque as shown in Equation (E6).

$$T = \int_0^{0.06} \rho(3200\rho)(10^6)(2\pi\rho)d\rho = (12{,}800\pi)(10^6)\frac{\rho^4}{4}\bigg|_0^{0.06} \qquad \text{(E6)}$$

ANS. $T = 130.3 \text{ kN}\cdot\text{m}$

Composite cross section: The integral in Equation (E5) can be written as the sum of two integrals:

$$T = \int_0^{0.06} \rho\tau_{x\theta}(2\pi\rho)d\rho = \underbrace{\int_0^{0.04} \rho\tau_{x\theta}(2\pi\rho)d\rho}_{T_{st}} + \underbrace{\int_{0.04}^{0.06} \rho\tau_{x\theta}(2\pi\rho)d\rho}_{T_{Cu/Zn}} \qquad \text{(E7)}$$

The internal torque in each material can be found by substituting Equation (E4) into Equations (E8) and (E9).

$$T_{st} = \int_0^{0.04} (\rho)(6400\rho)(10^6)(2\pi\rho)d\rho = (12{,}800\pi)(10^6)\frac{\rho^4}{4}\bigg|_0^{0.04} = 25.7 \text{ kN}\cdot\text{m} \qquad \text{(E8)}$$

$$T_{Cu/Zn} = \int_{0.04}^{0.06} (\rho)(3200\rho)(10^6)(2\pi\rho)d\rho = (6400\pi)(10^6)\frac{\rho^4}{4}\bigg|_{0.04}^{0.06} = 52.3 \text{ kN}\cdot\text{m} \qquad \text{(E9)}$$

Substituting Equations (E8) and (E9) into Equation (E7), we obtain the total internal torque on the composite section as shown in Equation (E10).

$$T = T_{st} + T_{Cu/Zn} = 25.7 + 52.3 \qquad \text{(E10)}$$

ANS. $T = 78 \text{ kN} \cdot \text{m}$

COMMENTS

1. The example demonstrates that the shear strain varies linearly across the cross section and that shear stress is piecewise linear.

2. The material models dictate the shear stress distribution across the cross section, but once the stress distribution is known, Equation (3.4-T) can be used for finding the equivalent internal torque. We emphasize, however, that Equation (3.4-T) does not depend upon the material model.

3. An alternative calculation for the internal torque for composites is as follows. We can find the torsional rigidity of the shaft as

$$\sum G_j J_j = (80)(10^9)\left(\frac{\pi}{2}\right)(0.04)^4 + (40)(10^9)\left(\frac{\pi}{2}\right)[(0.06)^4 - (0.04)^4]$$

$$= 975.15 \text{ kN} \cdot \text{m}^2$$

Comparing the torsional shear strain in Equation (3.2-T) to the shear strain given, we obtain:

$$\frac{d\phi}{dx} = 0.08 \text{ rad/m}$$

From Equation (4.7) we obtain the internal torque:

$$T = \frac{d\phi}{dx}\sum G_j J_j = (0.08)(975.15) = 78 \text{ kN} \cdot \text{m} \qquad \text{(E11)}$$

This result is same as that in Equation (E10).

EXAMPLE 4.6

A solid steel ($G_{st} = 80$ GPa) shaft 3 m long is securely fastened to a hollow bronze ($G_{Cu/Sn} = 40$ GPa) shaft that is 2 m long as shown in Figure 4.12. Determine the maximum value of shear stress in the shaft and the rotation of the right-hand end with respect to the wall.

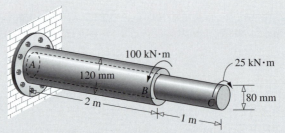

Figure 4.12 Composite shaft for Example 4.6.

PLAN

Section *BC* is homogeneous, and Equations (3.10-T) and (3.9-T) can be used in this section to find the relative rotation of section *C* with respect to *B* and the maximum shear stress in the section. Section *AB* is made from two materials, and we can use Equation (4.9) to find the relative rotation of section *B* with respect to *A* and Equation (4.8), to find the maximum shear stress in each material in section *AB*.

SOLUTION

We can find the polar moments and torsional rigidities as shown in Equations (E1) through (E5).

$$J_{st} = \frac{\pi}{32}(0.08)^4 = 4.02(10^{-6}) \text{ m}^4 \tag{E1}$$

$$J_{Cu/Sn} = \frac{\pi}{32}(0.12^4 - 0.08^4) = 16.33(10^{-6}) \text{ m}^4 \tag{E2}$$

$$G_{st}J_{st} = 321.6(10^3) \text{ N} \cdot \text{m}^2 \tag{E3}$$

$$G_{Cu/Sn}J_{Cu/Sn} = 653.2(10^3) \text{ N} \cdot \text{m}^2 \tag{E4}$$

$$\sum G_j J_j = G_{st}J_{st} + G_{Cu/Sn}J_{Cu/Sn} = 974.8(10^3) \text{ N} \cdot \text{m}^2 \tag{E5}$$

We make an imaginary cut in sections *BC* and *AB* and draw the free body diagrams as shown in Figure 4.13. By the equilibrium of moments, we obtain Equations (E6) and (E7).

$$T_{BC} = -25 \text{ kN} \cdot \text{m} \tag{E6}$$

$$T_{AB} - 100 + 25 = 0 \quad \text{or} \quad T_{AB} = 75 \text{ kN} \cdot \text{m} \tag{E7}$$

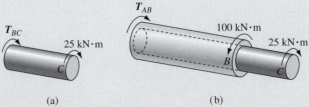

(a)   (b)

Figure 4.13 Free body diagrams for the calculation of internal torque in (a) section *BC* and (b) section *AB*.

We can use Equations (3.10-T) and (3.9-T) to find the relative rotation of the section at C with respect to B and the maximum shear stress in BC as shown in Equations (E8) and (E9).

$$\phi_C - \phi_B = \frac{T_{BC}(x_C - x_B)}{G_{st}J_{st}} = \frac{(-25)(10^3)(1)}{321.6(10^3)} = -0.0781 \text{ rad} \qquad (E8)$$

$$(\tau_{BC})_{max} = \frac{T_{BC}(\rho_{BC})_{max}}{J_{st}} = \frac{(-25)(10^3)(0.04)}{4.02(10^{-6})} = -250(10^6) \qquad (E9)$$

By using Equation (4.9), we can find the relative rotation of section B with respect to the section at A as shown in Equation (E10).

$$\phi_B - \phi_A = \frac{T_{AB}(x_B - x_A)}{\sum G_j J_j} = \frac{(75)(10^3)(2)}{974.8(10^3)} = 0.1538 \text{ rad} \qquad (E10)$$

Adding Equations (E8) and (E10), we obtain the relative rotation of the section at C with respect to the section at A. Noting that the rotation at A is zero, we obtain the rotation of the section at C as shown in Equation (E11).

$$\phi_C = 0.1538 - 0.0781 = 0.0757 \text{ rad} \qquad \text{or} \qquad (E11)$$

ANS. $\phi_C = 0.076 \text{ rad ccw}$

By using Equation (4.8), we can find the shear stress at any point on the imaginary cross section in AB as shown in Equation (E12).

$$\tau_i = \frac{(G_i \rho T_{AB})}{\sum G_j J_j} = \frac{G_i \rho (75)(10^3)}{974.8(10^3)} = (76.9)(10^{-3})G_i \rho \qquad (E12)$$

In AB, the shear stress in steel (τ_{st}) will be maximum at $\rho = 0.04$, and the shear stress in bronze ($\tau_{Cu/Sn}$) will be maximum at $\rho = 0.06$. Substituting the values of G and ρ in Equation (E13), we can find the maximum shear stress in each material in section AB as shown in Equations (E13) and (E14).

$$(\tau_{Cu/Sn})_{max} = (76.9)(10^{-3})(40)(10^9)(0.06) = 184.6(10^6) \qquad (E13)$$

$$(\tau_{st})_{max} = (76.9)(10^{-3})(80)(10^9)(0.04) = 246.2(10^6) \qquad (E14)$$

Comparing Equations (E9), (E13), and (E14), we see that the maximum shear stress occurs in section BC, and its magnitude is

ANS. $\tau_{max} = 250 \text{ MPa}$

COMMENTS

1. This example demonstrates that a similar analytical approach can be used for a composite shaft and a homogeneous shaft, provided one takes care to select the appropriate equations for the homogeneous and composite cross sections.

2. An alternative approach for calculation is to view the composite shaft as two homogeneous shafts; each shaft has an internal torque, but the two are constrained to rotate by the same amount, as depicted in Figure 4.14.

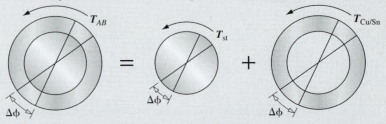

Figure 4.14 Composite shaft seen as two homogeneous shafts in Example 4.6.

We can write one equation that represents the internal torque as the sum of internal torque in steel and the internal torque in bronze. The second equation we generate by imposing the constraint that equal lengths of the shafts rotate by equal amount, and we use Equation (3.10-T) to write the equation. Substituting the torsional rigidities, we obtain a second equation for the unknown internal torques.

$$T_{AB} = T_{st} + T_{Cu/Sn} = 75 \text{ kN} \cdot \text{m} \tag{E15}$$

$$\Delta\phi = \frac{T_{st}\Delta x}{G_{st}J_{st}} = \frac{T_{Cu/Sn}\Delta x}{G_{Cu/Sn}J_{Cu/Sn}} \tag{E16}$$

$$T_{st} = 2.03 T_{Cu/Sn} \tag{E17}$$

Solving Equations (E15) and (E17) for the internal torques, we obtain $T_{st} = 24.75$ kN · m and $T_{Cu/Sn} = 50.25$ kN · m. Substituting these values in Equation (3.9-T), we obtain the values given in Equations (E13) and (E14). Similarly, by substituting $\Delta x = 2$ m and the internal torque values in Equation (E16), we can obtain the relative rotation in Equation (E10).

3. The approach outlined in comment 2 is more intuitive than the solution procedure used, but it is also more tedious. As the number of materials in a cross section grows, the approach shown in comment 2 becomes less attractive. Laminated composite shafts have many layers of different material that can be easily accounted for in the use of Equations (4.9) and (4.13).

EXAMPLE 4.7

A composite shaft with a titanium outer layer and an aluminum layer on the inside is to be designed for transmitting a torque of 1000 N · m. As shown in Figure 4.15, the shaft must be 2 m long and must have an outer diameter of 25 mm to fit existing attachments. The relative rotation of the two ends of the shaft is limited to 1.5 rad. The shear modulus of rigidity G, the allowable shear stress τ_{allow}, and the density γ are given in Table 4.2. Determine the amount, in grams, of each material to use if the lightest shaft is to be designed.

Figure 4.1 Composite shaft for Example 4.7.

TABLE 4.2 Material Properties for Example 4.7

Material	G (GPa)	τ_{allow} (MPa)	γ (Mg/m³)
Titanium alloy	36	450	4.4
Aluminum	28	180	2.8

PLAN

As the specific weight of aluminum is less than that of titanium, the more aluminum we use, the lighter will be the shaft. Thus, the problem is to determine the maximum diameter of the aluminum shaft without exceeding the specified limits. We can use Equation (4.13) to determine the maximum torsional shear stress in each shaft and Equation (4.9) to find the relative rotation of the two ends and determine the diameter of the aluminum shaft that satisfies all the limits.

SOLUTION

Let the diameter of the aluminum shaft be d mm. The polar moment for inertia for each material cross section is as shown in Equation (E1).

$$J_{\text{Ti}} = \frac{\pi}{32}(25^4 - d^4)(10^{-12}) \text{ m}^4 \quad \text{and} \quad J_{\text{Al}} = \frac{\pi}{32}d^4(10^{-12}) \text{ m}^4 \qquad \text{(E1)}$$

We can find the torsional rigidity of the shaft as shown in Equations (E2) and (E3).

$$\sum G_j J_j = (36)(10^9)\left(\frac{\pi}{32}\right)(25^4 - d^4)(10^{-12}) + (28)(10^9)\left(\frac{\pi}{32}\right)d^4(10^{-12}) \quad \text{or} \qquad \text{(E2)}$$

$$\sum G_j J_j = 1380.6 - 0.785d^4(10^{-3}) \qquad \text{(E3)}$$

The relative rotation of the two ends should be less than 1.5 rad over a length $x_2 - x_1 = 2$ m. From Equation (4.9), we obtain the limitation on d shown in Equation (E4).

$$\phi_2 - \phi_1 = \frac{T(x_B - x_A)}{\sum G_j J_j} = \frac{(1000)(2)}{1380.6 - 0.785d^4(10^{-3})} \le 1.5 \quad \text{or}$$

$$1380.6 - 0.785d^4(10^{-3}) \ge (2000/1.5) \quad \text{or}$$

$$0.785d^4(10^{-3}) \le (1380.6 - 2000/1.5) \quad \text{or} \quad d \le 15.7 \text{ mm} \qquad \text{(E4)}$$

The maximum torsional shear stress in titanium will occur at $\rho = 12.5(10^{-3})$ m. From Equation (4.13), we obtain the limitation on d shown in Equation (E5).

$$\tau_{Ti} = \frac{G_{Ti}\rho T}{\sum G_j J_j} = \frac{36(10^9)(12.5)(10^{-3})(1000)}{1380.6 - 0.785d^4(10^{-3})} \le 450(10^6) \quad \text{or}$$

$$1380.6 - 0.785d^4(10^{-3}) \ge \frac{450(10^3)}{450} \quad \text{or} \quad 0.785d^4(10^{-3}) \le 380.6 \quad \text{or}$$

$$d \le 26 \text{ mm} \tag{E5}$$

The maximum torsional shear stress in aluminum will occur at $\rho = (d/2)(10^{-3})$ m. From Equation (4.13) we obtain the limitation on d shown in Equation (E6).

$$\tau_{Al} = \frac{G_{Al}\rho T}{\sum G_j J_j} = \frac{28(10^9)(d/2)(10^{-3})(1000)}{1380.6 - 0.785d^4(10^{-3})} \le 180(10^6) \quad \text{or}$$

$$1380.6 - 0.785d^4(10^{-3}) \ge \frac{14d(10^3)}{180} \quad \text{or} \quad 0.785d^4(10^{-3}) + 77.8d \le 1380.6 \tag{E6}$$

To determine the limitation on d from Equation (E6), we must use a numerical method to find the roots of the equation. Before trying that, we check whether d obtained from Equations (E4) and (E5) would satisfy Equation (E6). Since $d = 15.7$ satisfies both Equations (E4) and (E5), we substitute this value in Equation (E6) and obtain $1269.1 \le 1380.6$, which implies that limitation on d from Equation (E6) is met by $d = 15.7$ mm. To ensure that we meet the limitation of Equation (E4), we round the value of d down to $d = 15$ mm. We multiply the density of each material by the volume to obtain the amount of titanium and aluminum to use in design of the shaft, as shown in Equations (E7) and (E8).

$$m_{Ti} = \left[\frac{\pi}{4}(0.025^2 - 0.015^2)\right](2)(4.4)(10^6) = 2764.6\text{g} \tag{E7}$$

ANS. $m_{Ti} = 2765\text{g}$

$$m_{Al} = \frac{\pi}{4}(0.015^2)(2)(2.8)(10^6) = 989.6\text{g} \tag{E8}$$

ANS. $m_{Al} = 990\text{g}$

COMMENT

If we constructed a hollow shaft made from titanium only, the inner diameter to the nearest millimeter would be 8 mm, to satisfy the conditions on the relative rotation and the maximum shear stress in titanium. The amount of titanium we would then need would be 3877 g. Because titanium is significantly more expensive than aluminum, this would result in a higher material cost. The total mass of the composite shaft is 3755 g, which is less than an all-titanium shaft mass by 122 g. Thus, if the shaft is used in an aircraft, the all-titanium shaft would result in higher fuel costs.

PROBLEM SET 4.2

4.10 The composite shaft shown in Figure P4.10 is constructed from aluminum (G_{Al} = 4000 ksi), bronze ($G_{Cu/Sn}$ = 6000 ksi), and steel (G_{st} = 12,000 ksi). (a) Determine the rotation of the free end with respect to the wall. (b) Plot the shear strain and shear stress across the cross section.

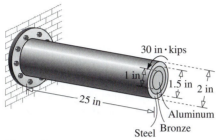

Figure P4.10

ANS. ϕ = 0.048 rad ccw

4.11 A composite shaft is constructed from aluminum (G_{Al} = 4000 ksi) and steel (G_{st} = 12,000 ksi) as shown in Figure P4.11. The allowable shear stress is 18 ksi in aluminum and 24 ksi in steel. (a) Determine the maximum torque T that can be applied. (b) Determine the rotation of the free end with respect to the wall when the maximum torque is applied.

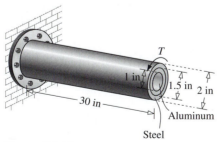

Figure P4.11

ANS. T_{max} = 28.9 in · kips
ϕ_B = 0.06 rad ccw

4.12 A solid steel (G = 80 GPa) shaft 3 m long is securely fastened to a hollow bronze (G = 40 GPa) shaft that is 2 m long, as shown in Figure P4.12. Determine (a) the magnitude of maximum shear stress in the shaft and (b) the rotation of the section at 1 m from the left wall.

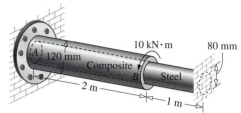

Figure P4.12

4.13 A shaft is to be designed to transmit a torque of 1000 N · m without twisting more than 0.5 rad over a length of 1 m. The mass of the shaft cannot exceed 1500 g, and the outside diameter must be 30 mm to fit existing attachments. You can design the shaft to be all aluminum, all titanium, or a composite of the two materials. The shear modulus of rigidity G, the allowable shear stress τ_{allow}, and the density γ of titanium and aluminum are given in Table P4.13. Determine to the nearest millimeter the diameters of the lightest shaft and the corresponding mass.

TABLE P4.13 Material Properties of Titanium Alloy and Aluminum

Material	G (GPa)	τ_{allow} (MPa)	γ (Mg/m³)
Titanium alloy	36	300	4.4
Aluminum	28	180	2.8

4.14 A 1 m long shaft is to be designed to transmit a torque of 3300 N · m. The outside diameter of the shaft must be 40 mm to fit existing attachments. The shaft can be all aluminum, all titanium, or a composite of the two materials. The shear modulus of rigidity G, the allowable shear stress τ_{allow}, and the density γ of titanium and aluminum are given in Table P4.13. Determine to the nearest millimeter the diameters of the *lightest* shaft and W, the corresponding mass, in grams.

ANS.   Composite Shaft, d_1 = 20 mm, d_2 = 30 mm W = 3519 g

4.15 A composite circular shaft is made from two nonlinear materials as shown in Figure P4.15. The stress–strain relationship for material 1 is given by $\tau_1 = G\gamma^{0.5}$ and for material 2 it is given by $\tau_2 = 2G\gamma^{0.5}$. Determine the rotation of the section at B in terms of T_{ext}, L, G, and R.

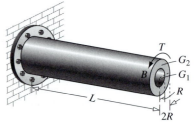

Figure P4.15

4.3   COMPOSITE SYMMETRIC BEAMS

Figure 4.16 shows a laminated composite cross section. We assume that all material cross sections are symmetric about the y axis and that the loading in still in the plane of symmetry; that is, the assumptions of symmetric bending about the z axis are applicable. We assume that all materials are securely bonded to each other: that is, there is no relative movement at any interface of the material. This assumption of secure bonding is necessary to preserve our kinematic assumptions. We further assume that we still have small strains, that all materials are linear-elastic-isotropic, and that there are no inelastic strains. In other words, all assumptions prior to Assumption 8 in Table 3.3 are still valid. Thus, Equation (3.6-B), which precedes Assumption 8, is still valid and forms our starting point for deriving the formulas for composite beams. Equation (3.5-B) will yield the location of the origin (neutral axis), while Equation (3.6-B) can be developed to give the bending formulas.

4.3.1   Normal Bending Stress in Composite Beams

Each material in the laminated cross section shown in Figure 4.16 has a modulus of elasticity E_i that is constant over the material cross-sectional area A_i. Suppose there are n materials in the cross section. The integral in Equation (3.6-B) can be written as the sum of the integrals over each material, as shown in Equation (4.10a).

$$M_z = \frac{d^2 v}{dx^2}\left[\int_{A_1} E_1 y^2 dA + \int_{A_2} E_2 y^2 dA + \cdots + \cdots + \int_{A_n} E_n y^2 dA \right] \qquad (4.10a)$$

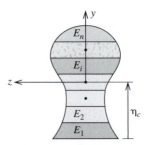

Figure 4.16 Cross section of a composite symmetric beam.

Here E_i is a constant in each integral and can be taken outside the integral. The remaining integral is the second area moment of inertia $(I_{zz})_i$ of the ith material. We thus obtain Equation (4.10b).

$$\boldsymbol{M}_z = \frac{d^2v}{dx^2}[E_1(I_{zz})_1 + E_2(I_{zz})_2 + \cdots + \cdots + E_n(I_{zz})_n]\qquad(4.10b)$$

Or, we can write the relation more compactly as Equation (4.11).

$$\boldsymbol{M}_z = \frac{d^2v}{dx^2}\left[\sum_{j=1}^{n} E_j(I_{zz})_j\right]\qquad(4.11)$$

Equation (4.11) shows that the bending rigidity of the composite beam is the sum of all the bending rigidities of the materials in the composite. We can write Equation (3.3-B) for the ith material as Equation (4.12),

$$(\sigma_{xx})_i = -E_iy\frac{d^2v}{dx^2}(x)\qquad(4.12)$$

where $(\sigma_{xx})_i$ is the bending normal stress in the ith material. Substituting Equation (4.11) into Equation (4.12), we obtain Equation (4.13).

$$(\sigma_{xx})_i = -\left[\frac{E_iy\boldsymbol{M}_z}{\sum_{j=1}^{n} E_j(I_{zz})_j}\right]\qquad(4.13)$$

Equations 4.11 and 4.13, which are applicable to composite beams, now replace Equations (3.8-B) and (3.9-B) for the homogeneous cross section. If we consider a homogeneous material, then $E_1 = E_2 = \cdots = E_i \cdots = E_n = E$. Substituting this in Equations (4.11) and (4.13), we obtain Equations (3.8-B) and (3.9-B) for homogeneous material, as expected.

The internal bending moment $\boldsymbol{M}_z$ in Equations (4.11) and (4.13) represents the statically equivalent moment *over the entire cross section,* as it did for the homogeneous cross section. The analysis techniques for finding the internal torque $\boldsymbol{M}_z$ at a cross section remain the same as before.

We know that the bending normal *strain* varies linearly across the composite cross section. But the normal stress in each material, $(\sigma_{xx})_i$ given by Equation (4.13), is a piecewise linear function that changes its slope with each material, as demonstrated shortly in Example 4.8.

4.3.2 Location of the Neutral Axis (origin) in Composite Beams

In our discussion of the location of the origin for homogeneous beams, we chose the origin such that the total axial force on a cross section was zero, which resulted in decoupling the axial problem from the bending problem. The location of the origin for a nonhomogeneous section is determined from Equation (3.5-B), which is identical to

Equation (3.5-A). Hence, the result in Equation (4.5), which gives the location of the origin for multiple materials, is valid here also and is rewritten as Equation (4.14) for convenience,

$$
\eta_c = \frac{\sum_{j=1}^{n} \eta_j E_j A_j}{\sum_{j=1}^{n} E_j A_j}
\tag{4.14}
$$

where η_j is the location of the centroid of the jth material as measured from a common datum line.

The bending normal stress varies linearly with y in each material, and the maximum bending normal stress in each material will be where the y coordinate is the greatest. With y measured from the neutral axis, we obtain the following observation.

- The maximum bending normal stress in each material is at the point on the material that is furthest from the neutral axis of a composite beam.

4.3.3 Bending Shear Stress in Composite Beams

Equation (3.15-B), relating shear and normal stresses in the beam, was derived from the equilibrium of a beam element and is independent of the material model, provided the kinematic assumptions remain valid, as is the case for our composite beams. Substituting Equation (4.13) into Equation (3.15-B) and noting that the moment and bending rigidity do not change across the cross section, we obtain Equation (4.15),

$$
\tau_{sx} t = -\frac{d}{dx} \int_{A_s} \frac{-E y M_z}{\sum_{j=1}^{n} E_j (I_{zz})_j} dA = \frac{d}{dx} \left[\frac{M_z}{\sum_{j=1}^{n} E_j (I_{zz})_j} \int_{A_s} E y \, dA \right]
$$
$$
= \frac{d}{dx} \left[\frac{M_z Q_{\text{comp}}}{\sum_{j=1}^{n} E_j (I_{zz})_j} \right]
\tag{4.15}
$$

where

$$
Q_{\text{comp}} = \int_{A_s} E y \, dA
\tag{4.16}
$$

Assuming that the beam is not tapered and that the modulus of elasticity does not change with x, we have the implication that Q_{comp} and the bending rigidity in Equation (4.15) do not change with x and can be taken outside the derivative sign. Then, after noting that $V_y = -(dM_z / dx)$, we obtain Equation (4.17).

$$
\tau_{sx} t = \left[\frac{Q_{\text{comp}}}{\sum_{j=1}^{n} E_j (I_{zz})_j} \right] \frac{dM_z}{dx} = -\left[\frac{Q_{\text{comp}} V_y}{\sum_{j=1}^{n} E_j (I_{zz})_j} \right]
$$

$$\boxed{\tau_{sx} = \tau_{xs} = -\frac{Q_{comp}V_y}{\left[\sum_{j=1}^{n} E_j(I_{zz})_j\right]t}} \qquad (4.17)$$

Thus Q_{comp} is the new quantity that we need to evaluate in Equation (4.17). Let n_s represent the number of materials in A_s; that is, n_s = the number of materials between the free surface and the point at which we are evaluating the shear stress. The integral in Equation (4.16) can be written as the sum of integrals over each material as Equation (4.18a).

$$Q_{comp} = \int_{A_1} E_1 y \, dA + \int_{A_2} E_2 y \, dA + \cdots + \int_{A_{n_s}} E_{n_s} y \, dA \qquad (4.18a)$$

Here E_j is a constant in each integral and can be taken outside the integral. The remaining integral, designated $(Q_z)_j$, is the first area moment about of the jth material about the neutral axis. Thus, we obtain Equation (4.18b),

$$Q_{comp} = E_1(Q_z)_1 + E_2(Q_z)_2 + \cdots + E_{n_s}(Q_z)_{n_s} \qquad (4.18b)$$

or, written more compactly:

$$\boxed{Q_{comp} = \sum_{j=1}^{n_s} E_j(Q_z)_j} \qquad (4.19)$$

The geometric property $(Q_z)_j$ is independent of the material and can be found as in a homogeneous cross section. Then Equation (4.19) can be used to find Q_{comp}. If $n_s = n$ (i.e., if we consider the entire cross section), then from Equation (3.5-B) $Q_{comp} = 0$. Thus, once more, like Q_z for a homogeneous material, Q_{comp} starts and ends with zero as we move from one free surface and reach the other free surface and is maximum at the neutral axis. Thus in a material, as we move closer to the neutral axis, the value of Q_{comp} increases. We record the following observations.

- The maximum value of Q_{comp} is at the neutral axis.

- The maximum bending shear stress in each material is at the point on the material that is closest to the neutral axis of a composite beam.

4.3.4   Cross Section Transformation Method

For composite beam cross sections made from materials of rectangular cross section, an alternative approach is possible that often looks simpler, particularly if the number of materials involved is small. The process of solving by means of the cross section transformation method is described first and justified immediately afterward.

The dimensions of the cross section in the z direction (bending axis) are transformed by means of the equation $\tilde{z} = z(E_j/E_{ref})$, where E_{ref} is the modulus of elasticity of any material in the cross section that is used as the reference material and $\tilde{z}$ is the transformed dimension. Figure 4.17 shows a cross section made from three materials. If we

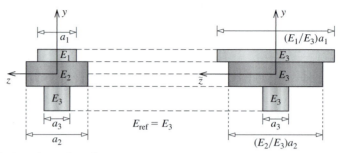

Figure 4.17 Transformation of a cross section. (a) Original composite cross section. (b) Transformed homogeneous cross section.

use material 3 as the reference material, the dimensions in the z direction are transformed by using the ratio of the modulus of elasticity of the material to the reference material. The transformed cross section is considered to be homogeneous with the modulus of elasticity of the reference material. The centroid, the second area moment of inertia $\tilde{I}_{zz}$, and the first area moment $\tilde{Q}_z$ are found for the transformed cross section as for any homogeneous material. The formulas in Equations (4.20a) and (4.20b) are then used for finding the bending normal and the shear stress.

$$(\sigma_{xx})_i = -\left(\frac{E_i}{E_{ref}}\right)\frac{M_z y}{\tilde{I}_{zz}} \tag{4.20a}$$

$$\tau_{sx} = \tau_{xs} = -\left(\frac{V_y \tilde{Q}_z}{\tilde{I}_{zz} t}\right) \tag{4.20b}$$

In these equations, $\tilde{I}_{zz}$ and $\tilde{Q}_z$ are, respectively, the second area moment of inertia and the first area moment about the axis through the centroid of the transformed homogeneous cross section, and t is the thickness perpendicular to the centerline of the original composite cross section. Note that the location of the centroid for the original composite cross section and the transformed homogeneous cross section will be the same, since the dimensions in the y direction are not being changed.

To justify the foregoing approach, we note that the differential area element can be written as $dA = dy\, dz$, which will now transform to $dA = (E_{ref}/E_j)dy\, d\tilde{z} = (E_{ref}/E_j)d\tilde{A}$, where $d\tilde{A}$ is the differential area element in the transformed cross section. Thus, the second moment of inertia transforms as shown in Equation (4.20c)

$$(I_{zz})_j = \int_{A_j} y^2 dA = (E_{ref}/E_j)\int_{A_j} y^2 dA \quad \text{or} \quad (I_{zz})_j = (E_{ref}/E_j)(\tilde{I}_{zz})_j \tag{4.20c}$$

and the total bending rigidity thus can be written as Equation (4.20d).

$$\sum_{j=1}^{n} E_j(I_{zz})_j = \sum_{j=1}^{n} E_j(E_{ref}/E_j)(\tilde{I}_{zz})_j = E_{ref}\sum_{j=1}^{n}(\tilde{I}_{zz})_j = E_{ref}\tilde{I}_{zz} \tag{4.20d}$$

Substituting the Equation (4.20d) into Equation (4.13), we obtain Equation (4.20a). Similarly the first moment of the area will transform as shown in Equation (4.20e).

$$(Q_z)_j = \int_{A_j} y \, dA = (E_{\mathrm{ref}}/E_j) \int_{\tilde{A}_j} y \, d\tilde{A} = (E_{\mathrm{ref}}/E_j)(\tilde{Q}_z)_j \qquad (4.20\mathrm{e})$$

Substituting Equation (4.20e) into Equation (4.19), we obtain Equation (4.20f).

$$Q_{\mathrm{comp}} = \sum_{j=1}^{n_s} E_j(E_{\mathrm{ref}}/E_j)(\tilde{Q}_z)_j = E_{\mathrm{ref}} \sum_{j=1}^{n_s} (\tilde{Q}_z)_j = E_{\mathrm{ref}}\tilde{Q}_z \qquad (4.20\mathrm{f})$$

Substituting Equations (4.20d) and (4.20f) into Equation (4.17), we obtain Equation (4.20b). The dimension t in Equation (4.20b) is not transformed because if the point is in the flange, then the direction perpendicular to the centerline is the y direction, which is not being transformed. Had we restricted ourselves to points in the web where the direction perpendicular to the centerline is the z direction and transformed the t dimension, the shear stress formula for the ith material would have been $(\tau_{xs})_i = -(E_i/E_{\mathrm{ref}})(V_y\tilde{Q}_z)/(\tilde{I}_{zz}\tilde{t})$. This formula reflects a similar modification in as much as the normal stress formula and the dimension $\tilde{t}$ is the dimension perpendicular to the centerline in the transformed homogeneous cross section, provided the point is in the web. We will use Equation (4.20b), which is valid at all points.

The advantage of the method described in this section is that the analytical process is the same as that for homogeneous cross sections. The method will become tedious, however, if the number of materials in the cross section is large or if the material cross sections are curvilinear instead of rectangular. Later, in Example 4.9, we shall demonstrate this method along with direct use of Equations (4.13) and (4.17).

EXAMPLE 4.8

A homogeneous wooden cross section and a cross section in which the wood is reinforced with steel are shown in Figure 4.18. The normal strain for both cross sections was found to vary as $\varepsilon_{xx} = -200y\,\mu$. The modulus of elasticity for steel is $E_{\mathrm{st}} = 30{,}000$ ksi, and for wood it is $E_{\mathrm{wood}} = 8000$ ksi. (a) Write expressions for the normal stress σ_{xx} as a function of y and plot the σ_{xx} distribution for each of the cross sections shown in Figure 4.18. (b) Calculate the equivalent internal moment M_z for each cross section.

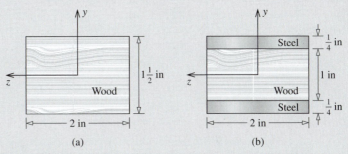

Figure 4.18 Cross sections for Example 4.8: (a) homogeneous and (b) laminated.

PLAN

(a) We can use Hooke's law to find the stress distribution from the given strain distribution. (b) The integral in Equation (3.4b-B) can be written as twice the integral for the top half because the stress distribution is symmetric about the center. The stress distribution can be substituted and the integration performed to obtain the equivalent internal moment.

SOLUTION

(a) From Hooke's law we can write Equations (E1) and (E2).

$$(\sigma_{xx})_{\text{wood}} = (8)(10^3)(-200y)(10^{-6}) = -1.6y \text{ ksi} \tag{E1}$$

$$(\sigma_{xx})_{\text{st}} = (30)(10^3)(-200y)(10^{-6}) = -6y \text{ ksi} \tag{E2}$$

For the homogeneous cross section, the stress distribution is as given in Equation (E1); for laminated, it switches from Equation (E1) to (E2) depending upon the value of y. We can write the stress distribution for both cross sections as a function of y as shown in Equations (E3) and (E4).

Homogeneous cross section

ANS. $\sigma_{xx} = -1.6y$ ksi $\quad -0.75 \le y < 0.75 \tag{E3}$

Laminated cross section

ANS. $\sigma_{xx} = \begin{cases} -6y \text{ ksi} & 0.5 < y \le 0.75 \\ -1.6y \text{ ksi} & -0.5 < y < 0.5 \\ -6y \text{ ksi} & -0.75 \le y < -0.5 \end{cases} \tag{E4}$

The strains and stresses can now be plotted as a function of y as shown in Figure 4.19.

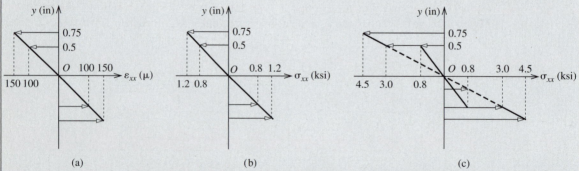

Figure 4.19 Strain and stress distribution for Example 4.8: (a) strain distribution, (b) stress distribution in a homogeneous cross section, and (c) stress distribution in a laminated cross section.

(b) Noting that $dA = 2\,dy$ and that the stress distribution is symmetric, we can write the integral in Equation (3.4b-B) as Equation (E5).

$$M_z = -\int_{-0.75}^{0.75} y\sigma_{xx}(2dy) = -2\left[\int_{0}^{0.75} y\sigma_{xx}(2dy)\right] \tag{E5}$$

Homogeneous Cross section: Substituting Equation (E3) into Equation (E5) and integrating, we obtain the equivalent internal moment as in Equation (E6).

$$M_z = -2\left[\int_{0}^{0.75} y(-1.6y \text{ ksi})(2dy)\right] = 6.4\frac{y^3}{3}\bigg|_{0}^{0.75} = 6.4\frac{(0.75)^3}{3} \tag{E6}$$

ANS. $M_z = 0.9 \text{ in} \cdot \text{kips}$

Laminated Cross section: Writing the integral in Equation (E5) as a sum of two integrals and substituting Equation (E4), we obtain the equivalent internal moment as shown in Equation (E7).

$$M_z = -2\left[\int_{0}^{0.5} y(-1.6y)(2dy) + \int_{0.5}^{0.75} y(-6y)(2dy)\right] \tag{E7}$$

$$= 4\left[1.6\frac{y^3}{3}\bigg|_{0}^{0.5} + 6\frac{y^3}{3}\bigg|_{0.5}^{0.75}\right]$$

ANS. $M_z = 2.64 \text{ in} \cdot \text{kips}$

COMMENTS

1. The example demonstrates that although the strain varies linearly across the cross section, the stress may not. In this example we considered material nonhomogeneity. In a similar manner, we can consider other models, such as elastic–perfectly plastic, or material models that have nonlinear stress–strain curves.

2. Figure 4.20 shows the stress distribution on the cross-sectional surface. The symmetry of stresses about the center results in a zero axial force.

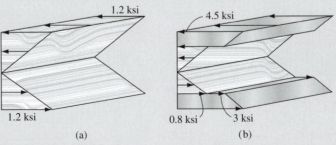

Figure 4.20 Stress distribution on the (a) homogeneous and (b) laminated cross sections for Example 4.8.

3. We can obtain the equivalent internal moment for a homogeneous cross section by replacing the triangular load with an equivalent load at the centroid of each triangle and then finding the equivalent moment as shown in Figure 4.21.

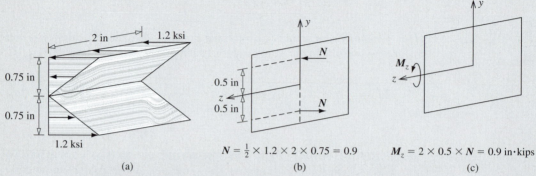

$$N = \tfrac{1}{2} \times 1.2 \times 2 \times 0.75 = 0.9 \qquad M_z = 2 \times 0.5 \times N = 0.9 \text{ in·kips}$$

    (a)   (b)   (c)

Figure 4.21 Statically equivalent internal moments for Example 4.8.

4. The approach outlined is very intuitive. However, for more complex cross-sectional shapes, this intuitive approach will become very tedious, and the generalization represented by Equation (3.4b-B) and the resulting formula can help simplify the calculations.

EXAMPLE 4.9

An aluminum ($E_{Al} = 10,000$ ksi) strip is securely fastened to a wooden ($E_{wood} = 1600$ ksi) beam as shown in Figure 4.22 to improve its load-carrying capacity. (a) On a section just to the right of support A, determine and plot the bending normal and shear stress across the cross section. (b) Determine the maximum bending normal and shear stress in aluminum and wood.

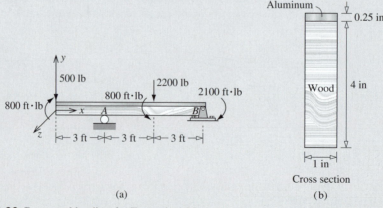

    (a)   (b)

Figure 4.22 Beam and loading for Example 4.9.

METHOD 1

PLAN

We can use the discontinuity functions to find the reactions at the supports and to write the moment equation. The shear force equation can be found by taking the derivative of the moment. The equations for shear force and bending moment can be plotted, and maximum values of shear force and moment obtained. We can use Equation (4.14) to determine the location of the neutral axis, to find I_{zz} for each material, and to calculate the bending rigidity for the cross section; we can find Q_{comp} as a function of y from Equation (4.19). (a) We can use Equations (4.13) and (4.17) to find and plot the bending normal stress and shear stress as a function of y. (b) The maximum bending normal stress in a material will exist at points that are furthest from the neutral axis in a material, while the maximum shear stress will exist on points in the material that are closest to the neutral axis.

SOLUTION

The reaction forces at A and B can be found from the free body diagram shown in Figure 4.23(a). The internal bending moment can be written as shown in Equation (E1) by using the discontinuity functions.

$$M_z = 800 - 500x + 1500\langle x - 3\rangle^1 - 800\langle x - 6\rangle^0 - 2200\langle x - 6\rangle^1$$
$$+ 2100\langle x - 9\rangle^0 - 1200\langle x - 9\rangle^1 \text{ ft}\cdot\text{lb}$$

(E1)

The shear force can be found by using Equation (3.11b-B) as shown in Equation (E2).

$$V_y = -\frac{dM_z}{dx} = 500 - 1500\langle x - 3\rangle^0 + 800\langle x - 6\rangle^{-1} + 2200\langle x - 6\rangle^0$$
$$- 2100\langle x - 9\rangle^{-1} + 1200\langle x - 9\rangle^0 \text{ lb}$$

(E2)

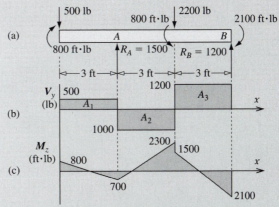

Figure 4.23 (a) Free body diagram and shear force (b) and bending moment (c) diagrams for Example 4.9.

Equations (E1) and (E2) can be plotted[1] as shown in Figure 4.23(b) and (c), and we can obtain the shear force and bending moment values shown in Equations (E3) and (E4).

$$V_{\text{max}} = 1200 \text{ lb} \qquad M_{\text{max}} = 2300 \text{ ft} \cdot \text{lb} \tag{E3}$$

$$V_A = -1000 \text{ lb} \qquad M_A = -700 \text{ ft} \cdot \text{lb} \tag{E4}$$

From Equation (4.14) and Figure 4.24, we can determine the location of the neutral axis as follows:

$$\eta_c = \frac{(1600)(4)(1)(2) + (10,000)(1)(0.25)(4.125)}{(1600)(4)(1) + (10,000)(1)(0.25)} = 2.597 \text{ in} \tag{E5}$$

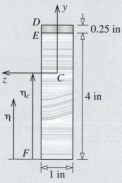

Figure 4.24 Location of the origin in Example 4.9.

We can use the parallel axis theorem to find the area moment of inertia about the z axis as shown in Equations (E6) and (E7).

$$(I_{zz})_{\text{Al}} = \frac{1}{12}(1)(0.25)^3 + (0.25)(2.597 - 4.125)^2 = 0.585 \text{ in}^4 \tag{E6}$$

$$(I_{zz})_{\text{wood}} = \frac{1}{12}(1)(4)^3 + (4)(2.597 - 2)^2 = 6.759 \text{ in}^4 \tag{E7}$$

We can find the total bending rigidity as shown in Equation (E8).

$$\sum E_j(I_{zz})_j = (1600)(6.759) + (10,000)(0.585) = 16.664(10^3) \text{ kips} \cdot \text{in}^2 \tag{E8}$$

We consider the material between the top (free) surface and any point y in aluminum as shown in Figure 4.25 and obtain the value of $(Q_{\text{comp}})_{\text{Al}}$ from Equation (4.19), as shown in Equation (E9).

$$(Q_{\text{comp}})_{\text{Al}} = E_{\text{Al}}(Q_z)_{\text{Al}} = (10,000)(1.653 - y)(1)\left(\frac{1.653 + y}{2}\right)$$

$$= 5000(1.653^2 - y^2) \tag{E9}$$

---

[1] Note that the moment and shear diagrams are obtained from Equations (E1) and (E2), and these may be different (flipped about the x axis) from what you saw in the introductory course, where you used a different methodology to obtain these diagrams. The values of shear force and moment, however, are independent of methodology.

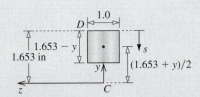

Figure 4.25 Calculation of Q_{comp} in aluminum.

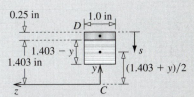

Figure 4.26 Calculation of Q_{comp} in wood.

Once more we consider the material between the top surface and any point y in wood as shown in Figure 4.26. There are two terms in the calculation of $(Q_{comp})_{wood}$: one over aluminum and the other over wood. The contribution from the aluminum can be found by substituting $y = 1.403$ in Equation (E9) as shown in Equation (E10).

$$(Q_{comp})_{wood} = [E_{Al}(Q_z)_{Al}]\big|_{y=1.403} + E_W(Q_z)_{wood}$$

$$(Q_{comp})_{wood} = 5000(1.653^2 - 1.403^2) + (1600)(1.403 - y)(1)\left(\frac{1.403 + y}{2}\right)$$

$$(Q_{comp})_{wood} = 3820 + 800(1.403^2 - y^2) \tag{E10}$$

(a) The bending normal stress as a function of y on a section just to the right of support A can be found by substituting Equations (E4) and (E8) into Equation (4.13) to obtain Equations (E11) and (E12).

$$(\sigma_{xx})_{Al} = -E_{Al}y\frac{M_A}{\sum E_j(I_{zz})_j} = -(10,000)(y)\frac{(-700)(12)}{16.664(10^3)} = 5041y \text{ psi} \tag{E11}$$

$$(\sigma_{xx})_{wood} = -E_{wood}y\frac{M_A}{\sum E_j(I_{zz})_j} = -(1600)(y)\frac{(-700)(12)}{16.664(10^3)} = 806.5y \text{ psi} \tag{E12}$$

We can write Equations (E11) and (E12) as follows:

$$\sigma_{xx} = \begin{cases} 5041y \text{ psi} & -2.597 \le y < 1.403 \\ 806.5y \text{ psi} & 1.403 < y \le 1.65 \end{cases} \tag{E13}$$

The bending shear stress τ_{xs} in each material, as a function of y on a section just to the right of support A, can be found by substituting Equations (E4), (E9), and (E10) into Equation (4.17). The directions y and s are opposite in Figures 4.25 and 4.26, hence $\tau_{xs} = -\tau_{xy}$ in each material and can be found as shown in Equations (E14) and (E15).

$$(\tau_{xs})_{Al} = -(\tau_{xy})_{Al} = -\frac{(Q_{comp})_{Al}V_A}{\left[\sum E_j(I_{zz})_j\right]t} = -\frac{5000(1.653^2 - y^2)(-1000)}{16.664(10^3)(1)}$$

$$(\tau_{xy})_{Al} = -300(1.653^2 - y^2) \tag{E14}$$

$$(\tau_{xy})_{\text{wood}} = -(\tau_{xy})_{\text{wood}} = -\frac{(Q_{\text{comp}})_{\text{wood}}V_A}{\left[\sum E_j(I_{zz})_j\right]t} = -\frac{[3820 + 800(1.403^2 - y^2)](-1000)}{16.664(10^3)(1)}$$

$$(\tau_{xy})_{\text{wood}} = -[229.2 + 48.01(1.403^2 - y^2)] \tag{E15}$$

We can write Equations (E14) and (E15) as Equation (E16).

$$\tau_{xy} = \begin{cases} -300\ (1.653^2 - y^2)\ \text{psi} & -2.597 \le y < 1.403 \\ -[229.2 + 48.01\ (1.403^2 - y^2)]\ \text{psi} & 1.403 < y \le 1.653 \end{cases} \tag{E16}$$

Equations (E13) and (E16) can be plotted as a function of y across the cross section as shown in Figure 4.27.

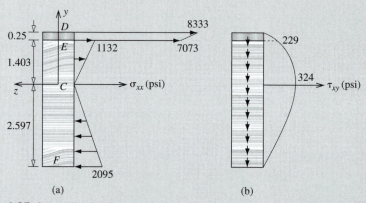

Figure 4.27 Stress distribution across the cross section for Example 4.9.

(b) The maximum bending normal stress will be at point D in aluminum and at point F in wood because in each material this point is furthest from the neutral axis. Substituting Equations (E3), (E8), and $y_D = 1.653$ and $y_F = -2.597$ into Equation (4.13), we obtain Equations (E17) and (E18).

$$(\sigma_{\text{max}})_{\text{Al}} = -E_{\text{Al}}y_D\frac{M_{\text{max}}}{\sum E_j(I_{zz})_j} = -(10,000)(1.653)\frac{(2300)(12)}{16.664(10^3)} = -27,378\ \text{psi} \tag{E17}$$

ANS. $(\sigma_{\text{max}})_{\text{Al}} = 27.4\ \text{ksi (C)}$

$$(\sigma_{\text{max}})_{\text{wood}} = -E_{\text{wood}}y_F\frac{M_{\text{max}}}{\sum E_j(I_{zz})_j} = -(1600)(-2.597)\frac{(2300)(12)}{16.664(10^3)} = 6882\ \text{psi} \tag{E18}$$

ANS. $(\sigma_{\text{max}})_{\text{wood}} = 6.9\ \text{ksi (T)}$

The maximum bending shear stress will be at point E in aluminum and at point C in wood because in each material this point is closest to the neutral axis. Substituting Equations (E3), (E9), and (E10) and $y_E = 1.403$ and $y_C = 0$ into Equation (4.17), we obtain Equations (E19) and (E20).

$$(\tau_{max})_{Al} = \left| \frac{(Q_{comp})_{Al}\big|_{y=1.403}\, V_{max}}{\left[\sum E_j(I_{zz})_j\right]t} \right| = \frac{5000(1.653^2 - 1.403^2)(1200)}{16.664(10^3)(1)} = 275 \text{ psi} \qquad \text{(E19)}$$

ANS. $(\tau_{max})_{Al} = 275$ psi

$$(\tau_{max})_{wood} = \left| \frac{(Q_{comp})_{wood}\big|_{y=0}\, V_{max}}{\left[\sum E_j(I_{zz})_j\right]t} \right| = -\frac{[3820 + 800(1.403^2 - 0^2)](1200)}{16.664(10^3)(1)} = 388 \text{ psi} \qquad \text{(E20)}$$

ANS. $(\tau_{max})_{wood} = 388$ psi

COMMENTS

1. The internal forces and moments were found from the shear and moment diagrams without regard to the homogeneity of the cross section. This once more emphasizes that the equilibrium equations that relate external forces to internal forces are independent of any material model.

2. The maximum bending shear stress in each material is an order of magnitude less than the maximum bending normal stress, which is consistent with the requirement for validity of our beam theory, as remarked in Section 3.2.3. But as the difference between the modulus of elasticity E drops by orders of magnitude (as in sandwich beams), then the normal and shear stresses for the material with the smaller E become nearly of the same order, and the theory is no longer valid.

3. Figure 4.27 shows that the bending normal stress σ_{xx} is discontinuous but the bending shear stress τ_{xy} is continuous at the aluminum–wood interface, even though τ_{xy} is represented by two different functions in Equation (E16). If examine Figure 4.26, we note that the calculation of Q_{comp} in wood is equal to the $(Q_{comp})_{Al}$ at the interface plus the additional value from wood. The additional value to Q_{comp} from the wood will be zero at the interface, implying that Q_{comp} is a continuous function at the interface, even though it is represented by different functions in each material. Since the rest of the terms in Equation (4.17) are the same for both materials, the continuity of the shear stress at the interface reflects the continuity of Q_{comp}.

METHOD 2

PLAN

With aluminum as the reference material, we note that the ratio $E_{wood}/E_{ref} = 0.16$. Thus, the dimension of the wood parallel to the z axis becomes $(1)(0.16) = 0.16$ in. We can now find the centroid and $\tilde{I}_{zz}$ for the transformed cross section. We can also find $\tilde{Q}_z$ as a function of y. (a) We can use Equations (4.20a) and (4.20b) to find and plot the bending normal stress and shear stress as a function of y. (b) The maximum bending normal stress in a material will exist at the points that are furthest from the neutral axis in a material, while the maximum shear stress will exist at the points that are closest to the neutral axis.

SOLUTION

Figure 4.28 shows the transformed cross section with aluminum as the reference material. The centroid of the transformed cross section can be found as shown in Equation (E21).

$$\eta_c = \frac{(4)(0.16)(2) + (1)(0.25)(4.125)}{(4)(0.16) + (1)(0.25)} = 2.597 \text{ in} \qquad (E21)$$

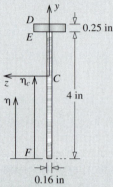

Figure 4.28 Transformed cross section for Example 4.9.

The area moment of inertia $\tilde{I}_{zz}$ for the transformed cross section can be found as shown in Equation (E22).

$$\tilde{I}_{zz} = \frac{1}{12}(1)(0.25)^3 + (0.25)(2.597 - 4.125)^2 + \frac{1}{12}(0.16)(4)^3 + (0.16)(4)(2.597 - 2)^2$$

$$\tilde{I}_{zz} = 1.6664 \text{ in}^4 \qquad (E22)$$

(a) Substituting Equations (E4) and (E22) into Equation (4.20a), we obtain the normal stress in aluminum and in wood as a function of y as shown in Equations (E23) and (E24).

$$(\sigma_{xx})_{Al} = -\left(\frac{E_{Al}}{E_{ref}}\right)\frac{M_A y}{\tilde{I}_{zz}} = -(1)(y)\frac{(-700)(12)}{1.6664} = 5041y \text{ psi} \qquad (E23)$$

$$(\sigma_{xx})_{wood} = -\left(\frac{E_{wood}}{E_{ref}}\right)\frac{M_A y}{\tilde{I}_{zz}} = -(0.16)(y)\frac{(-700)(12)}{1.6664} = 806.5y \text{ psi} \quad (E24)$$

Equations (E23) and (E24) are the same as Equations (E11) and (E12); hence the variation of σ_{xx} across the cross section is given by Equation (E12), and the plot is as given in Figure 4.27.

We can use Figure 4.25 to calculate $\tilde{Q}_z$ in aluminum because the dimensions are unchanged in the transformed cross section, as shown in Equation (E25).

$$\tilde{Q}_{Al} = (1.653 - y)(1)\left(\frac{1.653 + y}{2}\right) = \frac{(1.653^2 - y^2)}{2} \qquad (E25)$$

Figure 4.29 shows the calculation of $\tilde{Q}_z$ in wood. We consider the material between the top surface and any point y in wood. The contribution from the aluminum can be found by substituting $y = 1.403$ in Equation (E25) as shown in Equation (E26).

$$\tilde{Q}_{\text{wood}} = \frac{(1.653^2 - 1.403^2)}{2} + (1.403 - y)(0.16)\left(\frac{1.403 + y}{2}\right)$$

$$= 0.3820 + 0.08(1.403^2 - y^2) \qquad (E26)$$

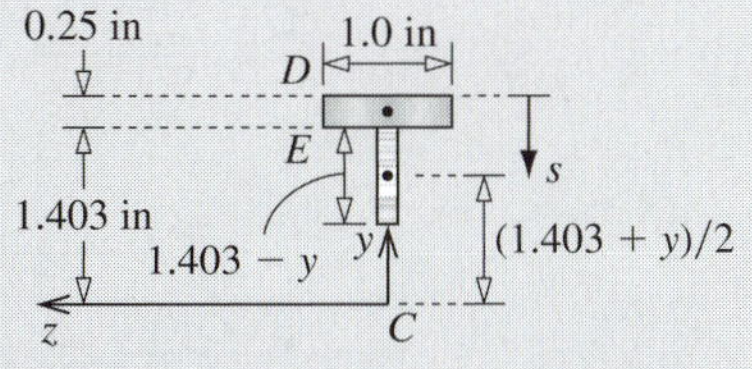

Figure 4.29 Calculation of Q_z in wood.

In each material, the bending shear stress τ_{xs} as a function of y on a section just to the right of support A can be found by substituting $t = 1$ and Equations (E3), (E25), and (E26) into Equation (4.20b). The directions y and s are opposite in Figures 4.25 and 4.29. Hence $\tau_{xs} = -\tau_{xy}$ in each material and can be found as shown in Equations (E27) and (E28).

$$(\tau_{xs})_{\text{Al}} = -(\tau_{xy})_{\text{Al}} = -\left(\frac{(V_y)_A \tilde{Q}_{\text{Al}}}{\tilde{I}_{zz} t}\right) = -\frac{(-1000)[(1.653^2 - y^2)/2]}{(1.6664)(1)}$$

$$(\tau_{xy})_{\text{Al}} = -300(1.653^2 - y^2) \qquad (E27)$$

$$(\tau_{xy})_{\text{wood}} = -(\tau_{xy})_{\text{wood}} = -\frac{(V_y)_A \tilde{Q}_{\text{wood}}}{\tilde{I}_{zz} t} - \frac{(-1000)[0.3820 + 0.08(1.403^2 - y^2)]}{(1.6664)(1)}$$

$$(\tau_{xy})_{\text{wood}} = -[229.2 + 48.01(1.403^2 - y^2)] \qquad (E28)$$

Equations (E27) and (E28) are the same as Equations (E14) and (E15); hence the plot of the shear stress is as given in Figure 4.27.

(b) The maximum bending normal stress will be at point D in aluminum and at point F in wood because in each material this point is furthest from the neutral axis. Substituting Equations (E2), (E18), and $y_D = 1.653$ and $y_F = -2.597$ into Equation (4.20a), we obtain the following:

$$(\sigma_{\text{max}})_{\text{Al}} = -\left(\frac{E_{\text{Al}}}{E_{\text{ref}}}\right)\frac{M_{\text{max}} y_D}{\tilde{I}_{zz}} = -(1)\frac{(2300)(12)(1.653)}{1.6664} = -27,378 \text{ psi}$$

ANS. $(\sigma_{\text{max}})_{\text{Al}} = 27.4 \text{ ksi (C)}$

$$(\sigma_{max})_{wood} = -\left(\frac{E_{wood}}{E_{ref}}\right)\frac{M_{max}y_F}{\tilde{I}_{zz}} = -(1.6)\frac{(2300)(12)(-2.597)}{1.6664} = 6882 \text{ psi}$$

ANS. $(\sigma_{max})_{wood} = 6.9 \text{ ksi (T)}$

The maximum bending shear stress will be at point E in aluminum and at point C in wood because this point in each material is closest to the neutral axis. Substituting Equations (E22), (E25), and (E26) and $y_E = 1.403$ and $y_C = 0$ into Equation (4.20b), we obtain Equations (E29) and (E30).

$$(\tau_{max})_{Al} = \left|\frac{(V_y)_{max}\tilde{Q}_{Al}}{\tilde{I}_{zz}t}\right| = \frac{(1200)[(1.653^2 - 1.403^2)/2]}{1.6664(1)} = 275 \text{ psi} \quad \text{(E29)}$$

ANS. $(\tau_{max})_{Al} = 275 \text{ psi}$

$$(\tau_{max})_{wood} = \left|\frac{(V_y)_{max}\tilde{Q}_{wood}}{\tilde{I}_{zz}t}\right| = \frac{(1200)[0.3820 + 0.08(1.403^2 - 0^2)]}{1.6664(1)} \quad \text{(E30)}$$

$$= 388 \text{ psi}$$

ANS. $(\tau_{max})_{wood} = 388 \text{ psi}$

COMMENTS

1. As in method 1, the calculation of internal forces and moments is independent of the nonhomogeneity of the material across the cross section.

2. The chief advantage of method 2 consists of the simplified equations. The calculation of $\tilde{I}_{zz}$ in Equation (E22) is simpler than computing the total bending rigidity $\sum E_j(I_{zz})_j$ in Equation (E8) of method 1. Similarly, finding $\tilde{Q}_{Al}$ and $\tilde{Q}_{wood}$ in Equations (E25) and (E26) is simpler than calculating $(Q_{comp})_{Al}$ and $(Q_{comp})_{wood}$ in Equations (E9) and (E10) in method 1. But as mentioned earlier, method 2 cannot be used for nonrectangular cross sections.

EXAMPLE 4.10

In reinforced concrete beams, the tension-carrying capacity of the concrete is ignored because the material's tensile strength is low, resulting in unpredictability of cracking on the tension side. Incorporate this information to solve the following problem.

The concrete beam shown in Figure 4.30 has been reinforced by embedding circular steel rods of diameter 20 mm. The maximum resisting (internal) moment in the beam was determined to be 40 kN · m. Determine the maximum bending normal stress in the concrete and steel. Use the following moduli of elasticity for steel and concrete: $E_{st} = 200$ GPa and $E_{conc} = 28$ GPa.

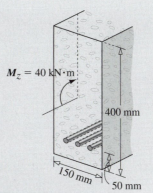

$M_z = 40$ kN·m

400 mm

150 mm

50 mm

Figure 4.30 Reinforced concrete beam cross section.

PLAN

We can use a modulus of elasticity of zero to simulate the zero tensile capacity of concrete. In other words, we view the cross section as if made of three materials: material 1 would be concrete from the top to the neutral axis with $E_1 = 28$ GPa; material 2 would be concrete below the neutral axis, with $E_2 = 0$; and material 3 would be steel, with $E_3 = 200$ GPa. Assuming that the neutral axis is η_c from the top, we can use Equation (4.14) to write and solve an equation in η_c. Once the location of neutral axis has been determined, the maximum bending normal stress in each material can be determined from Equation (4.13).

SOLUTION

We assume that the neutral axis is η_c from the top as shown in Figure 4.31. The area of concrete in compression is $A_1 = 0.15\eta_c$. The total area of steel is $A_3 = 3\pi(0.01)^2 = 0.942(10^{-3})$ m^2. Substituting these values and $E_1 = E_{conc} = 28$ GPa, $E_2 = 0$, and $E_3 = E_{st} = 200$ GPa in Equation (4.14) we obtain the following quadratic equation in η_c:

$$\eta_c = \frac{E_1 A_1 \eta_1 + E_2 A_2 \eta_3 + E_3 A_3 \eta_3}{E_1 A_1 + E_2 A_2 + E_3 A_3}$$

$$= \frac{(28)(0.15\eta_c)(\eta_c/2) + 0 + (200)[0.942(10^{-3})](0.35)}{(28)(0.15\eta_c) + 0 + (200)[0.942(10^{-3})]}$$

$$\eta_c^2 + 0.0897\eta_c - 0.0314 = 0 \qquad \text{(E1)}$$

Solving for the positive root in this quadratic equation, we obtain the value shown in Equation (E2).

$$\eta_c = 0.1379 \text{ m} \qquad \text{(E2)}$$

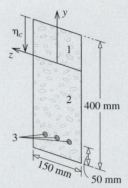

Figure 4.31 Neutral axis calculation for Example 4.10.

We can use the parallel axis theorem to find the area moment of inertia about the z axis as in Equations (E3) and (E4).

$$(I_{zz})_1 = \frac{1}{12}(0.15)(0.1379)^3 + (0.15)(0.1379)\left(\frac{0.1379}{2}\right)^2 = 131.1(10^{-6}) \text{ m}^4 \quad \text{(E3)}$$

$$(I_{zz})_3 = (3)\left[\frac{\pi}{4}(0.01)^4 + \pi(0.01)^2(0.35 - 0.1379)^2\right] = 42.42(10^{-6}) \text{ m}^4 \quad \text{(E4)}$$

We can find the total bending rigidity as shown in Equation (E5).

$$\sum E_j(I_{zz})_j = (28)(10^9)(131.1)(10^{-6}) + 0 + (200)(10^9)(42.42)(10^{-6})$$
$$= 12.15(10^6) \text{N} \cdot \text{m}^2 \quad \text{(E5)}$$

The maximum compressive bending stress in concrete will occur at the top surface. Substituting $y = 0.1379$ m, $M_z = 40$ kN $\cdot$ m, $E_1 = 28$ GPa, and Equation (E5) into Equation (4.13), we obtain Equation (E6).

$$(\sigma_{max})_{conc} = -E_1 y \frac{M_z}{\sum E_j(I_{zz})_j} = -(28)(10^9)(0.1379)\frac{(40)(10^3)}{12.15(10^6)}$$
$$= -12.7(10^6) \text{ N/m}^2 \quad \text{(E6)}$$

ANS. $(\sigma_{max})_{conc} = 12.7$ MPa (C)

The maximum tensile bending stress in steel will occur at the lowest point on the circular rod. Substituting $y = -(0.35 + 0.01 - \eta_c) = -0.2221$ m, $M_z = 40$ kN $\cdot$ m, $E_3 = 200$ GPa, and Equation (E5) into Equation (4.13), we obtain Equation (E7).

$$(\sigma_{max})_{st} = -E_3 y \frac{M_z}{\sum E_j(I_{zz})_j} = -(200)(10^9)(-0.2221)\frac{(40)(10^3)}{12.15(10^6)}$$
$$= 146.2(10^6) \text{ N/m}^2 \quad \text{(E7)}$$

ANS. $(\sigma_{max})_{st} = 146.2$ MPa (T)

COMMENT

If we analyze this problem, assuming that the cross section is made from two materials, then the location of the neutral axis from Equation (4.14) can be found as $\eta_c = 0.2153$ m. The corresponding bending rigidity will be $\sum E_j(I_{zz})_j = 26.2(10^6)\text{N} \cdot \text{m}^2$. The maximum normal stress in concrete in compression will be $(\sigma_{max})_{conc} = 9.42$ MPa (C), and the maximum normal stress in steel will be $(\sigma_{max})_{st} = 44.2$ MPa (T). Both these stress values are smaller than those obtained in Equations (E6) and (E7). Thus by neglecting the low capacity of concrete to support stresses in tension, we make a conservative approximation in our design.

PROBLEM SET 4.3

4.16 The cross section of a composite beam with a coordinate system that has its origin at C is shown in Figure P4.16. The normal strain at point A due to bending about the z axis is $\varepsilon_{xx} = -200 \, \mu$, and the moduli of elasticity of the materials are $E_1 = 8000$ ksi and $E_2 = 2000$ ksi (a) Plot the stress distribution across the cross section. (b) Determine the maximum bending normal stress in each material. (c) Determine the equivalent internal bending moment M_z by using Equation (3.4b-B). (d) Determine the equivalent internal bending moment M_z by using Equation (4.13).

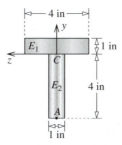

Figure P4.16

4.17 The cross section of a composite beam with a coordinate system that has an origin at C is shown in Figure P4.17. The normal strain at point A due to bending about the z axis is $\varepsilon_{xx} = -200 \, \mu$, and the moduli of elasticity of the materials are $E_1 = 200$ GPa and $E_2 = 70$ GPa. (a) Plot the stress distribution across the cross section. (b) Determine the maximum bending

normal stress in each material. (c) Determine the equivalent internal bending moment M_z by using Equation (3.4b-B). (d) Determine the equivalent internal bending moment M_z by using Equation (4.13).

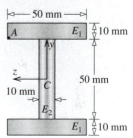

Figure P4.17

ANS. (b) $(\sigma_1)_{max} = 56$ MPa (T, C)
$(\sigma_2)_{max} = 14$ MPa (T, C) (c) and
(d) $M_z = 1512$ N $\cdot$ m

4.18 The cross section of a composite beam with a coordinate system that has an origin at C is shown in Figure P4.18. The normal strain at point A due to bending about the z axis is $\varepsilon_{xx} = 300 \, \mu$, and the moduli of elasticity of the materials are $E_1 = 30,000$ ksi and $E_2 = 20000$ ksi (a) Plot the stress distribution across the cross section. (b) Determine the maximum bending normal stress in each material. (c) Determine the equivalent internal bending moment M_z by using Equation (3.4b-B). (d) Determine the equivalent internal bending moment M_z by using Equation (4.13).

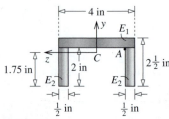

Figure P4.18

4.19 A simply supported, 3 m long beam has a uniform distributed load of 10 kN/m over its entire length. If the beam has the composite cross section shown in Figure P4.19, determine the maximum bending normal stress in each of the three materials. Aluminum, wood, and steel have moduli of elasticity of $E_{Al} = 70$ GPa, $E_{wood} = 10$ GPa, and $E_{st} = 200$ GPa, respectively.

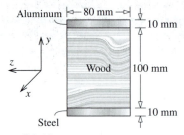

Figure P4.19

ANS. $(\sigma_{Al})_{max} = 102.1$ MPa (C)
$(\sigma_{wood})_{max} = 12.75$ MPa (C)
$(\sigma_{st})_{max} = 149.7$ MPa (T)

4.20 To improve the load-carrying capacity of a wooden beam ($E_{wood} = 2000$ ksi), a steel strip ($E_{st} = 30,000$ ksi) is securely fastened to it as shown in Figure P4.20. Determine (a) the maximum intensity of the load w if the allowable bending normal stresses in steel and in wood are 20 ksi and 4 ksi, respectively, and (b) the magnitude of the maximum shear stress in steel and in wood corresponding to the load in part (a).

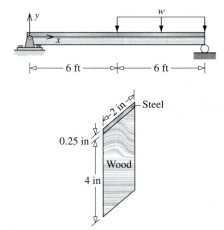

Figure P4.20

4.21 To reduce weight of a metal beam, the flanges are made of steel ($E_{st} = 200$ GPa) and a web of aluminum ($E_{Al} = 70$ GPa) as shown in Figure P4.21. Determine (a) the maximum bending normal stress in steel and in aluminum and (b) the magnitude of the maximum shear stress in steel and in aluminum.

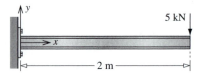

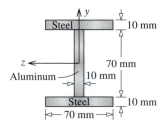

Figure P4.21

ANS. $|(\sigma_{st})_{max}| = 191.3$ MPa (T, C)
$|(\tau_{st})_{max}| = 5.95$ MPa
$|(\sigma_{Al})_{max}| = 52.1$ MPa (T, C)
$|(\tau_{Al})_{max}| = 6.41$ MPa

4.22 A steel (E_{st} = 200 GPa) tube of outside diameter 240 mm is attached to a brass ($E_{Cu/Zn}$ = 100 GPa) tube to form the cross section shown in Figure P4.22. Determine (a) the maximum bending normal stress in steel and in brass and (b) the magnitude of the maximum shear stress in the beam.

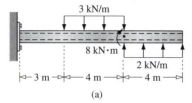

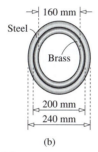

(b)

Figure P4.22

4.23 A concrete beam is reinforced by embedding circular steel rods of diameter 20 mm as shown in Figure P4.23. The maximum resisting (internal) moment in the beam was determined as 30 kN · m. Determine the maximum bending normal stress in the concrete and in steel given as E_{st} = 200 GPa and E_{conc} = 28 GPa. Neglect the low capacity of concrete to support stresses in tension.

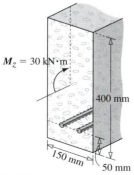

Figure P4.23

ANS. $(\sigma_{conc})_{max} = 10.92$ MPa (C)
$(\sigma_{st})_{max} = 160.2$ MPa (T)

4.24 A concrete beam is reinforced by embedding 2 in × 2 in square steel rods in the concrete as shown in Figure P4.24. The reinforced concrete beam is used as a 10 ft cantilever beam, with a force of 36 kips applied at the free end in the negative y direction. Determine the maximum compressive bending normal stress in the concrete. The moduli of elasticity for steel and concrete can be taken as E_{st} = 25,000 ksi and E_{conc} = 4,500 ksi. Neglect the low capacity of concrete to support stresses in tension.

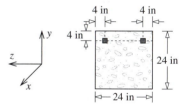

Figure P4.24

4.25 A concrete beam is reinforced by embedding 20 mm × 20 mm square steel rods in the concrete, loaded and supported as shown in Figure P4.25. The allowable tensile bending normal stress is 160 MPa, and the allowable compressive bending normal stress for concrete is 20 MPa. Determine the maximum intensity of the distributed load w. Take the moduli of elasticity for steel and concrete as E_{St} = 200 GPa and E_{conc} = 28 GPa. Neglect the low capacity of concrete to support stresses in tension.

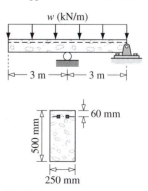

Figure P4.25

4.4 INTRODUCTION TO THE MICROMECHANICS OF COMPOSITES

The material constants obtained from the results of a tension test for a composite are macroscopic properties that in some manner depend upon its constituents and can vary significantly depending upon a host of factors, such as the manufacturing process, the manufacturing environment, the testing environment, and the history of loading. Thus, it is not surprising that experimental results of a tension test can show significant variations in the mechanical properties of seemingly similar composite materials. A theory intrinsically is an idealization of a complex reality, and the variability in mechanical properties of composites has spawned many theories. These theories attempt to predict macroscopic properties from the properties of individual constituents that may be microscopic in size. These theories constitute the field of micromechanics. In this section we will consider one such theory that can be obtained from the equations of mechanics of materials.[2]

A piece of the composite material, called a "unit cell" or a "representative volume element," is visualized in micromechanics to develop a theory. The properties of the unit cell are derived, and it is postulated that the properties of the entire body are similar to those of the unit cell. Clearly the postulates that the entire material behaves like the visualized unit cell and resembles it in geometry and material makeup of the unit cell are big approximations. The equations used to predict the properties of the unit cell are based on several assumptions and are yet another source of approximation in the theory. In spite of all these approximations, the theories improve our understanding of mechanics of composite materials. Experiments are the final arbiters of the accuracy of theoretical predictions.

The theories of micromechanics are theories of homogenization of constitutive materials to obtain macroscopic properties. Isotropic and transversely isotropic materials can be viewed as a subclass of orthotropic materials. We will develop equations that are applicable to orthotropic materials and use the relevant ones for isotropic and transversely isotropic materials. The orthotropic material that is easiest to visualize is the long-fiber composite lamina, discussed in the next section.

4.4.1 Orthotropic Composite Laminae

The discussion of orthotropic materials in Section 2.1.2 gave a brief description of a composite lamina. A composite orthotropic lamina is constructed by laying long fibers in a given direction and binding these fibers with an epoxy. A brief description of fibers and matrix materials follows.

The diameter of fibers is of the order of 10 μm. Short fibers (length on the order of 1 mm) are called whiskers. Fibers can have the corresponding tensile strength and stiffness (modulus of elasticity) orders of magnitude higher than those of bulk materials. The E-glass fibers have good electrical properties but lower strength and stiffness than the S-glass fibers. Glass fibers are extensively used in specialized sports equipment, automobiles, and aircraft structures. The terms "graphite fiber" and "carbon fiber" are often used interchangeably. There is a major difference, however: graphite fibers are subjected to very high temperatures. In comparison to glass fibers, carbon fibers are significantly more expensive and are usually used in advanced technological applications such as in aircraft structures. Aramid polymer fibers, the most popular being Kevlar, were originally

---

[2] Elasticity results are often used as an alternative to our mechanics of materials approach. See Gibson [1994], Hyer [1998], or Jones [1975] in Appendix D for additional details.

developed for use in radial tires but are now finding increased applications in structures. The list of fibers that can be used in structural application is continually growing. In Appendix C, Section C.10 gives the mechanical properties of a few fibers for use in this book.

The matrix holds the fibers together and protects them from abrasion and corrosion. Matrix material properties may often limit the application of a composite. Although metals and ceramics could be used as matrix materials, we will briefly consider only a few polymer materials, also called resins. There are two basic types of polymer resin in use in structural applications. The thermoset resins change irreversibly when cured (i.e., heated), and once cured, they do not melt if reheated. Thermoplastic polymers solidify when cooled and reversibly melt if reheated. Thermoset polyesters are inexpensive polymers, reinforced usually with glass fiber, for use as storage tanks, boat hulls, and public benches. Epoxy resins are usually used in advanced composite materials, and their properties are dependent upon the curing agent used. Polyimide resins are primarily used in high-temperature applications but are brittle. Section C.10 gives some mechanical properties of a few of the polymers for use in this book.

4.4.2 Unit Cell Approximation

Even though some of the results we will obtain are independent of the geometry of fiber and matrix, for the sake of simplicity, we will visualize a long-fiber composite lamina of rectangular geometry as shown in Figure 4.32. We assume the following:

E_f and E_m = modulus of elasticity of fiber and matrix

v_f and v_m = Poisson ratios of fiber and matrix

G_f and G_m = shear modulus of rigidity of fiber and matrix

V_f and V_m = volume of fiber and matrix

v_f and v_m = volumetric ratio of fiber and the matrix

The volumetric ratio is related to the volumes as $v_f = V_f/(V_f + V_m)$ and $v_m = V_m/(V_f + V_m)$, where the sum of the volumes of fiber and material is the total volume of the composite cell.

4.4.3 Longitudinal Modulus of Elasticity E_x

Consider a unit cell subjected to longitudinal stress σ_{xx} as shown in Figure 4.33. If A_f and A_m represent the cross-sectional areas of fiber and matrix perpendicular to the loading, then the axial rigidity of a cross section is $\sum E_j A_j = E_f A_f + E_m A_m$. The total axial

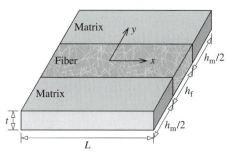

Figure 4.32 Unit cell approximation of a long-fiber composite.

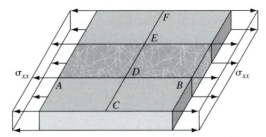

Figure 4.33 Longitudinal stress on a unit cell.

force that is applied is $\sigma_{xx}(A_f + A_m)$, which will be equal to the internal axial force N. From Equation (4.2) we obtain the axial strain at any cross section as in Equation (4.21).

$$\varepsilon_{xx} = \frac{N}{\sum E_j A_j} = \frac{\sigma_{xx}(A_f + A_m)}{E_f A_f + E_m A_m} = \frac{\sigma_{xx}(A_f + A_m)L}{(E_f A_f + E_m A_m)L} = \frac{\sigma_{xx}(V_f + V_m)}{(E_f V_f + E_m V_m)} \qquad (4.21)$$

Noting that $\varepsilon_{xx} = \sigma_{xx}/E_x$, we obtain Equation (4.22).

$$\frac{\sigma_{xx}}{E_x} = \frac{\sigma_{xx}(V_f + V_m)}{E_f V_f + E_m V_m} \qquad \text{or}$$

$$\boxed{E_x = E_f v_f + E_m v_m} \qquad (4.22)$$

Equation (4.21) is valid for any cross-sectional shape of the constituents because it depends only upon the value of the cross-sectional areas. The cross section perpendicular to the loading could belong to random short-fiber composite or to a long-fiber composite, as assumed here. Thus, Equation (4.22) is applicable to any composite body made of two constituents that have been mixed together. For this reason, Equation (4.22) is called the *rule of mixtures* for modulus of elasticity.

4.4.4   Poisson's Ratio v_{xy}

The transverse strain in the matrix and the fiber is as shown in Equation (4.23a).

$$(\varepsilon_{yy})_m = -v_m \varepsilon_{xx} \qquad (\varepsilon_{yy})_f = -v_f \varepsilon_{xx} \qquad (4.23a)$$

The reduction in the length of the line CF in Figure 4.33 can be found as shown in Equation follows:

$$\delta_{CF} = (\varepsilon_{yy})_m(CD + EF) + (\varepsilon_{yy})_f DE = (-v_m \varepsilon_{xx})\left(\frac{h_m}{2} + \frac{h_m}{2}\right) + (-v_f \varepsilon_{xx})(h_f) \qquad (4.23b)$$

The transverse strain of line CF is as shown in Equation (4.23c).

$$\varepsilon_{yy} = \frac{\delta_{CF}}{CF} = \frac{\delta_{CF}}{(h_m + h_f)} = \frac{(-v_m \varepsilon_{xx})(h_m) + (-v_f \varepsilon_{xx})(h_f)}{(h_m + h_f)}$$

$$= -\varepsilon_{xx}\frac{[v_m h_m + v_f h_f](tL)}{[h_m + h_f](tL)} \qquad (4.23c)$$

Noting that $v_{xy} = -\varepsilon_{yy}/\varepsilon_{xx}$ and $V_f = th_f L$, $V_m = th_m L$, we obtain Equation (4.24).

$$-v_{xy}\varepsilon_{xx} = -\varepsilon_{xx}\frac{[v_m V_m + v_f V_f]}{[V_m + V_f]} \qquad \text{or}$$

$$\boxed{v_{xy} = v_m V_m + v_f V_f} \qquad (4.24)$$

Equation (4.24) is the rule of mixtures for Poisson's ratio. However, unlike Equation (4.22) the assumption that the matrix and fiber are rectangular is used in Equation (4.23c). Thus, the use of Equation (4.24) for fibers of other cross-sectional shape is an approximation.

4.4.5 Transverse Modulus of Elasticity E_y

Figure 4.34 shows a unit cell subjected to a transverse stress σ_{yy}. Thus, the axial force applied in the y direction is $\sigma_{yy}Lt$, which would be equal to the internal axial force N on a cross section perpendicular to the load. The cross-sectional area of the imaginary cut is tL irrespective of whether the cut is in the matrix or the fiber, and the cross section is homogeneous. The total deformation of the line CF is the sum of deformations of lines CD, DE, and EF (i.e., $\delta_{CF} = \delta_{CD} + \delta_{DE} + \delta_{EF}$). The deformation of each line can be found by using Equation (3.10-A) as shown in Equation (4.25a).

$$\delta_{CF} = \frac{(\sigma_{yy}Lt)(h_m/2)}{E_m(Lt)} + \frac{(\sigma_{yy}Lt)(h_f)}{E_f(Lt)} + \frac{(\sigma_{yy}Lt)(h_m/2)}{E_m(Lt)} = \frac{\sigma_{yy}h_m}{E_m} + \frac{(\sigma_{yy})h_f}{E_f} \qquad (4.25a)$$

Noting that deformation $\delta_{CF} = (h_m + h_f)\varepsilon_{yy} = (h_m + h_f)(\sigma_{yy}/E_y)$, we obtain Equation (4.25b).

$$\delta_{CF} = \frac{\sigma_{yy}(h_m + h_f)}{E_y} = \frac{\sigma_{yy}h_m}{E_m} + \frac{\sigma_{yy}h_f}{E_f} \qquad \text{or} \qquad \frac{h_m + h_f}{E_y} = \frac{h_m}{E_m} + \frac{h_f}{E_f} \qquad (4.25b)$$

Multiplying the equation by (Lt) and noting that $V_f = Lth_f$ and $V_m = Lth_m$, we obtain Equation (4.26).

$$\frac{V_m + V_f}{E_y} = \frac{V_m}{E_m} + \frac{V_f}{E_f} \qquad \text{or}$$

$$\boxed{\frac{1}{E_y} = \frac{v_m}{E_m} + \frac{v_f}{E_f}} \qquad (4.26)$$

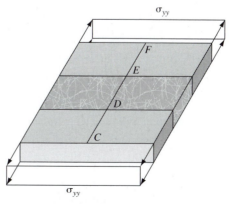

Figure 4.34 Transverse stress on a unit cell.

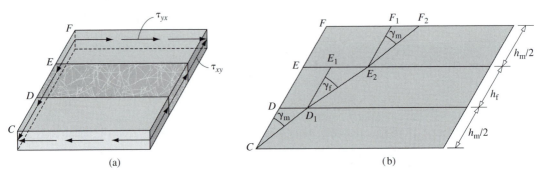

Figure 4.35 (a) Shear stress on unit cell. (b) Exaggerated shear deformation of unit cell.

Equation (4.26) is called the *inverse rule of mixtures*. We made use of rectangular geometry in Equation (4.25a) and in moving from Equation (4.25b) to Equation (4.26). Thus, the use of Equation (4.26) for fibers of other cross-sectional shapes is an approximation.

4.4.6 Shear Modulus G_{xy}

Figure 4.35(a) shows a unit cell subjected to uniform shear stress. Figure 4.35(b) shows the shear deformation due to the applied shear stress. The shear strain in the matrix and fiber are different, and hence the change in angle (i.e., from a right angle) is different for the matrix and the fiber as shown in Figure 4.35(b). Assuming small shear strain, we can write Equations (4.27a), (4.27b), and (4.27c).

$$\tan\gamma_m \approx \gamma_m = \frac{DD_1}{CD} \quad \text{or} \quad DD_1 = \gamma_m\left(\frac{h_m}{2}\right) \tag{4.27a}$$

$$\tan\gamma_f \approx \gamma_f = \frac{E_1E_2}{DE} \quad \text{or} \quad E_1E_2 = \gamma_f h_f \tag{4.27b}$$

$$\tan\gamma_m \approx \gamma_m = \frac{F_1F_2}{EF} \quad \text{or} \quad F_1F_2 = \gamma_m\left(\frac{h_m}{2}\right) \tag{4.27c}$$

The total shear displacement FF_2 is the shear strain γ_{xy} multiplied by CF; that is, $FF_2 = \gamma_{xy}(h_f + h_m)$. It can also be written as the sum of the shear displacements: $FF_2 = DD_1 + E_1E_2 + F_1F_2$. Equating the two expressions for FF_2, we obtain Equation (4.27d),

$$\gamma_{xy}(h_f + h_m) = \gamma_f h_f + \gamma_m h_m \quad \text{or} \quad \frac{\tau_{xy}}{G_{xy}}(h_f + h_m) = \left(\frac{\tau_f}{G_f}\right)h_f + \left(\frac{\tau_m}{G_m}\right)h_m \tag{4.27d}$$

where τ_f and τ_m are the shear stresses in the fiber and the matrix, respectively. The shear force at any cross section perpendicular to line $CDEF$ is $(\tau_{xy})(Lt)$. With the cross-sectional area that is perpendicular to $CDEF$ equal to (Lt), the shear stress in the fiber and in the matrix is equal to the applied shear stress (i.e., $\tau_f = \tau_m = \tau_{xy}$). Multiplying the resulting equation by Lt and noting the definition of volumes, we obtain Equation (4.28).

$$\frac{\tau_{xy}}{G_{xy}}(V_f + V_m) = \left(\frac{\tau_{xy}}{G_f}\right)V_f + \left(\frac{\tau_{xy}}{G_m}\right)V_m \quad \text{or}$$

$$\boxed{\frac{1}{G_{xy}} = \frac{v_f}{G_f} + \frac{v_m}{G_m}} \tag{4.28}$$

Equation (4.28) is the inverse rule of mixtures for shear modulus of elasticity. Once more, we made use of rectangular geometry in obtaining the results. Thus the rule of mixtures for modulus of elasticity is the only one that is independent of the shape of the fiber; for the rest of the material constants, the rule of mixtures and the inverse rule of mixtures are approximations, which are justified as being of the same order of approximation one makes in deriving the equations of mechanics of materials.

EXAMPLE 4.11

(a) What is the effective modulus of elasticity in the axial direction of the reinforced concrete bar in Example 4.4, assuming that 95 kg of steel was used to construct the bar?
(b) Calculate the contraction of the bar under a compressive axial force of 1000 N.

PLAN

(a) Knowing the density of the steel from Table 4.1, we can find the volume of steel used. Knowing the total volume, we can find the volume ratios of steel and concrete. Knowing the modulus of elasticity of each material, we can find the effective axial modulus of elasticity by using Equation (4.22). (b) From the effective modulus of elasticity, we can find use Equation (3.10-A) to find the elongation.

SOLUTION

(a) The total volume of the concrete bar can be found as in Equation (E1).

$$V = (0.2)(0.2)(2.5) = 0.1 \ \text{m}^3 = 100(10^{-3}) \ \text{m}^3 \qquad \text{(E1)}$$

From Table 4.1 we know the density of steel is 7870 kg/m^3. Knowing that the mass of steel used is 95 kg, we can find the volume of steel used and volume ratios of steel and concrete as shown in Equation (E2).

$$V_{st} = \frac{95}{7870} = 12.07(10^{-3}) \qquad v_{st} = \frac{V_{st}}{V} = 0.1207$$

$$v_{conc} = 1 - v_{st} = 0.8793 \qquad \text{(E2)}$$

From Table 4.1, the moduli, of elasticity of steel and concrete are $E_{st} = 200$ GPa and $E_{conc} = 20$ GPa. From the rule of mixtures, Equation (4.22), we obtain the effective axial modulus of elasticity as shown in Equation (E3).

$$E_{effective} = E_{st}v_{st} + E_{conc}v_{conc} = (200)(0.1207) + (20)(0.8793)$$
$$= 41.726 \qquad \text{(E3)}$$

ANS. $E_{effective} = 41.7$ GPa

(b) The cross-sectional area of the reinforced concrete bar is $A = 0.04$ m^2. The axial force is $N = -1000$ N. The axial contraction from Equation (3.10-A) can be written as Equation (E4).

$$u_2 - u_1 = \frac{N(x_2 - x_1)}{E_{\text{effective}}A} = \frac{(-1000)(2.5)}{(41.7)(10^9)(0.04)} = -1.5(10^{-6})\text{m} \qquad (E4)$$

ANS. $u_2 - u_1 = -0.0015$ mm

COMMENTS

1. The rule of mixtures, Equation (4.22), can be used for determining the effective modulus of elasticity for any two constituents, not just long-fiber composites. Depending upon a variety of factors, however, the accuracy of the effective modulus may be very different from that obtained from a tension test.

2. The cross-sectional area of the steel bars is $A_{\text{st}} = V_{\text{st}}/(2.5) = 4.828(10^{-3})\text{m}^2$. The cross-sectional area of the concrete is $A_{\text{conc}} = 0.04 - A_{\text{st}} = 35.172(10^{-3})\text{m}^2$. The axial rigidity of the composite bar $E_{\text{st}}A_{\text{st}} + E_{\text{conc}} A_{\text{conc}} = 1668(10^6)$ N, which is the same as $E_{\text{effective}}A = (41.7)(10^9)(0.04) = 1668(10^6)$ N. This is not surprising, in as much as the rule of mixtures was derived from the equations for composite axial bars. But the perspective used in the rule of mixtures is very different from that for composite axial bars.

EXAMPLE 4.12

A composite specimen has fibers aligned in the x direction as symbolically shown in Figure 4.36. The composite has 40% of S-glass fibers by volume in a polyester matrix. The composite is subjected to the normal and shear stresses. From the data given in Section C.10, determine the strains ε_{xx}, ε_{yy}, and γ_{xy}.

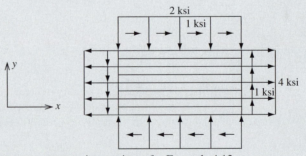

Figure 4.36 Stresses on a composite specimen for Example 4.12.

PLAN

By using the rule of mixtures and the inverse rule mixtures, we can determine the material constants E_x, E_y, ν_{xy}, and G_{xy}. The given stresses are $\sigma_{xx} = 4$ ksi, $\sigma_{yy} = -2$ ksi, and $\tau_{xy} = 1$ ksi. By using Equations (2.4a), (2.4b), (2.4c), and (2.4d) for orthotropic materials, the strains can be found.

SOLUTION

From Section C.10, we have the data for S-glass fiber and polyester matrix shown in Equation (E1).

$$E_f = 12,500 \text{ ksi} \qquad v_f = 0.22 \qquad E_m = 435 \text{ ksi} \qquad v_f = 0.38 \qquad \text{(E1)}$$

The volumetric ratios of the fiber and matrix are $v_f = 0.4$ and $v_m = 0.6$. From Equation (4.22) we obtain the longitudinal modulus of elasticity as shown in Equation (E2).

$$E_x = E_f v_f + E_m v_m = (0.4)(12,000) + (0.6)(435) = 5261 \text{ ksi} \qquad \text{(E2)}$$

From Equation (4.24) we obtain the major Poisson ratio as shown in Equation (E3).

$$v_{xy} = v_m v_m + v_f v_f = (0.4)(0.22) + (0.6)(0.38) = 0.308 \qquad \text{(E3)}$$

From Equation (4.26) we obtain the transverse modulus of elasticity as shown in Equation (E4).

$$\frac{1}{E_y} = \frac{v_f}{E_f} + \frac{v_m}{E_m} = \frac{0.4}{12,500} + \frac{0.6}{435} = 1.411(10^{-3}) \qquad \text{or} \qquad E_y = 708 \text{ ksi} \qquad \text{(E4)}$$

The shear moduli of elasticity for fiber and matrix can be found as shown in Equation (E5).

$$G_f = \frac{E_f}{2(1 + v_f)} = \frac{12,500}{2(1 + 0.22)} = 5122.9 \text{ ksi}$$

$$G_m = \frac{E_m}{2(1 + v_m)} = \frac{435}{2(1 + 0.38)} = 157.6 \text{ ksi} \qquad \text{(E5)}$$

From Equation (4.28) we obtain the shear modulus of elasticity for the composite as shown in Equation (E6).

$$\frac{1}{G_{xy}} = \frac{v_f}{G_f} + \frac{v_m}{G_m} = \frac{0.4}{5122.9} + \frac{0.6}{157.6} = 3.884(10^{-3}) \qquad \text{or} \qquad G_{xy} = 257.4 \text{ ksi} \qquad \text{(E6)}$$

From Equation (2.4d) we obtain the minor Poisson ratio as Equation (E7).

$$v_{yx} = \left(\frac{v_{xy}}{E_x}\right) E_y = \left(\frac{0.308}{5261}\right)(708) = 0.04145 \qquad \text{(E7)}$$

From Equation (2.4a) we obtain the normal strain in the x direction as shown in Equation (E8).

$$\varepsilon_{xx} = \frac{\sigma_{xx}}{E_x} - \frac{v_{yx}}{E_y}\sigma_{yy} = \frac{4}{5261} - \frac{0.04145}{708}(-2) = 877.4(10^{-6}) \qquad \text{(E8)}$$

ANS. $\varepsilon_{xx} = 877.4 \ \mu$

From Equation (2.4b) we obtain the normal strain in the y direction as in Equation (E9).

$$\varepsilon_{yy} = \frac{\sigma_{yy}}{E_y} - \frac{v_{xy}}{E_x}\sigma_{xx} = \frac{(-2)}{708} - \frac{0.308}{5261}(4) = 3059(10^{-6}) \qquad \text{(E9)}$$

ANS. $\varepsilon_{yy} = 3059 \ \mu$

From Equation (2.4c) we obtain the shear strain as shown in Equation (E10).

$$\gamma_{xy} = \frac{\tau_{xy}}{G_{xy}} = \frac{(1)}{257.4} = 3885(10^{-6}) \tag{E10}$$

ANS. $\gamma_{xy} = 3885\ \mu$

COMMENTS

1. The solution of the problem is based on calculations of material constants of an orthotropic material and on the use of stress–strain equations for orthotropic materials.

2. Suppose the fibers were at an angle to the applied stresses as shown in Figure 4.37. To solve the problem, we must transform the stresses into the material coordinate system x_1, y_1 in which we know the material constants as calculated earlier. The orthotropic stress–strain can be used to find the equations the strains in the coordinate system x_1, y_1. The strains then must be transformed to the coordinate system x, y. See Problems 4.34 through 4.37.

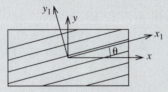

Figure 4.37   Fibers inclined to the loading.

PROBLEM SET 4.4

4.26 A reinforced concrete column was constructed for use in a building. The column contains 200 lb of steel bars, and it has a square cross section of 8 in × 8 in and a length of 75 in. Determine (a) the effective longitudinal modulus of elasticity and (b) the axial rigidity of the cross section of the bar. Use the data in Table P4.26.

TABLE P4.26   Material Properties of Steel and Concrete

	Concrete	Steel
Modulus of elasticity	4000 ksi	30,000 ksi
Specific weight	0.087 lb/in³	0.284 lb/in³

ANS. $E_x = 7822$ ksi

$E_x A = 500.6(10^3)$ kips

4.27 A reinforced concrete column is to be constructed to support a parking deck. The column will have a square cross section of 10 in × 10 in and a length of 96 in, and its longitudinal modulus of elasticity should be 8500 ksi. Determine to the nearest pound the amounts of steel and concrete needed to construct the reinforced concrete column. Use the data in Table P4.26.

4.28 A composite specimen has fibers aligned in the x direction as symbolically shown in Figure P4.28. The composite has 60% Kevlar fibers by volume in an epoxy matrix. The composite is subjected to normal and shear stresses. Use the data given in Section C.10 to determine the strains ε_{xx}, ε_{yy}, and γ_{xy}.

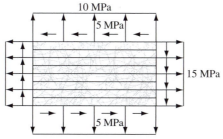

Figure P4.28

4.29 A composite is made from 60% carbon fibers in a matrix of epoxy. The composite is subjected to a 5 ksi axial stress in the direction of the fiber. Determine the axial and transverse strains. Use the data in Section C.10.

ANS. $\varepsilon_{xx} = 304.3\ \mu$ $\varepsilon_{yy} = -87.0\ \mu$

4.30 A composite is made from 60% carbon fibers in a matrix of epoxy. The fibers are aligned in the x direction, and the composite is subjected to a shear stress of $\tau_{xy} = 500$ psi. Determine the shear strain γ_{xy}. Use the data in Section C.10.

4.31 A carbon fiber–epoxy composite is to be constructed to have a longitudinal modulus of elasticity of $E_x = 15{,}000$ ksi. Determine the volumetric ratio of fiber and epoxy needed. Use the data in Section C.10.

4.32 A composite is to be constructed from Kevlar fiber and epoxy to have a longitudinal modulus of elasticity of $E_x = 8000$ ksi. Determine the volumetric ratio of fiber and epoxy needed. Use the data in Section C.10.

4.33 A composite is to be constructed from S-glass fiber and polyester to have a longitudinal modulus of elasticity of $E_x = 50$ MPa. Determine the volumetric ratio of fiber and epoxy needed. Use the data in Section C.10.

ANS. $v_f = 0.57$ $v_m = 0.43$

4.34 A composite specimen is made from 60% carbon fibers in a matrix of epoxy. The fibers are at an angle of $\theta = 15°$ counterclockwise to the x axis, as symbolically shown in Figure P4.34. Determine the axial and transverse strains when an axial stress of $\sigma_{xx} = 2$ ksi (T) is applied.

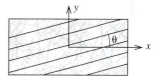

Figure P4.34

4.35 A composite specimen is made from 60% carbon fibers in a matrix of epoxy. The fibers are at an angle of $\theta = 15°$ counterclockwise to the x axis, as symbolically shown in Figure P4.34. The stress state is $\tau_{xy} = 500$ psi. Use the data in Section C.10 to determine the strains ε_{xx}, ε_{yy}, and γ_{xy}.

4.36 A composite specimen has fibers aligned at an angle of $\theta = 20°$ to the x direction as symbolically shown in Figure P4.36. The composite has 50% Kevlar fibers by volume in an epoxy matrix. The composite is subjected to an axial stress of $\sigma_{xx} = 25$ MPa (T). Use the data in Section C.10 to determine the strains ε_{xx}, ε_{yy}, and γ_{xy}.

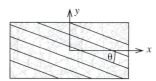

Figure P4.36

ANS. $\varepsilon_{xx} = 1200\ \mu$ $\varepsilon_{yy} = -622.8\ \mu$

$\gamma_{xy} = 1923.7\ \mu$

4.37 A composite specimen has fibers aligned at an angle of $\theta = 20°$ to the x direction as symbolically shown in Figure P4.36. The composite has 50% Kevlar fibers by volume in an epoxy matrix. The state of stress is $\sigma_{yy} = 20$ MPa (T). Use the data in Section C.10 to determine the strains ε_{xx}, ε_{yy}, and γ_{xy}.

4.38 A composite tube has an outer diameter 3 in, a thickness of 1/8 in, and a length of 36 in. The tube is made from S-glass fibers in a thermoset polyester in which the fibers are aligned parallel to the axis. The volumetric ratio of the fibers is 70%. Determine the change in length and in diameter when an axial force of 20 kips is applied.

4.39 A rectangular composite specimen of width 40 mm, thickness 4 mm, and length 200 mm is to carry an axial force of 10 kN in tension. The specimen is to be made from carbon fibers in a matrix of epoxy. To ensure adequate stiffness, the elongation is to be limited to 0.25 mm. Determine the volumetric ratio of the fibers.

ANS. $v_f = 0.338$

4.5   CLOSURE

In this chapter we saw that the kinematic equations, the constitutive equations, and the equilibrium equations do not change and that nonhomogeneity across a cross section requires us to write the integral in static equivalency as the sum of the integrals over each material for one-dimensional structural members. All the analytical techniques for finding internal forces and moments that are used for homogeneous cross sections can be used for composite cross sections.

For laminar composites, the properties of each lamina can be obtained by the rule of mixtures or by the inverse rule of mixtures. Once these properties have been obtained for each lamina, the formulas developed for composite one-dimensional structural members can be used for structural analysis.

5 | Inelastic Structural Behavior

5.0 OVERVIEW

Initial strain introduced by the turnbuckle in the wire attached to a traffic gate [Figure 5.1(a)] and the nonlinear stress–strain relationship of rubber in a stretch cord [Figure 5.1(b)] are examples of inelastic behavior that must be accounted for in stress analysis. Initial stresses introduced during assembly from manufacturing tolerances, concrete prestressed in compression to counter tensile stresses during loading, thermal stresses introduced from temperature changes, metals prestressed by loading into plastic regions to increase the load-carrying capacity of a structural member, the nonlinear and/or viscoelastic[1] response of rubber, plastic, biological materials—these are other examples of inelastic material behavior that must be accounted for in structural design and analysis. In this chapter we will account for initial strains, thermal strains, and viscoelastic strains in the context of axial members; we will also study the incorporation of nonlinear stress–strain equations in axial members, the torsion of circular shafts, and the symmetric bending of beams.

In Section 3.1, while developing the simplified theories on structural members, we made several assumptions regarding material behavior that permitted us to use Hooke's law. In this chapter we will consider the impact of dropping the following assumptions, listed earlier, in Table 3.3: Assumption 5 (of no inelastic strains), Assumption 6 (of elastic material behavior), and Assumption 7 (of a linear relationship between stress and strain). We will assume that the kinematic equations are still valid. The equations of static equivalency are unaffected by the material model. The equilibrium equations relating internal forces to external forces do not change with the material model. Thus, the primary impact of inelastic material behavior is on the stress distribution across the cross section. When the stress variation across the cross section is understood and accounted for, we can still use the analytical techniques we have used before.

---

[1]Many solid materials exhibit stress–strain behavior that has similarities to that of viscous materials.

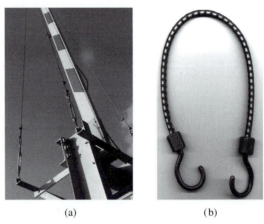

(a) (b)

Figure 5.1 Examples of inelastic structural behavior.

The learning objectives of this chapter are:

1. To understand the incorporation of thermal and initial strains into the theory and analysis of axial members.
2. To understand the incorporation of nonlinear material models into the basic simplified theories of axial members, torsion of circular shafts, and symmetric bending of beams.
3. To understand the phenomenon of viscoelasticity and to account for it in axial members.

5.1 EFFECTS OF TEMPERATURE

A material expands with an increase in temperature and contracts with a decrease in temperature. If the change in temperature is *uniform*, and if the material is *isotropic* and *homogeneous*, then all lines on the material will change dimensions by equal amounts, which will result in a normal strain. Since, however, there will be no change in the angles between any two lines, there will be no shear strain produced. Experimental observations confirm this deduction. Experiments also show that the change in temperature (ΔT) is related to the thermal normal strain (ε_T) as in Equation (5.1),

$$\varepsilon_T = \alpha \, \Delta T \tag{5.1}$$

where the Greek letter alpha (α) is called the linear coefficient of thermal expansion. The linear relationship given by Equation (5.1) is valid for metals in the region far from the melting point. In this linear region, the strains for most metals are small and the usual units for α are $\mu/°F$ or $\mu/°C$, where $\mu = 10^{-6}$. Throughout the discussion in this section we shall assume that we are in the linear region.

The tension test to determine material constants is conducted at an ambient temperature. We expect the stress–strain curve to have the same character at two different ambient temperatures. If we raise the temperature before we apply the force P on the specimen, then the specimen will expand and result in a thermal strain as given by Equation (5.1). But because there is no external force, there will be no resulting internal forces, hence no stresses. Thus, the increase in temperature before the application of the

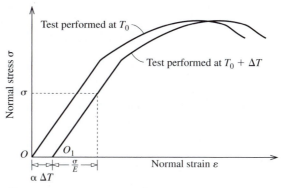

Figure 5.2 Effect of temperature on stress–strain curve.

force causes the starting point of the stress–strain curve to move from O to O_1 as shown in Figure 5.2. The total strain at any point is the sum of mechanical strain and thermal strains, which can be written as shown in Equation (5.2).

$$\varepsilon = \frac{\sigma}{E} + \alpha\,\Delta T \tag{5.2}$$

If there are no internal forces generated in a body owing to changes in temperature, then no stresses will be produced. Material nonhomogeneity, material anisotropy, non-uniform temperature distribution, or reaction forces from body constraints are the reasons for the generation of stresses from temperature changes. We record the following observation.

- No thermal stresses are produced in a homogeneous, isotropic, unconstrained body as a result of uniform temperature changes.

The generalized version of Hooke's law relates mechanical strains to stresses. The total normal strain, as seen from Equation (5.2), is the sum of mechanical and thermal strains. For isotropic materials with temperature changes, the generalized law can be written as shown in Equations (5.3). We do not have to comment on the homogeneity of the material or the uniformity of temperature change because Hooke's law is written for a point, not the whole body.

$$\varepsilon_{xx} = [\sigma_{xx} - \nu(\sigma_{yy} + \sigma_{zz})]/E + \alpha\,\Delta T \tag{5.3a}$$

$$\varepsilon_{yy} = [\sigma_{yy} - \nu(\sigma_{zz} + \sigma_{xx})]/E + \alpha\,\Delta T \tag{5.3b}$$

$$\varepsilon_{zz} = [\sigma_{zz} - \nu(\sigma_{xx} + \sigma_{yy})]/E + \alpha\,\Delta T \tag{5.3c}$$

$$\gamma_{xy} = \tau_{xy}/G \tag{5.3d}$$

$$\gamma_{yz} = \tau_{yz}/G \tag{5.3e}$$

$$\gamma_{zx} = \tau_{zx}/G \tag{5.3f}$$

Mechanical strain   Thermal strain

EXAMPLE 5.1

The circular bar shown in Figure 5.3 has a diameter of 100 mm. The bar is built into a rigid wall on the left, and a gap of 0.5 mm exists between the right-hand wall and the bar before the increase in temperature occurs. The temperature of the bar is uniformly increased by 80°C. Determine the average axial stress and the change in the diameter of the bar. Use $E = 200$ GPa, $\nu = 0.32$, and $\alpha = 11.7 \ \mu/°C$.

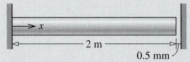

Figure 5.3 Bar for Example 5.1.

METHOD 1

PLAN

A reaction force in the axial direction will be generated to prevent expansion greater than the gap. This force, in turn, will generate σ_{xx}, but all other stress components will be zero. The total deformation is the gap from which the total axial strain for the bar can be found. The thermal strain can be calculated; thus in Equation (5.3a), the only unknown is σ_{xx}. The strain ε_{yy} can be found from Equation (5.3b) once σ_{xx} has been calculated. From ε_{yy}, the change in the diameter can be calculated.

SOLUTION

We can calculate thermal expansion as in Equation (E1).

$$\alpha \ \Delta T \ = \ 11.7(10^{-6})(80) \ = \ 936(10^{-6}) \tag{E1}$$

The total axial strain is the deformation (gap) divided by the length of the bar as shown in Equation (E2).

$$\varepsilon_{xx} \ = \ \frac{0.5(10^{-3})}{2} \ = \ 250(10^{-6}) \tag{E2}$$

Substituting Equations (E1), (E2), $\sigma_{yy} = 0$, and $\sigma_{zz} = 0$ into Equation (5.3a), we can obtain σ_{xx} as shown in Equation (E3).

$$\sigma_{xx} \ = \ E(\varepsilon_{xx} - \alpha \ \Delta T) \ = \ 200(10^9)(250 - 936)(10^{-6}) \ \text{N/m}^2 \qquad \text{or} \qquad \text{(E3)}$$

ANS. $\sigma_{xx} \ = \ 137.2$ MPa (C)

From Equation (5.3b), we can obtain ε_{yy} as shown in Equation (E4).

$$\varepsilon_{yy} \ = \ -\nu\frac{\sigma_{xx}}{E} + \alpha \ \Delta T \ = \ -0.25\left(\frac{-137.2(10^6)}{200(10^9)}\right) + 936(10^{-6}) \ = \ 1.107(10^{-3}) \quad \text{(E4)}$$

The change in diameter can be calculated as shown in Equation (E5).

$$\Delta D = \varepsilon_{yy}D = 1.107(10^{-3})(100) \tag{E5}$$

ANS. $\Delta D = 0.1107$ mm increase

METHOD 2

PLAN

We can think of the problem in two steps. (1) Ignore the restraining effect of the right-hand wall, think of the bar as free to expand, and find the thermal expansion δ_T. (2) Apply the force P to bring the bar back to the restraint position due to the right-hand wall and compute the corresponding stress.

SOLUTION

We draw an approximate deformed shape of the bar, assuming that there is no right-hand wall to restrain the deformation as shown in Figure 5.4.

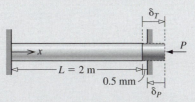

Figure 5.4 Approximate deformed shape of the bar in Example 5.1.

The thermal expansion δ_T is the thermal strain multiplied by the length of the bar as shown in Equations (E6).

$$\delta_T = (\alpha \, \Delta T)L = 11.7(10^{-6})(80)(2) = 1.872(10^{-3}) \tag{E6}$$

From Figure 5.4, we can see that by subtracting the gap from the thermal expansion, we will obtain the contraction δ_P we need to satisfy the restraint imposed by the right-hand wall. We can then find the mechanical strain and compute the corresponding stress, as shown in Equations (E7) and (E8).

$$\delta_P = \delta_T - 0.5(10^{-3}) = 1.372(10^{-3}) \qquad \varepsilon_P = \frac{\delta_P}{L} = \frac{1.372(10^{-3})}{2} = 0.686(10^{-3}) \tag{E7}$$

$$\sigma_P = E\varepsilon_P = 200(10^9)0.686(10^{-3}) \tag{E8}$$

ANS. $\sigma_P = 137.2$ MPa (C)

The change in diameter can be found as in method 1.

COMMENTS

1. If $\alpha \, \Delta T$ were less then ε_{xx} in Equation (E3), then σ_{xx} would come out as tension and our assumption that gap closes would be invalid. In such a case, there would be no stress σ_{xx} generated.

2. The increase in diameter is partly due to the Poisson effect and partly due to thermal strain in the y direction.

3. In method 1, we ignored the intermediate steps and conducted the analysis at equilibrium. In doing so we implicitly recognized that for a linear system, the process of reaching equilibrium is immaterial. In method 2 we conducted the thermal and mechanical strain calculations separately. Method 1 is more procedural, while method 2 is more intuitive.

EXAMPLE 5.2

The strains at a point on aluminum ($E = 70$ GPa, $G = 28$ GPa, $v = 0.25$, and $\alpha = 23$ $\mu/°C$) were found to be

$$\varepsilon_{xx} = 650 \, \mu \qquad \varepsilon_{yy} = 300 \, \mu \qquad \gamma_{xy} = 750 \, \mu$$

Assuming that the point is in plane stress, determine the stresses σ_{xx}, σ_{yy}, and τ_{xy} and the strain ε_{zz} if the temperature increases by 20°C.

PLAN

The shear strain can be calculated by using Equation (5.3d). In Equation (5.3a) and (5.3b), $\sigma_{zz} = 0$, $\varepsilon_{xx} = 650 \, \mu$, and $\varepsilon_{yy} = 300 \, \mu$ are known, and $\alpha \, \Delta T$ can be found and substituted to generate two equations in the two unknown stresses σ_{xx} and σ_{yy}, which can be found by solving the equations simultaneously. Then, from Equation (5.3c), the normal strain ε_{zz} can be found.

SOLUTION

From Equation (5.3d) we can find the shear stress as shown in Equation (E1)

$$\tau_{xy} = G\gamma_{xy} = 28(10^9)750(10^{-6}) \tag{E1}$$

ANS. $\tau_{xy} = 21 \, \text{MPa}$

Knowing $\Delta T = 20$, we can find the thermal strain as $\alpha \, \Delta T = 460(10^{-6})$. Equations (5.3a) and (5.3b) can be rewritten with $\sigma_{zz} = 0$ as shown in Equations (E2) and (E3).

$$\sigma_{xx} - v\sigma_{yy} = E(\varepsilon_{xx} - \alpha \, \Delta T) = 70(10^9)(650 - 460)(10^{-6}) \, \text{N/m}^2 \qquad \text{or}$$

$$\sigma_{xx} - 0.25\sigma_{yy} = 13.3 \, \text{MPa} \tag{E2}$$

$$\sigma_{yy} - \nu\sigma_{xx} = E(\varepsilon_{yy} - \alpha\,\Delta T) = 70(10^9) \times (300 - 460)(10^{-6}) \text{ N/m}^2 \qquad \text{or}$$

$$\sigma_{yy} - 0.25\sigma_{xx} = -11.2 \text{ MPa} \qquad \text{(E3)}$$

By solving Equations (E2) and (E3) we obtain the stresses.

ANS. $\sigma_{xx} = 11.2$ MPa (T) $\sigma_{yy} = 8.4$ MPa (C)

From Equation (5.3c) with $\sigma_{zz} = 0$, we obtain Equation (E4).

$$\varepsilon_{zz} = \frac{[-\nu(\sigma_{xx} + \sigma_{yy})]}{E} + \alpha\,\Delta T = \frac{[-0.25(11.2 - 8.4)(10^6)]}{70(10^9)} + 460(10^{-6}) \qquad \text{or} \qquad \text{(E4)}$$

ANS. $\varepsilon_{zz} = 450\,\mu$

COMMENT

Equations (E2) and (E3) have a very distinct structure. If we multiply either equation by ν and add the product to the other equation, the result will be to eliminate one of the unknowns.

5.2   INITIAL STRESS OR STRAIN IN AXIAL MEMBERS

Members in a statically indeterminate structure may have an initial stress or strain before loads are applied. These initial stresses or strains may be intentional or unintentional and can be caused by several factors. A good design must account for these factors by calculating the acceptable levels of prestress.

Nuts on a bolt are usually finger-tightened to hold an assembly in place. At this stage the assembly is usually stress free. The nuts are then given additional turns to pretension the bolts. When a nut is tightened by one full rotation, the distance it moves is called the *pitch*. Alternatively, pitch is the distance between two adjoining peaks on the threads. One reason for pretensioning is to prevent the nuts from becoming loose and falling off. Another reason is to introduce an initial stress that will be opposite in sign to that which will be generated by the loads. For example, cable in a bridge may be pretensioned by tightening the nut-and-bolt systems to counter any slackening in the cable due to wind or seasonal temperature changes.

If a member is shorter than is required during assembly, it will be forced to stretch, thus putting the entire structure into a prestress. Tolerances for manufacturing the members must be prescribed to ensure that the structure is not excessively prestressed.

In prestressed concrete, metal bars are initially stretched by applying tensile forces, and then concrete is poured over the bars. After the concrete has set, the applied tensile forces are removed. The initial prestress in the bar is redistributed, putting the concrete into compression. Concrete has good compressive strength but poor tensile strength. Prestressed concrete can be used in situations where it may be subjected to tensile stresses.

Structures are usually made of axial bars in different orientations, and for this reason the form of equation given in Equation (5.4) is preferred over Equation (3.10-A) in structural analysis,

$$\delta = \frac{NL}{EA}$$

(5.4)

where the $x_2 - x_1$ in Equation (3.10-A) is replaced by L, representing the length of the bar, and $u_2 - u_1$ is replaced by δ, representing the deformation of the bar in the undeformed direction. In Chapter 1 we used the component of deformation that is in the original direction of the bar in small-strain approximations, irrespective of movement of points on the bar. It should also be recognized that L, E, and A are positive; hence the sign of δ is the same as sign of N. We record the following observations.

- The deformation of the bar in the undeformed direction is represented by δ.

- If N is a tensile force, then δ is elongation; if N is a compressive force, then δ is contraction.

EXAMPLE 5.3

In the assembly of a machine, a subpart is modeled as shown in Figure 5.5. Bar A has to be pulled and attached to support at D before the force F is applied. Bars A and B are made of steel with modulus of elasticity $E = 200$ GPa, cross-sectional area $A = 100$ mm², and length $L = 2.5$ m. (a) Determine the initial axial stress in both bars before the force F is applied. (b) If the applied force $F = 10$ kN, determine the total axial stress in both bars.

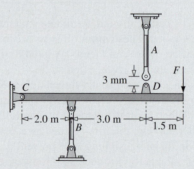

Figure 5.5 Two-bar model for Example 5.3.

PLAN

(a) After the gap has been closed, the two bars will be in tension. The degree of static redundancy for this problem is 1. To solve the problem, we can write one compatibility equation, and one equilibrium equation of moment about C. (b) We can consider the calculation of the internal forces with just F, assuming that the gap has closed and that the system is stress free before F is applied. Bar B will be in compression and bar A will be in tension owing to the force F. The internal forces in the bar can be found as in part (a). The initial stresses in part (a) can be superposed with the stresses due to just F to obtain the total axial stresses.

SOLUTION

(a) We draw the free body diagram of the rigid bars with bars A and B in tension as shown in Figure 5.6. By equilibrium of moment about point C, we obtain Equation (E1).

$$N_A(5) = N_B(2) \tag{E1}$$

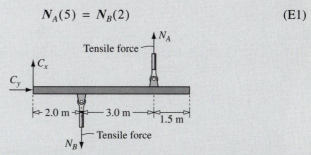

Figure 5.6 Free body diagram for part (a) of Example 5.3.

To get the compatibility equation, we draw the approximate deformed shape as shown in Figure 5.7. The movements of point E and point D on the rigid bar can be related by similar triangles. The movement of point E is equal to the deformation of the bar B, and the extension of bar A and the movement of point D equal the gap, as shown in Equations (E2) and (E3).

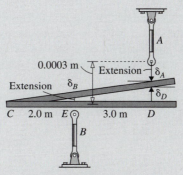

Figure 5.7 Deformed geometry for Example 5.3.

$$\frac{\delta_D}{5} = \frac{\delta_B}{2} \tag{E2}$$

$$\delta_D + \delta_A = 0.003 \tag{E3}$$

From Equations (E2) and (E3) we obtain Equation (E4).

$$2.5\delta_B + \delta_A = 0.003 \tag{E4}$$

From Equation (5.4) we obtain Equations (E5) and (E6).

$$\delta_A = \frac{N_A L_A}{E_A A_A} = \frac{N_A(2.5)}{20(10^6)} = 0.125 N_A(10^{-6}) \tag{E5}$$

$$\delta_B = \frac{N_B L_B}{E_B A_B} = \frac{N_B(2.5)}{20(10^6)} = 0.125 N_B(10^{-6}) \tag{E6}$$

Substituting Equations (E5) and (E6) into Equation (E4) we obtain Equation (E7).

$$2.5[0.125 N_B(10^{-6})] + 0.125 N_A(10^{-6}) = 0.003 \quad \text{or} \tag{E7}$$

$$2.5 N_B + N_A = 24{,}000$$

Solving Equations (E1) and Equation (E7), we obtain Equations (E8).

$$N_A = 3310.3 \text{ N} \quad \text{and} \quad N_B = 8275.9 \text{ N} \tag{E8}$$

The initial stresses in A and B can be found as shown in Equations (E9) and (E10).

$$\sigma_A = \frac{N_A}{A_A} = 33.1(10^6) \text{ N/m}^2 \quad \text{or} \tag{E9}$$

ANS. $\sigma_A = 33$ MPa (T)

$$\sigma_B = \frac{N_B}{A_B} = 82.7(10^6) \text{ N/m}^2 \quad \text{or} \tag{E10}$$

ANS. $\sigma_B = 83$ MPa (T)

(b) In the calculations that follow, the bars over the variables distinguish them from those in part (a). We draw the free body diagram of the rigid bars with bars A in tension and bar B in compression as shown in Figure 5.8. By equilibrium of moment about point C, we obtain Equation (E11).

$$F(6.5) - \overline{N}_A(5) - \overline{N}_B(2) = 0 \quad \text{or} \quad 5\overline{N}_A + 2\overline{N}_B = 65(10^3) \tag{E11}$$

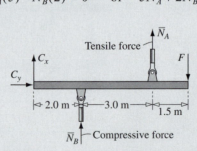

Figure 5.8 Free body diagram for part (b) of Example 5.3.

To get the compatibility equation, we draw the approximate deformed shape (Figure 5.9). For this part of the problem, the movements of points D and E are equal to the deformation of the bar. We can write the deformation relationship as follows:

$$\frac{\overline{\delta}_A}{5} = \frac{\overline{\delta}_B}{2} \tag{E12}$$

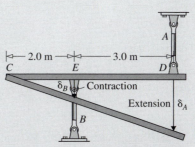

Figure 5.9 Deformed geometry for part (b) of Example 5.3.

The relation between deformation and internal forces is again as shown in Equations (E5) and (E6). Substituting Equations (E5) and (E6) in Equation (E12), we obtain Equation (E13).

$$0.125\overline{N_B}(10^{-6}) = 0.4[0.125\overline{N_A}(10^{-6})] \quad \text{or} \quad \overline{N_B} = 0.4\overline{N_A} \quad \text{(E13)}$$

Solving Equations (E11) and (E13) we obtain Equations (E14).

$$\overline{N_A} = 11.20(10^3) \quad \text{and} \quad \overline{N_B} = 4.48(10^3) \quad \text{(E14)}$$

The stresses in A and B can now be found as in Equations (E15) and (E16).

$$\overline{\sigma}_A = \frac{\overline{N_A}}{A_A} = 112(10^6) \text{ N/m}^2 = 112 \text{ MPa (T)} \quad \text{(E15)}$$

$$\overline{\sigma}_B = \frac{\overline{N_B}}{A_B} = 44.8(10^6) \text{ N/m}^2 = 45 \text{ MPa (C)} \quad \text{(E16)}$$

The total axial stress can be obtained by superposing stresses in Equations (E9) and (15) for bar A and in Equations (E10) and (E16) for bar B.

ANS. $(\sigma_A)_{\text{total}} = 145$ MPa (T) $\qquad (\sigma_B)_{\text{total}} = 38$ MPa (T)

COMMENTS

1. The importance of showing deformation (extension/contraction) in a deformed geometry that is consistent with the forces (tensile/compressive) shown in the free body diagram cannot be overstated. In Figure 5.6 both internal forces on the free body diagram were shown as tensile, and the corresponding deformations in the deformed geometry of Figure 5.7 were shown as extensions. In the free body diagram of Figure 5.8 the internal forces in bars A and B were shown as tensile and compressive, respectively. In the corresponding deformed geometry of Figure 5.9, the deformations of bars A and B were shown as extension and contractions.

2. We solved the problem twice, to incorporate the initial stress (strain) due to misfit and then to account for the external load. Since the problem is linear, it should not matter how we reach the final equilibrium position. In next section we will see that it is possible to solve the problem only once, but this requires an understanding of how initial strain is accounted for in the theory.

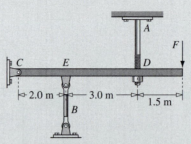

Figure 5.10 A problem similar to Example 5.3.

3. Consider the problem shown in Figure 5.10. After the nut has been finger-tightened, it is given an additional quarter-turn before the force F is applied. We are required to find the initial axial stress in both bars and the total axial stress. The pitch of the threads is 12 mm. Different mechanisms are used for introducing the initial strains for the problems in Figures 5.5 and 5.10, but the results are identical. To appreciate this, we recognize that the nut moves by pitch times the number of turns (i.e., 3 mm). Suppose that in both problems we initially ignore bar B and force F. Then in both problems the rigid bar at point D moves upward by 3 mm; but at this stage, both bar A and bar B are stress free. Now suppose that by applying a force, we stretch bar B and attach it to the rigid bar at point E. On removal of the force from bar B, the system would seek an equilibrium, which will be dictated by the properties of bar A and bar B, not by how we initially raised point D. The strain due to the tightening of a nut may be hard to visualize, but the analogous problem of strain due to misfit can be visualized and used as an alternative visualization aid.

5.3   TEMPERATURE EFFECTS IN AXIAL MEMBERS

Changes in length due to temperature variations introduce stresses due to the constraining effects of other members in a statically indeterminate structure. For purpose of analysis, there are a number of similarities between initial strain and thermal strain. Thus we shall rederive our theory to incorporate initial strain and see how this variable[2] affects our analysis. Clearly, Assumption 5 in Table 3.3 (absence of inelastic strains) is no longer valid. Hence, the formulas that follow this assumption are no longer valid. As we shall see, however, when the initial strain has been accounted for, the derivation of new formulas is the same as before.

We assume that our kinematic assumptions 2-A and 3 (on plane sections remaining plane and parallel and on small strain) are still valid. Hence the total strain at any cross section is uniform and only a function of x. Assuming that the material is isotropic and linearly elastic (Assumptions 4–7 are valid) but that temperature or any other

---

[2] In many numerical methods such as the finite element method (see Chapter 9), the initial strain approach is used not only for axial members but for all types of structural analysis, linear or nonlinear.

factor imposes an initial strain of ε_0 at a point, we obtain the stress–strain relationship of Equation (5.5).

$$\varepsilon_{xx} = \frac{du}{dx} = \frac{\sigma_{xx}}{E} + \varepsilon_0 \tag{5.5}$$

Substituting Equation (5.5) into Equation (3.4a-A) and assuming that the material is homogeneous and the initial strain ε_0 is uniform across the cross section, we obtain.

$$N = \int_A \left(E\frac{du}{dx} - E\varepsilon_0 \right) dA = \frac{du}{dx}\int_A E \, dA - \int_A E\varepsilon_0 dA = \frac{du}{dx}EA - EA\varepsilon_0 \tag{5.6}$$

or

$$\frac{du}{dx} = \frac{N}{EA} + \varepsilon_0 \tag{5.7}$$

Substituting Equation (5.7) into Equation (5.5), we obtain Equation (5.8).

$$\boxed{\sigma_{xx} = \frac{N}{A}} \tag{5.8}$$

If Assumptions 9 through 11 are still valid—that is, if N, E, and A are constant between x_1 and x_2 and ε_0 also does not change with x—then all quantities on the right-hand side of Equation (5.7) are constant between x_1 and x_2, and by integration we obtain Equation (5.9a).

$$u_2 - u_1 = \frac{N(x_2 - x_1)}{EA} + \varepsilon_0(x_2 - x_1) \tag{5.9a}$$

or alternatively

$$\boxed{\delta = \frac{NL}{EA} + \varepsilon_0 L} \tag{5.9b}$$

Equations (5.8) and (5.9b) imply that initial strain affects deformation but not stress. This seemingly paradoxical result has a different explanation for thermal strains and strains due to misfits, or from the pretensioning of bolts.

First we consider the strain ε_0 due to temperature changes. If a body is homogeneous and unconstrained, then no stresses are generated owing to temperature changes, as was observed in Section 5.1. This observation is equally true for statically determinate structures. A determinate structure simply expands or adjusts to account for the temperature changes. But in an indeterminate structure, the deformation of various members must satisfy the compatibility equations. The compatibility constraints cause internal forces to be generated, which in turn affect the stresses.

In thermal analysis $\varepsilon_0 = \alpha \, \Delta T$, an increase in temperature corresponds to extension, and a decrease in temperature corresponds to contraction. Equation (5.9b) assumes that N is positive in tension, and hence extensions due to ε_0 are positive and those due to contraction are negative. However, if on the free body diagram N is shown as a compressive force, and δ is shown as a contraction in the deformed shape, then consistency requires that the contraction due to ε_0 be treated as positive and the extension as negative in Equations (5.9b).

- The sign of $\varepsilon_0 L$ due to *temperature changes* must be consistent with the force N shown on the free body diagram.

We now consider the issue of initial strains caused by the factors discussed in Section 5.2. If we start our analysis with the undeformed geometry, then even when there is an initial strain or stress we have by implication imposed a *strain that is opposite in sign to the actual initial strain* before any external loads have been imposed. To elaborate this issue of sign, we put $\delta = 0$ in Equation (5.9b), to correspond to the undeformed state, and note that N and ε_0 must have opposite signs for the two terms on right-hand side to combine to yield a result of zero. But strain and internal forces must have same sign. For example, if a member is short and must be pulled to overcome the gap due misfit, then the bar has been extended at the undeformed state and is in tension before the application of external loads has begun. The problem can be corrected only if we think of ε_0 as negative with respect to the actual initial strain. Thus,

- *Prestrains (stresses) can be analyzed by using ε_0 as negative with respect to the actual initial strain in Equation (5.9b).*

A problem that has external forces in addition to the initial strain can be solved in two ways. We can find the stresses and deformation due to the initial strain and to the external forces individually, as we did in Section 5.2, and superpose the solution. The advantage of such an approach is that we have a good intuitive feel for the solution process. The disadvantage is that we are solving the problem twice. Alternatively, we can use Equation (5.9b) and solve the problem once, being careful with our signs, however, since the approach is less intuitive and more mathematical.

EXAMPLE 5.4

Bars A and B shown Figure 5.11 are made of steel with modulus of elasticity $E = 200$ GPa, coefficient of thermal expansion $\alpha = 12 \, \mu/°C$, cross-sectional area $A = 100 \, mm^2$, and length $L = 2.5$ m. If the applied force $F = 10$ kN and temperature of the bar A is decreased by 100°C, find the total axial stress in both bars.

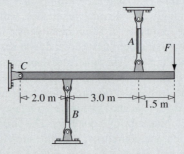

Figure 5.11 Two-bar mechanism for Example 5.4.

PLAN

The problem has one degree of static redundancy. We can write one compatibility equation and use Equation (5.9b) to get one equation relating the internal forces. By taking the moment about point C in the free body diagram of the rigid bar, we can obtain the remaining equation and solve the problem.

SOLUTION

The axial rigidity and the thermal strain can be calculated as shown in Equations (E1) and (E2).

$$EA = 200(10^9)(100)(10^{-6}) = 20(10^6) \text{ N} \tag{E1}$$

$$\varepsilon_0 = \alpha \, \Delta T = (12)(10^{-6})(-100) = -1200(10^{-6}) \tag{E2}$$

We draw the free body diagram of the rigid bar with bar A in tension and bar B in compression, as shown in Figure 5.12. By the moment equilibrium about point C, we obtain Equation (E3).

$$F(6.5) - N_A(5) - N_B(2) = 0 \quad \text{or} \quad 5N_A + 2N_B = 65(10^3) \tag{E3}$$

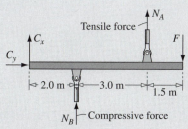

Tensile force

N_A

C_x

C_y

F

2.0 m   3.0 m   1.5 m

N_B Compressive force

Figure 5.12 Free body diagram for Example 5.4.

To get the compatibility equation, we draw the approximate deformed shape as shown in Figure 5.13. The movements of points D and E are equal to the deformation of the bar. The deformation relationship is given in Equation (E4).

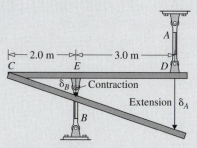

2.0 m   3.0 m

C E D A

δ_B Contraction

Extension δ_A

B

Figure 5.13 Deformed geometry for Example 5.4.

$$\frac{\delta_A}{5} = \frac{\delta_B}{2} \tag{E4}$$

From Equation (5.9b) we obtain Equations (E5) and E6).

$$\delta_A = \frac{N_A L_A}{E_A A_A} + \varepsilon_0 L_A = \frac{N_A(2.5)}{20(10^6)} - 1200(2.5)(10^{-6}) = (0.125 N_A - 3000)(10^{-6}) \tag{E5}$$

$$\delta_B = \frac{N_B L_B}{E_B A_B} = \frac{N_B(2.5)}{20(10^6)} = 0.125 N_B(10^{-6}) \qquad \text{(E6)}$$

Substituting Equations (E5) and (E6) in Equation (E2), we obtain Equation (E7).

$$0.125 N_B(10^{-6}) = 0.4[(0.125 N_A - 3000)(10^{-6})] \quad \text{or} \quad N_B = 0.4 N_A - 9600 \quad \text{(E7)}$$

Solving Equations (E3) and (E7), we obtain the internal forces shown in Equations (E8).

$$N_A = 14.52(10^3) \quad \text{and} \quad N_B = -3.79(10^3) \qquad \text{(E8)}$$

Noting that we assumed that bar B is in compression, we see that the sign of N_B in Equation (E6) implies that this variable is in tension. The stresses in A and B can now be found by dividing the internal forces by the cross-sectional area.

ANS. $\sigma_A = 145$ MPa (T) $\sigma_B = 38$ MPa (T)

COMMENTS

1. In Figures 5.5 and 5.10 the prestrain in member A is $0.003/2.5 = 1200(10^{-6})$ extension. This means that $\varepsilon_0 = -1200(10^{-6})$, which is the same as the thermal strain in Equation (E2). Thus, it is not surprising that the results of Examples 5.4 and 5.3 are identical. But unlike Example 5.3, this time we solved the problem only once.

2. It would be hard to intuit that bar B will be in tension because the initial strain is greater than the strain caused by the external force F. But this observation is obvious in the two solutions obtained in Example 5.3.

3. To calculate the initial strain by using the method in this example, it is recommended that the problem be formulated in terms of the force F initially. Then for the calculation of initial strain, substitute $F = 0$. This recommendation avoids some of the confusion that would be due to changing the sign of ε_0 in the initial strain calculations.

PROBLEM SET 5.1

5.1 An iron rim ($\alpha = 6.5$ μ/°F) of diameter 35.98 inches is to be placed on a wooden cask of diameter 36 inches. Determine the minimum temperature increase needed to slip the rim onto the cask.

5.2 Determine the angle by which the pointer shown in Figure P5.2 rotates from the vertical position if the temperature is increased by 60°C in both steel ($E_{st} = 200$ GPa, and

$\alpha_{st} = 12.0$ μ/°C) and aluminum ($E = 72$ GPa, and $\alpha = 23.0$ μ/°F.

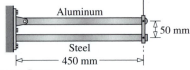

Figure P5.2

5.3 A plate ($E = 30,000$ ksi; $\nu = 0.25$; $\alpha = 6.5 \times 10^{-6}$/°F) cannot expand in the y direction and can expand at most by 0.005 inch in the *x* direction as shown in Figure P5.3. Assuming plane stress, determine the average normal stresses in the *x* and *y* directions due to a uniform temperature increase of 100°F.

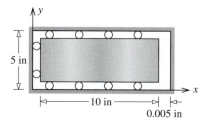

Figure P5.3

ANS. $\sigma_{xx} = 10$ ksi (C) $\sigma_{yy} = 22$ ksi (C)

5.4 Calculate ε_{xx}, ε_{yy}, γ_{xy}, ε_{zz}, and σ_{zz} (a) assuming plane stress and (b) assuming plane strain for the following stresses at a point, material properties, and change in temperature.

$$\sigma_{xx} = 100 \text{ MPa (T)} \qquad \sigma_{yy} = 150 \text{ MPa (T)}$$
$$\tau_{xy} = -125 \text{ MPa}$$

$$E = 200 \text{ GPa} \qquad \nu = 0.32 \qquad \alpha = 11.7 \ \mu/°C$$
$$\Delta T = -25°C$$

ANS. (a) $\varepsilon_{xx} = -32.5 \ \mu$, $\varepsilon_{yy} = 298 \ \mu$, $\gamma_{xy} = -1650 \ \mu$, $\varepsilon_{zz} = -693 \ \mu$ (b) $\varepsilon_{xx} = -254 \ \mu$, $\varepsilon_{yy} = 75.9 \ \mu$, $\sigma_{zz} = 138.5 \text{ MPa (T)}$

5.5 Calculate ε_{xx}, ε_{yy}, γ_{xy}, ε_{zz}, and σ_{zz} (a) assuming plane stress and (b) assuming plane strain for the following stresses at a point, material properties, and change in temperature.

$$\sigma_{xx} = 300 \text{ MPa (C)} \qquad \sigma_{yy} = 300 \text{ MPa (T)}$$
$$\tau_{xy} = 150 \text{ MPa}$$

$$G = 15 \text{ GPa} \qquad \nu = 0.2 \qquad \alpha = 26.0 \ \mu/°C$$
$$\Delta T = 75°C$$

5.6 Calculate ε_{xx}, ε_{yy}, γ_{xy}, ε_{zz}, and σ_{zz} (a) assuming plane stress and (b) assuming plane strain for the following stresses at a point, material properties, and change in temperature.

$$\sigma_{xx} = 22 \text{ ksi (C)} \qquad \sigma_{yy} = 25 \text{ ksi (C)}$$
$$\tau_{xy} = -15 \text{ ksi}$$

$$E = 30,000 \text{ ksi} \qquad \nu = 0.3 \qquad \alpha = 6.5 \ \mu/°F$$
$$\Delta T = 40°F$$

5.7 Calculate σ_{xx}, σ_{yy}, τ_{xy}, σ_{zz}, and ε_{zz} assuming that the point is in plane stress for the following strains at a point, material properties, and change in temperature.

$$\varepsilon_{xx} = -800 \ \mu \qquad \varepsilon_{yy} = -1000 \ \mu$$
$$\gamma_{xy} = -500 \ \mu$$

$$E = 30,000 \text{ ksi} \qquad \nu = 0.3 \qquad \alpha = 6.5 \ \mu/°F$$
$$\Delta T = 40°F$$

ANS. $\sigma_{xx} = 47.4$ ksi (C)
$\sigma_{yy} = 52.02$ ksi (C) $\tau_{xy} = -5.77$ ksi
$\varepsilon_{zz} = 1254 \ \mu$

5.8 Calculate σ_{xx}, σ_{yy}, τ_{xy}, σ_{zz}, and ε_{zz} assuming that the point is in plane stress for the following strains at a point, material properties, and change in temperature.

$$\varepsilon_{xx} = -3000 \ \mu \qquad \varepsilon_{yy} = 1500 \ \mu$$
$$\gamma_{xy} = 2000 \ \mu$$

$$E = 70 \text{ GPa} \qquad G = 28 \text{ GPa}$$
$$\alpha = 26.0 \ \mu/°C \qquad \Delta T = 75°C$$

5.9 Calculate σ_{xx}, σ_{yy}, τ_{xy}, σ_{zz}, and ε_{zz} assuming that the point is in plane stress for the following strains at a point, material properties, and change in temperature.

$$\varepsilon_{xx} = 1500 \ \mu \qquad \varepsilon_{yy} = -1200 \ \mu$$
$$\gamma_{xy} = -1000 \ \mu$$

$$E = 10,000 \text{ ksi} \qquad G = 3900 \text{ ksi}$$
$$\alpha = 12.8 \ \mu/°F \qquad \Delta T = -100°F$$

5.10 Derive the following relations of normal stresses in terms of normal strain from Equations (5.3a), (5.3b), and (5.3c).

$$\sigma_{xx} = [(1 - \nu)\varepsilon_{xx} + \nu\varepsilon_{yy} + \nu\varepsilon_{zz}]$$
$$\times \frac{E}{(1 - 2\nu)(1 + \nu)} - \frac{E\alpha \ \Delta T}{(1 - 2\nu)}$$

$$\sigma_{yy} = [(1 - \nu)\varepsilon_{yy} + \nu\varepsilon_{zz} + \nu\varepsilon_{xx}]$$
$$\times \frac{E}{(1 - 2\nu)(1 + \nu)} - \frac{E\alpha \ \Delta T}{(1 - 2\nu)} \quad (5.10)$$

$$\sigma_{zz} = [(1 - \nu)\varepsilon_{zz} + \nu\varepsilon_{xx} + \nu\varepsilon_{yy}]$$
$$\times \frac{E}{(1 - 2\nu)(1 + \nu)} - \frac{E\alpha \ \Delta T}{(1 - 2\nu)}$$

5.11 For a point in plane stress, show

$$\sigma_{xx} = [\varepsilon_{xx} + v\varepsilon_{yy}]$$

$$\times \frac{E}{(1 - v^2)} - \frac{E\alpha\,\Delta T}{(1 - v)}$$

$$\sigma_{yy} = [\varepsilon_{yy} + v\varepsilon_{xx}] \qquad (5.11)$$

$$\times \frac{E}{(1 - v^2)} - \frac{E\alpha\,\Delta T}{(1 - v)}$$

5.12 For a point in plane stress, show

$$\varepsilon_{zz} = -\left(\frac{v}{1 - v}\right)(\varepsilon_{xx} + \varepsilon_{yy})$$

$$\qquad (5.12)$$

$$+ \frac{(1 - 3v)}{(1 - v)}\alpha\,\Delta T$$

5.13 During assembly of the structure shown in Figure P5.13, a misfit between bar A and the attachment of the rigid bar was found. Determine the initial stress due to the misfit if bar A is pulled and attached. The modulus of elasticity of the circular bars A and B is $E = 10,000$ ksi, and the diameter is 1 inch.

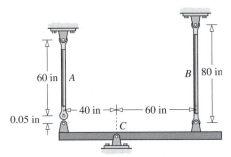

Figure P5.13

5.14 Bar A is 2 mm shorter than bar B owing to a manufacturing error. The attachment of these bars to the rigid bar would cause a misfit of 2 mm. Which of the two assembly configurations shown in Figure P5.14 would you recommend? Calculate the initial stress corresponding to your answer. Use a modulus of elasticity of $E = 70$ GPa and 25 mm as the diameter of the circular bars.

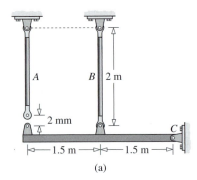

(a)

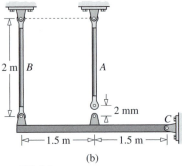

(b)

Figure P5.14

ANS. $\sigma_A = 14$ MPa (T) $\sigma_B = 28$ MPa (T)

5.15 A cross section of steel bolt passes through an aluminum sleeve as shown in Figure P5.15. After the unit has been assembled to the point of finger-tightening the nut (no deformation), the nut is given a quarter-turn. If the pitch of the threads is 3 mm, determine the initial axial stress developed in the sleeve and the bolt. The moduli of elasticity for steel and for aluminum are $E_{st} = 200$ GPa and $E_{Al} = 70$ GPa; the cross-sectional areas are $A_{st} = 500$ mm^2 and $A_{Al} = 1100$ mm^2.

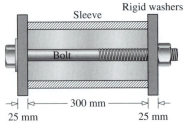

Figure P5.15

ANS. $\sigma_{Al} = 92.2$ MPa (C)
$\sigma_{st} = 202.8$ MPa (T)

5.16   The rigid bar in Figure P5.16 is horizontal when the unit is put together to the point of finger-tightening the nut. The pitch of the threads is 0.125 in. Use the material properties given in Table P5.16 to develop a table in steps of quarter-turn of the nut that can be used for prescribing the pretension in bar B. The maximum number of quarter-turns is limited by the yield stress.

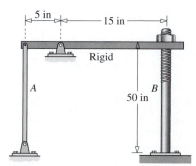

Figure P5.16

TABLE 5.16 Material Properties for Bars A and B

	Bar A	Bar B
Modulus of elasticity	10,000 ksi	30,000 ksi
Yield stress	24 ksi	30 ksi
Cross-sectional area	0.5 in²	0.75 in²

ANS. $N_{max} = 11$ quarter-turns
$\sigma_A = 22.4$ ksi $\sigma_B = 4.97$ ksi

5.17   The temperature increase in the bar in Figure P5.17 varies as $\Delta T = T_L(x^2/L^2)$. Determine the axial stress and movement of a point at $x = L/2$ in terms of length L, modulus of elasticity E, cross-sectional area A, coefficient of thermal expansion α, and increase of temperature at the end T_L.

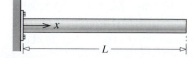

Figure P5.17

5.18   The temperature increase in the bar in Figure P5.17 varies as $\Delta T = T_L(x^2/L^2)$. Determine the axial stress and movement of a point at $x = L/2$ in terms of length L, modulus of elasticity E, cross-sectional area A, coefficient of thermal expansion α, and the increase of temperature at the end T_L.

Figure P5.18

5.19   The tapered bar shown in Figure P5.19 has a cross-sectional area that varies with x as $A = K(L - 0.5\ x)^2$. If the increase in temperature of the bar varies as $\Delta T = T_L(x^2/L^2)$, determine the axial stress at the midpoint in terms of the length L, modulus of elasticity E, cross-sectional area parameter K, coefficient of thermal expansion α, and increase of temperature at the end T_L.

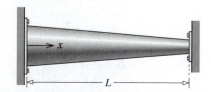

Figure P5.19

ANS. $\sigma_{xx} = 0.296\ \alpha T_L E$ (C)

5.20   Three metallic rods are attached to a rigid plate as shown in Figure P5.20. The temperature of the rods is lowered by 100°F after the forces described in Table P5.20 have been applied. Assuming that the rigid plate does not rotate, determine the movement of the rigid plate.

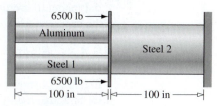

Figure P5.20

TABLE 5.20 Material Properties for Metallic Rods

	Area (in²)	E (ksi)	α (10⁻⁶/°F)
Aluminum	4	10,000	12.5
Steel 1	4	30,000	6.6
Steel 2	12	30,000	6.6

ANS. 0.0233 in

5.21 Solve Problem 5.15, assuming that in addition to the turning of the nut, the temperature of the assembled unit is raised by 40°C. The coefficients of thermal expansion for steel and aluminum are $\alpha_{st} = 12 \; \mu/°C$ and $\alpha_{Al} = 22.5 \; \mu/°C$.

ANS. $\sigma_{Al} = 104.7$ MPa (C)
   $\sigma_{st} = 230.4$ MPa (T)

5.22   Determine the axial stress in bar A of Problem 5.16, assuming that the nut is turned 1 full turn and the temperature of bar A is decreased by 50°F. The coefficient of thermal expansion for bar A is $\alpha_{st} = 22.5 \; \mu/°F$.

ANS. $\sigma_A = 25.7$ ksi (T)

5.4   NONLINEAR MATERIAL MODELS

Rubber, plastics, and muscles and other organic tissues exhibit nonlinearity in the stress–strain relationship, even at small strains. Metals also the exhibit nonlinearity after yield stress. In this section, we consider various nonlinear material models: that is, various forms of equations that are used for representing the stress–strain nonlinear relationship. The constants in the equations relating stress and strain are material constants. Material constants are found by matching the stress–strain equation to the experimental data in a least-squares sense.

A critical issue in nonlinear structural analysis is the determination of the origin of a given coordinate system such that the axial and bending problems are decoupled (i.e., separating axial load analysis from bending load analysis). We will take up this critical problem in nonlinear analysis a little later. In this section we bypass the problem by considering only symmetric cross sections, with symmetric material behavior in tension and compression. These limitations will ensure that the stress distribution in tension and compression will be symmetric about the centroid of the cross section. The centroid of the cross section will be the origin in all problems in this section.

We shall consider three material models:

1. Elastic–perfectly plastic, in which the nonlinearity is approximated by a constant.
2. Linear strain hardening, in which the nonlinearity is approximated by a linear function.
3. Power law, in which the nonlinearity is approximated by one term, a nonlinear function.

There are other material models described in Problem Set 5.2. The choice of material model to use depends not only upon the material stress–strain curve, but also upon the degree of accuracy needed and the resulting complexity of the analysis.

5.4.1   Elastic–Perfectly Plastic Material Models

Figure 5.14 shows the stress–strain curves describing an elastic–perfectly plastic behavior of a material. It is assumed that the material has the same behavior in tension and in

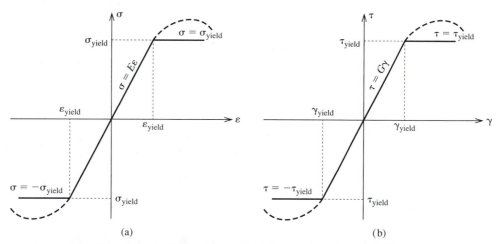

Figure 5.14 Elastic–perfectly plastic material behavior: (a) normal and (b) shear.

compression. Similarly for shear stress–strain, the material behavior is same for positive and negative stresses and strains.

Before yield stress, the stress–strain relationship is given by Hooke's law, after yield stress, the stress is a constant. Elastic–perfectly plastic material behavior is a simplifying approximation used to conduct an elastic–plastic analysis. The approximation is a conservative approximation, since it ignores the capacity of the material to carry stresses higher than the yield stress.

The equations describing the stress–strain curve are as follows:

$$\sigma = \begin{cases} \sigma_{yield} & \varepsilon \geq \varepsilon_{yield} \\ E\varepsilon & -\varepsilon_{yield} \leq \varepsilon \leq \varepsilon_{yield} \\ -\sigma_{yield} & \varepsilon \leq -\varepsilon_{yield} \end{cases} \quad \text{and} \quad \tau = \begin{cases} \tau_{yield} & \gamma \geq \gamma_{yield} \\ G\gamma & -\gamma_{yield} \leq \gamma \leq \gamma_{yield} \\ -\tau_{yield} & \gamma \leq -\gamma_{yield} \end{cases} \quad (5.13)$$

> **Definition 1** The set of points forming the boundary between the elastic and plastic regions on a body is called the elastic–plastic boundary.

Determining the location of the elastic–plastic boundary is one of the critical issues in elastic–plastic analysis. As our examples will demonstrate, the location of the elastic–plastic boundary is determined using the following observations:

1. On the elastic–plastic boundary, strain must be equal to yield strain, and stress equal to yield stress.

2. Deformations and strains are continuous at all points, including points at the elastic–plastic boundary.

Deformation that is not continuous implies that holes or cracks are being formed in the material. If strains, which are derivative displacements, are not continuous, then corners are being formed during deformation.

5.4.2   Linear Strain-Hardening Material Model

Figure 5.15, shows stress–strain curves for the linear strain-hardening model, also referred to as the *bilinear material*[3] *model*. It is assumed that the material has the same behavior in tension and in compression. Similarly for shear stress–strain, the material behavior is same for positive and negative stresses and strains.

This is another conservative, simplifying approximation of material behavior in which we once more ignore the ability of the material to carry stresses higher than those shown by the straight lines. Determining the location of the elastic–plastic boundary is once more a critical issue in the analysis and is determined as described in Section 5.4.1.

The equations describing the stress–strain curve are as follows:

$$
\sigma = \begin{cases}
\sigma_{\text{yield}} + E_2(\varepsilon - \varepsilon_{\text{yield}}) & \varepsilon \geq \varepsilon_{\text{yield}} \\
E_1\varepsilon & -\varepsilon_{\text{yield}} \leq \varepsilon \leq \varepsilon_{\text{yield}} \\
-\sigma_{\text{yield}} + E_2(\varepsilon + \varepsilon_{\text{yield}}) & \varepsilon \geq \varepsilon_{\text{yield}}
\end{cases}
$$

$$(5.14)$$

$$
\tau = \begin{cases}
\tau_{\text{yield}} + G_2(\gamma - \gamma_{\text{yield}}) & \gamma \geq \gamma_{\text{yield}} \\
G_1\gamma & -\gamma_{\text{yield}} \leq \gamma \leq \gamma_{\text{yield}} \\
-\tau_{\text{yield}} + G_2(\gamma + \gamma_{\text{yield}}) & \gamma \leq \gamma_{\text{yield}}
\end{cases}
$$

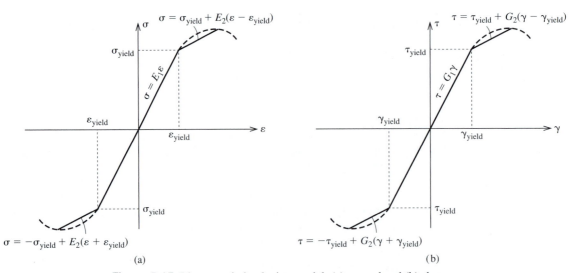

Figure 5.15 Linear strain hardening model: (a) normal and (b) shear.

---

[3] The numerical analysis technique called *incremental plasticity* uses a similar approximation. The major difference is that the nonlinear stress–strain curve is approximated not by a single straight line, but by a series of straight lines over small intervals.

5.4.3 Power Law Model

Figure 5.16 shows a power law representation of the nonlinear stress–strain curve. It is assumed that the material has the same behavior in tension and in compression. Similarly, for shear stress–strain, the material behavior is same for positive and negative stresses and strains. The equations describing the stress–strain relationship are as follows:

$$\sigma = \begin{cases} E\varepsilon^{n} & \varepsilon \geq 0 \\ -E(-\varepsilon)^{n} & \varepsilon < 0 \end{cases} \quad \text{and} \quad \tau = \begin{cases} G\gamma^{n} & \gamma \geq 0 \\ -G(-\gamma)^{n} & \gamma < 0 \end{cases} \tag{5.15}$$

The constants E and n, called the strength coefficient and the strain-hardening coefficient, are determined to fit the experimental stress–strain curve in a least-squares sense. Most metals in the plastic region and most plastics are represented by the solid curve, with a strain-hardening coefficient less than 1. Soft rubber, as well as muscles and other organic materials, represented by the dashed line in Figure 5.16, have a strain-hardening coefficient greater than 1.

From Equation (5.15) we note that when strain is negative, the term on the right becomes positive, permitting evaluation of the number to fractional powers. Furthermore, with negative strain we obtain negative stress, as we should.

In Section 5.4.2, we explained that the stress–strain relationship can be written by using different equations for different stress levels. We could, in a similar manner combine a linear equation for the linear part and nonlinear equation for the nonlinear part; or we could combine two nonlinear equations, thus creating additional material models. Several other material models are considered in Problem set 5.2.

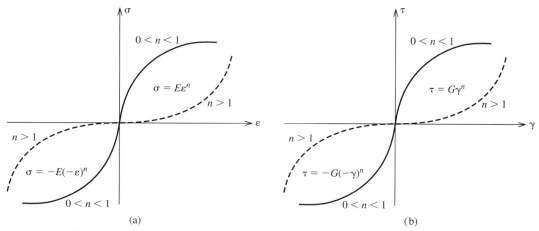

Figure 5.16 Nonlinear stress–strain curves: (a) normal and (b) shear.

EXAMPLE 5.5

Aluminum has a yield stress of $\sigma_{yield} = 40$ ksi in tension, a yield strain of $\varepsilon_{yield} = 0.004$, an ultimate stress of $\sigma_{ult} = 45$ ksi, and the corresponding ultimate strain of $\varepsilon_{ult} = 0.017$. Determine the material constants for (a) the elastic–perfectly plastic model, (b) the linear strain-hardening model, and (c) the nonlinear power law model, and (d) plot the stress–strain curves for each model.

PLAN

We have coordinates of three points on the curve: P_0 ($\sigma_0 = 0.00$, $\varepsilon_0 = 0.000$), P_1 ($\sigma_1 = 40.0$, $\varepsilon_1 = 0.004$), and P_2 ($\sigma_2 = 45.0$, $\varepsilon_2 = 0.017$). We can use these data to find the various constants in the material models.

SOLUTION

(a) The modulus of elasticity E is the slope between points P_0 and P_1 and can be found as in Equation (E1).

$$E_1 = \frac{\sigma_1 - \sigma_0}{\varepsilon_1 - \varepsilon_0} = \frac{40}{0.004} = 10{,}000 \text{ ksi} \tag{E1}$$

After yield stress, the stress is a constant. The stress–strain behavior can be written as shown in Equation (E2).

$$\sigma = \begin{cases} 10{,}000\varepsilon \text{ ksi} & |\varepsilon| \leq 0.004 \\ 40 \text{ ksi} & |\varepsilon| \geq 0.004 \end{cases} \tag{E2}$$

(b) In the linear strain-hardening model, the slope of the straight line before the yield stress is as calculated in Equation (E1). After the yield stress, the slope of the line can be found from the coordinates of points P_1 and P_2 as shown in Equation (E3).

$$E_2 = \frac{\sigma_2 - \sigma_1}{\varepsilon_2 - \varepsilon_1} = \frac{5}{0.013} = 384.6 \text{ ksi} \tag{E3}$$

The stress–strain behavior can be written as shown in Equation (E4).

$$\sigma = \begin{cases} 10{,}000\varepsilon \text{ ksi} & |\varepsilon| \leq 0.004 \\ 40 + 384.6(\varepsilon - 0.004) \text{ ksi} & |\varepsilon| \geq 0.004 \end{cases} \tag{E4}$$

(c) The two constants E and n in $\sigma = E\varepsilon^n$ can be found by substituting the coordinates of the two points P_1 and P_2, to generate the Equations (E5) and (E6).

$$40 = E(0.004)^n \tag{E5}$$

$$45 = E(0.017)^n \tag{E6}$$

Dividing Equation (E6) by Equation (E5) and taking the logarithm of both sides, we can solve for n as shown in Equation (E7).

$$\ln\left(\frac{0.017}{0.004}\right)^n = \ln\left(\frac{45}{40}\right) \quad \text{or} \quad n\ln(4.25) = \ln(1.125) \quad \text{or}$$

$$n = 0.0814 \tag{E7}$$

Substituting Equation (E7) into Equation (E5), we can obtain the value of E as follows:

$$E = 40/(0.004)^{0.0814} = 62.7 \text{ ksi} \tag{E8}$$

We can now write the stress–strain equations for the power law model as shown in Equation (E9).

$$\sigma = \begin{cases} 62.7\varepsilon^{0.0814} \text{ ksi} & \varepsilon \geq 0 \\ -62.7(-\varepsilon)^{0.0814} \text{ ksi} & \varepsilon < 0 \end{cases} \tag{E9}$$

(d) Stresses at different strains can be found by using Equations (E2), (E4), and (E9) and plotted as shown in Figure 5.17.

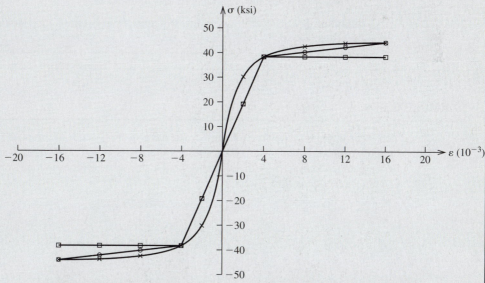

Figure 5.17 Stress–strain curves for the different models in Example 5.5: $\square$, elastic–perfectly plastic; $\mathbf{O}$, linear strain hardening; $\mathbf{\times}$, power law.

5.5 ELASTIC–PERFECTLY PLASTIC AXIAL MEMBERS

Stress is uniform across the cross section of an axial member. Thus, when an axial member in a structure reaches yield stress, the entire member will become plastic, assuming that the axial member has no distributed load. In accordance with Figure 5.14(a), the member that has turned plastic can carry stress equal to yield stress and can elongate or compress to any amount needed to satisfy the compatibility equations. It is the compatibility equations (continuity of displacement) that are used in determining the elongation or strain in an axial member that has turned plastic.

If we plot the applied load vs the displacement of the point at which the load is applied, we obtain the load deflection curve. In the elastic region, the load deflection curve is a straight line. When one member in a structure goes plastic, the deformation is controlled by the other members. But since the stress is constant in the plastic member, there is a change in the slope of the load deflection curve. The slope of the load deflection curve changes each time a member goes plastic. But if a sufficient number of members go plastic, we may have uncontrolled deformation, whereupon the structure is said to have collapsed. The load at which the structure collapses is called the collapse load. For the simple structures we will consider, "collapse load" corresponds to all members just turning plastic. But more complex structures can collapse without requiring all members to turn plastic. The analytical technique called *limit analysis* can be used for determining the collapse load for complex structures, but it is beyond the scope of this book.

We record the following for future use.

Definition 2 The plot of the applied force vs the deflection at that point in the direction of the applied force is called the load deflection curve.

Definition 3 The load at which the structure exhibits unbounded deformation is called the collapse load.

EXAMPLE 5.6

The steel members shown in Figure 5.18, have a cross-sectional area $A = 100$ mm^2, a modulus of elasticity $E = 200$ GPa, and a yield stress $\sigma_{yield} = 250$ MPa. The three members are attached to a roller that can slide in the slot shown. Plot the load deflection curve and determine the collapse load.

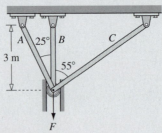

Figure 5.18 Geometry for Example 5.6.

PLAN

We can use the small-strain approximation to write the compatibilty equations relating the deformation of each bar to the displacement of the roller δ_P. By using Equations (5.4), we can relate the internal forces to the deformation of each bar and hence to δ_P. The free body diagram of the roller allows us to relate the internal forces in the members to the external forces. Since the yield stress and cross-sectional areas are the same for all members, the member that reaches yield stress first will be the one that has the greatest internal force. At the yield stress of that member, the corresponding F and δ_P can be found. We repeat the foregoing calculations for each member until all have turned plastic.

SOLUTION

The length of bar B is given and the lengths of bars A and C can be found from geometry as follows:

$$L_B = 3 \text{ m} \qquad L_A = 3/\cos 25 = 3.31 \text{ m} \qquad L_C = 3/\cos 55 = 5.23 \text{ m} \qquad \text{(E1)}$$

The axial rigidity for all three members is

$$EA = 200(10^9)100(10^{-6}) = 20(10^6) \text{ N} \qquad \text{(E2)}$$

Compatibility equations: We draw an exaggerated deformed shape as shown in Figure 5.19 and use the small-strain approximation to relate the displacement of the pin to the deformation of the bars as follows:

$$\delta_A = \delta_P \cos 25 = 0.9063\delta_P \qquad \text{(E3)}$$

$$\delta_B = \delta_P \qquad \text{(E4)}$$

$$\delta_C = \delta_P \cos 55 = 0.5736\delta_P \qquad \text{(E5)}$$

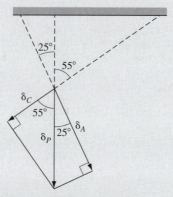

Figure 5.19 Deformation relationship for Example 5.6.

Equilibrium equations: We can draw the free body diagram as shown in Figure 5.20 and relate the internal forces in the members to the force F by equilibrium of forces in the y direction as follows:

$$N_A \cos 25 + N_B + N_C \cos 55 - F = 0 \qquad \text{or}$$

$$0.9063 \, N_A + N_B + 0.5736 \, N_C = F \qquad (E6)$$

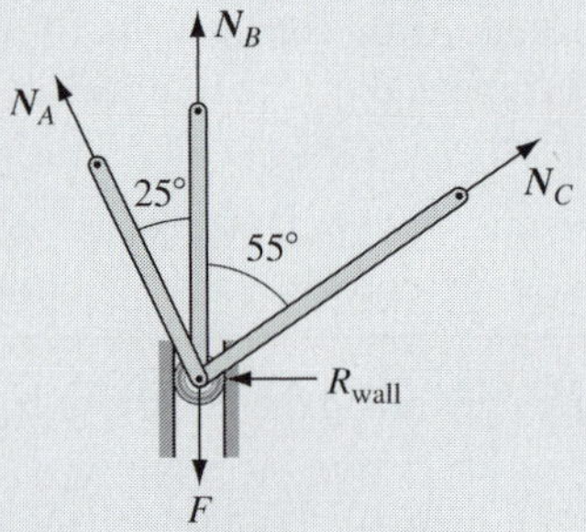

Figure 5.20 Equilibrium of forces in Example 5.6.

Assuming linear-elastic behavior for all members, we can write the internal force in terms of the deformation of the member by using Equations (5.4). Next we substitute Equations (E3), (E4), and (E5) and obtain the internal force in terms of δ_P as follows:

$$N_A = \left(\frac{E_A A_A}{L_A}\right)\delta_A = \left[\frac{20(10^6)}{3.31}\right](0.9063\delta_P) = 5.476(10^6)\delta_P \qquad (E7)$$

$$N_B = \left(\frac{E_B A_B}{L_B}\right)\delta_B = \left[\frac{20(10^6)}{3}\right](\delta_P) = 6.667(10^6)\delta_P \qquad (E8)$$

$$N_C = \left(\frac{E_C A_C}{L_C}\right)\delta_C = \left[\frac{20(10^6)}{5.23}\right](0.5736\delta_P) = 2.1934(10^6)\delta_P \qquad (E9)$$

First member turns plastic: From Equations (E7), (E8), and (E9), we see that the internal force in B is maximum. With all members having the same cross-sectional area and yield stress, we know bar B will be the first to turn plastic. The internal force in B at the yield stress can be obtained as follows:

$$N_B = \sigma_{\text{yield}}A_B = (250)(10^6)(100)(10^{-6}) = 25(10^3) \text{ N} \qquad (E10)$$

At yield stress, Equation (E8) is still valid, and we can find the value of δ_{P1} at which member B turns plastic as follows:

$$\delta_{P1} = \frac{25(10^3)}{6.667(10^6)} = 3.75(10^{-3}) \text{ m} \qquad (E11)$$

Substituting Equation (E11) into Equations (E7) and (E9), we obtain the internal forces in members A and C:

$$N_A = (5.476)(10^6)(3.75)(10^{-3}) = 20.53(10^3) \tag{E12}$$

$$N_C = (2.1934)(10^6)(3.75)(10^{-3}) = 8.225(10^3) \tag{E13}$$

Substituting Equations (E10), (E12), and (E13) into Equation (E6), we obtain the force F_1, at which member B turns plastic, as follows:

$$0.9063(20.53)(10^3) + 25(10^3) + 0.5736(8.225)(10^3) = F_1 \quad \text{or}$$

$$F_1 = 53.4 \text{ kN} \tag{E14}$$

Second member turns plastic: From Equations (E7) and (E9), and the fact that the cross-sectional area and the yield stress are the same, we know that A will turn plastic before C. The internal force in A will be as follows:

$$N_A = \sigma_{\text{yield}} A_A = (250)(10^6)(100)(10^{-6}) = 25(10^3) \text{ N} \tag{E15}$$

At yield stress, Equation (E7) is still valid and we can find the value of δ_{P2} at which member A turns plastic as follows:

$$\delta_{P2} = \frac{25(10^3)}{5.476(10^6)} = 4.57(10^{-3}) \text{ m} \tag{E16}$$

Substituting Equation (E16) into Equations (E7) and (E9), we obtain the internal forces in member C as follows:

$$N_C = (2.1934)(10^6)(4.57)(10^{-3}) = 10.01(10^3) \text{ N} \tag{E17}$$

Substituting Equations (E10), (E15), and (E17) into Equation (E6), we obtain the force F_2, at which member A turns, plastic:

$$0.9063(25)(10^3) + 25(10^3) + 0.5736(10.01)(10^3) = F_2 \quad \text{or}$$

$$F_2 = 62.0 \text{ kN} \tag{E18}$$

Third member turns plastic: The internal force in C at the yield stress can be obtained as follows:

$$N_C = \sigma_{\text{yield}} A_C = (250)(10^6)(100)(10^{-6}) = 25(10^3) \text{ N} \tag{E19}$$

At yield stress, Equation (E11) is still valid, and we can find the value of δ_{P3}, at which member C turns plastic, as follows:

$$\delta_{P3} = \frac{25(10^3)}{2.1934(10)^6} = 11.4(10^{-3}) \text{ m} \tag{E20}$$

Substituting Equation (E10), (E15), and (E19) into Equation (E6), we obtain the force at which member C turns plastic. The structure now cannot support any load. Thus the force at which C turns plastic is the collapse load and can be found as follows:

$$0.9063(25)(10^3) + 25(10^3) + 0.5736(25)(10^3) = F_{collapse} \qquad \text{(E21)}$$

ANS. $F_{collapse} = 48.3$ kN

We can use Equations (E11), (E14), (E16), (E18), (E20), and (E21) to plot F vs. δ_P; we can obtain the load deflection curve for the structure as shown in Figure 5.21

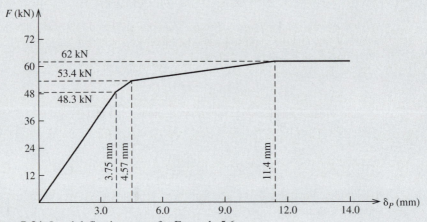

Figure 5.21 Load deflection curve for Example 5.6.

COMMENTS

1. The compatibility Equations (E3), (E4), and (E5) and the equilibrium Equation (E6) are independent of material behavior and do not change as the members turn plastic.

2. There is a change in the slope of the load deflection curve at loads at which a member turns plastic. This is the nonlinearity in the problem.

3. Calculating load deflection curves can be tedious as the number of members increases. The collapse load, however, corresponds to the equilibrium condition when all members are at yield stress, and its calculation is simple.

4. If all the members were made from different materials and had different cross-sectional areas, then, from the internal forces in Equations (E7), (E8), and (E9), the stress in each member would have to be found and compared to the yield stress of the respective member before it could be decided which member turned plastic first.

5.6   ELASTIC–PERFECTLY PLASTIC CIRCULAR SHAFTS

In Section 5.4.1, we studied the approximation of elastic–perfectly plastic material behavior. With reference to Figure 5.22, we recall the following.

1. Before yield stress, the material stress–strain relationship is represented by Hooke's law, and after yield stress the stress is assumed to be constant.

2. To determine the strain (deformation) in the horizontal portion *AB* of a curve, we use the requirement that deformation be continuous.

3. Unloading (elastic recovery) from a point in the plastic region is along line *BC*, which is parallel to the linear portion of the stress–strain curve *OA*.

We make use of the foregoing observations in our analysis, as described next.

The torsional shear stress varies linearly across a homogeneous, linear, elastic cross section and has its maximum at the outer radius of the shaft. Thus, the plastic zone will start from the outside and move inward for a homogeneous material, as shown in Figure 5.23(a).

The linear variation of torsional shear strain with the radial coordinate ρ was derived on the basis of geometry. If the kinematic assumptions[4] remain valid during yielding, then the linear variation of shear strain is unaffected by the fact that part of the cross section is now plastic, as shown in Figure 5.23(b). At the elastic–plastic boundary, we know that the shear strain is γ_{yield}. Thus Equation (3.6-T) can be rewritten as

$$\gamma_{\text{yield}} = \rho_y \frac{d\phi}{dx} \tag{5.16}$$

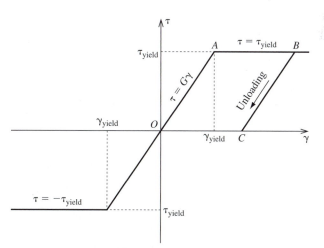

Figure 5.22   Elastic–perfectly plastic material behavior.

---

[4] If strains in the plastic region are very large, then the analysis is not valid because the assumption of small strains has been violated.

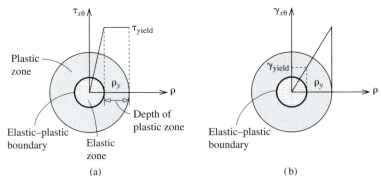

Figure 5.23 Elastic-perfectly plastic behavior of a circular shaft in torsion.

where ρ_y is the radial coordinate of the elastic–plastic boundary. If we assume[5] that ρ_y is not a function of x, then $d\phi/dx$ is constant and can be replaced by $(\phi_2 - \phi_1)/(x_2 - x_1)$. We obtain the following formula for the relative rotation:

$$\phi_2 - \phi_1 = \left(\frac{\gamma_{yield}}{\rho_y}\right)(x_2 - x_1) \tag{5.17}$$

Shear stress varies linearly from a value of zero at the center to a value of yield stress at the elastic–plastic boundary. After the elastic–plastic boundary, the shear stress is equal to the yield stress. The shear stress distribution can be written as follows:

$$\tau = \begin{cases} \dfrac{\tau_{yield}\rho}{\rho_y} & \rho \le \rho_y \\[2mm] \tau_{yield} & \rho \ge \rho_y \end{cases} \tag{5.18}$$

We can find the equivalent internal torque by substituting Equation (5.18) into Equation (3.4-T). We can relate the internal torque to external torque through equilibrium. There are two kinds of problem we will consider:

1. problems in which we know ρ_y, such as Example 5.8 and Example 5.9,

2. problems in which we need to find ρ_y, such as Example 5.10.

5.6.1 Residual Shear Stress

By prestressing the shaft into the plastic region, we can increase the torque-carrying capacity of the shaft, but now we need to compute the residual stresses in the shaft. During the elastic recovery (i.e., unloading along line *BC* in Figure 5.22), the assumption of material linearity is valid, and Equation (3.9-T) can be used for calculating the elastic shear stress. The internal torque for calculating the elastic shear stress corresponds to the torque that was equivalent to the stress distribution for a given location of the elastic–plastic boundary. We subtract the elastic stresses from the elastic–plastic stress distribution and obtain the residual stresses as demonstrated shortly (see Example 5.8).

---

[5] This implies that the shaft is not tapered and that external torque is not a function of x between x_1 and x_2.

EXAMPLE 5.7

In a hollow circular shaft, the shear strain at a section in polar coordinates was found to be $\gamma_{x\theta} = 3\rho(10^{-3})$, where ρ is the radial coordinate measured in inches. (a) Write expressions for $\tau_{x\theta}$ as a function of ρ and plot the distribution of the shear strain $\gamma_{x\theta}$ and shear stress $\tau_{x\theta}$ across the cross section. (b) Determine the equivalent internal torque acting at the cross section. Assume that the shaft is made from an elastic–perfectly plastic material that has a yield stress of $\tau_{yield} = 24$ ksi and a shear modulus $G = 6000$ ksi.

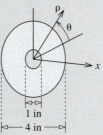

Figure 5.24 Hollow shaft for Example 5.7.

PLAN

(a) We can find the yield strain in shear γ_{yield} from the given yield stress τ_{yield} and the shear modulus G. We can then find the location of the elastic–plastic boundary by finding ρ_y at which the shear strain reaches the value of γ_{yield}. The shear stress at points before ρ_y can be found from Hooke's law, and the yield stress will come after ρ_y. (b) The internal torque can be calculated by substituting the expression for $\tau_{x\theta}$ in Equation (3.4-T).

SOLUTION

(a) The location of the elastic–plastic boundary can be found as follows:

$$\gamma_{yield} = \frac{\tau_{yield}}{G} = \frac{24(10^3)}{6000(10^3)} = 0.004 = 0.003\rho_y \quad \text{or}$$

$$\rho_y = \frac{0.004}{0.003} = 1.33 \text{ in} \tag{E1}$$

Up to ρ_y, the stress and strain are related by Hooke's law; hence

$$\tau_{x\theta} = G\gamma_{x\theta} = 6(10^6)(3\rho)(10^{-3}) = 18\rho(10^3) \tag{E2}$$

After ρ_y the stress is equal to τ_y, and the shear stress can be written as

$$\tau_{x\theta} = \begin{cases} 18\rho \text{ ksi} & 0.5 \le \rho \le 1.333 \\ 24 \text{ ksi} & 1.333 \le \rho \le 2.0 \end{cases} \tag{E3}$$

The distribution of the shear strain and shear stress across the cross section is shown in Figure 5.25.

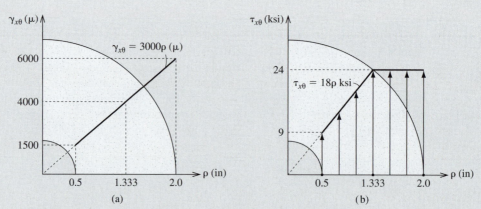

Figure 5.25 (a) Strain and (b) stress distribution in Example 5.7.

(b) The internal torque can be found from Equation (3.4-T) as follows:

$$T = \int_A \rho \tau_{x\theta} dA = \int_{A_E} \rho \tau_{x\theta} dA + \int_{A_P} \rho \tau_{x\theta} dA \qquad (E4)$$

$$\boxed{T_E} \qquad \boxed{T_P}$$

The differential area dA is the area of a ring of radius ρ and thickness $d\rho$ [i.e., $dA = (2\pi\rho)d\rho$]. Substituting the stress distribution from Equation (E1) into Equation (E2) and performing integration, we obtain the internal torque:

$$T_E = \int_{0.5}^{1.333} (\rho)(18\rho)(2\pi\rho)d\rho = (36\pi) \left. \frac{\rho^4}{4} \right|_{0.5}^{1.333} = 87.50 \text{ in} \cdot \text{kips} \qquad (E5)$$

$$T_P = \int_{1.333}^{2.0} (\rho)(24)(2\pi\rho)d\rho = (48\pi) \left. \frac{\rho^3}{3} \right|_{1.333}^{2.0} = 283.07 \text{ in} \cdot \text{kips} \qquad (E6)$$

$$T = T_E + T_P = 370.6 \text{ in} \cdot \text{kips} \qquad \textbf{ANS.} \quad T = 370.6 \text{ in} \cdot \text{kips}$$

COMMENTS

1. In this problem we knew the strain distribution and hence could locate the elastic–plastic boundary easily. In most problems we do not know the strains due to a load, and finding the elastic–plastic boundary is significantly more difficult.

2. The relation of internal to external torque can be established by drawing the appropriate free body diagram for a particular problem. The relationship depends upon the free body diagram and is independent of the elasticity or plasticity of the material.

EXAMPLE 5.8

A 100 mm diameter solid steel (G = 80 GPa, τ_{yield} = 160 MPa) circular shaft has a plastic zone that is 30 mm deep. Assuming an elastic–perfectly plastic material, determine (a) the equivalent internal torque at the cross section and (b) the residual stresses in the cross section when the shaft is unloaded.

PLAN

The radius of the shaft is 50 mm, and the plastic zone starting from outside is 30 mm deep. Thus the elastic–plastic boundary is located at ρ_y = 20 mm. (a) The shear stress varies linearly, starting from zero at ρ = 0 and reaching yield stress at ρ_y = 20 mm. After 20 mm the shear stress is constant and equal to the yield stress. We can write the equation for the shear stress in terms of ρ and use Equation (3.4-T) to determine the equivalent internal torque. (b) By using the internal torque obtained in part (a) and Equation (3.9-T), we can determine the shear stress distribution during unloading and subtract it from the stress distribution in part (a) to obtain the residual stress distribution.

SOLUTION

(a) The shear stress varies linearly from zero at the center to yield stress at the elastic–plastic boundary. Thus for $0 \leq \rho \leq 0.02$ we have

$$\tau = \frac{\tau_{yield}\rho}{\rho_y} = \left[\frac{160(10^6)}{20(10^{-3})}\right]\rho = 8000(10^6)\rho \tag{E1}$$

Beyond the elastic–plastic boundary, the shear stress is equal to the yield stress. We can write the equation for shear stress as

$$\tau = \begin{cases} 8000\rho \text{ MPa} & 0 \leq \rho \leq 0.02 \\ 160 \text{ MPa} & 0.02 \leq \rho \leq 0.05 \end{cases} \tag{E2}$$

Substituting Equation (E1) into Equation (3.4-T), we can obtain the equivalent internal torque as follows:

$$T = \int_A \rho\tau_{x\theta}dA = \int_0^{0.02} \rho[8000(10^6)\rho](2\pi\rho)d\rho + \int_{0.02}^{0.05} \rho[160(10^6)](2\pi\rho)d\rho \quad \text{or}$$

$$T = [16,000\pi](10^6)\frac{\rho^4}{4}\Big|_0^{0.02} + [320\pi](10^6)\frac{\rho^3}{3}\Big|_{0.02}^{0.05} \tag{E3}$$

$$= [2.01 + 39.2](10^3) = 41.3(10^3) \text{ N} \cdot \text{m}$$

ANS. T = 41.3 kN · m

(c) During unloading we can use Equation (3.9-T) to obtain the elastic rebound shear stress. The polar moment for the shaft is

$$J = (\pi/2)(0.05^4) = 9.8175(10^{-6}) \qquad (E4)$$

$$\tau_{\text{elastic}} = \frac{T\rho}{J} = \frac{41.3(10^3)}{9.8175(10^{-6})}\rho \quad \text{or} \quad \tau_{\text{elastic}} = 4207\rho \text{ MPa} \qquad (E5)$$

The residual shear stresses are $\tau_{\text{residual}} = \tau - \tau_{\text{elastic}}$. Substituting Equations (E2) and (E5), we obtain

$$\textbf{ANS.} \quad \tau_{\text{residual}} = \begin{cases} (3793\rho) \text{ MPa} & 0 \le \rho \le 0.02 \\ (160 - 4207\rho) \text{ MPa} & 0.02 \le \rho \le 0.05 \end{cases}$$

COMMENTS

1. The calculation of residual stress can be shown graphically as in Figure 5.26.

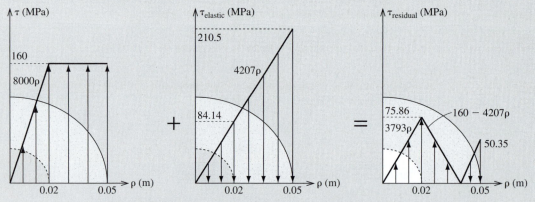

Figure 5.26 Residual stress calculation for Example 5.8.

2. The torque the shaft could carry without any plastic deformation can be calculated by using yield stress as the maximum elastic stress, that is, as $160(10^6) = T(0.5)/[7.8715(10^{-6})]$ or $T = 25.2$ kN · m. With the residual stresses, the maximum torque the shaft can carry without any additional plastic deformation is now 41.3 kN · m, which would be equivalent to a shaft of the same dimensions but made from a material with a yield stress of 210.5 MPa.

EXAMPLE 5.9

The steel ($G = 80$ GPa, $\tau_{\text{yield}} = 160$ MPa) shaft shown in Figure 5.27 has a plastic zone 30 mm deep in section AB. Determine (a) the magnitude of the applied torque T_{ext} and (b) the rotation of section C with respect to the wall.

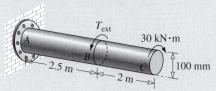

Figure 5.27 Shaft for Example 5.9.

PLAN

The cross section in AB is the same as in Example 5.8, and hence the internal torque in AB is $T_{AB} = 41.3$ kN · m. (a) After making an imaginary cut in AB, we can draw a free body diagram and determine the external torque T_{ext} by equilibrium. (b) We can find the rotation of section B with respect to section at A by using Equation (5.17). To find the rotation of the section at C with respect to B, we first check to see whether section BC is elastic or parts of it have gone plastic. If it is elastic, we can use Equation (3.10-T) to find the relative rotation of C with respect to B and then compute the rotation of the section at C with respect to A.

SOLUTION

(a) After making an imaginary cut in AB and taking the right-hand part to avoid calculating the wall reaction, we draw the free body diagram as shown in Figure 5.28.
By equilibrium of the moment we obtain

$$T_{\text{ext}} = 41.3 + 30 \qquad \text{or} \qquad \textbf{ANS.} \quad T_{\text{ext}} = 71.3 \ \text{kN} \cdot \text{m}$$

Figure 5.28 Free body diagram of section AB for Example 5.9.

(b) The strain at yield is

$$\gamma_{\text{yield}} = \frac{\tau_{\text{yield}}}{G} = \frac{160(10^6)}{80(10^9)} = 2(10^{-3}) \tag{E1}$$

We can use Equation (5.17) to find the relative rotation of section at B with respect to section at A as follows:

$$\gamma_{\text{yield}} = \rho_y \frac{d\phi}{dx} = 20(10^{-3})\frac{d\phi}{dx} = 2(10^{-3}) \quad \text{or} \quad \frac{d\phi}{dx} = 0.1 \quad \text{or} \tag{E2}$$

$$\phi_B - \phi_A = 0.1(x_B - x_A) = 0.1(2.5) = 0.25$$

After making an imaginary cut in BC as shown in Figure 5.29, we can draw the free body diagram. By equilibrium of the moment, we obtain the internal torque T_{BC} as

$$T_{BC} = -30 \text{ kN} \cdot \text{m} \tag{E3}$$

Figure 5.29 Free body diagram of section BC for Example 5.9.

and find the maximum shear stress in section BC from Equation (3.9-T):

$$\tau_{\text{max}} = \frac{T_{BC}\rho_{\text{max}}}{J} = \frac{-30(10^3)(50)(10^{-3})}{\pi(0.1)^4/32} = -153(10^6) \tag{E4}$$

The maximum shear stress in BC of 153 MPa does not exceed the yield stress of 160 MPa; hence we can use Equation (3.10-T) to find the rotation of the section at C with respect to B, as follows:

$$\phi_C - \phi_B = \frac{-30(10^3)(2)}{80(10^9)(9.817)(10^{-6})} = -0.0764 \tag{E5}$$

Adding Equations (E2) and (E5), we obtain the rotation of the section at C with respect to the section at A. Noting that the rotation at A is zero, we obtain the rotation at C as $\phi_C - \phi_A = 0.25 - 0.0764 = 0.1736$.

ANS. $\phi_C = 0.1736$ ccw

COMMENTS

1. If the maximum shear stress in section BC as given by Equation (E4) is greater than the yield stress, then Equation (E5) cannot be used because it is valid for the fully elastic segment of the shaft. In such cases we must find the location of elastic–plastic boundary; the relative rotation $\phi_C - \phi_B$ will have to be found as was done for section AB.

2. In Example 5.8 and this example, the location of the elastic–plastic boundary was known, simplifying the calculations significantly. In the next example we consider the problem of locating the elastic–plastic boundary for applied torques.

EXAMPLE 5.10

The steel (G = 80 GPa, τ_{yield} = 160 MPa) shaft is loaded as shown in Figure 5.30. Determine the location of the elastic plastic boundary in section AB.

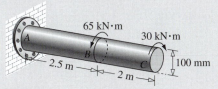

Figure 5.30 Shaft for Example 5.10.

PLAN

We can assume that the elastic–plastic boundary is located at a radial distance of ρ_y. We can write the stress distribution in terms of ρ_y and obtain the internal torque in AB in terms of ρ_y, which we can equate to the internal torque obtained from equilibrium to determine ρ_y.

SOLUTION

The shear stress distribution across the cross section can be written as follows:

$$\tau = \begin{cases} \dfrac{\tau_{yield}\rho}{\rho_y} & 0 \le \rho \le \rho_y \\[2mm] \tau_{yield} & \rho_y \le \rho \le 0.05 \end{cases} \tag{E1}$$

Substituting Equation (E1) in Equation (3.4-T) to obtain the equivalent internal torque, we write

$$T_{AB} = \int_A \rho\tau\,dA = \int_0^{\rho_y} \rho\left[\frac{\tau_{yield}\rho}{\rho_y}\right](2\pi\rho)\,d\rho + \int_{\rho_y}^{0.05} \rho\tau_{yield}(2\pi\rho)d\rho \qquad \text{or} \tag{E2}$$

$$T_{AB} = \left[2\pi\frac{\tau_{yield}}{\rho_y}\right]\frac{\rho^4}{4}\Bigg|_0^{\rho_y} + [2\pi\tau_{yield}]\frac{\rho^3}{3}\Bigg|_{\rho_y}^{0.05} = \frac{\pi\tau_{yield}}{6}[3\rho_y^3 + 4(0.05^3 - \rho_y^3)]$$

We make an imaginary cut in AB and draw the free body diagram shown in Figure 5.31. By equilibrium of the moment, we obtain the internal torque as follows:

$$T_{AB} - 65 + 30 = 0 \qquad \text{or} \qquad T_{AB} = 35 \text{ kN} \cdot \text{m} = 35(10^3) \text{ N} \cdot \text{m} \tag{E3}$$

Figure 5.31 *Free body diagram of section AB for Example 5.10.*

Equating Equation (E2) and (E3), we obtain

$$[0.5(10^{-3}) - \rho_y^3] = \frac{35(10^3)}{83.78(10^6)} = 0.4178(10^{-3}) \qquad \text{or}$$

$$\rho_y = [(0.5 - 0.4178)(10^{-3})]^{1/3} = 0.0435$$

The radius of the elastic–plastic boundary is

ANS. $\rho_y = 43.5$ mm

COMMENTS

1. The calculation of the elastic–plastic boundary will involve finding the roots of a cubic equation. For a hollow shaft, an iterative technique may be needed.

2. Once the location of the elastic–plastic boundary is known, the relative rotation and/or the residual stresses upon removal of the external loads can be found as was demonstrated in Examples 5.8 and 5.9.

5.7   ELASTIC–PERFECTLY PLASTIC BEAMS

Figure 5.32 shows the stress–strain curve for an elastic–perfectly plastic material. We recall the following observations.

1. The bending normal strain is linear across the cross section even if part of the cross section becomes plastic.

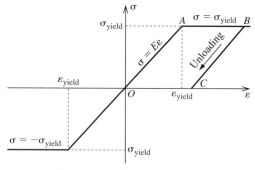

Figure 5.32 *Elastic–perfectly plastic material behavior.*

2. For pure bending, the normal stress distribution of σ_{xx} should result in zero axial force even if part of the cross section becomes plastic.

3. Before yield stress, the material stress–strain relationship is represented by Hooke's law; after yield stress, the stress is assumed to be constant.

4. At the elastic–plastic boundary, the bending normal stress should equal the yield stress, and the bending normal strain must equal the yield strain.

5. A plastic region will start growing from either the top or the bottom of the cross section because the maximum bending normal stress occurs at the point furthest from the neutral axis.

6. Unloading (elastic recovery) from a point in the plastic region is along line BC, which is parallel to the linear portion of the stress–strain curve OA shown in Figure 5.32.

We shall make use of these observations in our analysis. The location of the elastic–plastic boundary on the cross section is critical. We start with the following definition.

Definition 4 a = the distance of the elastic–plastic boundary from the neutral axis.

Consider a symmetric cross section of a beam that is symmetric not only about the y axis but also about the z axis, as shown in Figure 5.33(a). Points A and B will reach yield stress simultaneously at the same load. As the load increases, the plastic zone will move inward in a symmetric manner, and the neutral axis will continue to be the centroid of the cross section. The bending normal stress varies linearly with y in the elastic region and is equal to yield stress (tension or compression) in the plastic region, as shown in Figure 5.33(a). If we know a, we can write the stress distribution as a function of y and use Equation (3.4b-B) to find the equivalent internal moment at the cross section. Alternatively, if the internal moment is known, and we need to find the location of the elastic–plastic boundary, we can find a by writing the stress distribution in terms of a and then calculating the internal moment in terms of a and equating it to the given value moment value. In Example 5.11, given shortly, the location of the elastic–plastic boundary is provided (a is known) and the corresponding

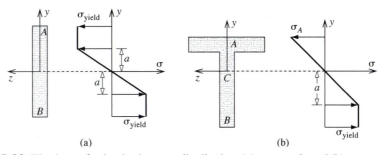

Figure 5.33 Elastic–perfectly plastic stress distribution: (a) symmetric and (b) unsymmetric.

calculation of the internal moment M_z is shown. In Example 5.12, the internal moment M_z is given and the calculation for the location of the elastic–plastic boundary a is shown.

Now consider a cross section that is not symmetric about the z axis, as shown in Figure 5.33(b). The centroid of the cross section at C is the location of the neutral axis if the entire cross section is elastic. Point B is further away than point A from the neutral axis at C. Thus, point B will reach yield stress before point A as the loads are increased. The plastic zone will start from point B and move upward. To satisfy Equation (3.4a-B), that is, for the total axial force on the cross section to be zero, the neutral axis will shift from the centroid. So with cross sections that are unsymmetric with respect to the z axis, we will need to find the location of the neutral axis also. Once more we will start by writing the stress distribution in terms of a across the cross section. We will first determine the location of the neutral axis in terms of a by using Equation (3.4a-B). Once a is known, the neutral axis as well as the location of the elastic–plastic boundary will be known. The equivalent internal moment M_z can be determined by using Equation (3.4b-B) once a is known. In Example 5.13 we will demonstrate the calculation of a and M_z for cross sections that have only one plane of symmetry.

The concept of shape factors, which is important in the elastic–perfectly plastic analysis of beams, requires the following definitions.

Definition 5 The ratio of the plastic moment to the elastic moment is called the shape factor for the cross section and will be designated by f.

$$f = \frac{M_p}{M_e} \qquad (5.19)$$

Definition 6 The moment at which the maximum bending normal stress on a cross section just reaches the yield stress is called the elastic moment and will be designated by M_e.

Definition 7 The internal moment for which the entire cross section becomes fully plastic is called the plastic moment and will be designated by M_p.

As the name suggests, the shape factor depends only upon the shape of the cross section and is independent of the material from which the beam is made. When the cross section becomes fully plastic, the beam can no longer support any load at the cross section and the corresponding moment represents the collapse moment. For an elastic design, the shape factor thus reflects the margin of safety. Example 5.14 will demonstrate the calculation of a shape factor.

In all examples that follow, we will be evaluating integrals over the cross section. Calculation errors can be significantly reduced by drawing an approximate stress distribution across the cross section, and then recording the y coordinate of points (sometimes in terms of a) at which either the stress σ_{xx} or the differential area dA changes.

EXAMPLE 5.11

The normal strain at a cross section in a beam due to bending about the z axis was found to vary as $\varepsilon_{xx} = -0.0125y$, with y measured in meters. (a) Write expressions for normal stress σ_{xx} as a function of y and plot the σ_{xx} distribution across the cross section. (b) Determine the equivalent internal moment. Assume that the beam is made from an elastic–perfectly plastic material that has a yield stress of $\sigma_{yield} = 250$ MPa in tension and compression and a modulus of elasticity $E = 200$ GPa.

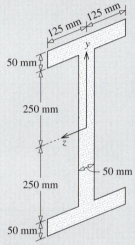

Figure 5.34 Beam cross section for Example 5.11.

PLAN

(a) Points furthest from the origin will be the most strained, and the plastic zone will start from the top and bottom and move inward symmetrically. We can find the yield strain ε_{yield} from the given yield stress σ_{yield} and the modulus of elasticity E. We can then find the location of elastic–plastic boundary by finding y_y at which the normal strain reaches the value of ε_{yield}. The normal stress before y_y can be found from the Hooke's law; after y_y it will be the yield stress. (b) Substituting the stress expression in Equation (3.4b-B), we can find the internal moments M_z.

SOLUTION

(a) The location of the elastic–plastic boundary can be found as follows:

$$\varepsilon_{yield} = \frac{\sigma_{yield}}{E} = \frac{(\pm 250)(10^6)}{200(10^9)} = (\pm 1.25)(10^{-3}) = -0.0125y_y \qquad \text{or}$$

(E1)

$$y_y = \frac{(\pm 1.25)(10^{-3})}{(-0.0125)} = (\mp 0.1) \text{ m}$$

Up to the elastic–plastic boundary (i.e., y_y), the material is in the linear range and Hooke's law applies. Thus

$$\sigma_{xx} = 200(10^9)(-0.0125y) = -2500y \text{ MPa} \qquad |y| \le 0.1 \text{ m}$$

The normal stress as a function of y can be written as follows:

$$\sigma_{xx} = \begin{cases} -250 \text{ MPa} & 0.1 \le y \le 0.3 \\ -2500y \text{ MPa} & -0.1 \le y \le 0.1 \\ 250 \text{ MPa} & -0.3 \le y \le -0.1 \end{cases} \qquad \text{(E2)}$$

The normal strain and stress as a function of y can be plotted as shown in Figure 5.35.

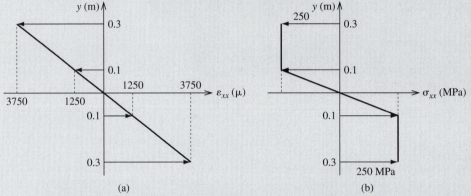

(a) (b)

Figure 5.35 (a) Strain and (b) stress distribution in Example 5.11.

(b) The internal moment M_z can be calculated by substituting Equation (E2) into Equation (3.4b-B). The area dA changes from the web to the flange, and the integral in Equation (3.4b-B) must account for changes in stress distribution as well as changes in the differential area. To better appreciate this statement, we plot the stress distribution across the entire cross section as shown in Figure 5.36.

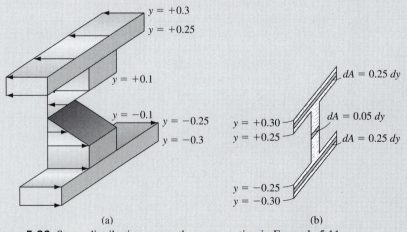

(a) (b)

Figure 5.36 Stress distribution across the cross section in Example 5.11.

We note that the stress distribution is symmetric. Thus we could calculate the moment from the top half and double it to obtain the total equivalent moment as in Equation (E3).

$$M_z = -\int_A y\sigma_{xx}\,dA = -2\left[\int_0^{0.1} y\sigma_{xx}\,dA + \int_{0.1}^{0.25} y\sigma_{xx}\,dA + \int_{0.25}^{0.3} y\sigma_{xx}\,dA\right] \quad \text{(E3)}$$

Substituting for σ_{xx} and dA in equation (E3), we obtain

$$M_z = -2(10^6)\left[\int_0^{0.1} y(-125y)dy + \int_{0.1}^{0.25} y(-12.5)dy + \int_{0.25}^{0.3} y(-62.5)dy\right] \quad \text{or}$$

$$M_z = -2(10^6)\left[(-125)\frac{y^3}{3}\Big|_0^{0.1} + (-12.5)\frac{y^2}{2}\Big|_{0.1}^{0.25} + (-62.5)\frac{y^2}{2}\Big|_{0.25}^{0.3}\right] \quad \text{or}$$

$$M_z = 83.33(10^3) + 656.25(10^3) + 1718.75(10^3) \quad \textbf{ANS.} \quad M_z = 2458 \text{ kN} \cdot \text{m}$$

COMMENTS

1. The total internal axial force is zero because the tensile stress behavior below the centroid is symmetrically duplicated by the compressive stress behavior above it.

2. The largest contribution toward the moment (third term) is from the stresses in the flange, as expected.

3. If on a beam we knew the cross section at which the calculated internal moment existed, then by free body diagram we could relate it to the external moment. If the beam is subjected to transverse loads, the internal moment will change with x and hence the location of the elastic–plastic boundary will change with x, making the problem significantly more difficult.

EXAMPLE 5.12

The internal moment at a section of the beam was found to be 21 kN · m. The beam material has a yield stress of 200 MPa. Determine the location of the elastic–plastic boundary.

PLAN

By symmetry, the neutral axis will stay at the centroid even as the plastic zone grows. We assume that the elastic–plastic boundary is a units from the centroid in the web. We can write the stress expression across the cross section in terms of a and the coordinate y. We can substitute the stress expression in Equation (3.4b-B), find the internal moment M_z in terms of a, and equate it to 21 kN to determine a.

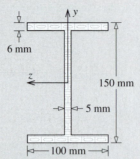

Figure 5.37 Cross section for Example 5.12.

SOLUTION

We draw the stress distribution, assuming that the location of the elastic–plastic boundary is in the web, and record the value of y at each point where either stress or area changes, as shown in Figure 5.38. The stress σ and dA in each interval can be written as follows:

$$\sigma = -200 \text{ MPa} \qquad dA = 0.100 \, dy \qquad 0.069 \le y \le 0.075$$

$$\sigma = -200 \text{ MPa} \qquad dA = 0.005 \, dy \qquad a \le y \le 0.069$$

$$\sigma = -\left(\frac{200}{a}y\right)\text{MPa} \qquad dA = 0.005 \, dy \qquad -a \le y \le a \qquad \text{(E1)}$$

$$\sigma = 200 \text{ MPa} \qquad dA = 0.005 \, dy \qquad -0.069 \le y \le -a$$

$$\sigma = 200 \text{ MPa} \qquad dA = 0.100 \, dy \qquad -0.075 \le y \le -0.069$$

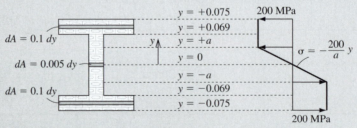

Figure 5.38 Stress distribution in Example 5.12.

We note that moment from the bottom half ($y < 0$) will be the same as that from the top half ($y > 0$). Thus, we need only find the moment from the top half and double it. From Equation (3.4b-B), we find the moment in terms of a as follows:

$$M_z = -2(10^6)\left[\int_0^a y\left(-\frac{200}{y_y}y\right)(0.005 \, dy) + \int_a^{0.069} y(-200)(0.005 \, dy) + \int_{0.069}^{0.075} y(-200)0.100 \, dy\right] \qquad \text{or}$$

$$M_z = 400(10^3)\left[\frac{0.005}{a}\left(\frac{y^3}{3}\right)\Big|_0^a + 0.005\left(\frac{y^2}{2}\right)\Big|_a^{0.069} + 0.1\left(\frac{y^2}{2}\right)\Big|_{0.069}^{0.075}\right] = 22.04(10^3) - \frac{a^2}{3}(10^6) \quad \text{(E2)}$$

Equating the moment in Equation (E2) to 21 kN, we can find a as follows:

$$22.04(10^3) - \frac{a^2}{3}(10^6) = 21(10^3) \quad \text{or} \quad a = 0.0548 \text{ m}$$

ANS. $a = 54.8$ mm

COMMENT

Since $a < 69$ mm, our assumption about the location of the elastic–plastic boundary is correct. If it had turned out to be incorrect, the problem would have had to be re-solved assuming that the elastic–plastic boundary was in the flange.

EXAMPLE 5.13

A beam of an elastic–perfectly plastic material has a yield stress of 48 ksi, and point A on the beam cross section shown in Figure 5.39 *just* reaches yield stress. Determine (a) the location of the neutral axis and (b) the moment required to produce the given state of stress.

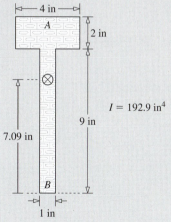

Figure 5.39 Cross section for Example 5.13.

PLAN

We draw the stress distribution across the cross section, assuming that the neutral axis is in the web at a distance a from the top. We write the stress distribution in terms of y and a. Substituting the stress distribution into Equation (3.4a-B) and equating it to zero, we determine a. Substituting the stress distribution into Equation (3.4b-B), we obtain the internal moment.

SOLUTION

We can draw the stress distribution as shown in Figure 5.40 and write the stress expression and value of dA for each interval as follows.

$$\sigma = -\left(\frac{48}{a}\right)y \qquad dA = 4\,dy \qquad (a-2) \le y \le a$$

$$\sigma = -\left(\frac{48}{a}\right)y \qquad dA = dy \qquad -a \le y \le (a-2) \qquad \text{(E1)}$$

$$\sigma = 48 \text{ ksi} \qquad dA = dy \qquad -(11-a) \le y \le -a$$

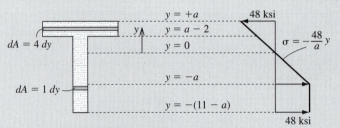

Figure 5.40 Stress distribution for Example 5.13.

We can substitute Equation (E1) in Equation (3.4a-B) and obtain the equation for determining the value of a as follows:

$$\int_A \sigma_{xx}\,dA = \int_{-(11-a)}^{-a} 48\,dy + \int_{-a}^{(a-2)} -\left(\frac{48}{a}\right)y\,dy + \int_{(a-2)}^{a} -\left(\frac{48}{a}\right)y(4dy) = 0 \qquad \text{or}$$

$$\text{(E2)}$$

$$48\left[(y)\Big|_{-(11-a)}^{-a} - \left(\frac{y^2}{2a}\right)\Big|_{-a}^{(a-2)} - 4\left(\frac{y^2}{2a}\right)\Big|_{(a-2)}^{a}\right] = 0 \qquad \text{or} \qquad 10a - 4a^2 + 12 = 0$$

The two roots of the quadratic Equation (E2) are $a = 3.386$ and $a = -0.886$. Only the positive root is admissible. Thus the location of the neutral axis is

ANS. $a = 3.386$ in   (E3)

We can substitute Equations (E1) and (E3) in Equation (3.4b-B) and obtain the equivalent internal moment as follows:

$$M_z = -\left[\int_{-(11-a)}^{-a} y48 \, dy + \int_{-a}^{(a-2)} y\left(-\frac{48}{a}y\right) dy + \int_{(a-2)}^{a} y\left(-\frac{48}{a}y\right)(4dy) \right] \quad \text{or}$$

$$M_z = -48\left[\left(\frac{y^2}{2}\right)\Big|_{-7.614}^{-3.386} - \left(\frac{y^3}{3(3.386)}\right)\Big|_{-3.386}^{1.386} - 4\left(\frac{y^3}{3(3.386)}\right)\Big|_{1.386}^{3.386} \right] = 1995.6 \quad \text{or}$$

ANS. $M_z = 1996$ in · kips

COMMENT

Once stress at a point on a cross section has been given, we can assume the location of the elastic–plastic boundary when drawing the approximate stress distribution. However, in some problems it may be necessary to check the assumption about the location of the elastic–plastic boundary.

EXAMPLE 5.14

Determine the shape factor for the beam cross section shown in Figure 5.39 (Example 5.13). The area moment of inertia of the cross section about the axis passing through the centroid is $I_{zz} = 192.9$ in⁴.

PLAN

Point B is further than point A from the neutral axis, and hence the elastic stress will be maximum at point B. We can find the elastic moment by using Equation (3.9-B) and equating it to yield stress. To find the plastic moment, we can assume that the neutral axis is a distance a from the bottom. We can write the stress distribution in terms of a and the coordinate y. We can use Equation (3.4a-B) to find the value of a. From Equation (3.4b-B), we can find the plastic moment and then obtain the shape factor from Equations (5.19).

SOLUTION

From Equation (3.9-B) we find the bending normal stress at point B and equate it to the yield stress to obtain the elastic moment:

$$\sigma_{\text{yield}} = -\left[\frac{M_e(-7.09)}{192.9}\right] \quad \text{or} \quad M_e = 27.21\sigma_{\text{yield}} \tag{E1}$$

We draw the stress distribution assuming that the location of the neutral axis is at a distance a from the bottom and record the value of y at each point where either stress or area changes, as shown in Figure 5.41. The stress σ and dA in each interval can be written as follows:

$$\begin{aligned}
\sigma &= -\sigma_{\text{yield}} & dA &= 4dy & (9-a) &\le y \le (11-a) \\
\sigma &= -\sigma_{\text{yield}} & dA &= dy & 0 &\le y \le (9-a) \\
\sigma &= \sigma_{\text{yield}} & dA &= dy & -a &\le y \le 0
\end{aligned} \tag{E2}$$

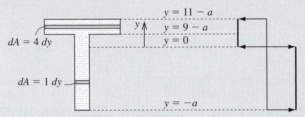

Figure 5.41 Stress distribution for Example 5.14.

Substituting Equation (E2) in Equation (3.4a-B), we can find the value of a as follows:

$$\int_{-a}^{0} \sigma_{\text{yield}}\,dy + \int_{0}^{(9-a)} -\sigma_{\text{yield}}\,dy + \int_{(9-a)}^{(11-a)} -\sigma_{\text{yield}}4\,dy = \sigma_{\text{yield}}[a - (9-a) - 4(2)] = 0 \quad \text{or} \tag{E3}$$

$$a = 8.5 \text{ in}$$

Substituting Equations (E2) and (E3) in Equation (3.4b-B), we obtain

$$M_p = -\int_A y\sigma_{xx}dA = -\left[\int_{-a}^{0} y(\sigma_{\text{yield}})dy + \int_{0}^{(9-a)} y(-\sigma_{\text{yield}})dy + \int_{(9-a)}^{(11-a)} y(-\sigma_{\text{yield}})4dy\right] \quad \text{or} \tag{E4}$$

$$M_p = (-\sigma_{\text{yield}})\left[\left(\frac{y^2}{2}\right)\Big|_{-8.5}^{0} - \left(\frac{y^2}{2}\right)\Big|_{0}^{0.5} - 4\left(\frac{y^2}{2}\right)\Big|_{0.5}^{2.5}\right] = 48.25\sigma_{\text{yield}}$$

Substituting equations (E1) and (E4) into Equations (5.19), we find the shape factor:

$$f = \frac{M_p}{M_e} = \frac{48.25\sigma_{\text{yield}}}{27.21\sigma_{\text{yield}}} \qquad \textbf{ANS.} \quad f = 1.77$$

COMMENT

Note in the preceding calculations of a shape factor that the yield stress could have any value. It will appear in both numerator and denominator and will cancel out. This emphasizes that the shape factor is independent of the material and is a geometric property of the cross section.

5.8   MATERIAL NONLINEARITY IN STRUCTURAL MEMBERS

In this section we consider the power law model in representing the stress–strain relationship. We assume that the behavior of a material is the same in tension and compression. Similarly, the behavior of a material is the same for positive and negative shear stress–shear strain.

We will assume that the kinematics for the structural members is still valid, and thus the strain distribution across the cross section is as before. The equations of equivalency and equilibrium do not change with the material model. Thus, the change with the use of the power law model in place of Hooke's law is in the stress distribution across the cross section. The new stress distribution can be integrated across the cross section to obtain new expressions for internal forces and moments that can be related to external forces and moments through equilibrium equations. In the case of bending, the material nonlinearity may cause the shifting of the neutral axis from the centroid, and the new location would be determined by ensuring that the total axial force is zero. The ideas of this paragraph are elaborated by Examples 5.15 through 5.19.

EXAMPLE 5.15

Re-solve Example 5.7, assuming that the shaft material has a stress–strain relationship given by $\tau = 450\gamma^{0.75}$ ksi.

PLAN

(a) By substituting the strain expression into the stress–strain equation, we can obtain stress as a function of ρ and plot it. (b) We can use Equation (3.4-T) to find the internal torque.

SOLUTION

(a) Substituting the strain distribution into the stress–strain relation, we obtain:

$$\tau_{x\theta} = 450(0.003)^{0.75}\rho^{0.75} \qquad \text{or} \qquad \tau_{x\theta} = 5.768\rho^{0.75} \text{ ksi} \qquad \text{(E1)}$$

The shear stress can be found at several points and plotted as shown in Figure 5.42.

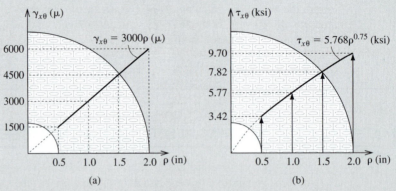

Figure 5.42 Strain and stress distribution for Example 5.15.

(b) For the internal torque calculation, we note that the differential area dA is the area of a ring of radius ρ and thickness $d\rho$ [i.e., $dA = (2\pi\rho)d\rho$]. Substituting the stress distribution from Equation (E1) into Equation (3.4-T) and performing integration, we obtain the internal torque.

$$T = \int_A \rho\tau_{x\theta}dA = \int_{0.5}^{2.0} (\rho)(5.768\rho^{0.75})(2\pi\rho)d\rho = (36.24)\int_{0.5}^{2.0}(\rho^{2.75})d\rho \quad \text{or}$$

$$T = (36.24)\left.\frac{\rho^{3.75}}{3.75}\right|_{0.5}^{2.0} = (9.664)[2^{3.75} - (0.5)^{3.75}]$$

ANS. $T = 129.3 \text{ in} \cdot \text{kips}$

COMMENT

On a shaft with loading, if we know the section at which the internal torque exists, then by making an imaginary cut and drawing free body diagram, we can relate the internal torque to the external torque.

EXAMPLE 5.16

The circular shaft shown in Figure 5.43 is made from a material that has a stress–strain relationship given by $\tau = 3400\gamma^{0.75}$ MPa. Determine the maximum torsional shear stress and the rotation of the cross section at C.

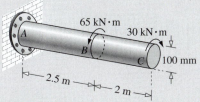

Figure 5.43 Shaft for Example 5.16.

PLAN

Starting with the kinematics of linear shear strain, we can derive the expression of shear stress and relative rotation in terms of the internal torque. The shaft and the loading are the same as in Example 5.10, and we can use the values of the internal torques to determine the maximum torsional shear strain and the relative rotations.

SOLUTION

From Equation (3.2-T), the kinematic equation for shear strain is

$$\gamma_{x\theta} = \rho \frac{d\phi}{dx} \tag{E1}$$

From the given material, the shear stress is

$$\tau_{x\theta} = 3400 \left(\rho \frac{d\phi}{dx} \right)^{0.75} (10^6) \; \text{N/m}^2 \tag{E2}$$

Noting that $d\phi/dx$ is only a function of x and does not change across the cross section and that $dA = 2\pi\rho \, d\rho$, we can write the static equivalency equation (3.4-T) as

$$T = \int_A \rho \tau_{x\theta} dA = \int_A \rho \left[3400 \left(\rho \frac{d\phi}{dx} \right)^{0.75} (10^6) \right] dA$$

$$= 3400(10^6) \left(\frac{d\phi}{dx} \right)^{0.75} \int_0^{0.05} \rho^{1.75} \, (2\pi\rho) d\rho \qquad \text{or}$$

$$T = 6800\pi(10^6) \left(\frac{d\phi}{dx} \right)^{0.75} \left. \frac{\rho^{3.75}}{3.75} \right|_0^{0.05} = 75.295(10^3) \left(\frac{d\phi}{dx} \right)^{0.75} \qquad \text{or}$$
$$\tag{E3}$$

$$\left(\frac{d\phi}{dx} \right)^{0.75} = \frac{T}{75.295(10^3)}$$

Substituting Equation (E3) into Equation (E2), we obtain

$$\tau_{x\theta} = 3400\rho^{0.75}\left[\frac{T}{75.295(10^3)}\right](10^6) = 45.156(10^3)T\rho^{0.75} \text{ N/m}^2 \qquad \text{(E4)}$$

Equation (E3) can be written as

$$\frac{d\phi}{dx} = \left[\frac{T}{75.295(10^3)}\right]^{4/3} \quad \text{or} \quad \phi_2 - \phi_1 = \left[\frac{T}{75.295(10^3)}\right]^{4/3}(x_2 - x_1) \qquad \text{(E5)}$$

From Example 5.10 and by inspection, we have the magnitudes of internal torque:

$$|T_{AB}| = 35(10^3) \text{ N} \cdot \text{m} \quad \text{and} \quad |T_{BC}| = 30(10^3) \text{ N} \cdot \text{m} \qquad \text{(E6)}$$

The maximum torsional shear stress will exist in AB at the outer surface. From Equations (E4) and (E6), we have

$$|\tau_{max}| = 45.156(10^3)|T_{AB}|\rho_{max}^{0.75} = 45.156(10^3)35(10^3)(0.05)^{0.75} = 167.1(10^6) \text{ N/m}^2$$

ANS. $|\tau_{max}| = 167.1$ MPa

By using Equations (E5) and (E6), we can find the magnitudes of the relative rotations. By inspection of Figure 5.43, we can find the directions of the relative rotations to obtain

$$\phi_B - \phi_A = \left[\frac{|T_{AB}|}{75.295(10^3)}\right]^{4/3}(x_B - x_A)$$

$$= \left[\frac{35(10^3)}{75.295(10^3)}\right]^{4/3}(2.5) = 0.9002 \text{ rad ccw} \qquad \text{(E7)}$$

$$\phi_C - \phi_B = \left[\frac{|T_{BC}|}{75.295(10^3)}\right]^{4/3}(x_B - x_A)$$

$$= \left[\frac{30(10^3)}{75.295(10^3)}\right]^{4/3}(2.0) = 0.5864 \text{ rad cw} \qquad \text{(E8)}$$

The relative rotation of C with respect to A can be found from Equations (E7) and (E8). Noting that rotation at A is zero, we obtain the rotation at C as $\phi_C - \phi_A = 0.9002 - 0.5864 = 0.3138$ or

ANS. $\phi_C = 0.3138$ rad ccw

COMMENTS

1. In writing Equation (E2) we assumed that $d\phi/dx$ is positive. Following the sign conventions for internal torque and rotations, we would get a negative value for $d\phi/dx$. We could have also developed (written) Equations (E4) and (E5) for negative values for torque, and this may be necessary if we are trying to write a computer program for such problems. For hand calculations it is easier to determine the directions of shear stress and rotation by inspection.

2. The problem once more emphasizes the modular character of the mechanics equations. Except for the stress distribution, the fundamental equations of kinematics, equivalency, and equilibrium did not change.

EXAMPLE 5.17

Re-solve Example 5.11, assuming that the stress–strain relationship is given by $\sigma = 9000\varepsilon^{0.6}$ MPa in tension and compression.

PLAN

The stress distribution can be obtained from the stress–strain equation and the strain distribution. We can use Equation (3.4b-B) to find the internal moment.

SOLUTION

(a) Substituting the strains in the stress–strain relation, we obtain

$$\sigma_{xx} = \begin{cases} E\varepsilon_{xx}^{0.6} & \varepsilon_{xx} \geq 0 \\ -E(-\varepsilon_{xx})^{0.6} & \varepsilon_{xx} \leq 0 \end{cases} \quad \text{or} \quad \sigma_{xx} = \begin{cases} 649.2(-y)^{0.6} & y \leq 0 \\ -649.2(y)^{0.6} & y \geq 0 \end{cases} \quad \text{(E1)}$$

The strains and stresses can be found at different values of y and plotted as shown in Figure 5.44.

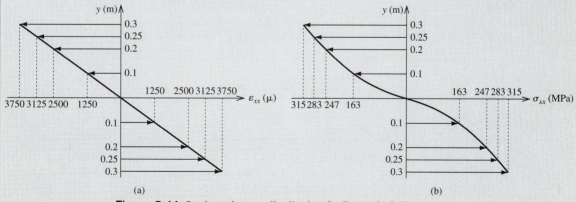

(a)   (b)

Figure 5.44 Strain and stress distribution for Example 5.17.

(b) The internal moment M_z can be found by substituting the stress expressions of Equation (E1) into Equation (3.4b-B). We note that the stresses are symmetric in tension and compression, and hence we can compute the moment from the top half and double it.

$$M_z = -\int_A y\sigma_{xx}dA = -2\left[\int_0^{0.25} y\sigma_{xx}(0.05\ dy) + \int_{0.25}^{0.3} y\sigma_{xx}(0.25\ dy)\right] \quad \text{(E2)}$$

Substituting for σ_{xx} in Equation (E2), we obtain the internal moment as follows:

$$M_z = -2\left[\int_0^{0.25} y(-649.2y^{0.6})(0.05\ dy) + \int_{0.25}^{0.3} y(-649.2y^{0.6})(0.25\ dy)\right] \quad \text{or}$$

$$M_z = -2(10^6)\left[(-32.46)\frac{y^{2.6}}{2.6}\Big|_0^{0.25} + (-162.3)\frac{y^{2.6}}{2.6}\Big|_{0.25}^{0.3}\right] = 679.3(10^3) + 2059.8(10^3)$$

ANS. $M_z = 2739\ \text{kN}\cdot\text{m}$

COMMENT

To better appreciate the stress distribution, we can plot it across the entire cross section as shown in Figure 5.45.

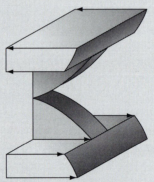

Figure 5.45 Stress distribution across the cross section in Example 5.17.

EXAMPLE 5.18

The beam cross section shown in Figure 5.46 is made from a material that has a stress–strain curve given by $\sigma = 400\varepsilon^{0.4}$ ksi. Determine (a) the location of the neutral axis and (b) the bending normal stress in terms of y and the internal moment M_z.

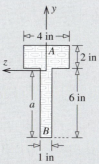

Figure 5.46 Beam cross section for Examples 5.18 and 5.19.

PLAN

The bending strain is still a linear function of y as given by Equation (3.2-B). An expression for the stress can be written by using the given stress–strain equation. The neutral axis can be determined from Equation (3.4a-B). By using Equation (3.4b-B), curvature can be written in terms of the internal moment, which on substitution into the stress expression will give the desired result.

SOLUTION

From Equation (3.2-B) the bending normal strain is given by:

$$\varepsilon_{xx} = -y\frac{d^2v}{dx^2}(x) \tag{E1}$$

From the given stress–strain relationship, and assuming that the curvature is positive, the bending normal stress can be written as

$$\sigma_{xx} = \begin{cases} 400\ \varepsilon_{xx}^{0.4} & \varepsilon_{xx} \geq 0 \\ -400\ (-\varepsilon_{xx})^{0.4} & \varepsilon_{xx} \leq 0 \end{cases} \quad \text{or} \quad \sigma_{xx} = \begin{cases} 400\left(-y\dfrac{d^2v}{dx^2}\right)^{0.4} & y < 0 \\ -400\left(y\dfrac{d^2v}{dx^2}\right)^{0.4} & y > 0 \end{cases} \tag{E2}$$

(a) From Equation (3.4a-B), the zero axial force equation $\int_A \sigma_{xx}dA$ can be written as

$$\int_{-a}^{0} 400\left(-y\frac{d^2v}{dx^2}\right)^{0.4}(1)dy + \int_{0}^{(6-a)}\left[-400\left(y\frac{d^2v}{dx^2}\right)^{0.4}\right](1)dy$$

$$+ \int_{(6-a)}^{(8-a)}\left[-400\left(y\frac{d^2v}{dx^2}\right)^{0.4}\right](4)dy = 0 \quad \text{or}$$

$$\left[400\left(\frac{d^2v}{dx^2}\right)^{0.4}\right]\left[\int_{-a}^{0}(-y)^{0.4}(1)dy+\int_{0}^{(6-a)}[-(y)^{0.4}](1)dy\right.$$

$$\left.+\int_{(6-a)}^{(8-a)}[-(y)^{0.4}](4)dy\right]=0 \quad \text{or}$$

$$\frac{-(-y)^{1.4}}{1.4}\bigg|_{-a}^{0}-\frac{y^{1.4}}{1.4}\bigg|_{0}^{6-a}-4\frac{y^{1.4}}{1.4}\bigg|_{6-a}^{8-a}=0 \quad \text{or}$$

$$a^{1.4}+3(6-a)^{1.4}-4(8-a)^{1.4}=0 \quad \text{(E3)}$$

The root of Equation (E3) can be found numerically to obtain

ANS. $a=5.804$ in

(b) From Equation (3.4b-B) the internal moment $M_z=-\int_A y\sigma_{xx}dA$ can be written as

$$M_z=-\left[\int_{-a}^{0}\left[400y\left(-y\frac{d^2v}{dx^2}\right)^{0.4}\right](1)dy+\int_{0}^{(6-a)}y\left[-400\left(y\frac{d^2v}{dx^2}\right)^{0.4}\right](1)dy\right.$$

$$\left.+\int_{(6-a)}^{(8-a)}y\left[-400\left(y\frac{d^2v}{dx^2}\right)^{0.4}\right](4)dy\right] \quad \text{or}$$

$$M_z=-\left[400\left(\frac{d^2v}{dx^2}\right)^{0.4}\right]\left[\int_{-a}^{0}y[(-y)^{0.4}]dy-\int_{0}^{(6-a)}y^{1.4}dy-4\int_{(6-a)}^{(8-a)}y^{1.4}dy\right] \quad \text{or}$$

$$M_z=-\left[400\left(\frac{d^2v}{dx^2}\right)^{0.4}\right]\left[\frac{(-y)^{2.4}}{2.4}\bigg|_{-a}^{0}-\frac{y^{2.4}}{2.4}\bigg|_{0}^{6-a}-4\frac{y^{2.4}}{2.4}\bigg|_{6-a}^{8-a}\right] \quad \text{or} \quad \text{(E4)}$$

$$M_z=-\left[\frac{400}{2.4}\left(\frac{d^2v}{dx^2}\right)^{0.4}\right][-a^{2.4}-(6-a)^{2.4}-4(8-a)^{2.4}+4(6-a)^{2.4}]$$

Substituting the value of $a=5.804$ into Equation (E4) we obtain

$$M_z=15{,}738\left(\frac{d^2v}{dx^2}\right)^{0.4} \text{ in}\cdot\text{kips} \quad \text{or} \quad \left(\frac{d^2v}{dx^2}\right)^{0.4}=\frac{M_z}{15{,}738} \quad \text{(E5)}$$

Substituting Equation (E5) into Equation (E2), we obtain the bending normal stress distribution:

$$\textbf{ANS.} \quad \sigma_{xx} = \begin{cases} 0.0254 M_z(-y)^{0.4} \text{ ksi} & y < 0 \\ -0.0254 M_z(y)^{0.4} \text{ ksi} & y > 0 \end{cases} \tag{E6}$$

COMMENTS

1. This example highlights that the location of the neutral axis is not at the centroid for nonlinear materials.

2. Though the algebra is more tedious than for linear elastic materials, the basic ideas of calculating the neutral axis and the internal bending moment do not change.

EXAMPLE 5.19

Determine the maximum bending normal stress for the two beams and loading shown in Figure 5.47. The beam cross section and material are the same as in Example 5.18.

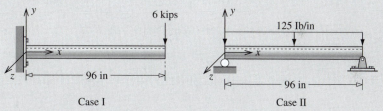

Case I   Case II

Figure 5.47 Beam and loading for Example 5.19.

PLAN

The maximum internal moment for the cantilever beam will be at the wall and in the center for the simply supported beam. The maximum bending normal stress for the cross section shown in Figure 5.46 will be at the bottom of the beam. The maximum bending normal stress can be found from the expression obtained in Example 5.18.

SOLUTION

Free body diagrams can be drawn after making imaginary cuts at the wall for case I and in the center for case II, as shown in Figure 5.48.

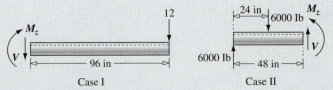

Case I   Case II

Figure 5.48 Beam and loading in Example 5.19.

The stress expression from Example 5.18 is

$$\sigma_{xx} = \begin{cases} 0.0254 M_z (-y)^{0.4} \text{ ksi} & y < 0 \\ -0.0254 M_z (y)^{0.4} \text{ ksi} & y > 0 \end{cases} \tag{E1}$$

Case I: The internal moment from Figure 5.48(a) is

$$M_z = -(6)(96) = (-576) \text{ in} \cdot \text{kips} \tag{E2}$$

The point furthest from the neutral axis on the cross section in Figure 5.46 was point B. The coordinate of point B is $y_B = -5.804$ in. Substituting the y coordinate and Equation (E2) into Equation (E1), we obtain the maximum bending normal stress as follows:

$$\sigma_{max} = 0.0254(-576)(5.804)^{0.4} = -29.56 \quad \text{or} \quad \textbf{ANS.} \quad \sigma_{max} = 29.6 \text{ ksi (C)}$$

Case II: The internal moment from Figure 5.48(a) is

$$M_z = (24)(6000) = 144,000 \text{ in} \cdot \text{lb} = 144 \text{ in} \cdot \text{kips} \tag{E3}$$

Substituting the y coordinate of point B ($y_B = -5.804$ in) and Equation (E2) into Equation (E1), we obtain the maximum bending normal stress as follows:

$$\sigma_{max} = 0.0254(144)(5.804)^{0.4} = 7.39 \quad \textbf{ANS.} \quad \sigma_{max} = 7.4 \text{ ksi (T)}$$

COMMENTS

1. The example once more emphasizes that the equilibrium equations relating the internal moment to the external forces are not dependent upon material nonlinearity.

2. Once a formula for normal stress and moment has been developed for a nonlinear material, it can be used like the flexure formula for a linear elastic material. The primary difference from the flexure formula for a linear elastic material is that the internal moment–stress formula is applicable to a specific geometry and a specific material nonlinearity.

PROBLEM SET 5.2

Nonlinear material models

5.23 Bronze has a yield stress of $\sigma_{\text{yield}} =$ 18 ksi in tension, a yield strain of $\varepsilon_{\text{yield}} = 0.0012$, and ultimate stress of $\sigma_{\text{ult}} = 50$ ksi; the corresponding ultimate strain is $\varepsilon_{\text{ult}} = 0.50$. Determine the material constants and plot the resulting stress–strain curve for (a) the elastic–perfectly plastic model, (b) the linear strain-hardening model, and (c) the nonlinear power law model.

5.24 Cast iron has a yield stress of $\sigma_{\text{yield}} =$ 220 MPa in tension, a yield strain of $\varepsilon_{\text{yield}} =$ 0.00125, and ultimate stress of $\sigma_{\text{ult}} = 340$ MPa; the corresponding ultimate strain is $\varepsilon_{\text{ult}} = 0.20$. Determine the material constants and plot the resulting stress–strain curve for (a) the elastic–perfectly plastic model, (b) the linear strain-hardening model, (c) and the nonlinear power law model.

Elastic–plastic response of axial members

5.25 A force F is applied to the roller that slides inside a slot as shown in Figure P5.25. Both bars have a cross-sectional area of $A =$ 100 mm², a modulus of elasticity of $E =$ 200 GPa, and a yield stress of 250 MPa. Bars AP and BP have lengths of $L_{AP} = 200$ mm and $L_{BP} = 250$ mm, respectively. Draw the load deflection curve and determine the collapse load.

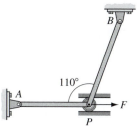

Figure P5.25

5.26 A force F is applied to the roller that slides inside a slot as shown in Figure P5.26. Both bars have a cross-sectional area of $A = 100$ mm², a modulus of elasticity of $E =$ 200 GPa, and a yield stress of 250 MPa. Bars AP and BP have lengths of $L_{AP} = 200$ mm and $L_{BP} = 250$ mm, respectively. Draw the load deflection curve and determine the collapse load.

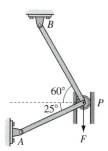

Figure P5.26

5.27 A force F is applied to the roller that slides inside a slot as shown in Figure P5.27. Both bars have a cross-sectional area of $A = 100$ mm², a modulus of elasticity of $E =$ 200 GPa, and a yield stress of 250 MPa. Bars AP and BP have lengths of $L_{AP} = 200$ mm and $L_{BP} = 250$ mm, respectively. Draw the load deflection curve and determine the collapse load.

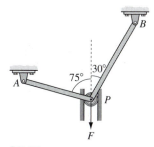

Figure P5.27

ANS. $P_{\text{collapse}} = 28.1$ kN
$\delta_{\text{collapse}} = 0.966$ mm

5.28 The three steel members shown in Figure P5.28 have a cross-sectional area of $A = 100$ mm^2, a modulus of elasticity $E = 200$ GPa, and a yield stress 250 MPa. The three members are attached to a roller that can slide in the slot as shown. Draw the load deflection curve and determine the collapse load.

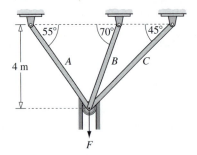

Figure P5.28

ANS. $P_{collapse} = 61.65$ kN
$\delta_{collapse} = 10.0$ mm

5.29 The three steel ($E = 200$ GPa, $\sigma_{yield} = 200$ MPa) bars shown in Figure P5.29 have lengths of $L_A = 4$ m, $L_B = 3$ m, and $L_C = 2$ m, respectively. All bars have the same cross-sectional area of 500 mm^2. Draw the load deflection curve and determine the collapse load.

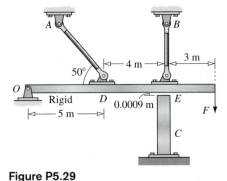

Figure P5.29

5.30 Draw the load deflection curve and determine the collapse load for the two-bar mechanism shown in Figure P5.30. Use the data given in Table P5.30.

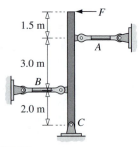

Figure P5.30

TABLE P5.30 Properties of Bars A and B

Property	Bar A	Bar B
Modulus of elasticity	200 GPa	170 GPa
Yield stress	250 MPa	220 MPa
Cross-sectional area	100 mm^2	75 mm^2
Length	1.5 m	1.5 m

ANS. $P_{collapse} = 24.31$ kN
$\delta_{collapse} = 6.3$ mm

5.31 Three poles are pin-connected to a ring at D and to the supports on the ground. The ring slides on a vertical rigid pole. The coordinates of the four points are as given in Figure P5.31. All poles have a cross-sectional area of $A = 1$ in^2, a modulus of elasticity of $E = 10,000$ ksi, and a yield stress of $\sigma_{yield} = 40$ ksi. Draw the load deflection curve and determine the collapse load.

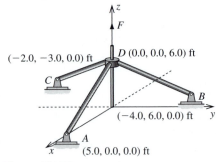

Figure P5.31

Problems on the nonlinear response of circular shafts

5.32 A solid circular shaft of diameter 3 inches has a shear strain at a section in polar coordinates that was found to be $\gamma_{x\theta} = 0.002\rho$, where ρ is the radial coordinate measured in inches. The shaft is made from elastic–perfectly plastic material that has a yield stress of $\tau_{\text{yield}} = 18$ ksi and a shear modulus $G = 12{,}000$ ksi. (a) Write expressions for $\tau_{x\theta}$ as a function of ρ and plot the shear strain $\gamma_{x\theta}$ and shear stress $\tau_{x\theta}$ distribution across the cross section. (b) Determine the equivalent internal torque acting at the cross section.

ANS. $T = 123.25$ in · kips.

5.33 A solid circular shaft of diameter 3 inches has a shear strain at a section in polar coordinates was found to be $\gamma_{x\theta} = 0.002\rho$, where ρ is the radial coordinate measured in inches. The shaft is made from a bilinear material that has a yield stress of $\tau_{\text{yield}} = 18$ ksi and shear modulus $G_1 = 12{,}000$ ksi and $G_2 = 4{,}800$ ksi. (a) Write expressions for $\tau_{x\theta}$ as a function of ρ and plot the shear strain $\gamma_{x\theta}$ and shear stress $\tau_{x\theta}$ distribution across the cross section. (b) Determine the equivalent internal torque acting at the cross section.

ANS. $T = 150.3$ in · kips.

5.34 A solid circular shaft of diameter 3 inches has a shear strain at a section in polar coordinates was found to be $\gamma_{x\theta} = 0.002\rho$, where ρ is the radial coordinate measured in inches. The shaft material has a stress strain relationship given by $\tau = 243\gamma^{0.4}$ ksi. (a) Write expression for $\tau_{x\theta}$ as a function of ρ and plot the shear strain $\gamma_{x\theta}$ and shear stress $\tau_{x\theta}$ distribution across the cross section. (b) Determine the equivalent internal torque acting at the cross section.

5.35 A solid circular shaft of diameter 3 inches has a shear strain at a section in polar coordinates was found to be $\gamma_{x\theta} = 0.002\rho$, where ρ is the radial coordinate measured in inches. The shaft material has a stress

strain relationship given by $\tau = (12{,}000\gamma - 120{,}000\gamma^2)$ ksi. (a) Write expression for $\tau_{x\theta}$ as a function of ρ and plot the shear strain $\gamma_{x\theta}$ and shear stress $\tau_{x\theta}$ distribution across the cross section. (b) Determine the equivalent internal torque acting at the cross section.

ANS. $T = 186.3$ in · kips

5.36 A hollow circular shaft has an inner diameter of 50 mm and an outside diameter of 100 mm. The shear strain at a section in polar coordinates was found to be $\gamma_{x\theta} = 0.2\rho$, where ρ is the radial coordinate measured in meters. The shaft is made from an elastic–perfectly plastic material that has a shear yield stress of $\tau_{\text{yield}} = 175$ MPa and a shear modulus $G = 26$ GPa. (a) Write expressions for $\tau_{x\theta}$ as a function of ρ and plot the shear strain $\gamma_{x\theta}$ and shear stress $\tau_{x\theta}$ distribution across the cross section. (b) Determine the equivalent internal torque acting at the cross section.

5.37 A hollow circular shaft has an inner diameter of 50 mm and an outside diameter of 100 mm. The shear strain at a section in polar coordinates was found to be $\gamma_{x\theta} = 0.2\rho$, where ρ is the radial coordinate measured in meters. The shaft is made from a linear strain-hardening material that has a shear yield stress of $\tau_{\text{yield}} = 175$ MPa and shear moduli $G_1 = 26$ GPa and $G_2 = 14$ GPa. (a) Write expressions for $\tau_{x\theta}$ as a function of ρ, and plot the shear strain $\gamma_{x\theta}$ and shear stress $\tau_{x\theta}$ distribution across the cross section. (b) Determine the equivalent internal torque acting at the cross section.

5.38 A hollow circular shaft has an inner diameter of 50 mm and an outside diameter of 100 mm. The shear strain at a section in polar coordinates was found to be $\gamma_{x\theta} = 0.2\rho$, where ρ is the radial coordinate measured in meters. The shaft material has a stress–strain relationship given by $\tau = 3435\gamma^{0.6}$ MPa. (a) Write expressions for $\tau_{x\theta}$ as a function of ρ, and plot the shear strain $\gamma_{x\theta}$ and shear stress $\tau_{x\theta}$ distribution across the cross section. (b) Determine the equivalent internal torque acting at the cross section.

5.39 A hollow circular shaft has an inner diameter of 50 mm and an outside diameter of 100 mm. The shear strain at a section in polar coordinates was found to be $\gamma_{x\theta} = 0.2\rho$, where ρ is the radial coordinate measured in meters. The shaft material has a stress–strain relationship given by $\tau = (26{,}000\gamma - 208{,}00\gamma^2)$ MPa. (a) Write expressions for $\tau_{x\theta}$ as a function of ρ, and plot the distribution of the shear strain $\gamma_{x\theta}$ and the shear stress $\tau_{x\theta}$ across the cross section. (b) Determine the equivalent internal torque acting at the cross section.

5.40 A 5 ft long hollow shaft with an outside diameter of 4 inches and an inside diameter of 2 inches is twisted through an angle of 12°. The shaft material has a shear yield stress of 24 ksi and a shear modulus of $G = 4000$ ksi, Assuming elastic–perfectly plastic material behavior, determine (a) the magnitude of the applied torque and (b) the residual shear stress when the applied torque is removed. (c) Show that the residual stresses are self-equilibrating (i.e., produce no internal torque).

ANS. $T = 316.4$ in · kips

5.41 The elastic–perfectly plastic shaft shown in Figure P5.41 has a plastic zone 30 mm deep in section BC. The shear yield stress of shaft is 200 MPa and its shear modulus $G = 80$ GPa. Determine (a) the magnitude of the applied torque T_{ext}, (b) the rotation of section A with respect to the wall, and (c) the residual stresses in the shaft when all the external torques are removed.

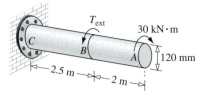

Figure P5.41

ANS. $T_{ext} = 117.7$ kN · m
$\phi_A = 0.1715$ rad ccw

5.42 The solid, elastic–perfectly plastic shaft shown in Figure P5.42 has a diameter of 4 inches, a shear yield stress of 30,000 psi, and a shear modulus of $G = 12{,}000$ ksi. The plastic zone in AB is 0.5 inch deep. Determine (a) the magnitude of the applied torque T_{ext}, (b) the rotation of section C with respect to the wall, and (c) the residual stresses in the shaft when all the external torques are removed.

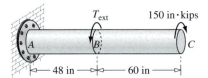

Figure P5.42

5.43 The shaft shown in Figure P5.43 is made from elastic–perfectly plastic material; it has a shear yield stress of 200 MPa and a shear modulus of $G = 80$ GPa. The plastic zone in section AB is 25 mm deep. Determine (a) the torque T_{ext}, (b) the rotation of section B.

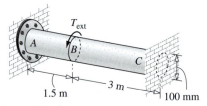

Figure P5.43

ANS. $T = 90.0$ kN · m $\phi_B = 0.15$ rad ccw

5.44 Determine the equivalent internal bending moment M_z for the cross section shown in Figure P5.44, the normal stress on the cross section is as follows:

$$\sigma_{xx} = \begin{cases} 24 \text{ ksi} & -5 \le y < -4 \\ -6y \text{ ksi} & -4 < y < 4 \\ -24 \text{ ksi} & 4 < \ \le 5 \end{cases}$$

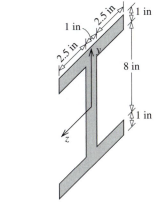

Figure P5.44

5.45 A rectangular beam has the dimensions shown in Figure P5.45. The normal strain due to bending about the z axis was found to vary as $\varepsilon_{xx} = -0.01y$, with y measured in meters. Determine the equivalent internal moment that produced the given state of strain. The beam is made from an elastic–perfectly plastic material that has a yield stress of $\sigma_{yield} = 250$ MPa and a modulus of elasticity of $E = 200$ GPa.

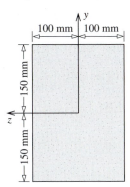

Figure P5.45

ANS. $M_z = 864.5$ kN · m

5.46 A rectangular beam has the dimensions shown in Figure P5.45. The normal strain due to bending about the z axis was found to vary as $\varepsilon_{xx} = -0.01y$, with y measured in meters. Determine the equivalent internal moment that would produce the given strain. The beam is made from a bilinear material that has a yield stress of

$\sigma_{yield} = 200$ MPa and moduli of elasticity $E_1 = 250$ GPa and $E_2 = 80$ GPa.

ANS. $M_z = 765$ kN · m

5.47 A rectangular beam has the dimensions shown in Figure P5.45. The normal strain due to bending about the z axis was found to vary as $\varepsilon_{xx} = -0.01y$, with y measured in meters. Determine the equivalent internal moment that would produce the given strain. The beam material has a stress–strain relationship given by $\sigma = 952\varepsilon^{0.2}$ MPa.

ANS. $M_z = 1061$ kN · m

5.48 A rectangular beam has the dimensions shown in Figure P5.45. The normal strain due to bending about the z axis was found to vary as $\varepsilon_{xx} = -0.01y$, with y measured in meters. Determine the equivalent internal moment that would produce the given strain. The beam material has a stress–strain relationship given by $\sigma = (200\varepsilon - 2,000\varepsilon^2)$ MPa.

ANS. $M_z = 890$ N · m

5.49 A wide-flange beam has the dimensions shown in Figure P5.49. The normal strain due to bending about the z axis was found to vary as $\varepsilon_{xx} = -0.4y(10^{-3})$, with y measured in inches. Determine the equivalent internal moment that would produce the given strain. The beam is made from an elastic–perfectly plastic material that has a yield stress of $\sigma_{yield} = 40$ ksi, and a modulus of elasticity of $E = 10,000$ ksi.

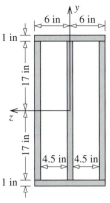

Figure P5.49

5.50 A beam has the dimensions shown in Figure P5.49. The normal strain due to bending about the z axis was found to vary as $\varepsilon_{xx} = -0.4y(10^{-3})$, with y measured in inches. Determine the equivalent internal moment that would produce the given strain. The beam is made from a bilinear material that has a yield stress of $\sigma_{yield} = 40$ ksi and moduli of elasticity $E_1 = 10,000$ ksi and $E_2 = 2000$ ksi.

5.51 A wide-flange beam has the dimensions shown in Figure P5.49. The normal strain due to bending about the z axis was found to vary as $\varepsilon_{xx} = -0.4y(10^{-3})$, with y measured in inches. Determine the equivalent internal moment that would produce the given strain. The beam material has a stress–strain relationship given by $\sigma = 70\varepsilon^{0.1}$ ksi.

5.52 A wide-flange beam has the dimensions shown in Figure P5.49. The normal strain due to bending about the z axis was found to vary as $\varepsilon_{xx} = -0.4y(10^{-3})$, with y measured in inches. Determine the equivalent internal moment that would produce the given strain. The beam material has a stress–strain relationship given by $\sigma = 10,000\varepsilon - 90,000\varepsilon^2$ ksi.

5.53 A beam made from an elastic–perfectly plastic material has a yield stress of 30 ksi. Determine the internal bending moment if the elastic–plastic boundary is 0.5 inch from the top and bottom on the rectangular cross section shown in Figure P5.53.

Figure P5.53

ANS. $M_z = 48.75$ in · kips

5.54 Determine the shape factor for the rectangular cross section shown in Figure P5.53.

ANS. $f = 1.5$

5.55 An elastic–perfectly plastic material has a yield stress of $\sigma_{yield} = 40$ ksi. Point A in Figure P5.55 is at yield stress as a result of bending of the beam. Determine (a) the location of the neutral axis, assuming that it is in the web, and (b) the applied moment that produced the state of stress.

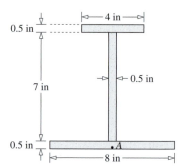

Figure P5.55

5.56 For the cross section shown in Figure P5.55 determine the shape factor.

5.57 The uniformly loaded, simply supported beam shown in Figure P5.57 is made of an elastic–perfectly plastic material that has a yield stress of 30 ksi. The beam has a hollow square cross section as shown. If point A is at yield stress, determine the intensity w of the uniform load.

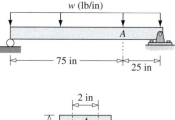

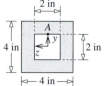

Figure P5.57

ANS. $w = 427$ lb/in

5.58 A beam of an elastic–perfectly plastic material has a yield stress of 50 ksi and the cross section shown in Figure P5.58. If point A just reaches yield stress, determine (a) the location of the neutral axis and (b) the applied moment that produced the state of stress.

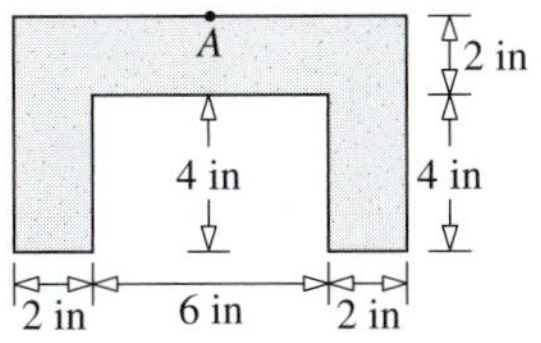

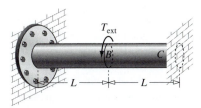

Figure P5.58

ANS. $a = 2.186$ in $\quad M_z = 2091$ in · kips

5.59 A circular solid shaft of radius R is made from a nonlinear material that has a shear stress–shear strain relationship given by $\tau = K\gamma^{0.4}$. Show that $\tau_{max} = 0.5411(T/R^3)$ and $\phi_2 - \phi_1 = 0.2154 (x_2 - x_1) (T/KR^{3.4})^{2.5}$, where τ_{max} is the maximum shear stress at a section, T is the internal torque at the section, and ϕ_2 and ϕ_1 are the sectional rotations at x_1 and x_2.

5.60 The circular solid shaft of radius R shown in Figure P5.60 is made from a nonlinear material that has a shear stress–shear strain relationship given by $\tau = K\gamma^{0.6}$. Assuming small deformation, determine (a) the rotation of section B with respect to the left-hand wall and (b) the maximum shear stress in the shaft.

Figure P5.60

ANS. $\phi_B = 0.124\left(\dfrac{T_{ext}}{KR^{3.6}}\right)^{1.667} L$

$\tau_{max} = 0.2865\dfrac{T_{ext}}{R^3}$

5.61 The hollow square beam shown in Figure P5.61 is made from a material that has a stress–strain relation given by $\sigma = K\varepsilon^{0.4}$. Assume that the material behaves identically in tension and in compression. In terms of K, L, a, and M_{ext}, determine the bending normal stress at point A.

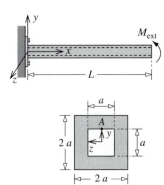

Figure P5.61

5.62 The stress–strain curve for a material is given by $\sigma = K\varepsilon^{0.5}$. Assume that the material behaves identically in tension and in compression. For the rectangular cross section shown in Figure P5.62, show that the bending normal stress is given by

$$\sigma_{xx} = \begin{cases} \dfrac{-5\sqrt{2}}{bh^2}\left(\dfrac{y}{h}\right)^{0.5} M_z & y > 0 \\[3mm] \dfrac{5\sqrt{2}}{bh^2}\left(\dfrac{y}{h}\right)^{0.5} M_z & y < 0 \end{cases}$$

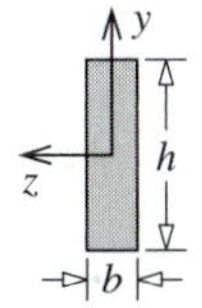

Figure P5.62

5.63 The cantilever beam shown in Figure P5.63 is made from a material that has a stress–strain relation given by $\sigma = K\varepsilon^{0.7}$. Assume that the material behaves identically in tension and in compression. In terms of K, w, a, and L, determine the maximum bending normal stress at point A, just above the bottom flange.

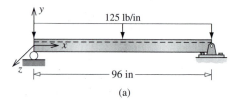

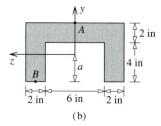

(a)

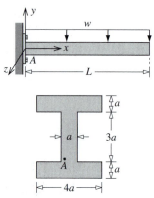

Figure P5.63

ANS. $\sigma_A = 0.0233(wL^2/a^3)$ (C)

Figure P5.64

5.64 The simply supported beam in Figure P5.64 is made from a material having the stress–strain curve given by $\sigma = 400\varepsilon^{0.4}$ ksi. Determine (a) the neutral axis location and (b) the bending normal stress at points A and B on the midsection of the beam.

5.65 The strain displacement relationship for a large axial strain is given by

$$\varepsilon_{xx} = \frac{du}{dx} + \frac{1}{2}\left(\frac{du}{dx}\right)^2$$

where we recognize that u is only a function of x, and hence the strain from the equations is uniform across the cross section. For a linear, elastic, homogeneous material, show

$$\frac{du}{dx} = \sqrt{1 + \frac{2N}{EA}} - 1 \quad \text{and} \quad \sigma_{xx} = \frac{N}{A}$$

5.9   VISCOELASTICITY

Solids exhibiting over time stress–strain behavior that has similarities to that of a viscous fluid are said to be viscoelastic. The phenomenon of viscoelasticity is seen in one of the two ways: a solid under constant stress (load) continues to deform over time, or a solid under constant strain (deformation) shows a decrease in stress over time. Metals that exhibit viscoelasticity when used at elevated temperatures are found in turbine blades, boiler walls, and pipes in power plants. By "elevated temperatures," we mean temperatures that are about a third of melting point temperature on an absolute scale. Warming up to loosen the muscles before vigorous exercise is recognition of the viscoelasticity of the muscle tissues. Wooden beams and floors in old houses sag owing to viscoelasticity. Plastics, polymer composites, tar, putty, solid rocket propellant, and skin all exhibit the phenomenon of viscoelasticity[6] for axial members.

---

[6] For additional details, see Shames and Cozzarelli [1997], Flugge [1967], and Morrison [2001] in Appendix D.

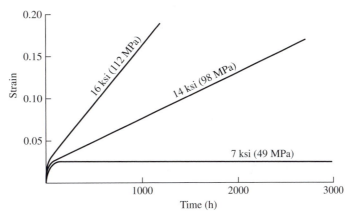

Figure 5.49 Strain vs time for a typical metal.

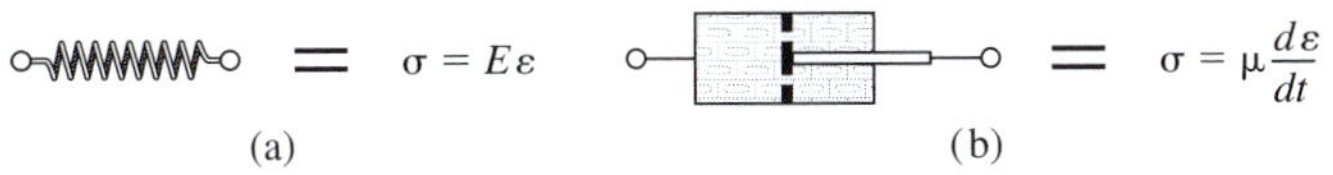

$$\multimap\!\!\!\text{WWWW}\!\!\!\multimap \quad = \quad \sigma = E\varepsilon \qquad \multimap\!\!\boxed{}\!\!\multimap \quad = \quad \sigma = \mu\frac{d\varepsilon}{dt}$$

$$\text{(a)} \hspace{10em} \text{(b)}$$

Figure 5.50 Spring and dashpot representation for (a) elastic and (b) viscous material response.

Figure 5.49 shows results exhibited at elevated temperature by a typical metal in a tension test in which a specimen is stressed to a certain constant level and the strain is observed over time. As seen from the graph, the strain grows over time. This growth in strain over time is called *creep,* and the strain is referred to as *creep strain.* The graph also shows that the material's initial response to stress is instantaneous and is followed by creep, which persists over long periods of time. The inertial forces (dynamic effects) can be neglected in the analysis of creep because of the slowness of the process. *Quasistatic* analysis is the term for such time analyses in which the inertial forces can be neglected.

We need an equation that models elastic as well as viscous responses to incorporate the empirical information relating stress, strain, and time that is obtained by experiments and shown in graphs like Figure 5.49. Linear springs are used to symbolically represent Hooke's law for linear material behavior as shown in Figure 5.50. Nonlinear springs are used to represent nonlinear material behavior such as discussed in Section 5.4. Dashpots[7] are used to symbolically represent linear viscous behavior as shown in Figure 5.50, where μ is a viscosity parameter relating stress and strain rates as in a Newtonian viscous fluid. A nonlinear stress–strain rate relationship is used for non-Newtonian viscous fluids. Springs and dashpots in series and in parallel are used to create mathematical equations with elasticity parameters E_i and viscosity parameters μ_i. These parameters are then determined in a least-squares manner from experimental data. In Chapter 2 it was emphasized that the choice of a material model is dictated both by the experimental data and by the degree of accuracy needed for the analysis. A material model that does not fit

---

[7] A piston with holes moving through a viscous fluid is one way of visualizing a dashpot.

the experimental data well will produce a high error rate in the theoretical predictions. A material model that fits the experimental data very accurately may well be so complex that no analytical model (theory) can be built. In the context of viscoelasticity, the material model selected influences the type (linear or nonlinear), the number, and the configuration in which the springs and dashpots are used.

We will restrict ourselves to linear systems in which the spring and dashpot responses can be superposed. Sections 5.9.1 and 5.9.2 discuss the simplest linear viscoelastic models, with one spring and one dashpot. Section 5.9.5 describes briefly more complex linear models. It should be recognized that even though we are considering only linear systems, the systems will be nonconservative because dashpots (viscous materials) always dissipate energy.

5.9.1 Maxwell Model

The Maxwell model is shown symbolically as a spring and dashpot in series in Figure 5.51. In this model both the spring and the dashpot are subjected to the same force (stress), and the total deformation (strain) is the sum of the deformations. The constitutive stress–strain equation for the Maxwell model is obtained by adding the strain rates for the spring and the dashpot to obtain total strain rates as shown in Equation (5.20),

$$\frac{d\varepsilon}{dt} = \frac{1}{E}\frac{d\sigma}{dt} + \frac{\sigma}{\mu}$$ (5.20)

where μ is the viscosity parameter. If $\mu \rightarrow \infty$ in Equation (5.20), we get an elastic behavior. If $E \rightarrow \infty$, then we get a viscous behavior.

Figure 5.51 Maxwell model representation.

5.9.2 Kelvin Model

In the Kelvin model, the spring and the dashpot are attached in parallel as shown in Figure 5.52. In this model both the spring and the dashpot are subjected to the same deformation (strain), and the total force (stress) transmitted is the sum of forces carried by the two components. The constitutive stress–strain equation for the Kelvin model is obtained by adding the stresses for the spring and the dashpot to obtain the total stress as shown in Equation (5.21).

$$\sigma = \mu\frac{d\varepsilon}{dt} + E\varepsilon$$ (5.21)

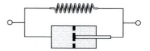

Figure 5.52 Kelvin model representation.

If $\mu \to 0$ in Equation (5.21), we get an elastic behavior; if $E \to 0$, then we get a viscous behavior.

5.9.3 Creep Test

In the creep test, the most widely used tension test for obtaining viscoelastic properties, the material is subjected to a sudden stress, which is then held constant. That is, the material is subjected to a step change in stress. We can model the step change by using the discontinuity functions studied in Section 3.3 with time t replacing coordinate x as shown in Equation (5.22),

$$\sigma = \sigma_0 \langle t \rangle^0 \tag{5.22}$$

where σ_0 is the initial stress applied at time $t = 0$ as shown in Figure 5.53(a). Equation (5.22) can be substituted in the Maxwell model constitutive equation (5.20) and integrated by using the properties of the discontinuity functions (discussed in Section 3.3) to obtain Equation (5.23).

$$\varepsilon(t) = \frac{\sigma_0}{E} \langle t \rangle^0 + \frac{\sigma_0}{\mu} \langle t \rangle^1 \tag{5.23}$$

We note that $t > 0$. Hence the discontinuity functions can be replaced by the arguments, and Equation (5.23) can be written as Equation (5.24).

$$\boxed{\varepsilon(t) = \frac{\sigma_0}{E} + \frac{\sigma_0}{\mu} t} \tag{5.24}$$

In Figure 5.53(b), which plots the strain obtained in Equation (5.24), the Maxwell material shows an immediate elastic response followed by a linear increase in strain with time. As $t \to \infty$ the strain $\varepsilon \to \infty$; that is, deformation and strain grow continuously, like a fluid. For this reason, Maxwell materials are often called *Maxwell fluids*.

Equation (5.22) can be substituted in the Kelvin model constitutive Equation (5.21) and integrated to obtain the solution (see Problem 5.66) given by Equation (5.25).

$$\boxed{\varepsilon(t) = \frac{\sigma_0}{E}[1 - e^{-(E/\mu)t}]} \tag{5.25}$$

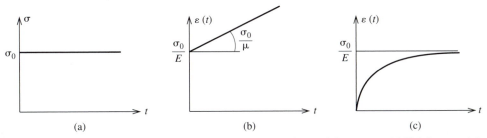

Figure 5.53 (a) Applied stress in creep test. (b) Maxwell material response. (c) Kelvin material response.

Figure 5.53(c) shows the plot of the strain obtained in Equation (5.25). Unlike the Maxwell material, the strain does not jump instantaneously when the stress is applied; instead, it grows toward the elastic strain value. As $t \to \infty$ the strain $\varepsilon \to \sigma_0/E$. That is, the material is elastically deformed, and hence Kelvin materials behave like solids and are often referred to as *Kelvin solids*.

In Equation (5.25) the ratio μ/E has dimensions of time and may be considered to be a material constant. This ratio will dictate the rate at which Kelvin material reaches the elastic solution. In a creep test, this ratio is referred to as the *retardation time*.

5.9.4 Relaxation Test

The relaxation test is another widely used test for determining viscoelastic properties. In this test strain is suddenly applied and held constant as shown in Equation (5.26),

$$\varepsilon(t) = \varepsilon_0 \langle t \rangle^0 \tag{5.26}$$

where ε_0 is the initial applied strain applied at time $t = 0$ as shown in Figure 5.54(a). Equation (5.26) can be substituted in the Maxwell model constitutive Equation (5.20) and integrated to obtain the solution (see Problem 5.67) given by Equation (5.27).

$$\boxed{\sigma(t) = E\varepsilon_0 e^{-(E/\mu)t}} \tag{5.27}$$

Figure 5.54(b) shows the plot of the stress obtained in Equations (5.27). The Maxwell material initially shows an elastic stress that continuously decreases with time. As $t \to \infty$ the stress $\sigma \to 0$. That is, the material becomes unstressed with time. Once more we note that the ratio μ/E in Equation (5.27) has the dimension of time, and this ratio will control the rate at which the stress is relaxed. In a relaxation test, this ratio is referred to as *relaxation time*. Thus, both retardation and relaxation time are represented by the same ratio, with the difference being that retardation time is used in the context of creep testing and relaxation time in the context of relaxation testing.

Equation (5.26) can be substituted in the Kelvin model constitutive Equation (5.21) to obtain Equation (5.28).

$$\sigma = \mu\varepsilon_0 \langle t \rangle^{-1} + E\varepsilon_0 \langle t \rangle^0 \tag{5.28}$$

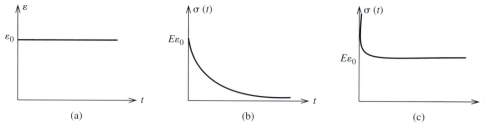

Figure 5.54 (a) Applied strain in relaxation test. (b) Maxwell material response. (c) Kelvin material response.

Equation (5.28) shows that the stress initial response is that of a delta function,[8] which then reduces to the elastic solution. Real materials are unlikely to show such behavior, and this emphasizes that Equation (5.28) is an idealization for analytical purposes. If predictions based on the analysis do not correlate well with experimental data, other models should be considered.

5.9.5 Generalized Viscoelastic Linear Models

More springs and more dashpots can be put together in series and parallel to create more complex models. Figure 5.55 shows the next level of complexity, where we have increased the number of parameters from two to three. In this manner we can continue adding springs and dashpots to add more parameters that can be used to get a better fit between the experimental data and the constitutive model. In Maxwell and Kelvin models we see that Equations (5.20) and (5.21) are differential equations in time that represent the constitutive relationship between stress and strain. The general form of these differential equation is shown as Equation (5.29),

$$\sum_{i=0}^{n} p_i \frac{d^i \sigma}{dt^i} = \sum_{j=0}^{m} q_j \frac{d^j \varepsilon}{dt^j} \tag{5.29}$$

where p_i and q_j are related to the E's and μ's in some manner dictated by the configuration of springs and dashpots; n and m are integers that depend upon the number of springs and dashpots. Comparing Equation (5.29) with Equation (5.20), we see that $n = 1$, $m = 0$, $p_0 = 1/\mu$, $p_1 = 1/E$, and $q_0 = 1$ for the Maxwell model. Comparing Equation (5.29) with Equation (5.21), we see that $n = 0$, $m = 1$, $p_0 = 1$, $q_0 = E$, and $q_0 = \mu$ for the Kelvin model. Solutions of Equation (5.29) are generally found by means of Laplace transformation,[9] and the reader is referred to the books cited in footnote 6 of this chapter for additional details.

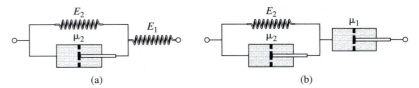

Figure 5.55 Three parameter for (a) a standard solid and (b) a standard fluid.

---

[8] See Section 3.3 for the definition of the delta function $\langle t \rangle^{-1}$.

[9] See Kreyszig [1979] in Appendix D.

EXAMPLE 5.20

The stepped axial metal bar shown Figure 5.56 is an approximation of a rod used in transmitting load in the machinery at a power station, where temperatures are high. The bar is made from a material having the viscoelastic properties shown in Figure 5.49. The modulus of elasticity of the bar is $E = 14,000$ ksi, and the cross-sectional areas of AB and BC are 1 in^2 and 0.5 in^2. Determine the movement of the section at C at times (a) $t = 0$ and (b) $t = 1000$ hours.

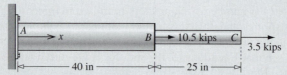

Figure 5.56 Viscoelastic axial bar for Example 5.20.

PLAN

The relative displacements of the ends of segments AB and BC can be found by using Equation (3.10-A) and added to get the movement of the section at C at time $t = 0$. The stress in each segment can also be found, and by using Figure 5.49 we can obtain the strain at 1000 hours for each segment. By multiplying the strain by the length, we can obtain the relative movement of segment ends and add them to find the movement of the section at C that is due to creep.

SOLUTION

The internal force in each segment can be obtained as given by Equation (E1).

$$N_{AB} = 14 \text{ kips} \quad \text{and} \quad N_{BC} = 3.5 \text{ kips} \tag{E1}$$

The relative movements of the segment ends due to elastic deformation can be obtained by using Equation (3.10-A) as shown in Equations (E2) and (E3).

$$(u_B)_e - (u_A)_e = \frac{N_{AB}(x_B - x_A)}{E_{AB}A_{AB}} = \frac{(14)(40)}{(14,000)(1)} = 0.040 \text{ in} \tag{E2}$$

$$(u_C)_e - (u_B)_e = \frac{N_{BC}(x_C - x_B)}{E_{BC}A_{BC}} = \frac{(3.5)(25)}{(14,000)(0.5)} = 0.0125 \text{ in} \tag{E3}$$

The stress in each segment can be found as shown in Equations (E4).

$$\sigma_{AB} = \frac{N_{AB}}{A_{AB}} = \frac{14}{1} = 14 \text{ ksi} \quad \text{and} \quad \sigma_{BC} = \frac{N_{BC}}{A_{BC}} = \frac{3.5}{0.5} = 7 \text{ ksi} \tag{E4}$$

(a) Adding Equations (E2) and (E3), we obtain Equation (E5), and noting that the section at A cannot move, we obtain the movement at point C at $t = 0$.

$$(u_C)_e - (u_A)_e = 0.040 + 0.0125 \qquad \text{or} \qquad \text{(E5)}$$

ANS. $(u_C)_e = 0.0525$ in

(b) From the stresses in Equation (E4) and the curves in Figure 5.49, we estimate the normal strain at $t = 1000$ hours as shown in Equations (E6).

$$\varepsilon_{AB} = 0.075 \qquad \text{and} \qquad \varepsilon_{BC} = 0.02 \qquad \text{(E6)}$$

The relative movements of the segment ends due to creep can be found as shown in Equations (E7) and (E8).

$$(u_B)_c - (u_A)_c = \varepsilon_{AB}(x_B - x_A) = (0.075)(40) = 3.0 \text{ in} \qquad \text{(E7)}$$

$$(u_C)_c - (u_B)_c = \varepsilon_{BC}(x_C - x_B) = (0.02)(25) = 0.5 \text{ in} \qquad \text{(E8)}$$

Adding Equations (E7) and (E8) and noting that the section at A cannot move, we obtain the movement of the section at C that is due to creep.

$$(u_C)_c = 3 + 0.5 = 3.5 \text{ in} \qquad \text{(E9)}$$

COMMENTS

1. The elastic deformation is only $0.0525/3.5 \times 100 = 1.5\%$ of the creep deformation. A similar analysis at 2000 hours shows that the elastic deformation is only $0.0525/5.7 \times 100 = 0.92\%$. Thus, as time passes, elastic deformation relative to creep becomes negligible.

2. At 7 ksi the creep strain (ε_{BC}) is 0.025. At 14 ksi (i.e., at double the stress value), the creep strain is not double but even greater: ε_{AB} is 0.075. In other words, creep strain and stress are not linearly related. Increasing stress increases the creep strain disproportionately. Thus, the material with the viscoelastic properties shown in Figure 5.49 cannot be modeled by any of the linear models discussed in this section.

EXAMPLE 5.21

A beam is held up by a 1/2 inch diameter circular rod AB made from a Maxwell material ($E = 15{,}000$ ksi and $\mu = 12 \times 10^9$ ksi $\cdot$ s) as shown in Figure 5.57. Model the beam as a rigid beam with a specific weight of 500 lb/ft. Determine the rotation of the beam from the horizontal position 10,000 hours after assembly.

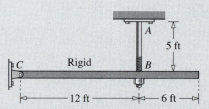

Figure 5.57 Viscoelastic rod supporting the rigid beam in Example 5.21.

PLAN

The internal axial force in bar AB can be found by taking the moment about C on a free body diagram of the rigid beam. The internal axial force divided by area gives us the stress σ_0 at $t = 0$. Viewing this stress as a step function in time, the strain at 10,000 hours can be found from Equation (5.24). From this strain, the deformation of AB can be obtained, and by using geometry, the rotation of the bar can be determined.

SOLUTION

The total weight of the beam is length times the specific weight, as shown in Equation (E1).

$$W = (500)(18) = 9000 \text{ lb} = 9 \text{ kips} \tag{E1}$$

Figure 5.58(a) shows the free body diagram of the rigid bar. By equilibrium of moment about C, we obtain Equation (E2).

$$12(N_{AB}) - 9(9) = 0 \quad \text{or} \quad N_{AB} = 6.75 \text{ kips} \tag{E2}$$

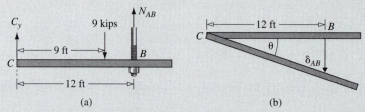

(a)   (b)

Figure 5.58 (a) Free body diagram and (b) exaggerated deformed geometry for Example 5.21.

The axial stress in bar AB just after assembly can be found as shown in Equation (E3).

$$\sigma_{AB} = \frac{N_{AB}}{A_{AB}} = \frac{6.75}{\pi(1/4)^2} = 34.377 \text{ ksi} \tag{E3}$$

The strain in AB after 10,000 hours can be found from Equation (5.24) as shown in Equation (E4).

$$\varepsilon_{AB} = \frac{\sigma_{AB}}{E} + \frac{\sigma_{AB}}{\mu}t = \frac{34.377}{15,000} + \frac{34.377}{12(10^9)}(10,000)(3600) = 0.1054 \tag{E4}$$

The deformation of bar AB can be found as follows:

$$\delta_{AB} = \varepsilon_{AB} L_{AB} = (0.1054)(5) = 0.5271 \text{ ft} \qquad (E5)$$

From the exaggerated deformed geometry shown in Figure 5.58(b), the angle of rotation for the rigid bar can be found as shown in Equation (E6).

$$\tan \theta = \frac{\delta_{AB}}{CB} = \frac{0.5271}{12} = 0.0439 \qquad \text{or} \qquad (E6)$$

ANS. $\theta = 2.52°$

COMMENT

In this example the initial analysis (at $t = 0$) is conducted like a regular elastic problem. The viscoelastic response then is found from the viscoelastic model. This is possible because we are considering only linear systems. Nonlinear viscoelastic problems depend upon the history of loading, and in such cases the elastic and viscoelastic problems may not be separable.

EXAMPLE 5.22

Obtain the constitutive equation for the three-parameter solid shown in Figure 5.59.

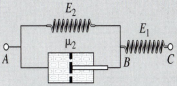

Figure 5.59 Three-parameter solid for Example 5.22.

PLAN

We can consider a tensile stress σ applied to ends A and C. Segments AB and BC will be subjected to the same stress. The sum of the strains (deformation) in AB and BC is equal to the total strain ε. The strain in BC is given by Hooke's law. The strain in AB can be written as the total strain minus the elastic strain in BC. The elastic stress and the viscous stress in AB can be written in terms of the strain in AB and added and equated to the applied stress to obtain the constitutive equation.

SOLUTION

The stress in AB and BC is equal to the applied stress as shown in Equation (E1).

$$\sigma_{AB} = \sigma_{BC} = \sigma \qquad \text{(E1)}$$

The sum of the strains in AB and BC is equal to the total strain. The strain in BC can be written as Hooke's law, as follows:

$$\varepsilon = \varepsilon_{AB} + \varepsilon_{BC} = \varepsilon_{AB} + \frac{\sigma_{BC}}{E_1} = \varepsilon_{AB} + \frac{\sigma}{E_1} \qquad \text{or} \qquad \varepsilon_{AB} = \varepsilon - \frac{\sigma}{E_1} \qquad \text{(E2)}$$

The elastic stress in AB is $E_2\varepsilon_{AB}$. The viscous stress in AB is $\mu_2\,(d\varepsilon_{AB}/dt)$. The sum of the elastic and viscous stresses in AB should equal the applied stress, as shown in Equation (E3).

$$\sigma = E_2\varepsilon_{AB} + \mu_2\frac{d\varepsilon_{AB}}{dt} \qquad \text{(E3)}$$

Substituting Equation (E2) into Equation (E3), we obtain Equation (E4).

$$\sigma = E_2\left(\varepsilon - \frac{\sigma}{E_1}\right) + \mu_2\frac{d}{dt}\left(\varepsilon - \frac{\sigma}{E_1}\right) = E_2\varepsilon - E_2\frac{\sigma}{E_1} + \mu_2\frac{d\varepsilon}{dt} - \frac{\mu_2 d\sigma}{E_1 dt} \qquad \text{or} \qquad \text{(E4)}$$

$$\sigma\left(1 + \frac{E_2}{E_1}\right) + \frac{\mu_2 d\sigma}{E_1 dt} = E_2\varepsilon + \mu_2\frac{d\varepsilon}{dt}$$

COMMENTS

1. This example demonstrates how one can add complexity (additional parameters) to the constitutive equation.

2. Comparing Equation (E4) with Equation (5.29), we see that $n = 2$, $m = 2$, $p_0 = (1 + E_2/E_1)$, $p_1 = \mu_2/E_1$, $q_0 = E_2$, and $q_0 = \mu_2$.

PROBLEM SET 5.3

5.66 Show that the solution in Equation (5.25) satisfies the differential equation given by Equation (5.21) for $\sigma = \sigma_0\langle t\rangle^0$.

5.67 Show that the solution in Equation (5.27) satisfies the differential equation given by Equation (5.20) for $\varepsilon(t) = \varepsilon_0\langle t\rangle^0$.

5.68 A 100 kg mass is suspended by cables as shown in Figure P5.68. The cable material has viscoelastic properties given by the graphs shown in Figure 5.49. The cable cross-sectional area is 20 mm². Determine how far

down the mass has moved, 2000 hours after being suspended.

Figure P5.68

ANS. $\delta = 37.5$ mm

5.69 Two rods, each having a diameter of 15 mm, support a rigid platform of mass

5000 kg as shown in Figure P5.69. The rods are made from a material that conforms to the Maxwell model, with $E = 200$ GPa and $\mu = 100\,(10^6)$ GPa · s. Determine the rotation of the rod from the horizontal after 3000 hours.

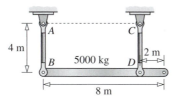

Figure P5.69

ANS. $\theta = 3.99°$ cw

5.70 Two rods identical to those in Problem 5.69 support a rigid platform of mass 5000 kg as shown in Figure P5.69. Determine how far down pins B and D will have moved after (a) 100 hours and (b) 10,000 hours.

5.71 During the assembly of a joint, a rod must be pulled to close a gap. The joint is modeled as shown in Figure P5.71. The rod is made from a Maxwell material with $E = 200$ GPa and $\mu = 100\,(10^6)$ GPa · s. Determine the stress in the rod (a) just after assembly, (b) after 500 hours, and (c) after 5000 hours.

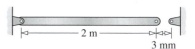

Figure P5.71

5.72 A Kelvin material is subjected to a linearly varying strain $\varepsilon = (\varepsilon_0/a)\langle t\rangle^1$, $0 \le t \le a$. Determine the stress response in terms of E, μ, a, and t, and plot $\sigma(t)$ vs t.

5.73 A Maxwell material is subjected to a linearly varying stress $\sigma = (\sigma_0/a)\langle t\rangle^1$, $0 \le t \le a$. Determine the strain response in terms of E, μ, a, and t, and plot $\varepsilon(t)$ vs t.

5.74 Obtain the constitutive equation of the three-parameter fluid model shown in Figure P5.74. Compare your results with Equation (5.29) and determine the values of m, n, p's, and q's in Equation (5.29).

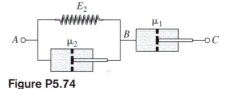

Figure P5.74

5.10 CLOSURE

In this chapter we saw that the kinematic equations, the static equivalency equations, and the equilibrium equations do not change. Rather, the change is in the material model. The initial strain, including temperature effects and the power law stress–strain relationship, results in a new set of formulas, but the process for obtaining the formulas remains unchanged. In elastic–perfectly plastic materials, the critical issue in the analysis is the determination of the location of the elastic–plastic boundary. In axial problems, the issue of the elastic–plastic boundary is the identification of the member that turns plastic. In bending problems, the determination of the elastic–plastic boundary is coupled with the determination of the neutral axis. In torsion, we saw that residual stresses can be obtained by subtracting the stresses obtained from the elastic formulas from the total stresses. This idea can also be applied to determine the residual stresses in a beam cross section. Prestressing (residual stresses) or prestraining (initial strains) are effective ways of increasing the capacity of a material to carry higher loads. In viscoelasticity the constitutive equations relating stress and strain are constructed by using springs and dashpots to symbolize the behavior of elastic and viscous materials. In the context of axial members, we did not need to develop new formulas. In shafts and beams, however, the stress–strain equations can be incorporated, as was done with power law nonlinearity. The concepts of nonlinear and/or viscoelastic material behavior can be used in analysis and in applications of plastics, rubber, and biological materials.

6 Thin-Walled Structural Members

6.0 OVERVIEW

The pole of a street light, the metal beams of a gantry crane, and the pipes of a highway sign-post (Figure 6.1) are all thin-walled structural members. Most metal beams have cross-sectional dimensions much greater than the thickness of the flange or the web and can be said to be thin walled. Box beams and columns made from thin metal plates and concrete can also be described as thin-walled members. These thin-walled structural members are used extensively in bridges, buildings, ships, aircraft, gantry cranes, and so on. The theory encompassing all thin-walled structures[1] can fill several graduate-level courses. Our focus is limited to the bending and torsion of long thin-walled structural members. In the introductory course on the mechanics of materials, and in Chapter 3, we limited ourselves to symmetric bending of beams and torsion of circular shafts. In this chapter we will study the unsymmetric bending of beams and the torsion of noncircular cross sections that have thin walls.

We developed the theory for the symmetric bending of beams under two limitations: the beam had to have a plane of symmetry, and the loading had to be in the plane of symmetry. We will drop these limitations in this chapter to develop the theory for the unsymmetric bending of beams. The theory will be initially developed under the following limitation: the external forces must be applied such that there is no twist or axial deformation. In symmetric bending, we saw the condition for decoupling the axial and bending problems was that the loads must pass through the centroid of the cross section. In this chapter we will see that the condition for decoupling torsion from unsymmetric bending is that the loads must pass through a point called the *shear center* of the cross section.

We will also develop the theory associated with the torsion of thin-walled tubes and consider problems in which the bending loads do not pass through the shear center, resulting in combined bending–torsion problems.

---

[1] See Szilard [1974], Kraus [1967], Murray [1984], and Vinson [1989] in Appendix D.

(a)   (b)   (c)

Figure 6.1 Examples of thin-walled structural members. (*Photograph of gantry crane courtesy of North American Industries.*)

We will follow the logic described in Chapter 3. The primary change will be in the kinematic equations for strains, which will now include deformations in the y and z directions. Though we will carry a new set of variables, the process of moving from one step to next remains the same as in the derivation of the elementary theories highlighted in Table 3.3. Once more we will try to develop the simplest possible theory for unsymmetric bending, making assumptions as we did for symmetric bending in Table 3.3. These assumptions are points at which we can add complexities, as was demonstrated for symmetric bending in Chapters 3, 4, and 5.

The learning objectives of this chapter are:

1. To understand the theory, its limitations, and its application in the design and analysis of unsymmetric bending of beams.

2. To understand the concept of shear center, how to determine its location, and how to use it in structural analysis.

3. To understand the theory of the torsion of thin-walled tubes and combined loading problems.

6.1   THEORY OF UNSYMMETRIC BENDING OF BEAMS

The theory of the unsymmetric bending of beams will be developed subject to limitations similar to those described in Section 3.2.1, namely:

1. The length of the member is significantly greater (approximately 10 times) than the greatest dimension in the cross section.

2. Points considered are away from regions of stress concentration

3. The variation of external loads or changes in cross-sectional area is gradual except in regions of stress concentration.

We will assume that the external forces are not functions of time; that is, we assume a static problem. Figure 6.2 shows a segment of a beam that may be loaded by transverse forces P_y and P_z, bending moments $(M_y)_{ext}$ and $(M_z)_{ext}$, and transverse distributed forces $p_y(x)$ and $p_z(x)$.

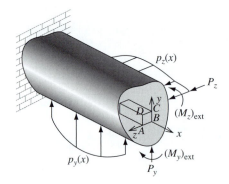

Figure 6.2 Possible loading in unsymmetric bending of beams.

The objectives of the theoretical derivations in this section are:

1. To obtain formulas for bending normal stress in terms of internal bending M_y and M_z.
2. To obtain the boundary value problem for calculating beam deflection $v(x)$ in the y direction and $w(x)$ in the z direction.

The formulas for bending shear stress to internal shear forces V_y and V_z are derived in Section 6.2.

The logic shown in Figure 6.3 and discussed in Section 3.1 will be used to develop the simplest theory for the unsymmetric bending of beams. Assumptions will be identified as we move from one step to the next. These assumptions are points at which complexities can be added to the theory, as described in Chapters 3, 4, and 5.

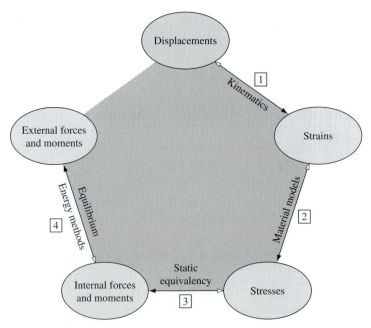

Figure 6.3 Logic in structural analysis.

We introduce the following sign conventions.

Definition 1 The displacements v and w are positive in the positive y and z direction, respectively.

Definition 2 The external distributed forces per unit length $p_y(x)$ and $p_z(x)$ are positive in the positive y and z direction, respectively.

6.1.1 Deformation

Restricting ourselves to static problems, we make the following assumption.

Assumption 1 Deformations are not functions of time.[2]

To decouple the bending from the axial and torsion problems, we make the follow-ing assumption[3] and obtain the requisite conditions from the theory later.

Assumption 2a The loads are such that there is no axial or torsional deformation.

Figure 6.2 shows a section of an unsymmetric beam with an undeformed rectangle *ABCD*. If there is no twist during bending, then the right angles of the rectangle *ABCD* cannot change. The right angles are between the y and z directions; hence the shear strain γ_{yz} is zero. The definition of $\gamma_{yz} = \partial v/\partial z + \partial w/\partial y = 0$ is satisfied if v is not a function of z and w is not a function of y as shown in Equations (6.1a) and (6.1b).

$$v(x, y, z) = v(x, y) \tag{6.1a}$$

$$w(x, y, z) = w(x, z) \tag{6.1b}$$

Assumption 2b Changes in cross-sectional dimensions are significantly smaller than bending displacements.

Assumption 2b implies that dimensional changes in the cross section are much smaller then the movement of the cross section as a whole. The longer the beam, the higher the validity of Assumption 2b. Neglecting dimensional changes in the y and z directions[4] implies that the normal strain in the y and z directions can be neglected.

---

[2] See Problem 6.17 for inclusion of time.

[3] The numbering on the assumptions corresponds to Table 3.3: for example, Assumptions 2a, 2b, and 2c are related to Assumption 2 of the table; assumptions displayed later (e.g., Assumptions 3–7) are the same as in the table. Thus the addition of complexities when the assumptions are not valid follows logic similar to that used for symmetric bending discussed in Chapters 3, 4, and 5.

[4] Dimensional changes due to Poisson's effect can be obtained once stresses have been found. The neglect of small-dimensional changes in the cross section permits simplification in the kinematics.

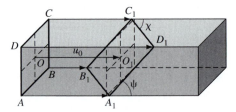

Figure 6.4 Plane sections remain plane during bending.

For our immediate calculations, that is, $\varepsilon_{yy} = \partial v / \partial y \approx 0$ and $\varepsilon_{zz} = \partial w / \partial z \approx 0$, which implies that v cannot be a function of y and w cannot be a function of z as shown in Equations (6.2a) and (6.2b).

$$v = v(x) \tag{6.2a}$$

$$w = w(x) \tag{6.2b}$$

Equations (6.2a) and (6.2b) imply that if we know the bending curve $v(x)$ and $w(x)$ of one line on the beam, we know the curves for all the lines on the beam.

Assumption 2c Plane sections before deformation remain plane after deformation.

Assumption 2c implies that the displacement u in the x direction is a linear function of y and z. From Figure 6.4 we obtain the displacement in the x direction of any point on plane $ABCD$ as shown in Equation (6.3),

$$u = u_0 - \psi y - \chi z \tag{6.3}$$

where u_0 is the axial displacement of a cross section, ψ is the rotation about the z axis and χ is the rotation about the y axis as shown in Figure 6.4. A plus or minus sign with ψ and χ does not change the linear approximation, but the reason for choosing the sign becomes clear with the next assumption. To develop a theory for unsymmetric bending, in which we do not have any axial deformation ($u_0 = 0$), we shall assume that the cross section undergoes rotation about the y and z axes only.

At present we have four unknown variables: the deflections v and w and the rotations of the cross section ψ and χ. With the next assumption we relate v to ψ and w to χ, hence reducing the problem to determination of v and w only.

Assumption 2d Planes perpendicular to the axis remain nearly perpendicular after deformation.

The assumption implies that the shear strains γ_{xy} and γ_{xz} are *nearly* zero. We cannot use this assumption in building a theoretical model if shear is important, as it is in sandwich beams and short beams. We note from Figure 6.5 that the beam deflection in the positive y and z directions causes a point with positive coordinates to move in the negative x direction, and this was the reason for using the minus signs in Equation (6.3). We can approximate sine and tangent functions by the arguments of the functions for small strains and relate these to the slope of the deflection curves as shown in

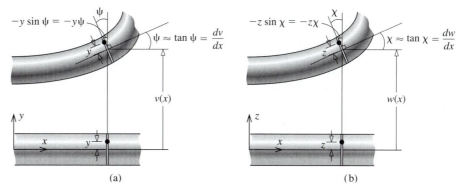

Figure 6.5 Displacement in the x direction due to bending.

Figure 6.5. Substituting the values of ψ, χ, and $u_0 = 0$ in Equation (6.3), we obtain Equation (6.4).

$$u = -y\frac{dv}{dx} - z\frac{dw}{dx} \tag{6.4}$$

6.1.2 Strain Distribution

Assumption 3 The strains are small.

From the definition of strain ε_{xx} and Equation (6.4) we obtain Equation (6.5),

$$\boxed{\varepsilon_{xx} = \frac{du}{dx} = -y\frac{d^2v}{dx^2} - z\frac{d^2w}{dx^2}} \tag{6.5}$$

where d^2v/dx^2 and d^2w/dx^2 are the curvatures[5] of the beam during bending about the z and y axis, respectively. Both curvatures are functions of x only. We thus have the following observation:

- The normal strain ε_{xx} in bending is a linear function of y and z. It will be maximum at the point that is the farthest from the origin.

6.1.3 Material Model

Our motivation is to develop a simple theory for unsymmetric bending. We therefore make assumptions regarding material behavior that will permit us to use the simplest material model, namely, Hooke's law.

---

[5] From calculus, the curvature of a curve defined by $v(x)$ is equal to

$$\frac{d^2v}{dx^2}\bigg/\left[1 + \left(\frac{dv}{dx}\right)^2\right]^{3/2}.$$

For small slopes, the denominator is nearly 1, and we obtain d^2v/dx^2 for the curvature.

> **Assumption 4** The material is isotropic.
>
> **Assumption 5** There are no inelastic strains.
>
> **Assumption 6** The material is elastic.
>
> **Assumption 7** Stress and strain are linearly related.

Substituting Equation (6.5) in Hooke's law,[6] $\sigma_{xx} = E\varepsilon_{xx}$, we obtain Equation (6.6).

$$\sigma_{xx} = -Ey\frac{d^2v}{dx^2} - Ez\frac{d^2w}{dx^2} \tag{6.6}$$

Though the strain is a linear function of y, we cannot say the same for stress at this stage because the modulus of elasticity E could change across the cross section, as in laminated structures.

6.1.4 Internal Forces and Moments

Figure 6.6(a) shows the stresses on a differential element with the outward normal in the x direction. Figure 6.6(b) shows the statically equivalent internal forces and moments in unsymmetric bending. The static equivalency of normal stress σ_{xx} to the internal forces and moments shown in Figure 6.6 results in the Equations (6.7a), (6.7b), and (6.7c).

$$N = \int_A \sigma_{xx}\, dA = 0 \tag{6.7a}$$

$$M_z = -\int_A y\sigma_{xx}\, dA \tag{6.7b}$$

$$M_y = -\int_A z\sigma_{xx}\, dA \tag{6.7c}$$

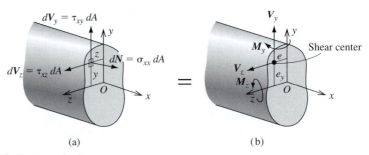

(a)   (b)

Figure 6.6 Static equivalency.

---

[6] We are assuming that σ_{yy} and σ_{zz} are negligible.

In symmetric bending we saw that the equilibrium of external transverse forces requires internal shear forces, which implies that the shear stresses τ_{xy} and τ_{xz} are nonzero. By Hooke's law this implies that the shear strains γ_{xy} and γ_{xz} cannot be zero. Assumption 2d implied that the shear strain was small but not zero. In beam bending, one of the checks on the validity of the analysis is to compare the maximum shear stresses τ_{xy} and τ_{xz} to the maximum normal stress σ_{xx}. If the maximum bending normal stress is comparable to either of the maximum bending shear stress components, then the shear strain cannot be neglected in kinematic considerations and our theory is not valid.

- The maximum bending normal stress σ_{xx} in the beam should be nearly an order of magnitude (factor of 10) greater than the maximum bending shear stress τ_{xy} and τ_{xz}.

From statics we know that any distributed force can be replaced by a force and a moment at any point, or by a single force (and no moment) at a specific point. The specific point at which the shear stress can be represented by shear forces V_y and V_z (components of a single force) alone, and no internal torque, is called the *shear center*. The concept of shear center is discussed in detail in Section 6.2.1, where its coordinates e_y and e_z are determined. The static equivalency of shear stresses to the internal forces and moments shown in Figure 6.6 results in Equations (6.8a), (6.8b), and (6.8c).

$$V_y = \int_A \tau_{xy}\, dA \tag{6.8a}$$

$$V_z = \int_A \tau_{xz}\, dA \tag{6.8b}$$

$$T = \int_A [(y - e_y)\tau_{xz} - (z - e_z)\tau_{xy}]\, dA = 0 \tag{6.8c}$$

The zero on the right-hand side in Equations (6.7a) and (6.8c) is a consequence of Assumption 2a. Equation (6.7a) is used for determining the origin, as discussed in Section 6.1.7. Equation (6.8c) is used for determining the shear center, as discussed in Section 6.2.1.

6.1.5 Sign Convention

Figure 6.7 shows the sign convention for the internal bending moments and internal shear forces.

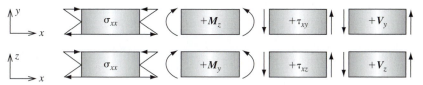

Figure 6.7 Sign convention for internal bending moments and shear forces.

The sign convention for internal bending moments and shear forces consists of the following definitions.

Definition 3 On a free body diagram, the positive internal shear force V_y is shown in the same direction as positive τ_{xy}.

Definition 4 On a free body diagram, the positive internal shear force V_z is shown in the same direction as positive τ_{xz}.

Definition 5 On a free body diagram, the positive internal bending moment M_z is shown to put the positive y face in compression.

Definition 6 On a free body diagram, the positive internal bending moment M_y is shown to put the positive z face in compression.

In drawing the positive internal moments and internal shear forces on a free body diagram, we do not consider the external forces and moments. The sign convention for internal forces and moments is tied to the coordinate system used in describing the beam geometry. Equilibrium equations will give the proper sign for the internal bending moments and internal shear forces.

6.1.6   Bending Formulas

Substituting σ_{xx} from Equation (6.6) into Equations (6.7b) and (6.7c), and noting that d^2v/dx^2 and d^2w/dx^2 are functions of x only, while the integration is with respect to y and z ($dA = dy\ dz$), we obtain Equations (6.9a) and (6.9b).

$$M_z = \frac{d^2v}{dx^2}\int_A Ey^2\ dA + \frac{d^2w}{dx^2}\int_A Eyz\ dA \qquad (6.9a)$$

$$M_y = \frac{d^2v}{dx^2}\int_A Eyz\ dA + \frac{d^2w}{dx^2}\int_A Ez^2\ dA \qquad (6.9b)$$

To simplify further, we would like to take E outside the integral, which implies that E cannot change across the cross section. We make the following assumption:

Assumption 8 The material is homogeneous across the cross section.[7]

From Equations (6.9a) and (6.9b) with a constant E, we obtain the moment curvature formulas in unsymmetric bending shown in Equations (6.10a) and (6.10b),

$$M_z = EI_{zz}\frac{d^2v}{dx^2} + EI_{yz}\frac{d^2w}{dx^2} \qquad (6.10a)$$

---

[7] See Problem 6.19 for laminated beams.

$$M_y = EI_{yz}\frac{d^2v}{dx^2} + EI_{yy}\frac{d^2w}{dx^2} \qquad (6.10b)$$

where I_{yy}, I_{zz}, I_{yz} are the second area moments of inertia defined in Equation (6.11) and discussed briefly in Appendix C.

$$I_{zz} = \int_A y^2 \, dA \qquad I_{yy} = \int_A z^2 \, dA \qquad I_{yz} = \int_A yz \, dA \qquad (6.11)$$

Equations (6.10a) and (6.10b) can be solved for the curvatures d^2v/dx^2 and d^2w/dx^2 in terms of the internal moments M_y and M_z by using Cramer's rule (see Section B.6) to obtain

$$\frac{d^2v}{dx^2} = \frac{1}{E}\left(\frac{I_{yy}M_z - I_{yz}M_y}{I_{yy}I_{zz} - I_{yz}^2}\right) \qquad (6.12a)$$

$$\frac{d^2w}{dx^2} = \frac{1}{E}\left(\frac{I_{zz}M_y - I_{yz}M_z}{I_{yy}I_{zz} - I_{yz}^2}\right) \qquad (6.12b)$$

For statically determinate beams,[8] we can integrate Equations (6.12a) and (6.12b) and use boundary conditions on v, w, dv/dx, and dw/dx to determine the integration constants, thereby obtaining the deflection of the beam at any x.

Substituting Equations (6.12a) and (6.12b) into Equation (6.6), we obtain the *bending normal stress formula* for unsymmetric bending as shown in Equation (6.13).

$$\sigma_{xx} = -\left(\frac{I_{yy}M_z - I_{yz}M_y}{I_{yy}I_{zz} - I_{yz}^2}\right)y - \left(\frac{I_{zz}M_y - I_{yz}M_z}{I_{yy}I_{zz} - I_{yz}^2}\right)z \qquad (6.13)$$

If axis y or z is an axis of symmetry, then $I_{yz} = 0$, and Equation (6.13) will simplify to

$$\sigma_{xx} = -\left(\frac{M_z}{I_{zz}}\right)y - \left(\frac{M_y}{I_{yy}}\right)z$$

which is the formula for symmetric bending. Similar simplification can be achieved in unsymmetric bending by using principal coordinates for the area moments of inertia discussed in Section A.6 of Appendix A.

6.1.7 Location of Origin

The internal axial force N should be zero because we assumed that there are no external axial forces. This implies that the stress distribution must be such that the internal axial force N in Equation (6.7a) is zero, as shown in Equation (6.14).

$$\int_A \sigma_{xx} \, dA = 0 \qquad (6.14)$$

---

[8] For statically indeterminate beams we could carry the reaction force as an unknown in the moments and determine the reaction from additional boundary conditions on the deflection or slope. Alternatively, we could solve a fourth-order differential equation. See Problem 6.16.

Equation (6.14) does not depend upon the material models and implies the following, irrespective of the material model:

- The resultant axial force on a cross section in bending is zero. Thus the total compressive force must equal the total tensile force in bending.

We can use Equation (6.14) to determine the origin for inelastic material behavior as we did in symmetric bending. For our simple material model, we substitute Equation (6.6) into Equation (6.14) and note that d^2v/dx^2 and d^2w/dx^2 are functions of x only, while the integration is with respect to y and z ($dA = dy\,dz$) to obtain Equation (6.15).

$$-\frac{d^2v}{dx^2}\int_A Ey\,dA - \frac{d^2w}{dx^2}\int_A Ez\,dA = 0 \qquad (6.15)$$

We note that d^2v/dx^2 and d^2w/dx^2 cannot be zero in general, since that would imply that there is no bending. Therefore the integrals in Equation (6.15) must be zero, as shown in Equations (6.16a) and (6.16b).

$$\int_A yE\,dA = 0 \qquad (6.16a)$$

$$\int_A zE\,dA = 0 \qquad (6.16b)$$

Equations (6.16a) and (6.16b) are used for determining the origin in laminated beams as discussed in symmetric bending (see Section 4.3.2). However, if Assumption 8 is valid, then E is a constant across the cross section, and we obtain Equations (6.17a) and (6.17b).

$$\int_A y\,dA = 0 \qquad (6.17a)$$

$$\int_A z\,dA = 0 \qquad (6.17b)$$

Equations (6.17a) and (6.17b) are satisfied if y and z are measured from the centroid of the cross section. That is, the origin of a homogeneous cross section is at the centroid of the cross section.

6.1.8   Neutral Axis

If the normal stress changes from compressive to tensile to satisfy the condition in Equation (6.14), then it must be zero about some line.

Definition 7 The line where the bending normal stress σ_{xx} is zero is called the *neutral axis*.

Setting $\sigma_{xx} = 0$ in Equation (6.13), we obtain the following equations for the neutral axis:

$$y = (\tan\beta)z \qquad \tan\beta = \frac{I_{zz} - I_{yz}(M_z/M_y)}{I_{yz} - I_{yy}(M_z/M_y)} \qquad (6.18)$$

In symmetric bending, the neutral axis was always the z axis passing through the centroid, and it was unaffected by the loading. Equation (6.18), however, shows that the orientation of the axis depends not only upon the shape of the cross section that affects the area moments of inertia but also upon the ratios of the internal moments that would in turn depend upon the external loading. If the ratios of the internal moments in a segment of a beam do not change, then the orientation of neutral axis will remain fixed for that segment of the beam. This situation will occur in sections of the beam in which the character of the external load does not change.

In symmetric bending about the z axis the orientation of the neutral axis was fixed, and the deflection that was in the y direction was perpendicular to the neutral axis. In unsymmetric bending the orientation of the neutral axis can change, but the deflection continues to be perpendicular to it, as will be demonstrated in Example 6.2.

We record the following observations for a beam made from a linear, elastic, homogeneous material.

- The origin of the coordinate system must be the centroid.
- The neutral axis passes through the centroid of the cross section.
- The bending normal stress σ_{xx} varies linearly with y and z.
- The bending normal stress σ_{xx} is zero at the neutral axis and maximum at point furthest from the neutral axis on a cross section.
- The bending normal stress σ_{xx} is tensile on one side of the neutral axis and compressive on the other side.
- To prevent axial deformation, the external bending forces must pass through the centroid.
- The orientation of the neutral axis depends upon the shape of cross section as well as the external loading.
- The displacement of the beam is always perpendicular to the neutral axis.

6.1.9   Equilibrium Equations

We consider a differential element dx of the beam as shown in Figure 6.8. By equilibrium of the forces in the y and z directions, we obtain Equations (6.19a) and (6.19b).

$$-V_y + (V_y + dV_y) + p_y\, dx = 0 \qquad \text{or} \qquad \frac{dV_y}{dx} = -p_y \tag{6.19a}$$

$$\frac{dV_z}{dx} = -p_z \tag{6.19b}$$

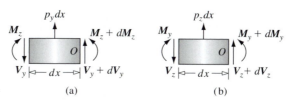

(a)   (b)

Figure 6.8 A differential beam element with forces in the y and z directions.

By equilibrium of the moment about point O in Figure 6.8(a), we obtain Equation (6.19c).

$$-M_z + (M_z + dM_z) + V_y \, dx + (p_y \, dx) \frac{dx}{2} = 0 \qquad \text{or} \qquad \frac{dM_z}{dx} = -V_y \qquad (6.19c)$$

Similarly, by equilibrium of the moment about point O in Figure 6.8(b), we obtain Equation (6.19d).

$$\frac{dM_y}{dx} = -V_z \qquad (6.19d)$$

By substituting Equations (6.10a) and (6.10b) into Equations (6.19c) and (6.19d) and then substituting the results in Equations (6.19a) and (6.19b), we can obtain the fourth-order differential equations on the deflections v and w (see Problem 6.16). For statically indeterminate problems, these fourth-order differential equations must be solved in a manner similar to that in symmetric bending. We shall confine our attention to statically determinate problems in which the internal moments can be determined by static equilibrium and deflection can be obtained by integrating the second-order equations (6.12a) and (6.12b).

EXAMPLE 6.1

A steel cantilever beam 2 m long has a modulus of elasticity $E = 200$ GPa; its cross section and loading are as shown in Figure 6.9. Determine (a) the orientation of the neutral axis at the cross section of the maximum bending normal stress and (b) the value of the maximum bending normal stress. The second area moments of inertia for the cross section are

$$I_{yy} = 0.4125(10^6) \text{ mm}^4 \qquad I_{zz} = 1.5125(10^6) \text{ mm}^4 \qquad I_{yz} = 0.45(10^6) \text{ mm}^4$$

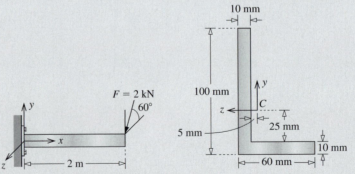

Figure 6.9 Beam and loading for Example 6.1.

PLAN

(a) The maximum internal moments M_y and M_z will occur on the cross section at $x = 0$ and can be determined by drawing a free body diagram. We can use Equation (6.13) to determine an expression for σ_{xx} as a function of y and z. By setting the stress expression to zero, we can obtain the straight-line equation representing the neutral axis. (b) We can identify the points that look furthest from the neutral axis, determine σ_{xx} at these points, and obtain the maximum value for σ_{xx}.

SOLUTION

(a) We can make an imaginary cut at $x = 0$ and draw free body diagrams as shown in Figure 6.10, using the sign convention for internal shear forces and bending moments. To simplify visualization, we draw two free body diagrams corresponding to viewing down the positive z and the negative y axes. We note that the external load has a component in the negative y direction and in the positive z direction.

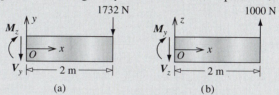

Figure 6.10 Internal moment calculations for Example 6.1: (a) $F \sin 60 = 1732$ N and (b) $F \cos 60 = 1000$ N.

By equilibrium of the moment at point O in Figure 6.10, we obtain Equations (E1) and (E2).

$$M_z = -3464 \text{ N} \cdot \text{m} \qquad \text{(E1)}$$

$$M_y = 2000 \text{ N} \cdot \text{m} \qquad \text{(E2)}$$

From the given area moments of inertia we obtain (E3).

$$I_{yy}I_{zz} - I_{yz}^2 = 0.4214(10^{-12}) \text{ m}^8 \qquad \text{(E3)}$$

Substituting moments from Equations (E1) and (E2) and the area moments of inertia in Equation (6.13), we obtain Equation (E4).

$$\sigma_{xx} = -\left[\frac{(0.4125)(-3464) - (0.45)(2000)}{0.4214(10^{-12})}\right](10^{-6})y$$

$$-\left[\frac{(1.5125)(2000) - 0.45(-3464)}{0.4214(10^{-12})}\right](10^{-6})z \qquad \text{or}$$

$$\sigma_{xx} = [5.527y - 10.878z](10^9) \qquad \text{(E4)}$$

We set $\sigma_{xx} = 0$ in Equation (E4) and obtain the straight-line equation for the neutral axis as shown in Equation (E5).

$$y = \left(\frac{10.878}{5.527}\right)z = \tan 63.1° \ z \qquad \text{(E5)}$$

From Equation (E5) we see that the neutral axis is 63.1° from the z axis, as shown in Figure 6.11.

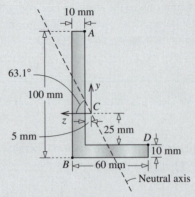

Figure 6.11 Orientation of the neutral axis in Example 6.1.

(b) By inspection of Figure 6.11, we identify points A, B, and D as points that are likely to be the furthest from the neutral axis. We record the coordinates of these points and substitute these coordinates in Equation (E4) to obtain the bending normal stress at the three points, as shown in Equations (E6), (E7), and (E8).

$$y_A = 0.065 \ \text{m} \qquad z_A = 0.005 \ \text{m} \qquad \sigma_A = 0.350(10^9) \ \text{N/m}^2 \qquad \text{(E6)}$$

$$y_B = -0.035 \ \text{m} \qquad z_B = 0.015 \ \text{m} \qquad \sigma_B = -0.357(10^9) \ \text{N/m}^2 \qquad \text{(E7)}$$

$$y_D = -0.025 \ \text{m} \qquad z_D = -0.045 \ \text{m} \qquad \sigma_D = 0.351(10^9) \ \text{N/m}^2 \qquad \text{(E8)}$$

From Equations (E6), (E7), and (E8) we see that the stress at point B is the largest in magnitude. Thus the maximum bending normal stress exists at point B on a cross section at $x = 0$ and its value is

ANS. $\sigma_{max} = 357$ MPa (C)

COMMENTS

1. Points A and D are on one side of the neutral axis and both are in tension, while point B is on the other side of neutral axis and is in compression, which is consistent because the neutral axis separates the compressive side from the tensile side.

2. We had to find σ_{xx} at several points before we could determine the maximum bending stress in the cross section. In symmetric bending, we would have calculated the stresses at the top and/or bottom only.

3. In this problem we were able to determine the maximum moments by inspection. For more complex loading we will have to draw bending moment diagrams to find the location and value of the maximum internal moments. If the moments M_y and M_z are maximum at different locations, the preceding calculations will have to be performed at each of the locations separately.

EXAMPLE 6.2

For the steel beam ($E = 200$ GPa) shown in Example 6.1, (a) determine the deflection at the midpoint in the y and z directions and (b) show that the direction of the deflection at the midpoint is perpendicular to the neutral axis.

PLAN

(a) We can obtain internal moments M_y and M_z as functions of x by drawing a free body diagram. The differential equations are given by Equations (6.12a) and (6.12b), the boundary conditions are the deflections v and w, and their first derivatives are zero at $x = 0$. The boundary value problems can be solved and the deflections obtained at $x = 1$ m. (b) Since the loading does not change across the beam, the ratio of M_y to M_z does not change across the beam; hence the orientation of the neutral axis at the midpoint is the same as at $x = 0$, which we determined in Example 6.1. Knowing the direction of the displacement vector and the neutral axis, we can show that the displacement vector is perpendicular to the neutral axis.

SOLUTION

(a) We make an imaginary cut at x and take the right-hand part to avoid calculating the wall reactions. To simplify visualization, we draw two free body diagram corresponding to viewing down the positive z and negative y axes, as shown in Figure 6.12. We note that the external load has a component in the negative y and positive z directions. We can balance the moment at point O and obtain the internal moments as shown in Equations (E1) and (E2).

$$M_z = -1732(2 - x) \text{ N} \cdot \text{m} \tag{E1}$$

$$M_y = 1000(2 - x) \text{ N} \cdot \text{m} \tag{E2}$$

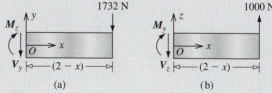

(a)   **(b)**

Figure 6.12 Internal moment calculations for Example 6.2: (a) $F \sin 60 = 1732$ N and (b) $F \cos 60 = 1000$ N.

The area moments of inertia from Example 6.1 are as shown in Equations (E3) and (E4).

$$I_{yy} = 0.4125(10^6) \text{ mm}^4 \quad I_{zz} = 1.5125(10^6) \text{ mm}^4 \quad I_{yz} = 0.45(10^6) \text{ mm}^4 \tag{E3}$$

$$I_{yy}I_{zz} - I_{yz}^2 = 0.4214(10^{-12}) \text{ m}^8 \tag{E4}$$

Substituting Equations (E1) through (E4) into Equation (6.12a) and (6.12b), we obtain Equations (E5) and (E6).

$$\frac{d^2v}{dx^2} = \frac{1}{200(10^9)} \left[\frac{(0.4125)(-1732) - (0.45)(1000)}{0.4214(10^{-12})} \right] (2 - x)(10^{-6}) \quad \text{or}$$

$$\frac{d^2v}{dx^2} = -0.0276 + 0.0138x \tag{E5}$$

$$\frac{d^2w}{dx^2} = \frac{1}{200(10^9)}\left[\frac{(1.5125)(1000) - (0.45)(-1732)}{0.4214(10^{-12})}\right](2-x)(10^{-6}) \quad \text{or}$$

$$\frac{d^2w}{dx^2} = 0.0544 - 0.0272x \tag{E6}$$

Noting that the slope and deflections at $x = 0$ are zero, we obtain the boundary conditions shown in Equations (E7) through (E10).

$$v(0) = 0 \tag{E7}$$

$$\frac{dv}{dx}(0) = 0 \tag{E8}$$

$$w(0) = 0 \tag{E9}$$

$$\frac{dw}{dx}(0) = 0 \tag{E10}$$

Integrating Equation (E5), we obtain Equation (E11).

$$\frac{dv}{dx} = -0.0276x + 0.0138\frac{x^2}{2} + c_1 \tag{E11}$$

Using the boundary condition of Equation (E8), we obtain Equation (E12):

$$\frac{dv}{dx}(0) = 0 + 0 + c_1 = 0 \quad \text{or} \quad c_1 = 0 \tag{E12}$$

Integrating Equation (E11), we obtain Equation (E13).

$$v = -0.0276\frac{x^2}{2} + 0.0138\frac{x^3}{6} + c_2 \tag{E13}$$

Using the boundary condition of Equation (E7), we obtain Equation (E14).

$$v(0) = 0 + 0 + c_2 = 0 \quad \text{or} \quad c_2 = 0 \tag{E14}$$

Thus we obtain Equation (E15).

$$v(x) = (-13.82x^2 + 2.30x^3)(10^{-3}) \text{ m} \tag{E15}$$

Similarly, by integrating Equation (E6) and using Equations (E9) and (E10), we can obtain Equation (E16).

$$w(x) = (27.20x^2 - 4.53x^3)(10^{-3}) \text{ m} \tag{E16}$$

The deflection at the midpoint can be obtained by substituting $x = 1$ m in Equations (E15) and (E16).

$$v(1) = (-13.82 + 2.30)(10^{-3}) \text{ m} = -11.52(10^{-3}) \text{ m} \tag{E17}$$

ANS. $v(1) = -11.52$ mm

$$w(1) = (27.20 - 4.53)(10^{-3}) \text{ m} = 22.67(10^{-3}) \text{ m} \qquad \text{(E18)}$$

ANS. $w(1) = 22.67$ mm

(b) Let $\bar{\mathbf{j}}$ and $\bar{\mathbf{k}}$ be the unit vectors in the y and z direction, respectively. Let $\bar{\mathbf{D}}_1$ be the displacement vector at $x = 1$ m. Thus from Equations (E17) and (E18), we obtain Equation (E19).

$$\bar{\mathbf{D}}_1 = [v(1)\bar{\mathbf{j}} + w(1)\bar{\mathbf{k}}] = (-11.52\bar{\mathbf{j}} + 22.67\bar{\mathbf{k}}) \text{ mm} \qquad \text{(E19)}$$

Let $\bar{\mathbf{i}}_{NA}$ be a unit vector in the direction of the neutral axis. The orientation of the neutral axis does not change across the beam because the loading does not change. Hence from Example 6.1 we obtain Equation (E20).

$$\bar{\mathbf{i}}_{NA} = \cos 63.1 \; \bar{\mathbf{j}} + \sin 63.1 \quad \bar{\mathbf{k}} = 0.8915\,\bar{\mathbf{j}} + 0.4531\,\bar{\mathbf{k}} \qquad \text{(E20)}$$

We take the dot product of Equations (E19) and (E20) to obtain Equation (E21).

$$\bar{\mathbf{D}}_1 \cdot \bar{\mathbf{i}}_{NA} = (-11.52)(0.8915) + (22.67)(0.4531) = 0.00 \qquad \text{(E21)}$$

Equation (E21) shows that the vector $\bar{\mathbf{D}}_1$ and the vector $\bar{\mathbf{i}}_{NA}$ are perpendicular, which proves that the displacement is perpendicular to the neutral axis at the midpoint in the beam.

COMMENT

We could have proved that the displacement is perpendicular to the neutral axis by calculating the angle (slope) made by the displacement vector with the z axis as $\tan^{-1}(-11.52/22.67) = -26.9°$, which is 90° to the neutral axis.

PROBLEM SET 6.1

6.1 The internal bending moments on the cross section shown in Figure P6.1 were determined to be $M_y = 10$ in · kips and $M_z = -15$ in · kips. Determine (a) the orientation of the neutral axis and (b) the maximum bending normal stress.

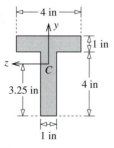

Figure P6.1

6.2 The internal bending moments on the cross section shown in Figure P6.2 were determined to be $M_y = -2$ kN · m and $M_z = -2.5$ kN · m. Determine (a) the orientation of the neutral axis and (b) the maximum bending normal stress.

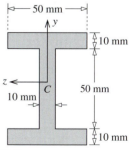

Figure P6.2

ANS. $\beta_{NA} = -75.3°$ $\sigma_{max} = 321.7$ ksi (T, C)

6.3 The internal bending moments on the cross section shown in Figure P6.3 were determined to be $M_y = 10$ in · kips and $M_z = 15$ in · kips. Determine (a) the orientation of the neutral axis and (b) the maximum bending normal stress.

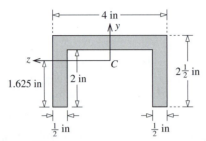

Figure P6.3

6.4 The internal bending moments on the cross section shown in Figure P6.4 were determined to be $M_y = 20$ in · kips and $M_z = 25$ in · kips. Determine (a) the orientation of the neutral axis and (b) the maximum bending normal stress.

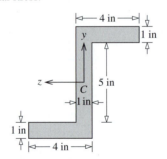

Figure P6.4

ANS. $\beta_{NA} = -60.54°$
$\sigma_{max} = 5.66$ ksi (T, C)

6.5 The internal bending moments on the cross section shown in Figure P6.5 were determined to be $M_y = -10$ in · kips and $M_z = -12$ in · kips. Determine (a) the orientation of the neutral axis and (b) the maximum bending normal stress.

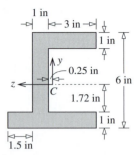

Figure P6.5

6.6 The internal bending moments on the cross section shown in Figure P6.6 were determined to be $M_y = 5$ in · kips and $M_z = -5$ in · kips. Determine (a) the orientation of the neutral axis and (b) the maximum bending normal stress.

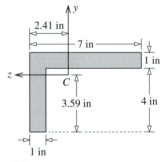

Figure P6.6

ANS. $\beta_{NA} = 4.78°$
$\sigma_{max} = 0.809$ ksi (C)

6.7 A force of 2 kN acts in the negative y direction on the beam shown in Figure P6.7 such that there is no twist. Determine (a) the maximum bending normal stress and (b) the orientation of neutral axis on the section of maximum bending normal stress.

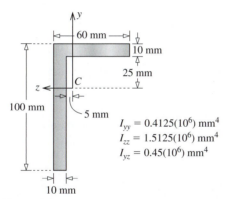

$I_{yy} = 0.4125(10^6)$ mm^4
$I_{zz} = 1.5125(10^6)$ mm^4
$I_{yz} = 0.45(10^6)$ mm^4

Figure P6.7

6.8 The modulus of elasticity for the beam in Problem 6.7 is $E = 200$ GPa. (a) Determine the direction and magnitude of the maximum deflection. (b) Show that the direction of the deflection at the midpoint is perpendicular to the neutral axis.

6.9 The cross section shown in Figure P6.9(a) has a uniform thickness of $t = 0.05$ inch. (a) Determine the intensity of the distributed load w if the maximum bending normal stress is not to exceed 30 ksi. (b) Corresponding to your answer in part (a), determine the deflection of the free end. The area moments of inertia of the cross section are given in Figure P6.9(b).

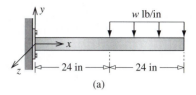

(a)

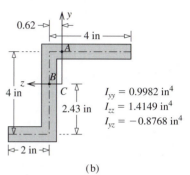

$I_{yy} = 0.9982$ in^4
$I_{zz} = 1.4149$ in^4
$I_{yz} = -0.8768$ in^4

(b)

Figure P6.9

6.10 An 8 ft long beam that is simply supported in the y and z directions has the cross section shown in Figure P6.10. The uniform load acts in the y-z plane at an angle of 30° to the z axis. Assuming no twisting of the cross section, determine (a) the orientation of the neutral axis at the cross section of the maximum bending normal stress and (b) the value of the maximum normal stress.

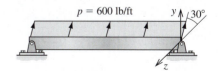

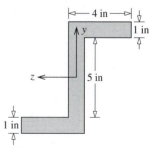

Figure P6.10

ANS. $\theta_{NA} = -72.9°$ from the z axis
$\sigma_{max} = 7.6$ ksi (T or C)

6.11 The modulus of elasticity for the beam in Problem 6.10 is $E = 30,000$ ksi. (a) Determine the direction and magnitude of the maximum deflection. (b) Show that the direction of the deflection at the mid-point is perpendicular to the neutral axis.

ANS. $v_{max} = -0.0273$ in
$w_{max} = -0.0889$ in

6.12 A cantilever beam is loaded such that there is no twist. The distributed load acts in the y-z plane at an angle of 24° from the x-y plane as shown in Figure P6.12. On a section at $x = 60$ in, determine (a) the orientation of the neutral axis and (b) the bending normal stress at points A and B.

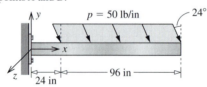

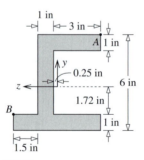

Figure P6.12

6.13 The modulus of elasticity for the beam in Problem 6.12 is $E = 30,000$ ksi. Determine the deflection of the beam at $x = 60$ inches and show that it is perpendicular to the neutral axis.

6.14 A cantilever beam is loaded as shown in Figure P6.14 such that there is no twist in the cross section. Determine (a) the orientation of the neutral axis at the section containing points A and B and (b) the bending normal stress at points A and B.

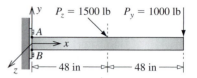

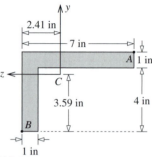

Figure P6.14

ANS. $\sigma_A = 5.1$ ksi (T) $\sigma_B = 16$ ksi (C)

6.15 The modulus of elasticity for the beam in Problem 6.14 is $E = 10,000$ ksi. Determine the deflection of the beam at the free end.

6.16 Show that the fourth-order differential equations governing the deflection of beams are

$$\frac{d^2}{dx^2}\left[EI_{zz}\frac{d^2v}{dx^2} + EI_{yz}\frac{d^2w}{dx^2}\right] = p_y \quad (6.20a)$$

$$\frac{d^2}{dx^2}\left[EI_{yz}\frac{d^2v}{dx^2} + EI_{yy}\frac{d^2w}{dx^2}\right] = p_z \quad (6.20b)$$

6.17 Show that the differential equations governing unsymmetric free vibrations of a beam are

$$\frac{\partial^2}{\partial x^2}\left[EI_{zz}\frac{\partial^2v}{\partial x^2} + EI_{yz}\frac{\partial^2w}{\partial x^2}\right]$$

$$+ \rho_m A\frac{\partial^2v}{\partial t^2} = 0 \quad (6.21a)$$

$$\frac{\partial^2}{\partial x^2}\left[EI_{yz}\frac{\partial^2v}{\partial x^2} + EI_{yy}\frac{\partial^2w}{\partial x^2}\right]$$

$$+ \rho_m A\frac{\partial^2w}{\partial t^2} = 0 \quad (6.21b)$$

where ρ_m is the material density and A is the cross-sectional area.

6.18 A beam resting on an elastic foundation has a distributed spring force that depends on the deflection at a point. For unsymmetric beams, show that the differential equations governing the deflection of the beam are

$$\frac{d^2}{dx^2}\left[EI_{zz}\frac{d^2v}{dx^2}+EI_{yz}\frac{d^2w}{dx^2}\right]$$
$$+k_yv=p_y \qquad (6.22a)$$

$$\frac{d^2}{dx^2}\left[EI_{yz}\frac{d^2v}{dx^2}+EI_{yy}\frac{d^2w}{dx^2}\right]$$
$$+k_zw=p_z \qquad (6.22b)$$

where k_y and k_z are the foundation moduli (spring forces per unit length) in the y and z direction, respectively.

6.19 For a laminated (different E_i) beam in unsymmetric bending, show that the moment–curvature relationships are given by

$$M_z=\left(\sum_{i=1}^{n}E_i(I_{zz})_i\right)\frac{d^2v}{dx^2}$$
$$+\left(\sum_{i=1}^{n}E_i(I_{yz})_i\right)\frac{d^2w}{dx^2} \qquad (6.23a)$$

$$M_y=\left(\sum_{i=1}^{n}E_i(I_{yz})_i\right)\frac{d^2v}{dx^2}$$
$$+\left(\sum_{i=1}^{n}E_i(I_{yy})_i\right)\frac{d^2w}{dx^2} \qquad (6.23b)$$

where $(I_{yy})_i$, $(I_{zz})_i$, and $(I_{yz})_i$ are the area moments of inertia of the ith material about the origin determined by Equations (6.16a) and (6.16b).

6.20 For the laminated unsymmetric bending of the beam described in Problem 6.19, obtain an expression for the bending normal stress in the jth material.

6.2   SHEAR STRESS IN THIN OPEN SECTIONS

The development of theory for shear stresses in unsymmetric bending is very similar to the development of theory for shear stresses in symmetric bending discussed in Section 3.2.3. The primary difference is the use of the Equation (6.13) for σ_{xx} in place of the expression for it in symmetric bending.

Consider an open cross section of a beam with thickness t as shown in Figure 6.13. Our objective is to get a formula for the shear stress τ_{sx} at an arc distance s measured

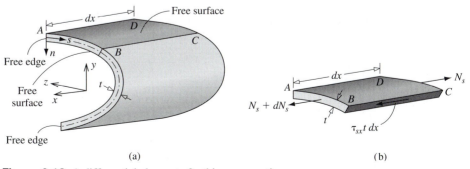

(a)   (b)

Figure 6.13 A differential element of a thin open section.

from a free edge. We note that the shear stress τ_{nx} is zero on the free surfaces; hence the symmetric counterpart τ_{xn} is also zero on the free surfaces. Thus, for thin sections we can approximate τ_{xn} as zero through the thickness. We can also approximate τ_{sx} as constant across the thickness because the section is thin.

The total axial force across the cross section is zero in accordance with Equation (6.7a). However, if we consider only the part of the cross section *ABCD* shown in Figure 6.13(a), then an axial force will result from the stress σ_{xx}. We thus have Equation (6.24a),

$$N_s = \int_{A_s} \sigma_{xx} \, dA \qquad (6.24a)$$

where

Definition 8 The direction of the *s* coordinate is from the free surface toward the point at which the shear stress is being calculated.

Definition 9 The area A_s is the area between free edge and the point at which the shear stress is being evaluated.

It should be noted that in an open section there are two (or more) free edges, and either edge could be used in the definition of A_s. But the choice of the free edge defines the origin of *s* and hence the direction of *s*. If we take a differential element *dx*, then the stress σ_{xx} will vary across the element; hence the resulting axial force N_s in Equation (6.24a) will vary across the element and must be balanced by the shear force, as shown in Figure 6.13(b). Note that no forces are shown on side *AD* because this is a free edge. By equilibrium of forces, we obtain Equation (6.24b).

$$(N_s + dN_s) - N_s + \tau_{sx} t \, dx = 0 \qquad \text{or} \qquad \tau_{sx} t = -\frac{d}{dx}(N_s) = -\frac{d}{dx} \int_{A_s} \sigma_{xx} \, dA \qquad (6.24b)$$

Equation (6.24b) is identical to the equilibrium Equation (3.15-B) in symmetric bending. The differences from symmetric bending start with the substitution of σ_{xx}. Substituting Equation (6.13) into Equation (6.24b) and noting that the internal moments and the area moments of inertia do not vary over the cross section (across A_s), we obtain Equation (6.24c).

$$\tau_{sx} t = \frac{d}{dx}\left[\left(\frac{I_{yy}M_z - I_{yz}M_y}{I_{yy}I_{zz} - I_{yz}^2} \right) \int_{A_s} y \, dA + \left(\frac{I_{zz}M_y - I_{yz}M_z}{I_{yy}I_{zz} - I_{yz}^2} \right) \int_{A_s} z \, dA \right] \qquad (6.24c)$$

We define the first moment of the area A_s as in Equations (6.24d) and (6.24e).

$$Q_z = \int_{A_s} y \, dA \qquad (6.24d)$$

$$Q_y = \int_{A_s} z \, dA \qquad (6.24e)$$

To simplify further, we make the following assumption:

Assumption 9 The beam is not tapered.

Assumption 9 implies that the area moments of inertia and the first moment of the area A_s are not functions of x, and hence during the differentiation in Equation (6.24c) will not be differentiated. We obtain

$$\tau_{sx}t = \left(\frac{I_{yy}\dfrac{dM_z}{dx} - I_{yz}\dfrac{dM_y}{dx}}{I_{yy}I_{zz} - I_{yz}^2}\right)Q_z + \left(\frac{I_{zz}\dfrac{dM_y}{dx} - I_{yz}\dfrac{dM_z}{dx}}{I_{yy}I_{zz} - I_{yz}^2}\right)Q_y \qquad (6.24f)$$

Substituting the relationships of Equations (6.19c) and (6.19d), we obtain Equation (6.25),

$$\boxed{q = \tau_{sx}t = -\left(\frac{I_{yy}Q_z - I_{yz}Q_y}{I_{yy}I_{zz} - I_{yz}^2}\right)V_y - \left(\frac{I_{zz}Q_y - I_{yz}Q_z}{I_{yy}I_{zz} - I_{yz}^2}\right)V_z} \qquad (6.25)$$

where q is the shear flow at a point on the cross section. The thickness t can vary across the cross section; that is, it can be a function of s. If y or z is an axis of symmetry, then $I_{yz} = 0$, and Equation (6.25) simplifies to

$$q = \tau_{sx}t = -\left(\frac{V_yQ_z}{I_{zz}}\right) - \left(\frac{V_zQ_y}{I_{yy}}\right) \qquad (6.26)$$

Equation (6.26) represents the shear stress due to combined bending about the y and z axes when either axis is an axis of symmetry. It can be derived from symmetric bending by superposition but because of the limitation in the theory of symmetric bending that the load must lie in the axis of symmetry, this would require that both y and z be symmetric axes.

6.2.1 Shear Center

The location of the shear center is an important design consideration. If weight (bending load) causes the wing of an aircraft to twist in an unpredictable manner, the angle of attack that impacts the aerodynamic characteristics is changed in unpredictable manner. A tall building may twist from improper accounting for dead weight and the shear center.

From statics we know that any distributed force can be replaced by a force and a moment at any point, or by a *single force (and no moment) at a specific point*. The specific point at which the shear stress (shear flow) in Equation (6.25) can be represented by just shear forces V_y and V_z (components of a single force) and no internal torque is called the shear center. Figure 6.14 shows the static equivalency between the shear stress τ_{xs} and shear forces V_y and V_z. Clearly if we take a moment at any point other than the shear center, these forces will result in a moment. Hence a cross section can have only one shear center.

It should be emphasized that we developed the theory of unsymmetric bending under the limitation of a cross section that does not twist. If the internal forces and

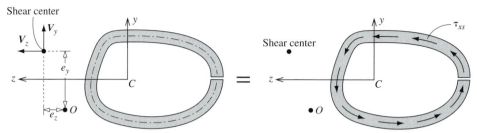

Figure 6.14 Static equivalency of shear force.

moments representing stresses must be in equilibrium with the external forces, the implication is that the external forces must also pass through the shear center. Thus we can define the shear center in one of two ways:

Definition 10a The shear center is a point in space at which the shear stress due to bending can be replaced by statically equivalent internal shear forces and no internal torque.

Definition 10b The shear center is a point in space such that if the line of action of the external forces passes through the point, the cross section will not twist.

To determine the location of the shear center, we consider one shear force at a time. We discuss the method with just V_y, but the same steps can be repeated with V_z. We use Equation (6.25) [or (6.26) if applicable] with just V_y to find the shear stress τ_{xs} in terms of V_y at any arbitrary location defined by using a convenient variable. In other words, τ_{xs} *will be a function of a variable that defines the location of a point on the cross section.* Since the two systems shown in Figure 6.14 are statically equivalent, the two systems should have the same resultant moment at *any point O.* We take the moment about point O and find e_z. Since V_y will appear on both sides of the moment equation, it cancels out and we obtain the location e_z, which depends only on the geometry. We repeat these steps with V_z and find e_y. The point O around which we take the moment for static equivalency is completely arbitrary. *We shall choose point O such that it reduces algebra.*

It should be emphasized that the shear center may not lie on the cross section and may not even be enclosed by the perimeter of the cross section. As long as the line of action of the external forces passes through the shear center, we meet the requirement that the external forces not twist the cross section. If the external forces do not pass through the shear center, then the shear center can be used for separating the shear stresses due to bending from those due to torque.

If the shear stresses are symmetric about a line, then clearly the moment from the shear stresses on one side of the line will balance the moments from the other side, and we may conclude that the shear center lies on this line of symmetry.

In Section 6.1.7 we observed that external bending forces must pass through the centroid if axial deformation is to be avoided. In this section we conclude that the external bending forces must pass through the shear center if torsional deformation is to be avoided. Thus, if we want no axial or torsional deformation, the external forces must be along the line joining the centroid and the shear center. If the bending (transverse) forces do not pass through either, we will have axial and torsional deformations. If we prevent axial and torsional deformations, we will have axial and torsional stresses. Recall that in symmetric bending, the transverse forces are in the plane of symmetry, which contains both the shear center and the centroid of the cross section.

We record the following observations regarding the shear center.

- Associated with each cross section is a unique shear center.

- The shear center depends only on the geometry and is independent of the loading.

- The shear center lies on the axis around which the shear stress distribution is symmetric.

- The shear center decouples the shear stresses due to bending from the shear stresses due to torsion.

- If bending forces are not to produce any axial or torsional deformation, the external forces must be along the line joining the centroid and the shear center of the cross section.

EXAMPLE 6.3

Determine the direction of shear flow in a cross section of a beam for the two cases shown in Figure 6.15, assuming that the forces are passing through the shear center (SC), as illustrated.

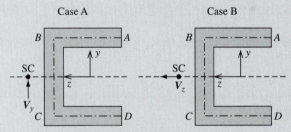

Figure 6.15 Cross sections for Example 6.3.

METHOD 1

PLAN

The shear flow, which is always tangential to the centerline, is in the vertical direction in *BC* and in the horizontal direction in *AB* and *CD*. It must be zero on the free edges at *A* and *D*. We can derive the flow direction in case A by noting that the resultant force in the *y* direction must be same as the direction of V_y and zero in the *z* direction. We can derive the flow direction in case B by noting that the resultant force in the *y* direction must be zero; in the *z* direction it must be in the same direction as V_z, and the flow must be symmetric about the *z* axis.

SOLUTION

Case A: The resultant force in the z direction must be zero. Hence the shear flow in section AB must be opposite to the shear flow in CD. The condition will be met if the shear flow is either clockwise or counterclockwise and does not change directions anywhere on the cross section. Since CD is the only part that can yield a component in the y direction, the shear flow in CD must be in the same direction as V_y for static equivalency to hold. Thus the flow in case A must be clockwise, as shown in Figure 6.16(a).

Case B: The resultant force in the y direction must be zero. Hence the shear flow in BC must change direction (i.e., must go to zero at some point in BC). Since the z axis is the axis of symmetry with respect to geometry and loading, the shear flow must be zero at point E as shown in Figure 6.16(b). We assume that the shear flow that is zero at A and E does not change direction between A and E (i.e., does not go to zero between A and E). The direction of the flow in AB must be in the direction of V_z for static equivalency to hold, hence the flow direction in AE is counterclockwise. Similar reasoning then results in a clockwise direction of shear flow in DE as shown in Figure 6.16(b).

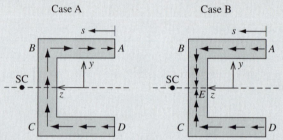

Figure 6.16 Shear flow for Example 6.3.

METHOD 2

PLAN

We can use Equation (6.26) for determining the direction of the flow without performing any calculations.

SOLUTION

Case A: From Equation (6.26) we have $\tau_{sx}t = -(V_y Q_z/I_{zz})$, with I_{zz} and t always positive. Since V_y is positive, we conclude that the shear stress and Q_z will have opposite signs. If we start from point A and move in the direction $ABCD$, then Q_z always will be positive, and hence the shear stress τ_{xs} or τ_{sx} always will be negative. With s measured from A toward B as positive, the shear stress flow will be in the direction shown in Figure 6.16(a) to yield a negative τ_{xs}. If we start from point D and move in the direction $DCBA$, then Q_z will always be negative, and hence τ_{xs} will always be positive. With s measured from D toward C as positive, we once more obtain the shear flow shown in Figure 6.16(a).

Case B: From Equation (6.26) we have $\tau_{sx}t = -(V_z Q_y / I_{yy})$, with I_{yy} and t always positive. Since V_z is positive, we conclude that the shear stress and Q_y will have opposite signs. Starting from point A or point D, Q_y will be negative and will go to zero at point E. Thus the shear stress τ_{xs} is positive whether we start from point A or point D and hence is in the direction shown in Figure 6.16(b).

COMMENTS

1. In case A, the moment about the shear center due to shear flow in AB and CD is opposite to the direction of moment due to shear flow in BC. This holds true as long as the shear center is to the left of BC. In case B, the symmetry of the shear flow about the z axis ensures that the moment about the shear center will be zero irrespective of the location of the shear center on the z axis. Thus if the location of the shear center had not been known, we could have intuitively deduced the direction in which the shear center lay.

2. Method 1 is more intuitive than method 2, which relies on a formula. We may need both methods at times to deduce the complete flow.

3. In this problem we had the z axis as the axis of symmetry, which we used in method 1 to deduce the direction of the flow. In method 2 the symmetry resulted in I_{yz} being zero, and we could use the simplified Equation (6.26). When there is no axis of symmetry, it may not be possible to deduce the direction of flow by either of the two methods. In such cases, we must use Equation (6.25), follow the sign conventions, and determine the direction of shear stress from the subscripts.

EXAMPLE 6.4

A beam that is simply supported in the y and the z directions is shown in Figure 6.17. The cross section has a uniform thickness t, and the area moments of inertia are as follows:

$$I_{yy} = 0.7928R^3 t \qquad I_{zz} = 2.2375R^3 t \qquad I_{yz} = 0$$

(a) Determine the shear stress at points A and B, which are on a section just to the left of the applied loads. (b) Sketch the shear flow on the section in part (a).

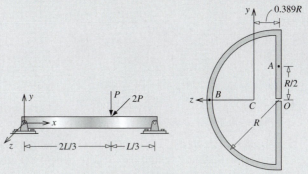

Figure 6.17 Beam and cross section for Example 6.4.

PLAN

(a) We can find the reactions at the left-hand support by drawing free body diagrams of the entire beam. We can then make an imaginary cut just to the left of the applied load and find the shear forces on the cross section containing points A and B. We can find Q_y and Q_z at points A and B and use Equation (6.26) [or Equation (6.25)] to find the shear stresses at points A and B. (b) We can draw the shear flow direction, as demonstrated in Example 6.3.

SOLUTION

(a) To simplify visualization, we draw two free body diagrams of the entire beam corresponding to viewing down the positive z and negative y axes, as shown in Figure 6.18. We determine the reaction at the left-hand support by equilibrium of the moment at the right-hand support, as shown in Equations (E1) and (E2).

$$R_{Ly}(L) - P\left(\frac{L}{3}\right) = 0 \quad \text{or} \quad R_{Ly} = \frac{P}{3} \tag{E1}$$

$$R_{Lz}(L) - (2P)\left(\frac{L}{3}\right) = 0 \quad \text{or} \quad R_{Lz} = \frac{2P}{3} \tag{E2}$$

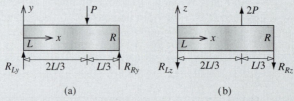

(a)   (b)

Figure 6.18 Free body diagrams for the reaction calculations in Example 6.4.

We make a cut just to the left of the applied load and draw free body diagrams as shown in Figure 6.19. The internal shear forces and bending moment are drawn in accordance with our sign convention. By force equilibrium we obtain Equations (E3) and (E4).

$$V_y = -R_{Ly} = -\frac{P}{3} \tag{E3}$$

$$V_z = R_{Lz} = \frac{2P}{3} \tag{E4}$$

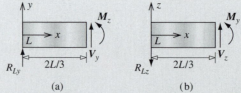

(a)   (b)

Figure 6.19 Free body diagrams for the calculation of shear forces in Example 6.4.

Point A: The area A_s is the area between the free edge at O and point A as shown in Figure 6.20(a). We find the first moment of the area by multiplying the area by the distance of the centroid of A_s to point C as shown in Equations (E5) and (E6).

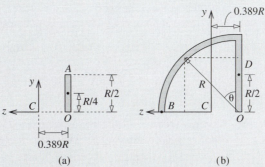

(a)   (b)

Figure 6.20 First area moments for (a) point A and (b) point B in Example 6.4.

$$Q_{yA} = \left(\frac{R}{2}\right)(t)(-0.389R) = -0.1945R^2 t \qquad \text{(E5)}$$

$$Q_{zA} = \left(\frac{R}{2}\right)(t)\left(\frac{R}{4}\right) = 0.125R^2 t \qquad \text{(E6)}$$

Substituting Equations (E3), (E4), (E5), and (E6) in Equation (6.26), we obtain the shear stress at point A as shown in Equation (E7).

$$\tau_A t = -\frac{\left(-\dfrac{P}{3}\right)(0.125R^2 t)}{2.2375R^3 t} - \frac{\left(\dfrac{2P}{3}\right)(-0.1945R^2 t)}{0.7928R^3 t} = \frac{0.0186P}{R} + \frac{0.1636P}{R} \qquad \text{or} \qquad \text{(E7)}$$

ANS. $\tau_A = \dfrac{0.1822P}{Rt}$

Point B: The area A_s is the area between the free edge at O and point B as shown in Figure 6.20(b). Because of the circular arc, we start our calculations of Q_y and Q_z from the definitions in Equations (6.24d) and (6.24e) and write the integral as the sum of two integrals. The integral over the straight edge OD can be computed by multiplying the area of the straight part by the location of the centroid with respect to point C. The integral over the curved part DB must be evaluated by integration. To perform the integration, we can write the y and z coordinates as $y = R \cos \theta$ and $z = R \sin \theta - 0.389R$ and the differential area as $dA = t\,ds = tR\,d\theta$, and integrate from $\theta = 0$ to $\theta = \pi/2$ as shown in Equation (E8).

$$Q_{zB} = \int_{A_{OD}} y\,dA + \int_{A_{DB}} y\,dA = (Rt)\left(\frac{R}{2}\right) + \int_0^{\pi/2} R \cos \theta(tR\,d\theta) \qquad \text{or}$$

$$Q_{zB} = \frac{R^2 t}{2} + \left[R^2 t \sin \theta\right]\Big|_0^{\pi/2} = 1.5R^2 t \qquad \text{(E8)}$$

In a similar manner, Q_{yB} can be calculated and is as shown in Equation (E9).

$$Q_{yB} = \int_{A_{OD}} z\,dA + \int_{A_{DB}} z\,dA = (Rt)(-0.389R) + \int_0^{\pi/2} (R \sin \theta - 0.389R)(tR\,d\theta) \qquad \text{or}$$

$$Q_{yB} = -0.389R^2 t + R^2 t(-\cos \theta - 0.389\theta)\Big|_0^{\pi/2} = 0 \qquad \text{(E9)}$$

Substituting Equations (E3), (E4), (E8), and (E9) in Equation (6.26), we obtain the shear stress at point B as shown in Equation (E10).

$$\tau_B t = -\frac{\left(-\dfrac{P}{3}\right)(1.5R^2 t)}{2.2375R^3 t} - \frac{\left(\dfrac{2P}{3}\right)(0)}{0.7928R^3 t} = \frac{0.2235P}{R} + 0 \qquad \text{or} \qquad \text{(E10)}$$

ANS. $\quad \tau_B = \dfrac{0.2235P}{Rt}$

(b) We can sketch the shear flow as we did in Example 6.2, as shown in Figure 6.21. The flow was constructed for each shear force as follows.

Just V_z: The shear force $V_z = +2P/3$ was drawn as shown in Figure 6.21(a). The shear flow must be symmetric with respect to the z axis because the geometry is symmetric with respect to V_z. The shear flow must be such that the resultant vertical force in the y direction is zero. The resultant vertical force due to the flow in OD will cancel the resultant vertical force in OE because the flows are symmetric about the z axis, hence must be in opposite directions. In the arc section DBE, the flow will reverse direction at B to ensure symmetry, and the resultant force in the y direction from the flow in DB will cancel that in EB. The flow in DB and EB must be such that it produces a horizontal component in the direction V_z, and we obtain the flow as shown in Figure 6.21(a).

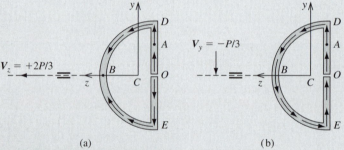

(a)   (b)

Figure 6.21 Shear flow in Example 6.4.

Just V_y: The shear force $V_y = -P/3$ was drawn as shown in Figure 6.21(b). The shear flow must be such that the resultant horizontal force in the z direction is zero. The flow in OD and OE will not result in any horizontal force. The shear flow at a point in the arc DBE has a horizontal component, but the resultant from the top half will cancel the bottom half if the flow does not change direction. Thus we conclude that the shear flow is either clockwise or counterclockwise over the entire section. However, the vertical component in OD and OE will be opposite to the vertical component due to the flow in DBE. The difference between the two should be in the direction of the force shown in Figure 6.21(b). This can be resolved in two ways. (1) The shear flow starts from zero at point O and increases as we move toward point D (or E), and hence the shear flow in DBE will be of greater magnitude than in OD and OE, which means that the flow direction is likely to be as shown in Figure 6.21. (2) With V_y negative, the sign of shear stress from Equation (6.26) is that of Q_z. If we start from O and take the path $ODBE$, then Q_z will always be positive. Hence the shear stress will be in the positive s direction as shown. The second argument is more definitive than the first.

COMMENTS

1. The shear stress (τ_{xs}) at A and B turned out to be positive. The cross section has an outward normal in the positive x direction, and thus shear stress will be positive in the positive s direction. The direction of positive s is from O toward A for point A and from point O toward point B. Thus at point A the shear stress is upward, and at point B the shear stress is positive downward.

2. We calculated Q_{yB} and found it to be zero in Equation (E9). We could have avoided this calculation, had we noted that z is measured from the centroid and therefore the following is true: $\int_A z \, dA = 0$, where A is the total cross-sectional area. Because the geometry is symmetric with respect to the z axis, the integral over the top part is half that of the integration over the total area and hence is zero.

3. Notice that the two flows shown in Figure 6.21 add at point A, since these are in the same direction, which is consistent with our calculations.

EXAMPLE 6.5

Determine the shear center for the cross section shown in Figure 6.22.

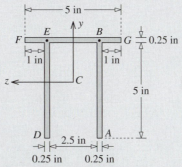

Figure 6.22 Cross section for Example 6.5.

PLAN

The shear center will lie on the y axis because it is the axis of symmetry. We need to find e_y only. Hence, we consider the shear force V_z placed at an assumed location of the shear center. If we take the moment at point E, the resultant of the shear flow in DE and $GBEF$ will pass through point E and hence will not produce any moment about E. We can find the shear flow in AB by using Equation (6.26), and then we can take the moment about point E to find e_y.

SOLUTION

We can find $I_{yy} = 7.344$ in^4. The area A_s in calculation of Q_y is defined from the free edge A to some location defined by parameter s measured from A as shown in Figure 6.23. The first moment of the area A_s can be found by multiplying A_s by the distance of the centroid of A_s to C. The shear flow q as a function of s can then be found as shown in Equations (E1) and (E2).

$$Q_y = (s)(0.25)(-1.375) = -0.34375s \text{ in}^3 \tag{E1}$$

$$q = -\frac{V_z Q_y}{I_{yy}} = -\frac{V_z(-0.34375s)}{7.344} = 0.1872s V_z \tag{E2}$$

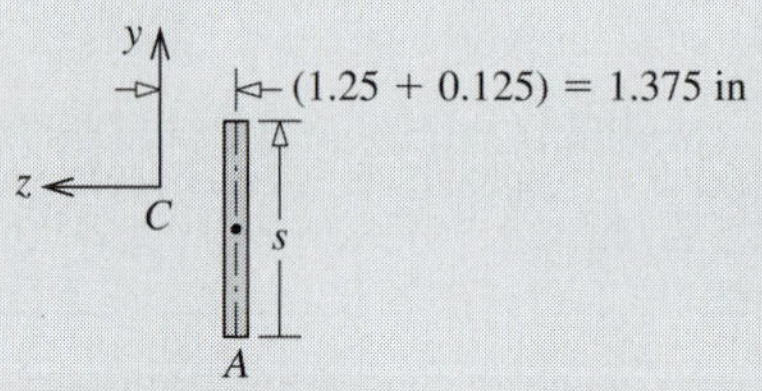

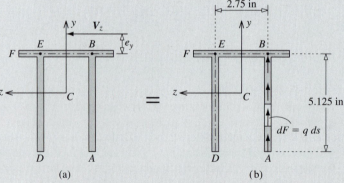

Figure 6.23 First area moment in Example 6.5.

Figure 6.24 shows the two statically equivalent systems with the shear flow drawn in the positive s direction only in AB. We equate the moment about point E for the two systems. To obtain the moment of the shear flow in AB, we find the moment due to the differential element ds and integrate from point A to point B, as shown in Equation (E3).

$$V_z e_y = \int_{s_A=0}^{s_B=5.125} (2.75)(q\ ds) = 2.75 \int_0^{5.125} (0.1872s V_z)\ ds$$

$$= 0.1287 V_z \left.\frac{s^2}{2}\right|_0^{5.125} = 1.69 V_z \qquad \text{or} \tag{E3}$$

ANS. $e_y = 1.69$ in

Figure 6.24 Static equivalency in Example 6.5.

COMMENTS

1. In the final results V_z cancels out. Thus we may assume that V_z in either the positive or negative direction in Figure 6.24(a), but the flow direction in Figure 6.24(b) must correspond to the assumed direction for V_z. If the flow direction and the shear force directions are not consistent, the location of the shear center may be calculated erroneously as that on the other side of the point at which we are taking the moment.

2. In this problem the moment arm was a constant value of 2.75 in. This may not always be the case. If sections *BA* and *DE* were at an angle other than 90° to *FEBG*, the moment arm would be a function of *s*.

3. In all shear center calculations, the shear flow must be found as a function of a variable, and the moment is found by integration.

EXAMPLE 6.6

Determine the shear center for the cross section shown in Figure 6.25.

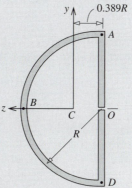

Figure 6.25 Cross section for Example 6.6.

PLAN

The shear center will lie on the *z* axis because it is the axis of symmetry. We need to find e_z only. Hence we consider the shear force V_y placed at an assumed location of the shear center. The shear flow in *OA* and *OD* passes through *O* and will not produce any moment. We can find the shear flow in *ABD* by using Equation (6.26) and take moment about point *O* to find e_z.

SOLUTION

In the calculation of Q_y, the area A_s is between the free edge at *O* and a location defined by parameter θ measured from *A*, as shown in Figure 6.26.

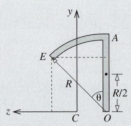

Figure 6.26 First area moment for Example 6.6.

We start our calculations of Q_z from the definition in Equations (6.24e) and write the integral as the sum of two integrals. The integral over the straight edge OA can be computed by multiplying the area of the straight part by the location of the centroid with respect to point C. The integral over the curved part must be evaluated by integration. To perform the integration, we can write the y coordinate as $y = R\cos\theta$ and the differential area as $dA = t\,ds = tR\,d\theta$ and integrate from $\theta = 0$ to some location θ, as shown in Equation (E1).

$$Q_z = \int_{A_{OA}} y\,dA + \int_{A_{AE}} y\,dA = (Rt)\left(\frac{R}{2}\right) + \int_0^\theta R\cos\theta(tR\,d\theta) \qquad \text{or}$$

$$Q_z = \frac{R^2 t}{2} + \left[R^2 t\,\sin\theta\right]\Big|_0^\theta = (0.5 + \sin\theta)(R^2 t) \tag{E1}$$

From Example 6.4 we have $I_{zz} = 2.2375R^3 t$. Noting that I_{yz} and V_z are zero, we obtain from Equation (6.26) the shear flow q as a function of θ as shown in Equation (E2).

$$q = -\left[\frac{V_y Q_z}{I_{zz}}\right] = -\left[\frac{V_y(0.5 + \sin\theta)(R^2 t)}{2.2375R^3 t}\right] = -\left[\frac{0.4469 V_y(0.5 + \sin\theta)}{R}\right] \tag{E2}$$

Figure 6.27 shows the two statically equivalent systems with the shear flow drawn in the positive s direction, which corresponds to OAB. We equate the moment about point O for the two systems in Figure 6.27. To obtain the moment of the shear flow in ABD, we find the moment due to the differential element ds and integrate from point A to point D as shown in Equation (E3).

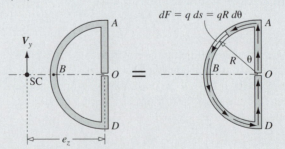

Figure 6.27 Static equivalency in Example 6.6.

$$V_y e_z = -\int_{\theta_A=0}^{\theta_D=\pi} R(qR\,d\theta) = R^2\int_0^\pi \left(\frac{0.4469\,V_y(0.5+\sin\theta)}{R}\right)d\theta$$

$$= V_y(0.4469R)(0.5\theta - \cos\theta)\big|_0^\pi \qquad (E3)$$

ANS. $e_z = 1.6R$

COMMENTS

1. In Equation (E3), V_y produces a clockwise moment about point O, while the shear flow shown in Figure 6.27 produces a counterclockwise moment about point O, which is reflected in the minus sign before the integral.

2. The shear flow in Equation (E2) is negative; yet in Figure 6.27 it is shown in the positive direction. This is permissible because we are relating two positive variables, and Equation (E3) shows that the moments due to a positive V_y and a positive q will be opposite in sign. On substituting q in Equation (E3), we account for the sign in Equation (E2).

3. We can show that the shear flow in the negative direction (i.e., the reverse of the direction shown in Figure 6.27) corresponds to the sign in Equation (E2). In such a case, there will be no minus sign in Equation (E3); but now a positive value of q must be substituted because the sign in Equation (E2) has already been counted.

4. The integration in Equation (E1) is on an open interval (indefinite integral) to find the shear flow at an arbitrary location θ. The integration in Equation (E3) is on a closed interval (definite integral) to account for the contribution toward the moments from all points on the arc ABD.

5. We defined θ from point A, which is not a free edge; but A_s in Figure 6.26 is defined from the free edge at O. We can start a new variable from any point as long as the area A_s in the computation of Q_z (and Q_y) is defined from the free edge to the point where we are to find the shear stress (shear flow). To appreciate this, consider the variable s starting from the free edge at O. The area $dA = t\,ds$, and we can write the integral for Q_z as $Q_z = \int_{A_{OA}} y\,dA + \int_{A_{AF}} y\,dA = \int_{s_O=0}^{s_A=R} y(t\,ds) + \int_{s_A=R}^{s} y(t\,ds)$. In the first integral $y=s$, and its value turns out to be $R^2 t/2$ as in Equation (E1). In the second integral we let $s=(s_1-R)$, where s_1 now is defined from point A and is equal to $R\theta$. On substituting in the second integral, we obtain the second integral in Equation (E1). Thus, provided A_s is properly defined, the effect of starting a new variable is akin to that of changing the integration variable.

EXAMPLE 6.7

The cross section shown in Figure 6.28 has a uniform thickness of $t = 0.05$ in. The area moments of inertia are

$$I_{yy} = 0.9982 \text{ in}^4 \qquad I_{zz} = 1.4149 \text{ in}^4 \qquad I_{yz} = -0.8768 \text{ in}^4$$

Determine the shear center of the cross section.

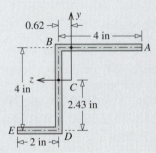

Figure 6.28 Cross section for Example 6.7.

PLAN

If we take the moment about point D, then we will not have to calculate the shear flow in sections BC and DE because the resultant of these flows will not produce any moment at D. The shear flow in AB can be found from Equation (6.25) by considering the shear forces V_y and V_z one at a time. The location of the shear center with respect to point D can then be found.

SOLUTION

From the given area moment of inertia, we obtain Equation (E1).

$$I_{yy}I_{zz} - I_{yz}^2 = 0.6436 \tag{E1}$$

The area A_s in the calculation of Q_y is defined from the free edge A to a location defined by the parameter s, measured from A as shown in Figure 6.29. The first moment of the area A_s can be found by multiplying the area A_s by the distance of the centroid of A_s to C as shown in Equations (E2) and (E3).

$$Q_y = (s)(0.05)[-(3.38 - 0.5s)] = 0.025s^2 - 0.169s \tag{E2}$$

$$Q_z = (s)(0.05)(1.57) = 0.0785s \tag{E3}$$

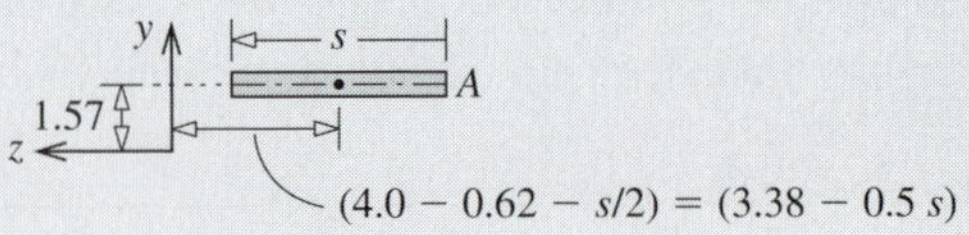

Figure 6.29 First area moment in Example 6.7.

From Equation (6.25) with just V_y, we obtain the shear flow in AB as shown in Equation (E4).

$$q = -\left[\frac{0.9982(0.0785s) - (-0.8768)(0.025s^2 - 0.169s)}{0.6436}\right]V_y$$

$$= -(0.0340s^2 - 0.1085s)V_y \tag{E4}$$

From Equation (6.25) with just V_z, we obtain the shear flow in AB as shown in Equation (E5).

$$q = -\left[\frac{(1.4149)(0.025s^2 - 0.169s) - (-0.8768)(0.0785s)}{0.6436}\right]V_z$$

$$q = -(0.0550s^2 - 0.2646s)V_z \tag{E5}$$

We consider *only* V_y and equate the moment about point D for the two systems in Figure 6.30. To obtain the moment of the shear flow in AB, we find the moment due to the differential element ds and integrate from point A to point B. We can then substitute Equation (E4) and obtain e_z as shown in Equation (E6).

$$V_y e_z = -\int_{s_A=0}^{s_B=4} 4(q\,ds) = 4\int_0^4 (0.0340s^2 - 0.1085s)V_y\,ds$$

$$= \left(0.136\frac{s^3}{3} - 0.434\frac{s^2}{2}\right)\Bigg|_0^4 V_y \quad \text{or} \tag{E6}$$

ANS. $e_z = -0.571$ in

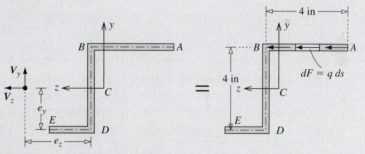

Figure 6.30 Static equivalency in Example 6.7.

We consider *only* V_z and equate the moment about point D for the two systems in Figure 6.30. To obtain the moment of the shear flow in AB, we find the moment due to the differential element ds and integrate from point A to point B. We can then substitute Equation (E4) and obtain e_y as shown in Equation (E7).

$$V_z e_y = \int_{s_A=0}^{s_B=4} 4(q\,ds) = -4\int_0^4 (0.0550s^2 - 0.2646s)V_z\,ds$$

$$= -\left(0.220\frac{s^3}{3} - 1.058\frac{s^2}{2}\right)\Bigg|_0^4 V_z \quad \text{or} \tag{E7}$$

ANS. $e_y = 3.77$ in

COMMENTS

1. The minus sign for e_z indicates that the shear center is to the right of point D.

2. We could have calculated e_y and e_z in a single step as follows. The shear flow due to V_y and V_z is the sum of the flows in Equations (E4) and (E5), as shown in Equation (E8).

$$q = -(0.0340s^2 - 0.1085s)V_y - (0.0550s^2 - 0.2646s)V_z \qquad \text{(E8)}$$

The moment at point D for both systems is as shown in Equation (E9).

$$V_z e_y - V_y e_z = \int_{s_A=0}^{s_B=4} 4(q\,ds) \qquad \text{(E9)}$$

On substituting Equation (E8) into Equation (E9) and integrating, we obtain Equation (E10).

$$V_z e_y - V_y e_z = 0.571 V_y + 3.77 V_z \qquad \text{or}$$

$$V_z(e_y - 3.77) - V_y(e_z + 0.571) = 0 \qquad \text{(E10)}$$

Since V_y and V_z are independent, the equation can be zero only if the coefficients of V_y and V_z are zero, which gives us the location of the shear center, as before.

6.3   SHEAR STRESSES IN THIN CLOSED SECTIONS

The primary difficulty in finding the shear stress in a thin closed section is that unlike the case of thin open sections, we have no free edge from which to start our calculations. If by inspection we could determine a point on the cross section where the shear flow is zero, we could use this point to start our variable s and define the area A_s. Calculations for the shear flow would then proceed in exactly the same manner as for thin open sections. Once shear flow is known, the shear center can be found, as was done for open sections. In this section we develop a method for finding the shear flow in sections of arbitrary shapes.

Figure 6.31(a) shows a thin closed section. We choose an arbitrary point on the cross section to start our variable s. At this point we make an imaginary cut as shown in Figure 6.31(b). Points A and B are two points on either side of the imaginary cut. We designate the shear flow at this imaginary cut as q_0. The total shear flow at any point is given by Equation (6.27),

$$q_c = q_0 + q \qquad \text{(6.27)}$$

where q_c is the shear flow in the closed section at any point, q is the shear flow given by Equation (6.25) for the open section, and q_0 is the unknown shear flow at the starting point, which is to be determined.

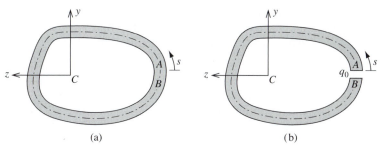

Figure 6.31 (a) Thin closed section. (b) An imaginary cut in the closed section.

On an open section, points A and B can move by different amounts. But for a closed section the two points A and B must have the same displacement. It is this continuity of displacement we next use to determine the starting value of q_0. The torsional shear strain can be written as

$$\gamma_{xs} = \frac{\partial u}{\partial s} + \frac{\partial v_s}{\partial x} = \frac{\tau_{xs}}{G} \tag{6.28a}$$

where u and v_s are displacements in the x and s direction, respectively, and G is the shear modulus of elasticity. Integrating Equation (6.28a) between points A and B, we obtain Equation (6.28b).

$$\int_{s_A}^{s_B} \frac{\partial u}{\partial s}\, ds = \oint \left[\frac{\tau_{xs}}{G} - \frac{\partial v_s}{\partial x} \right] ds \qquad \text{or}$$

$$u(s_B) - u(s_A) = \oint \left[\frac{\tau_{xs}}{G} - \frac{\partial v_s}{\partial x} \right] ds \tag{6.28b}$$

We assume that the shape and dimension of the cross section undergo negligible change. In fact, this is simply a restatement of Assumption 2b. We have also assumed that there is no twist of the cross section, in accordance with Assumption 2a. These two assumptions, along with Assumption 2c, imply that a cross section behaves like a rigid plate that rotates about the y and z axes as shown in Figure 6.4. Thus, the three assumptions imply that no point on the cross section moves relative to the other in the s direction (i.e., $v_s = 0$) in pure bending. Assuming that the material is homogeneous across the cross section in accordance with Assumption 8, and noting that shear stress is shear flow divided by thickness (i.e., $\tau_{xs} = q_c/t$), we obtain Equation (6.28c).

$$u(s_B) - u(s_A) = \frac{1}{G} \oint \left(\frac{q_c}{t} \right) ds \tag{6.28c}$$

Substituting Equation (6.27) and noting that $u(s_B) = u(s_A)$, we obtain Equation (6.29).

$$\boxed{ \oint \left(\frac{q_c}{t} \right) ds = \oint \left(\frac{q_0 + q}{t} \right) ds = 0 } \tag{6.29}$$

The starting value of the shear flow can be determined from Equation (6.29). A further simplification can be achieved if the thickness is uniform across the cross section. Noting that q_0 is a constant, we obtain Equation (6.30),

$$\boxed{q_0 = -\frac{1}{S}\oint q \, ds} \tag{6.30}$$

where S is the total path length of the perimeter of the cross section. Following are the steps for finding shear flow (shear stress) in a closed section:

Step 1. Choose a convenient point on the cross section to start the calculations.

Step 2. Use the point of step 1 like a free edge to calculate q from Equation (6.25).

Step 3. Use Equation (6.29) [or Equation (6.30) if applicable] to find q_0.

Step 4. Use Equation (6.27) to find the shear flow q_c at any point, and calculate the shear stress by dividing it by the thickness at that point.

6.3.1 Shear Centers of Thin Closed Sections

In open sections, we could avoid calculating the shear flow in many parts of the cross section by choosing the point such that the moment from the shear flow was zero. In closed sections, the shear flow must be found in *all* parts of the cross section to determine q_0. Except for this difference, the calculation for closed sections is essentially the same as for open sections.

EXAMPLE 6.8

Determine the shear flow at point D and sketch the flow on the cross section shown in Figure 6.32, assuming (a) just V_y and (b) just V_z. The cross section has a uniform thickness t with the centroid at C and the area moments of inertia as follows:

$$I_{yy} = 0.7928R^3t \qquad I_{zz} = 2.2375R^3t \qquad I_{yz} = 0$$

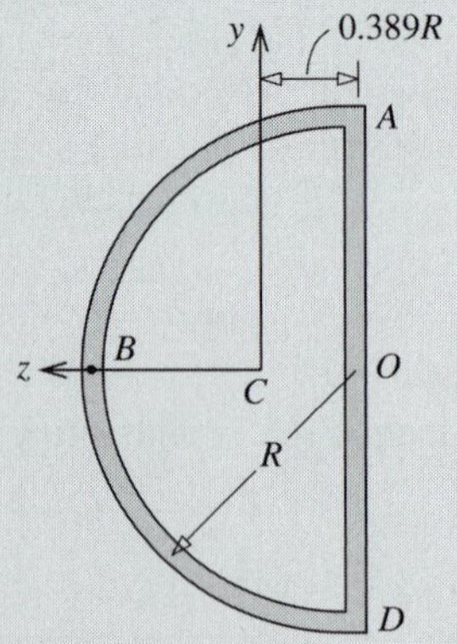

Figure 6.32 Cross section for Example 6.8.

PLAN

We can make an imaginary cut at D and determine the shear flow for each shear force by using Equation (6.26). Noting that thickness is uniform, we can determine the starting value of the shear flow at D from Equation (6.30). The total shear flow at any point can then be found by using Equation (6.27) and plotted for each shear flow.

SOLUTION

Step 1. We make an imaginary cut at D to start our computation.

Step 2. Figure 6.33 shows the area A_s used for computing Q_y and Q_z in the vertical and circular segments of the cross section.

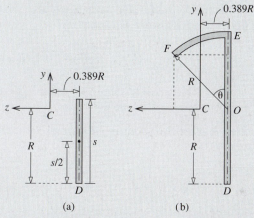

Figure 6.33 First area moment for Example 6.8 in (a) vertical and (b) circular segments of the cross section.

From Figure 6.33(a) we obtain the first moments of area A_s as shown in Equations (E1) and (E2).

$$Q_{y1} = (s)(t)(-0.389R) = -0.389sRt \qquad (E1)$$

$$Q_{z1} = (s)(t)[-(R - s/2)] = -sRt + 0.5s^2 t \qquad (E2)$$

From Figure 6.33(b) we obtain the first moments of area A_s as shown in Equations (E3) and (E4).

$$Q_{y2} = \int_{A_{DOE}} z\,dA + \int_{A_{EF}} z\,dA = Q_{y1}\big|_{s=2R} + \int_0^\theta (R\sin\theta - 0.389R)(tR\,d\theta) \qquad \text{or}$$

$$Q_{y2} = -0.778R^2 t + R^2 t(-\cos\theta - 0.389\theta)\big|_0^\theta$$

$$= -0.778R^2 t + R^2 t(1 - \cos\theta - 0.389\theta) \qquad \text{or}$$

$$Q_{y2} = R^2 t(0.222 - \cos\theta - 0.389\theta) \qquad (E3)$$

$$Q_{z2} = \int_{A_{DOE}} y\, dA + \int_{A_{EF}} y\, dA = Q_{z1}\Big|_{s=2R} + \int_0^\theta R\,\cos\theta(tR\,d\theta)$$

$$= 0 + R^2 t\,\sin\theta\Big|_0^\theta \quad \text{or} \quad Q_{z2} = R^2 t\,\sin\theta \tag{E4}$$

(a) Just V_y: Noting that $I_{yz} = 0$, and assuming a cut at D, we can find the shear flow in segment DOE from Equation (6.26) as shown in Equation (E5).

$$q_1 = -\left(\frac{V_y Q_{z1}}{I_{zz}}\right) = -(-sRt + 0.5s^2 t)\frac{V_y}{I_{zz}} \tag{E5}$$

Assuming a cut at D, the shear flow in segment EFD can be found from Equation (6.26) as shown in Equation (E6).

$$q_2 = -\left(\frac{V_y Q_{z2}}{I_{zz}}\right) = -(R^2 t\,\sin\theta)\frac{V_y}{I_{zz}} \tag{E6}$$

We note that thickness is uniform, and thus we can use Equation (6.30). The total path length is $S = (2R + \pi R)$.

Step 3. The shear flow at our starting point D can be found as shown in Equation (E7).

$$Sq_0 = -\oint q\, ds = -\left[\int_{S_{DOE}} q\, ds + \int_{S_{EFD}} q\, ds\right] = -\left[\int_0^{2R} q_1\, ds + \int_0^\pi q_2(R\,d\theta)\right] \quad \text{or}$$

$$R(\pi + 2)q_0 = \left(\frac{V_y}{I_{zz}}\right)\left[\int_0^{2R}(-sRt + 0.5s^2 t)ds + \int_0^\pi (R^2 t\,\sin\theta)(R\,d\theta)\right] \quad \text{or}$$

$$R(\pi + 2)q_0 = \left(\frac{V_y}{I_{zz}}\right)\left[\left(-\frac{s^2}{2}Rt + 0.5\frac{s^3}{3}t\right)\Big|_0^{2R} + (R^3 t)(-\cos\theta)\Big|_0^\pi\right]$$

$$= \left(\frac{V_y}{I_{zz}}\right)\left(\frac{4R^3 t}{3}\right) \quad \text{or}$$

$$q_0 = \left(\frac{V_y}{I_{zz}}\right)\left[\frac{4R^2 t}{3(\pi + 2)}\right] = \left(\frac{V_y}{2.2375R^3 t}\right)\left[\frac{4R^2 t}{3(\pi + 2)}\right] = 0.1159\frac{V_y}{R} \tag{E7}$$

ANS. $q_0 = 0.1159\dfrac{V_y}{R}$

Step 4. The total shear flow (q_{c1}) in segment *DOE* is sum of the flows in Equations (E7) and (E5), while in segment *EBF* the total shear flow (q_{c2}) is sum of the flows in Equations (E7) and (E6), as shown in Equations (E8) and (E9).

$$q_{c1} = 0.1159\frac{V_y}{R} - (-sRt + 0.5s^2t)\frac{V_y}{I_{zz}} \tag{E8}$$

$$q_{c2} = 0.1159\frac{V_y}{R} - (R^2t\sin\theta)\frac{V_y}{I_{zz}} \tag{E9}$$

The value of the shear flow at various points can be found from Equations (E8) and (E9) and plotted as shown in Figure 6.34(a). The flow has a zero value at 15.03° from the vertical in segment *ABD* and reaches a maximum value in each segment on the *z* axis.

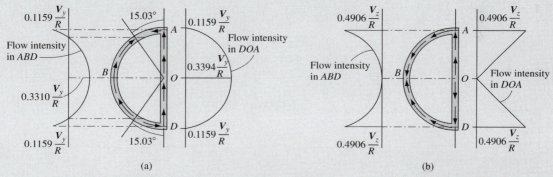

$$\text{(a)} \qquad\qquad\qquad\qquad \text{(b)}$$

Figure 6.34 Shear flow in Example 6.8 (a) due to V_y only and (b) due to V_z only.

(b) Just V_z: Noting that $I_{yz} = 0$, the shear flow in segment *DOE* can be found from Equation (6.26) as in Equation (E10).

$$q_1 = -\left(\frac{V_zQ_{y1}}{I_{yy}}\right) = 0.389\left(\frac{V_z}{I_{yy}}\right)(sRt) \tag{E10}$$

The shear flow in segment *EFD* can be found from Equation (6.26) as in Equation (E11).

$$q_2 = -\left(\frac{V_zQ_{y2}}{I_{yy}}\right) = -R^2t\left(\frac{V_z}{I_{yy}}\right)(0.222 - \cos\theta - 0.389\theta) \tag{E11}$$

We note that the thickness is uniform, and thus we can use Equation (6.30). The total path length is $S = (2R + \pi R)$.

Step 3. The shear flow at our starting point *D* can be found as shown in Equation (E12).

$$Sq_0 = -\oint q\,ds = -\left[\int_{S_{DOE}} q\,ds + \int_{S_{EFD}} q\,ds\right] = -\left[\int_0^{2R} q_1\,ds + \int_0^{\pi} q_2(R\,d\theta)\right] \qquad \text{or}$$

$$R(\pi + 2)q_0 = -\left(\frac{V_z}{I_{yy}}\right)\left[\int_0^{2R}(0.389\,sRt)ds - \int_0^{\pi}R^2t(0.222 - \cos\theta - 0.389\theta)(R\,d\theta)\right] \quad \text{or}$$

$$R(\pi + 2)q_0 = -\left(\frac{V_z}{I_{yy}}\right)\left[\left(0.389\frac{s^2}{2}Rt\right)\Big|_0^{2R} - (R^3t)\left(0.222\theta - \sin\theta - 0.389\frac{\theta^2}{2}\right)\Big|_0^{\pi}\right]$$

$$= -\left(\frac{V_z}{I_{yy}}\right)(2R^3t)$$

$$q_0 = -\left(\frac{V_z}{I_{yy}}\right)\left[\frac{2R^2t}{(\pi + 2)}\right] = -\left(\frac{V_z}{0.7928R^3t}\right)\left[\frac{2R^2t}{(\pi + 2)}\right] = -0.4906\frac{V_z}{R} \quad \text{(E12)}$$

ANS. $q_0 = -0.4906\dfrac{V_z}{R}$

Step 4. The total shear flow (q_{c1}) in segment *DOE* is sum of the flows in Equations (E12) and (E10), while in segment *EBF* the total shear flow (q_{c2}) is sum of the flows in Equations (E12) and (E11), as shown in Equations (E13) and (E14).

$$q_{c1} = -0.4906\frac{V_z}{R} + 0.389\left(\frac{V_z}{I_{yy}}\right)(sRt) \quad \text{(E13)}$$

$$q_{c2} = -0.4906\frac{V_z}{R} - R^2t\left(\frac{V_z}{I_{yy}}\right)(0.222 - \cos\theta - 0.389\theta) \quad \text{(E14)}$$

The value of the shear flow at various points can be found from Equations (E13) and (E14) and plotted as shown in Figure 6.34(b). The flow has zero value at points *O* and *B*, which are on the *z* axis.

COMMENTS

1. That the flow due to V_z is zero at the *z* axis is not surprising. We could have easily concluded that by inspection, since the geometry is symmetric about the shear force V_z. Had we done this, we could have saved much calculation. By using point *O* (zero flow point) as a starting point, we can define *s* as shown in Figure 6.35 and calculated Q_{yD} and the shear flow at *D* as shown in (E15) and (E16).

$$Q_{yD} = (R)(t)(-0.389R) = -0.389R^2t \quad \text{(E15)}$$

$$q_D = -\left(\frac{V_zQ_{yD}}{I_{yy}}\right) = \frac{V_z(0.389R^2t)}{0.7928R^3t} = 0.4907\frac{V_z}{R} \quad \text{(E16)}$$

The value of q_D in Equation (E16) is nearly the same magnitude as q_0 in Equation (E12). The difference in sign is because of the direction of *s*. In Figure 6.35 the positive direction of *s* is from *O* toward *D*, while in Figure 6.33(a) the direction is from *D* toward *O*.

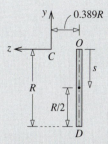

Figure 6.35 First area moment with O as starting point for Example 6.8.

2. With V_y there is no intuitive reasoning that would tell us that the zero flow point is 15.03° from the vertical. Thus, the method outlined for finding q_0 is a general method that should be used if we cannot determine the zero flow point by inspection.

EXAMPLE 6.9

Determine the location of the shear center of the cross section shown in Figure 6.32.

PLAN

The shear center will lie on the z axis because it is the axis of symmetry. We need to find e_z only. Hence, we consider the shear force V_y placed at an assumed location of the shear center. The shear flow due to V_y is known from the solution to Example 6.8. We can take the moment at point O and determine e_z.

SOLUTION

Figure 6.36 shows the two statically equivalent systems, with the shear flow drawn in the positive s direction, where the positive s is from point D toward the point under consideration. The shear flow that will produce a moment about point O is the shear flow in segment ABD, which is given by Equation (E9) in Example 6.8 and repeated as Equation (E1).

$$q_{c2} = 0.1159\frac{V_y}{R} - (R^2 t \sin \theta)\frac{V_y}{I_{zz}} = \frac{V_y}{R}[0.1159 - 0.4469 \sin \theta] \qquad \text{(E1)}$$

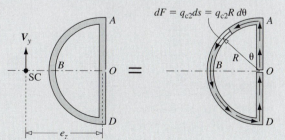

Figure 6.36 Static equivalency in Example 6.9.

We equate the moment about point O for the two systems in Figure 6.36. To obtain the moment of the shear flow in ABD, we find the moment due to the differential element ds and integrate from point A to point D as shown in Equation (E2).

$$V_y e_z = -\int_{\theta_A=0}^{\theta_D=\pi} R(q_{c2} R \, d\theta) = -R^2 \int_0^\pi \left(\frac{V_y}{R}[0.1159 - 0.4469 \, \sin \theta] \right) d\theta \quad \text{or} \quad \text{(E2)}$$

$$e_z = -(R)(0.1159\theta + 0.4469 \, \cos \theta)|_0^\pi = -R(0.1159\pi - 0.4469 - 0.4469)$$

ANS. $\quad e_z = 0.53R$

COMMENTS

1. In the calculation for the open section in Example 6.6, we needed to find the shear flow only in segment ABD, while in this example we had to find shear flow over the entire cross section before we could find q_0 as shown in Example 6.8.

2. The shear center is inside the perimeter of the closed section, while for the open section in Example 6.6 it was outside the perimeter of the cross section.

PROBLEM SET 6.2

6.21 A thin cross section of uniform thickness t is shown in Figure P6.21. If shear stresses were to be found at points A and B, what values of Q_y and Q_z are needed for the calculation? Assume $t \ll a$ and a gap at D of negligible thickness. Report the values of Q_y and Q_z in terms of t and a.

6.22 A thin cross section of uniform thickness t is shown in Figure P6.22. If shear stresses were to be found at points A and B, what values of Q_y and Q_z are needed for the calculation? Assume $t \ll a$ and a gap at D of negligible thickness. Report the values of Q_y and Q_z in terms of t and a.

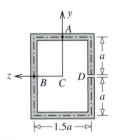

Figure P6.21

ANS. $(Q_y)_A = -1.031a^2 t$

$(Q_y)_B = 0 \qquad (Q_z)_A = 1.25ta^2$

$(Q_z)_B = 2.5ta^2$

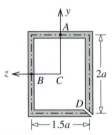

Figure P6.22

6.23 A thin cross section of uniform thickness t is shown in Figure P6.23. If shear stresses were to be found at points A and B, what values of Q_y and Q_z are needed for the calculation?

Report the values of Q_y and Q_z in terms of t and a. Assume $t \ll a$.

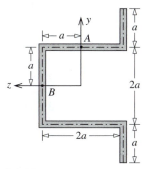

Figure P6.23

ANS. $(Q_y)_A = -1.5a^2t$ $(Q_y)_B = 0$
$(Q_z)_A = 2.5ta^2$ $(Q_z)_B = 4ta^2$

6.24 A thin cross section of uniform thickness t is shown in Figure P6.24. If shear stresses were to be found at points A and B, what values of Q_y and Q_z are needed for the calculation? Assume $t \ll a$ and a gap at D of negligible thickness. Report the values of Q_y and Q_z in terms of t and a.

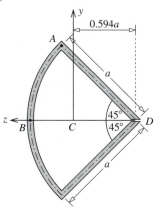

Figure P6.24

6.25 Shear forces on the cross section shown in Figure P6.25 were calculated as $V_y = 10$ kips and $V_z = -5$ kips. The cross section has a uniform thickness of 1/8 in. Determine the bending shear stresses at points A and B and report your answers as τ_{xy} or τ_{xz}.

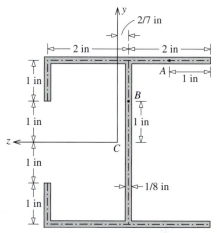

Figure P6.25

6.26 Shear forces on the cross section shown in Figure P6.26 were calculated as $V_y = 20$ kN and $V_z = 15$ kN. The cross section has a uniform thickness of 3 mm. Determine the bending shear stresses at points A and B and report your answers as τ_{xy} or τ_{xz}.

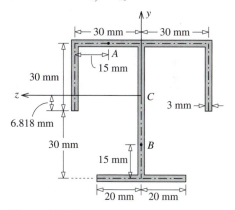

Figure P6.26

6.27 The cantilever beam shown in Figure P6.27 is loaded such that there is no twist. The cross section has a uniform thickness of 0.5 in. Calculations show that $I_{yy} = 80.25$ in^4 and $I_{zz} = 166.00$ in^4. Determine the bending normal stress and the bending shear stress at points A and B in the x-y-z coordinate system.

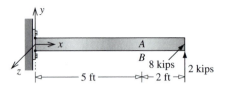

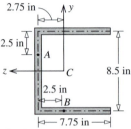

Figure P6.27

ANS. $(\sigma_{xx})_A = 5.03$ ksi (T)

$(\tau_{xy})_A = 0.98$ ksi $\qquad (\sigma_{xx})_B = 1.23$ ksi (T)

$(\tau_{xz})_B = -1.10$ ksi

6.28 The cross section shown in Figure P6.28 has a uniform thickness of $t = 0.05$ in. Assume that there is no twist of the cross section due to the applied load. Determine the bending normal stress and the bending shear stress at points A and B in the x-y-z coordinate system.

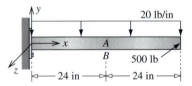

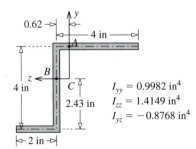

$I_{yy} = 0.9982$ in^4

$I_{zz} = 1.4149$ in^4

$I_{yz} = -0.8768$ in^4

Figure P6.28

6.29 Determine the location of the shear center for the cross section shown in Figure P6.24 with respect to point D.

6.30 Determine the location of the shear center for the cross section shown in Figure P6.25 with respect to point C.

6.31 Determine the location of the shear center for the cross section shown in Figure P6.26 with respect to point C.

6.32 The cross section shown in Figure P6.32 has a uniform thickness t. Assuming $t \ll a$, determine the location of the shear center with respect to point A.

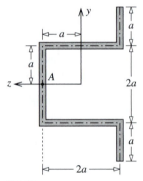

Figure P6.32

ANS. $e = 0.714a$

6.33 The cross section shown in Figure P6.33 has a uniform thickness t. Assuming $t \ll a$, determine the location of the shear center with respect to point A.

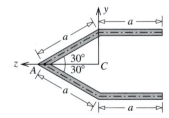

Figure P6.33

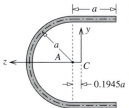

Figure P6.34

6.34 The cross section shown in Figure P6.34 has a uniform thickness t. Assuming $t \ll a$, determine the location of the shear center with respect to point A.

ANS. $e = 1.72a$

6.35 The cross section shown in Figure P6.35 has a uniform thickness t. Assuming $t \ll a$, determine the location of the shear center with respect to point A.

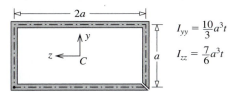

$$I_{yy} = \frac{10}{3}a^3 t$$

$$I_{zz} = \frac{7}{6}a^3 t$$

Figure P6.35

ANS. $e_y = 1.1a$ $e_z = 0.714a$

6.36 The cross section shown in Figure P6.36 has a uniform thickness t. Assuming $t \ll a$, determine the location of the shear center with respect to point A.

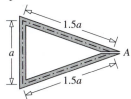

Figure P6.36

ANS. $e = 1.945a$

6.37 The cross section shown in Figure P6.37 has a uniform thickness t. Assuming $t \ll a$, determine the location of the shear center with respect to point A.

ANS. $e = a/(2\sqrt{3})$

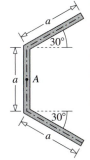

Figure P6.37

6.38 The cross section shown in Figure P6.38 has a uniform thickness t, and its boundaries are circular arcs. Assuming $t \ll a$, determine the location of the shear center with respect to point A in terms of radius a and angle α.

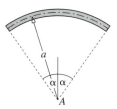

Figure P6.38

ANS. $e = 2a\dfrac{(\sin \alpha - \alpha \cos \alpha)}{\alpha - \sin \alpha \cos \alpha}$

6.39 The cross section shown in Figure P6.39 has a uniform thickness t and boundaries made from circular arcs. Assuming $t \ll a$, determine the location of the shear center with respect to point A.

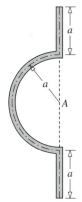

Figure P6.39

6.40 The thin cross section shown in Figure P6.40 has a thickness $t = 0.02a$. The cross section has a positive shear force V in the y direction. (a) Use Equation (6.29) to determine the shear flow at point A. (b) Determine the shear flow at points F and G.

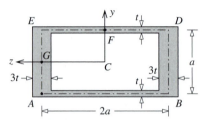

Figure P6.40

6.41 The cross section shown in Figure P6.41 has a uniform thickness t, and its curved boundaries are circular arcs. The thin cross section is subjected to a shear force $V_y = V$ acting through the shear center. Assuming $t \ll a$, determine the shear flow at points A, B, and D in terms of V, a, and t.

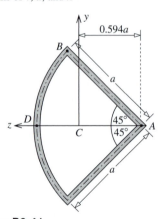

Figure P6.41

6.42 The cross section shown in Figure P6.41 has a uniform thickness t, and its curved boundaries are circular arcs. The thin cross section is subjected to a shear force $V_z = V$ acting through the shear center. Assuming $t \ll a$, determine the shear flow at points A, B, and D in terms of V, a, and t.

6.43 The cross section shown in Figure P6.41 has a uniform thickness t, and its curved boundaries are circular arcs. Assuming $t \ll a$, determine the location of the shear center with respect to point C.

6.44 The cross section shown in Figure P6.44 has a uniform thickness t, and its curved boundaries are circular arcs. The thin cross section is subjected to a shear force $V_y = V$ acting through the shear center. Assuming $t \ll a$, determine the shear flow at points A and B in terms of V, a, and t.

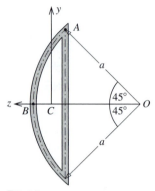

Figure P6.44

6.45 The cross section shown in Figure P6.44 has a uniform thickness t, and its curved boundaries are circular arcs. The thin cross section is subjected to a shear force $V_z = V$ acting through the shear center. Assuming $t \ll a$, determine the shear flow at points A and B in terms of V, a, and t.

6.46 The cross section shown in Figure P6.44 has a uniform thickness t, and its curved boundaries are circular arcs. Assuming $t \ll a$, determine the location of the shear center with respect to point O.

6.47 The thin cross section shown in Figure P6.47 has a uniform thickness t and is subjected to a shear force $V_y = V$ acting through the shear center. Determine the shear stress at points A and B in terms of V, a, and t.

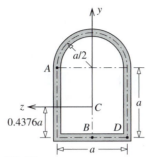

Figure P6.47

ANS. $(\tau_{xy})_A = 0.371\dfrac{V}{at}$

$(\tau_{xz})_B = 0$

6.48 The thin cross section shown in Figure P6.47 has a uniform thickness t and is subjected to a shear force $V_z = V$ acting through the shear center. Starting with point D, determine the shear stress at points A and B in terms of V, a, and t.

ANS. $(\tau_{xy})_A = -0.234\left(\dfrac{V}{at}\right)$

$(\tau_{xz})_B = 0.567\left(\dfrac{V}{at}\right)$

6.49 Determine the shear center of the cross section shown in Figure P6.47 relative to centroid C.

ANS. $e_y = 0.071a$ $e_z = 0$

6.50 The cross section shown in Figure P6.50 has a uniform thickness t, and its curved boundaries are circular arcs. The thin cross section is subjected to a shear force $V_y = V$ acting through the shear center. Assuming $t \ll a$, determine the shear flow at points A and B in terms of V, a, and t.

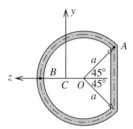

Figure P6.50

6.51 The cross section shown in Figure P6.50 has a uniform thickness t, and its curved boundaries are circular arcs. The thin cross section is subjected to a shear force $V_z = V$ acting through the shear center. Assuming $t \ll a$, determine the shear flow at points A and B in terms of V, a, and t.

6.52 The cross section shown in Figure P6.50 has a uniform thickness t, and its curved boundaries are circular arcs. Assuming $t \ll a$, determine the location of the shear center with respect to point O.

6.4   TORSION OF THIN-WALLED TUBES

The sheet metal skin on the fuselage or wing of an aircraft, and the shell of a tall building are examples of bodies that can be analyzed as thin-walled tubes. "Thin-walled" implies that the thickness t of the walls is smaller by a factor of at least 10 in comparison to the length b of the biggest line that can be drawn across two points on the cross section, as shown in Figure 6.37. The term "tube" implies that the length L is at least 10 times that of the cross-sectional dimension b. We assume that this thin-walled tube is subjected to torsional moments only.

The walls of the tube are bounded by two free surfaces; hence by symmetry of shear stresses, the shear stress in the normal direction τ_{xn} must go to zero on these bounding surfaces, as shown in Figure 6.38. This does not imply that τ_{xn} is zero in the interior, but

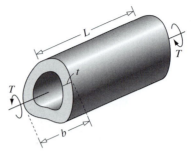

Figure 6.37 Torsion in a thin-walled tube.

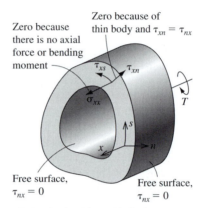

Figure 6.38 Deducing stress behavior in thin-walled tubes.

for thin walls an approximation of zero τ_{xn} is justified. A normal stress of σ_{xx} would imply either an internal axial force or an internal bending moment. Since there is no external axial force or bending moment, we approximate the value of σ_{xx} as zero.

Figure 6.38 shows that the only nonzero stress component is τ_{xs}, which can be assumed to be uniform in the n direction because the tube is thin. We next show that it is also uniform in the s direction by making imaginary cuts along the x direction through two A and B on the cross section, as shown in Figure 6.39.

We establish the following sign convention for the direction s.

Definition 11 The tangential s direction is positive counterclockwise with respect to the x direction.

By equilibrium of forces in the x direction in Figure 6.39, we obtain Equation (6.31a).

$$\tau_A(t_A\,dx) = \tau_B(t_B\,dx) \qquad \text{or} \qquad \tau_A t_A = \tau_B t_B \qquad (6.31\text{a})$$

Noting that the product of shear stress and thickness is the shear flow at a point, we obtain the following important conclusion from Equation (6.31a).

- Shear flow is constant in a cross section of a thin tube under torsion.

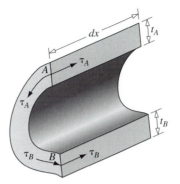

Figure 6.39 Deducing constant shear flow in a thin-walled tube.

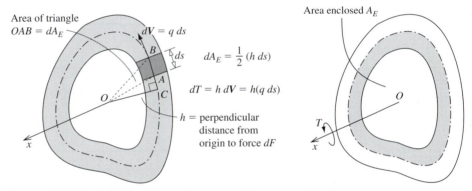

Figure 6.40 Equivalency of internal torque and shear stress (flow).

We can replace the shear stresses (shear flow) by an equivalent internal torque as shown in Figure 6.40. The line OC in Figure 6.40 is perpendicular to the line of action of the force dV, which is tangential to the arc at that point. The internal torque can be written, and q can be taken outside the integral as shown in Equation (6.31b).

$$T = \oint dT = \oint q(h\,ds) = q\oint(h\,ds) \tag{6.31b}$$

The area enclosed by the centerline of the tube can be written as in Equation (6.31c).

$$A_E = \oint d\,A_E = \oint \frac{1}{2}(h\,ds) \tag{6.31c}$$

Substituting Equation (6.31c) into Equation (6.31b), we obtain Equation (6.31d).

$$T = q\oint 2\,dA_E = 2qA_E \tag{6.31d}$$

We thus obtain Equation (6.32),

$$\boxed{\tau_{xs} = \frac{T}{2tA_E}} \tag{6.32}$$

where T = internal torque at the section containing the point at which the shear stress is to be calculated

A_E = area enclosed by the center-line of the tube

t = thickness at the point which the shear stress is to be calculated

---

Definition 12 The internal torque T is consider to be positive if it is counterclockwise with respect to the outward normal to an imaginary cut surface.

The thickness t in Equation (6.32) can vary along the cross section, provided the assumption of a thin wall is not violated. If the thickness varies, the shear stress will not be constant on the cross section even though the shear flow is constant.

6.4.1   Torsional Deformation

Assumption 10a The shape of the cross section does not change during torsional deformation.

Assumption 10a implies that all radial lines between the center of rotation O and a point on the cross section undergo the same angle of rotation ϕ as shown in Equation (6.33a).

$$\phi = \phi(x) \tag{6.33a}$$

Definition 13 The angle ϕ is positive counterclockwise with respect to the x axis.

The torsional shear strain can be written as

$$\gamma_{xs} = \frac{\partial u}{\partial s} + \frac{\partial v_s}{\partial x} = \frac{\tau_{xs}}{G} \tag{6.33b}$$

where u and v_s are the displacements in the x and s direction, respectively, and G is the shear modulus of elasticity. Integrating Equation (6.28a) between two arbitrary points s_1 and s_2 on the cross section, we obtain Equation (6.33c).

$$\int_{s_1}^{s_2} \frac{\partial u}{\partial s} ds = \int_{s_1}^{s_2} \left[\frac{\tau_{xs}}{G} - \frac{\partial v_s}{\partial x} \right] ds \quad \text{or}$$

$$u(s_2) - u(s_1) = \int_{s_1}^{s_2} \left[\frac{\tau_{xs}}{G} - \frac{\partial v_s}{\partial x} \right] ds = \int_{s_1}^{s_2} \left[\frac{q}{Gt} - \frac{\partial v_s}{\partial x} \right] ds \tag{6.33c}$$

The quantity $u(s_2) - u(s_1)$ represents relative warping of point at s_2 with respect to point at s_1.

The shear stress τ_{xs} is known from Equation (6.32). To obtain an expression for $\partial v_s/\partial x$, we consider Figure 6.41, where line OA rotates by an angle $\Delta\phi$. The displacement

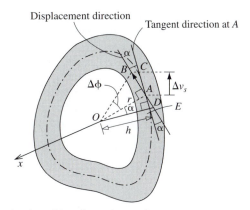

Figure 6.41 Deformation in a thin tube under torsion.

of point AB is shown by the vector AB, whose magnitude is $r\Delta\phi$. The displacement vector is at some angle α with the tangent at point A. Then the tangential component of the displacement vector is given by AC and can be written as in Equation (6.33d).

$$\Delta v_s = (r\Delta\phi)\cos\alpha \tag{6.33d}$$

Draw line OED in Figure 6.41 from the center of rotation point O such that the line is perpendicular to the tangent at A. The angle α in the triangle BAC is equal to angle EAD. Noting that angles OAD and OEA are right angles, we obtain from geometry that angle EOA is α. From the triangle AOD we obtain $r\cos\alpha = h$. Substituting $r\cos\alpha = h$ in Equation (6.33d), we obtain $\Delta v_s = h(\Delta\phi)$. Dividing by Δx and taking the limit as $\Delta x \to 0$, we obtain

$$\lim_{\Delta x \to 0}\left(\frac{\Delta v_s}{\Delta x}\right) = h\lim_{\Delta x \to 0}\left(\frac{\Delta\phi}{\Delta x}\right) \quad \text{or} \quad \frac{\partial v_s}{\partial x} = h\frac{d\phi}{dx}(x) \tag{6.33e}$$

where $d\phi/dx$ is the angle of twist per unit length. Substituting Equations (6.32) and (6.33e) into Equation (6.28b), we obtain Equation (6.33f).

$$u(s_2) - u(s_1) = \int_{s_1}^{s_2}\left[\frac{q}{Gt} - h\frac{d\phi}{dx}(x)\right]ds \tag{6.33f}$$

Noting that $d\phi/dx$ does not change with s, we obtain Equation (6.33g).

$$u(s_2) - u(s_1) = \int_{s_1}^{s_2}\frac{q}{Gt}\,ds - \frac{d\phi}{dx}(x)\int_{s_1}^{s_2}h\,ds \tag{6.33g}$$

Equation (6.33g) can be used for finding relative warping between two points on the cross section. Now consider going a round the entire cross section: $u(s_2) = u(s_1)$ for a closed cross section, to find Equation (6.33h):

$$\oint\frac{q}{Gt}\,ds - \frac{d\phi}{dx}(x)\oint h\,ds = 0 \quad \text{or} \quad \frac{d\phi}{dx}(x) = \frac{1}{2A_E}\oint\frac{q}{Gt}\,ds \tag{6.33h}$$

Substituting for q from Equation (6.31d) and assuming a homogeneous material across the cross section, we obtain *Bredte* formula in Equation (6.34).

$$\left| \frac{d\phi}{dx}(x) = \frac{T}{4A_E^2 G} \oint \frac{ds}{t} \right| \tag{6.34}$$

If $d\phi/dx$ is a constant between the two sections at x_1 and x_2, then we can write $d\phi/dx = (\phi_2 - \phi_1)/(x_2 - x_1)$ and obtain Equation (6.35).

$$\left| \phi_2 - \phi_1 = \left[\frac{T}{4A_E^2 G} \oint \frac{ds}{t} \right] (x_2 - x_1) \right| \tag{6.35}$$

If the cross section is of uniform thickness, then t can be taken outside the integral and Equation (6.35) can be written as

$$\left| \phi_2 - \phi_1 = \frac{TS(x_2 - x_1)}{4A_E^2 Gt} \right| \tag{6.36}$$

where S is the total path length of the closed section and T is an internal torque that must be determined by making an imaginary cut and drawing a free body diagram. The following possible ways in which T may be found are elaborated further in Example 6.11.

1. **T** is always drawn counterclockwise with respect to the outward normal of the imaginary cut. An equilibrium equation then is used to get a positive or negative value for **T**. In other words, we are following the sign convention for internal torque given by Definition 12. Therefore, the sign for relative rotation obtained from Equation (6.35) or (6.36) is positive counterclockwise with respect to x axis in accordance with Definition 13, and the direction of shear stress τ_{xs} in Equation (6.32) can be determined from the subscripts.

2. **T** is drawn at the imaginary cut in a direction to equilibrate the external torques. Since we are determining the direction of **T** by inspection, the direction of relative rotation in Equation (6.35) or (6.36) and the direction of shear stress τ_{xs} in Equation (6.32) must also be determined by inspection.

EXAMPLE 6.10

Show that Equations (6.32) and (6.36) give the same results as formulas for the torsion of circular shafts when the shaft is a thin circular tube (i.e., $t \ll R$, where t is the thickness and R is the radius of centerline of the circular cross section).

PLAN

The polar moment of inertia can be found in terms of t and R, and the condition $t \ll R$ can be used to simplify the expression. Substituting the polar moment into Equations (3.9-T) and (3.10-T), we find the shear stress and the relative rotation these results with and compare Equations (6.32) and (6.36).

SOLUTION

The polar moment of inertia for a thin circular tubular cross section with centerline radius R and thickness t is as calculated in Equation (E1).

$$J = \frac{\pi}{2}\left[\left(R + \frac{t}{2}\right)^4 - \left(R - \frac{t}{2}\right)^4\right] = \frac{\pi}{2}\left[\left(R + \frac{t}{2}\right)^2 - \left(R - \frac{t}{2}\right)^2\right]\left[\left(R + \frac{t}{2}\right)^2 + \left(R - \frac{t}{2}\right)^2\right] \quad \text{or}$$

$$J = \frac{\pi}{2}\left[\left(R + \frac{t}{2}\right) - \left(R - \frac{t}{2}\right)\right]\left[\left(R + \frac{t}{2}\right) + \left(R - \frac{t}{2}\right)\right]\left[2R^2 + 2\left(\frac{t}{2}\right)^2\right]$$

$$\approx \frac{\pi}{2}(2R)(t)(2R^2) = 2\pi R^3 t \tag{E1}$$

The shear stress at any point on the centerline ($\rho = R$) and the relative rotation of two sections on a circular shaft can be calculated from Equations (3.9-T) and (3.10-T) as given in Equations (E2) and (E3).

$$\tau_{x\theta} = \frac{T\rho}{J} = \frac{TR}{2\pi R^3 t} = \frac{T}{2\pi R^2 t} \tag{E2}$$

$$\phi_2 - \phi_1 = \frac{T(x_2 - x_1)}{GJ} = \frac{T(x_2 - x_1)}{2\pi GR^3 t} \tag{E3}$$

The area enclosed by the centerline of the circular tubular cross section is $A_E = \pi R^2$, and the perimeter length is $S = 2\pi R$. From Equations (6.32) and (6.36) we obtain Equations (E4) and (E5).

$$\tau_{xs} = \frac{T}{2tA_E} = \frac{T}{2\pi R^2 t} \tag{E4}$$

$$\phi_2 - \phi_1 = \frac{TS(x_2 - x_1)}{4A_E^2 Gt} = \frac{T(2\pi R)(x_2 - x_1)}{4(\pi R^2)^2 Gt} = \frac{T(x_2 - x_1)}{2\pi GR^3 t} \tag{E5}$$

The s direction in Equation (E4) is the tangential direction θ in Equation (E2). Thus Equations (E2) and (E3) for a circular shaft give the same results as Equations (E4) and (E5) for thin tubes.

EXAMPLE 6.11

A thin-walled tube with a shear modulus of $G = 4000$ ksi is subjected to torques as shown Figure 6.42(a).

Determine (a) the maximum shear stress in the tube, (b) the shear stress at point O, and (c) the relative rotation of section at D with respect to section at A. Show the results of part (b) on a stress cube.

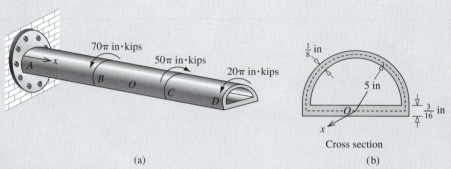

Figure 6.42 (a) Thin-walled tube for Example 6.11. (b) Cross section of the tube.

PLAN

(a) From Equation (6.32), we know that the maximum shear stress will exist in a section where internal torque is maximum and thickness is minimum. To determine maximum internal torque, we draw free body diagrams by making imaginary cuts in *AB*, *BC*, and *CD*. (b) In Equation (6.32) we substitute the internal torque in *BC* and the thickness at point *O* to obtain the shear stress at point *O*. (c) From Equation (6.35) we can obtain the relative rotation of the ends of segments *AB*, *BC*, and *CD*. By adding all the relative rotations, we obtain relative rotation of section at *D* with respect to the section at *A*.

SOLUTION

(a) Figure 6.43 shows the free body diagrams. The right-hand part of each cut is taken to avoid calculating the wall reaction that would appear to the left of the imaginary cut. The internal torques are drawn in accordance with the sign convention in Definition 12.

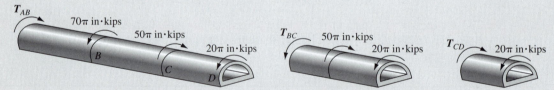

Figure 6.43 Internal torque calculations for Example 6.11.

By equilibrium of the torques in Figure 6.43, we obtain Equations (E2) through (E3).

$$T_{AB} + 50\pi - 70\pi - 20\pi = 0 \quad \text{or} \quad T_{AB} = 40\pi \text{ in} \cdot \text{kips} \quad \text{(E1)}$$

$$T_{BC} + 50\pi - 20\pi = 0 \quad \text{or} \quad T_{BC} = -30\pi \text{ in} \cdot \text{kips} \quad \text{(E2)}$$

$$T_{CD} - 20\pi = 0 \quad \text{or} \quad T_{CD} = 20\pi \text{ in} \cdot \text{kips} \quad \text{(E3)}$$

The maximum torque is in AB and minimum thickness is 1/8 in. The enclosed area is $A_E = \pi 5^2/2 = 12.5\pi$. From Equation (6.32) we obtain Equation (E4).

$$\tau_{max} = \frac{40\pi}{(12.5\pi)(1/8)} \tag{E4}$$

ANS. $\tau_{max} = 25.6$ ksi

(b) At point O the internal torque is T_{BC} and $t = 3/16$, and thus from Equation (6.32) we obtain Equation (E5).

$$\tau_0 = \frac{(-30)\pi}{(12.5\pi)(3/16)} \tag{E5}$$

ANS. $\tau_0 = -12.8$ ksi

The direction of the shear stress can be determined by using subscripts or intuitively as follows.

Stress direction by subscripts: We note that the shear stress at point O is negative. The outward normal in Figure 6.44(a) is in the positive x direction, and thus the force must be in the negative s direction, as shown in Figure 6.44(a).

(a)   (b)   (c)

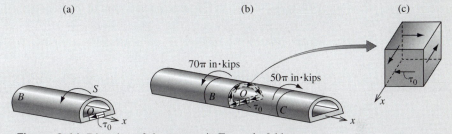

Figure 6.44 Direction of shear stress in Example 6.11.

Shear stress direction by inspection: Figure 6.44(b) shows a part of the tube between sections B and C. Section BO would rotate counterclockwise with respect to section OC. The shear stress must be opposite to this possible motion and hence is in the clockwise direction, as shown. The direction on the remaining surfaces can be drawn by using the observation that the members of the symmetric pair of shear stress components point either toward the corner or away from it, as shown in Figure 6.44(c).

(c) Since the cross-section of the tube is the same, we can calculate the following for the entire tube.

$$\oint \frac{ds}{t} = \frac{2(5)}{(3/16)} + \frac{\pi(5)}{(1/8)} = 179 \tag{E6}$$

Substituting the internal torques and Equation (E6) into Equation (6.35), we can obtain the relative rotation of the ends of each segment as shown in Equations (E7) through (E9).

$$\phi_B - \phi_A = \left[\frac{T_{AB}}{4A_E^2 G} \oint \frac{ds}{t} \right](x_B - x_A) = \frac{(40\pi)(179)(24)}{4(12.5\pi)^2(4000)} = 21.879(10^{-3}) \text{ rad} \tag{E7}$$

$$\phi_C - \phi_B = \left[\frac{T_{BC}}{4A_E^2 G}\oint\frac{ds}{t}\right](x_C - x_B) = \frac{(-30\pi)(179)(24)}{4(12.5\pi)^2(4000)} = -16.409(10^{-3})\text{ rad} \qquad \text{(E8)}$$

$$\phi_D - \phi_C = \left[\frac{T_{CD}}{4A_E^2 G}\oint\frac{ds}{t}\right](x_D - x_C) = \frac{(20\pi)(179)(20)}{4(12.5\pi)^2(4000)} = 9.1164(10^{-3})\text{ rad} \qquad \text{(E9)}$$

Adding Equations (E7), (E8), and (E9), we obtain the relative rotation of section at D with respect to section at A as shown in Equation (E10).

$$\phi_D - \phi_A = (21.879 - 16.409 + 9.1164)(10^{-3})\text{ rad} \qquad \text{(E10)}$$

ANS. $\phi_D - \phi_A = 0.0146$ rad ccw

COMMENTS

1. The shear flow in the cross section containing point O [Figure 6.42(b)] is constant over the entire cross section. The magnitude of torsional shear stress at point O, however, will be two-thirds that of the value of the shear stress in the circular part of the cross section because of the variation in thickness.

2. The methodology for calculating the relative rotations of thin tubes is similar to that for finding the relative rotations of circular shafts. The primary difference is the formulas used in finding the relative rotation of the ends of each segment.

6.4.2   Torsion of Multicell Tubes

Figure 6.45 shows a multicell cross section. In each cell there is a uniform shear flow, as was the case for a single cell. Equation (6.31d) gives the contribution from each single cell toward the internal torque. Thus, the total internal torque is as given in Equation (6.37),

$$T = \sum_{i=1}^{N} 2q_i A_{Ei} \qquad \text{(6.37)}$$

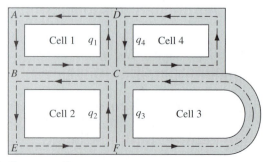

Figure 6.45 Multicell cross section.

where q_i is the shear flow in the ith cell, A_{Ei} is the area enclosed by the center line of the ith cell, and N is the total number of cells.

Equation (6.37) contains N unknown q_i's, which makes the problem statically indeterminate. As in all indeterminate problems, we generate the remaining $(N-1)$ equations from deformation considerations. Equation (6.33h) was obtained by starting from a point, integrating along a closed loop, and returning to the starting point. *The closed loop can be around any of the cell* or even about multiple cells. The value of $(d\phi/dx)$ is independent of the closed loop. We write Equation (6.33h) for the ith cell as shown in Equation (6.38).

$$\frac{d\phi}{dx}(x) = \left(\frac{1}{2A_{Ei}}\right)\oint_{\Gamma_i}\frac{q}{Gt}\,ds \qquad (6.38)$$

where Γ_i is the centerline path around the ith cell. By choosing N closed loops and equating the results to each other, we obtain the $N-1$ equations which, along with Equation (6.37), can be solved to obtain the Nq_i shear flows.

Some caution needs to be exercised in writing Equation (6.38) for each cell. Consider the integration loop around cell 1 in Figure 6.45. Along path BC the shear flow is $(q_1 - q_2)$, along path CD the shear flow is $(q_1 - q_4)$, while along AB and DA the shear flow is q_1. If we wrote an equation for cell 2 and integrated along path $EFCB$, then in going from C to B the shear flow would be $(q_2 - q_1)$ which is the negative of the value we used for cell 1 in going from B to C. We will follow the following sign convention:

- In writing the line integral in Equation (6.38), assume q to be positive in the direction of integration.

Example 6.12 demonstrates the analytical process for thin-walled, multicell tubes.

EXAMPLE 6.12

The cross section shown in Figure 6.46 is subjected to a torque of $T_{ext} = 900$ N · m. Calculate the maximum shear stress in the cross section and the angle of twist per unit length of the cross section. Use a shear modulus of $G = 28$ GPa.

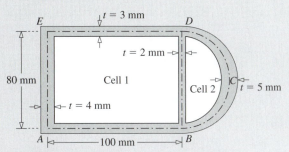

Figure 6.46 Cross section for Example 6.12.

PLAN

We can use Equation (6.37) to write one equation in the unknown shear flows of the two cells. From Equation (6.38), the angle of twist per unit length can be written for the two cells and equated to obtain the second equation for the two shear flows. By solving the two equations, the shear flows in each cell can be found. The shear stress is shear flow divided by thickness. The maximum shear stress will exist either in *AB*, *BD*, or *DE*. The angle of twist per unit length can be found from the equations written by using Equation (6.38).

SOLUTION

The enclosed area of the two cells can be written as in Equation (E1).

$$A_{E1} = (0.1)(0.08) = 8(10^{-3}) \text{ m}^2 \qquad A_{E2} = \left(\frac{\pi}{2}\right)\left(\frac{0.08}{2}\right)^2 = 2.513(10^{-3}) \text{ m}^2 \qquad (E1)$$

We obtain Equation (E2) from Equation (6.37) and the given torque.

$$T = 2[q_1 A_{E1} + q_2 A_{E2}] = 2[q_1(8)(10^{-3}) + q_2(2.513)(10^{-3})] = 900 \qquad \text{or}$$

$$8q_1 + 2.513q_2 = 450(10^3) \qquad (E2)$$

Figure 6.47 shows the shear flow in the cross section, assuming that the torque is counter-clockwise.

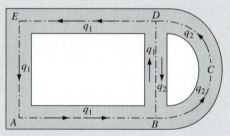

Figure 6.47 Shear flow for Example 6.12.

Equation (6.38) can be written for cell 1 as shown in Equation (E3).

$$\frac{d\phi}{dx}(x) = \left(\frac{1}{2A_{E1}}\right)\left[\int_{s_A}^{s_B}\frac{q_{AB}}{G_{AB}t_{AB}}ds + \int_{s_B}^{s_D}\frac{q_{BC}}{G_{BC}t_{BC}}ds + \int_{s_D}^{s_E}\frac{q_{CD}}{G_{CD}t_{CD}}ds + \int_{s_E}^{s_A}\frac{q_{DA}}{G_{DA}t_{DA}}ds\right] \qquad (E3)$$

Noting that *G*, *t*, and *q* are constant in each integral, we can rewrite Equation (E3) as follows.

$$\frac{d\phi}{dx}(x) = \left(\frac{1}{2A_{E1}}\right)\left[\frac{q_{AB}(s_B - s_A)}{G_{AB}t_{AB}} + \frac{q_{BD}(s_D - s_B)}{G_{BD}t_{BD}} + \frac{q_{DE}(s_E - s_D)}{G_{DE}t_{DE}} + \frac{q_{EA}(s_A - s_E)}{G_{EA}t_{EA}}\right] \qquad (E4)$$

From Figure 6.47 we obtain Equations (E5) and (E6).

$$q_{AB} = q_1 \qquad q_{BD} = q_1 - q_2 \qquad q_{DE} = q_1 \qquad q_{EA} = q_1 \tag{E5}$$

$$t_{AB} = 0.003 \text{ m} \qquad t_{BD} = 0.002 \text{ m} \qquad t_{DE} = 0.003 \text{ m} \qquad t_{EA} = 0.006 \text{ m} \tag{E6}$$

The shear modulus G is same for all parts and is a common factor. Substituting the value of the variables into Equation (E4), we obtain Equation (E7).

$$\frac{d\phi}{dx}(x) = \frac{1}{2(8)(10^{-3})(28)(10^9)}\left[\frac{q_1(0.1)}{0.003} + \frac{(q_1 - q_2)(0.08)}{0.002} + \frac{q_1(0.1)}{0.003} + \frac{q_1(0.08)}{0.004}\right]$$

$$\frac{d\phi}{dx}(x) = [0.2827q_1 - 0.0893q_2](10^{-6}) \tag{E7}$$

Equation (6.38) can be written for cell 2, and the equivalent form of Equation (E4) is Equation (E8).

$$\frac{d\phi}{dx}(x) = \left(\frac{1}{2A_{E2}}\right)\left[\frac{q_{BCD}(s_D - s_B)}{G_{BCD}t_{BCD}} + \frac{q_{DB}(s_D - s_B)}{G_{DB}t_{DB}}\right] \tag{E8}$$

From Figure 6.47 we obtain Equations (E9) and (E10).

$$q_{BCD} = q_2 \qquad q_{DB} = q_2 - q_1 \tag{E9}$$

$$t_{BCD} = 0.005 \text{ m} \qquad t_{DB} = 0.002 \text{ m} \tag{E10}$$

The shear modulus G is same for all parts and is a common factor. Substituting the value of the variables in Equation (E8), we obtain Equation (E11).

$$\frac{d\phi}{dx}(x) = \frac{1}{2(2.513)(10^{-3})(28)(10^9)}\left[\frac{q_2(0.1257)}{0.005} + \frac{(q_2 - q_1)(0.08)}{0.002}\right]$$

$$\frac{d\phi}{dx}(x) = [-0.2842q_1 + 0.4628q_2](10^{-6}) \tag{E11}$$

Equating Equations (E7) and (E11), we obtain Equation (E12).

$$[0.2827q_1 - 0.0893q_2](10^{-6}) = [-0.2842q_1 + 0.4628q_2](10^{-6}) \qquad \text{or}$$

$$q_1 = 0.9737q_2 \tag{E12}$$

Solving Equations (E2) and (E12), we obtain Equations (E13).

$$q_1 = 42.53(10^3) \text{ N/m} \qquad \text{and} \qquad q_2 = 43.675(10^3) \text{ N/m} \tag{E13}$$

Since the thickness of AB and DE is less than for EA, the shear stress in EA will be greater in AB and DE. We also evaluate shear stress in DB and BCD as shown in Equations (E14), (E15), and (E16).

$$\tau_{AB} = \frac{q_{AB}}{t_{AB}} = \frac{42.53(10^3)}{0.003} = 14.175(10^6) \text{ N/m}^2 \tag{E14}$$

$$\tau_{BCD} = \frac{q_2}{t_{BCD}} = \frac{43.675(10^3)}{0.005} = 8.735(10^6) \text{ N/m}^2 \tag{E15}$$

$$\tau_{DB} = \frac{q_2 - q_1}{t_{DB}} = \frac{(43.625 - 42.53)(10^3)}{0.002} = 0.5725(10^6) \ \text{N/m}^2 \qquad \text{(E16)}$$

Comparing Equations (E14), (E15), and (E16), we see that the maximum torsional shear stress is in *AB*. Its value is

ANS. $\tau_{max} = 14.2$ MPa

Substituting Equation (E13) into Equation (E8), we obtain Equation (E17).

$$\frac{d\phi}{dx}(x) = [0.2827(42.53) - 0.0893(43.625)](10^3)(10^{-6}) \qquad \text{(E17)}$$

ANS. $\frac{d\phi}{dx}(x) = 0.0081$ rad/m

COMMENTS

1. In Equation (E5) we wrote the shear flow in *BD* as $(q_1 - q_2)$, while in Equation (E9) we wrote it as $(q_2 - q_1)$ in accordance with our sign convention of assuming a positive value for the shear flow in the direction of integration.

2. Equation (E4) highlights that if q, t, and G are constants in each segment of the integration circuit, we can write Equation (6.38) as Equation (E18),

$$\frac{d\phi}{dx}(x) = \left(\frac{1}{2A_{Ei}}\right) \sum_{j=1}^{M} \frac{q_j(s_{j+1} - s_j)}{G_j t_j} \qquad \text{(E18)}$$

where M is the number of segments in the integration path of the ith cell and q_j is positive in the direction of increasing s.

6.5 COMBINED LOADING

Figure 6.48 shows a cross section of a beam on which the external loads P_y and P_z do not pass through the shear center. These loads bend the beams and, because the loads are not passing through the shear center, also subject the cross section to a torque. If we know the location of the shear center, we can calculate the torque on the cross section. The stresses at any point can then be found as follows.

1. The normal stress σ_{xx} due to bending can be found by using Equation (6.13).

2. The shear stress τ_{xs} due to bending can be found by using Equation (6.28b) to find the shear flow for a thin closed section and dividing the result by the thickness at that point.

3. The shear stress τ_{xs} due to torsion of thin tubes can be found by using Equation (6.32).

4. The shear stress due to bending and torsion can be superposed to obtain the total shear stress at a point.

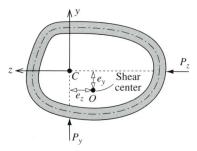

Figure 6.48 Loads not passing through the shear center.

EXAMPLE 6.13

A uniformly distributed load acts through the centroid of the cross section of a cantilever beam as shown in Figure 6.49. The cross section has a uniform thickness of 1/4 in. (a) Determine the normal and shear stresses at points A and B on a section at the built-in end. (b) Show the results of part (a) on stress cubes.

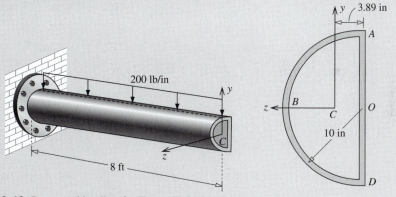

Figure 6.49 Beam and loading for Example 6.13.

PLAN

From Example 6.9 we know the location of the shear center of the cross section. We can make an imaginary cut at the built-in end and draw a free body diagram with internal bending moment M_z, torque T acting at the section, and internal shear force V_y acting at the shear center. The internal quantities can be determined by means of equilibrium equations. The shear flow values at points A and B due to V_y were calculated in Example 6.9, from which the shear stress due to bending can be found. The bending normal stresses and torsional shear stresses can be found from Equations (6.13) and (6.32), respectively. The total stresses at the points can be found from superposition and shown on a stress cube.

SOLUTION

From Example 6.9 we know the shear center is at $0.53R$. With $R = 10$ in, we obtain the shear center at 5.30 in from point O or 1.41 in from C. The total force from the distributed load is $P = (200)(96) = 19200$ lb = 19.2 kips. We make an imaginary cut at the wall and draw the free body diagram as shown in Figure 6.50. The internal quantities follow our sign convention.

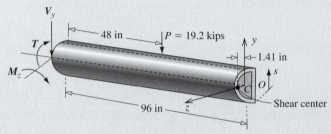

Figure 6.50 Free body diagram of the beam in Example 6.13.

By means of the equilibrium of forces and moments, we obtain Equations (E1), (E2), and (E3).

$$V_y = -19.2\,\text{kips} \tag{E1}$$

$$M_z = -(19.2)(48) = -921.6\,\text{in} \cdot \text{kips} \tag{E2}$$

$$T = -(19.2)(1.41) = -27.07\,\text{in} \cdot \text{kips} \tag{E3}$$

From Example 6.9, the area moment of inertia I_{zz} can be found as shown in Equation (E4).

$$I_{zz} = 2.2375R^3 t = 2.2375(10)^3 1/4 = 559.4\,\text{in}^4 \tag{E4}$$

Noting that $I_{yz} = 0$ and $V_z = 0$, we obtain the bending normal stress at point A from Equation (6.13) as in Equation (E5).

$$(\sigma_{xx})_A = -\left[\frac{M_z y_A}{I_{zz}}\right] = -\left[\frac{(-921.6)(10)}{559.4}\right] = 16.475 \tag{E5}$$

ANS. $(\sigma_{xx})_A = 16.5$ ksi (T)

Point B is on the neutral axis, and hence the bending normal stress is zero.

ANS. $(\sigma_{xx})_B = 0$

From Figure 6.34(a) the shear flows at points A and B are as given in Equation (E6).

$$q_A = 0.1159\frac{V_y}{R} \quad \text{and} \quad q_B = -0.3310\frac{V_y}{R} \tag{E6}$$

We can obtain the shear stress by dividing the shear flows in Equation (E6) by the thickness. Substituting $R = 10$, $t = 1/4$, and Equation (E1), we obtain the bending shear stresses at points A and B as shown in Equations (E7) and (E8).

$$(\tau_{xs})_A = \frac{q_A}{t_A} = \frac{(0.1159)(-19.2)}{(10)(1/4)} = -0.890 \text{ ksi} \qquad \text{(E7)}$$

$$(\tau_{xs})_B = \frac{q_B}{t_B} = \frac{(-0.3310)(-19.2)}{(10)(1/4)} = 2.542 \text{ ksi} \qquad \text{(E8)}$$

The positive direction of the tangential coordinate s is as shown in Figure 6.50. The enclosed area is $A_E = \pi(10)^2/2 = 157.1 \text{ in}^2$. From Equation (6.32) we obtain the torsional shear stresses at points A and B as shown in Equation (E9).

$$(\tau_{xs})_{A,\,B} = \frac{T}{2tA_E} = \frac{(-27.07)}{2(1/4)(157.1)} = -0.345 \text{ ksi} \qquad \text{(E9)}$$

The s direction in Figure 6.50 is counterclockwise. Hence by superposition we add the results in Equation (E9) to the results in Equations (E6) and (E7) to obtain the total shear stresses at points A and B as shown in Equation (E10).

$$(\tau_{xs})_A = -0.890 - 0.345 = -1.235 \text{ ksi} \qquad \text{and}$$

$$(\tau_{xs})_B = 2.542 - 0.345 = 2.197 \text{ ksi} \qquad \text{(E10)}$$

Noting that the s direction at point A is in the same direction as y, and point B is in the opposite direction to y, we obtain the shear stresses in the x and y coordinates:

ANS. $(\tau_{xy})_A = -1.24 \text{ ksi} \qquad (\tau_{xy})_B = -2.20 \text{ ksi}$

The results for the stresses at points A and B are as shown in Figure 6.51.

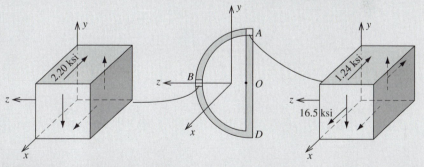

Figure 6.51 Stress cubes for Example 6.13.

COMMENTS

1. At point A there is a change in flow direction. The result shown in Figure 6.51 assumes that point A is on the straight side, DOA. If we assume that point A is on the curved side, ABD, then the shear stress at that point will be not τ_{xy} but τ_{xz} of the same magnitude.

2. To appreciate the full complexity of the problem solved in this example, suppose we had started with only the geometry of the cross section and the loading of the beam shown in Figure 6.49. We would have begun by finding the centroid of the cross section. Next we would have found the area moments of inertia about the y and z axes. We would then have determined the shear flow in a closed cross section due to V_y and V_z, as was done in Example 6.8. Next we would have found the shear center, as was done in Example 6.9. Only then could we have begun the calculations done in this example.

EXAMPLE 6.14

Determine the normal and shear stresses in Cartesian coordinates at points A and B for the beam and loading shown in Figure 6.52. The beam is made from sheet metal 5 mm thick. The second area moments of inertia are

$$I_{yy} = 1.086(10^{-6}) \text{ mm}^4 \qquad I_{zz} = 2.341(10^{-6}) \text{ mm}^4 \qquad I_{yz} = 0$$

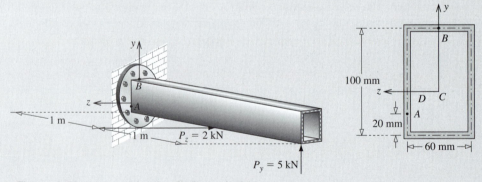

Figure 6.52 Beam and loading for Example 6.14.

PLAN

Owing to the double symmetry in the cross section, the shear center is at the origin. Thus, the applied loads will cause the beam to bend and twist. By making an imaginary cut at A and drawing a free body diagram, we can find the internal shear forces, bending moments, and torque. The bending normal stresses and torsional shear stresses can be found from Equations (6.13) and (6.32), respectively. The shear flow will be zero at point B due to just shear force V_y and at point D due to just shear force V_z. These zero points can be used for finding shear stresses due to bending from Equation (6.13). The superposition of the results will yield the stresses at points A and B.

SOLUTION

Figure 6.53 shows the free body diagram after an imaginary cut has been made at the wall. The internal forces and moments are drawn in accordance with our sign convention.

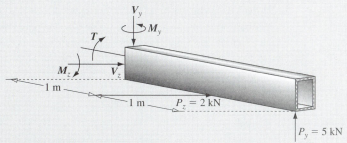

Figure 6.53 Free body diagram for Example 6.14.

By means of the equilibrium of forces and moments, we obtain the Equations (E1) through (E3). The equation for torque is obtained by taking the moment from the applied forces about the center of the cross section (shear center).

$$V_y = 5 \text{ kN} \quad \text{and} \quad V_z = -2 \text{ kN} \quad \text{(E1)}$$

$$M_y = -2 \text{ kN} \cdot \text{m} \quad \text{and} \quad M_z = 10 \text{ kN} \cdot \text{m} \quad \text{(E2)}$$

$$T + 5(10^3)(0.0325) - (10^3)2(0.0525) = 0 \quad \text{or} \quad T = -57.5 \text{ N} \cdot \text{m} \quad \text{(E3)}$$

1. **Bending normal stress:** The bending normal stress at points A and B can be found from Equation (6.13) as shown in Equations (E4) and (E5).

$$(\sigma_{xx})_A = -\left(\frac{M_z}{I_{zz}}\right)y_A - \left(\frac{M_y}{I_{yy}}\right)z_A = -\frac{10(10^3)}{2.341(10^{-6})}(-0.03) - \frac{-2(10^3)}{1.086(10^{-6})}(0.03)$$

$$= 183.4(10^6) \text{ N/m}^2 \quad \text{(E4)}$$

ANS. $(\sigma_{xx})_A = 183.4 \text{ MPa (T)}$

$$(\sigma_{xx})_B = -\left(\frac{M_z}{I_{zz}}\right)y_B - \left(\frac{M_y}{I_{yy}}\right)z_B = -\frac{10(10^3)}{2.341(10^{-6})}(0.05) - \frac{-2(10^3)}{1.086(10^{-6})}(0)$$

$$= -213.6(10^6) \text{ N/m}^2 \quad \text{(E5)}$$

ANS. $(\sigma_{xx})_B = 213.6 \text{ MPa (C)}$

2. **Torsional shear stress:** The enclosed area is $A_E = (0.1)(0.06) = 0.006 \text{ m}^2$. The torsional shear stress can be found from Equation (6.32) as shown in Equation (E6).

$$(\tau_{xs})_{A,B} = \frac{T}{2tA_E} = \frac{(-57.5)}{2(0.005)(0.006)} = -0.958(10^6) \text{ N/m}^2 \quad \text{(E6)}$$

Noting that the counterclockwise tangential (s) direction at A is in the negative y direction and at B is in the positive z direction, we obtain the torsional shear stresses in Cartesian coordinates as shown in Equation (E7).

$$(\tau_{xy})_A = 0.958 \text{ MPa} \quad \text{and} \quad (\tau_{xz})_B = -0.958 \text{ MPa} \quad (E7)$$

3. **Shear stress due to just V_y:** By symmetry about the y axis, we conclude that point B will have zero shear flow. With point B as the starting point, the area A_s, needed for the calculation of the shear stress at point A, can be seen in Figure 6.54(a). The first moment of the area about the z axis can be found as shown in Equation (E8).

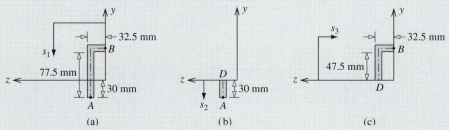

Figure 6.54 First area moments for Example 6.14.

$$(Q_z)_A = (32.5)(5)(50) + (78.5)(5)\left(\frac{78.5}{2} - 30\right) = 11.52(10^3) \text{ mm}^3$$

$$= 11.52(10^{-6}) \text{ m}^3 \quad (E8)$$

The bending shear stress due to just V_y at point A can be found by using Equation (6.26) as follows:

$$(\tau_{s_1 x})_A = -\left[\frac{V_y(Q_z)_A}{I_{zz}t}\right] = -\left[\frac{5(10^3)(11.52)(10^{-6})}{(2.341)(10^{-6})(0.005)}\right] = -4.919(10^6) \text{ N/m}^2 \quad (E9)$$

Noting that direction of s_1 at point A in Figure 6.54(a) is in the negative y direction, we obtain the shear stresses at points A and B in cartesian coordinate as in Equation (E10).

$$(\tau_{xy})_A = 4.919 \text{ MPa} \quad \text{and} \quad (\tau_{xz})_B = 0 \quad (E10)$$

4. **Shear stress due to just V_z:** By symmetry about the z axis, we conclude that point D will have zero shear flow. With point D as the starting point, the areas A_s, needed for calculation of shear stress at points A and B, can be seen in Figure 6.54(b) and (c), respectively. The first moment of the area about the y axis and the shear stresses can be found as in Equations (E11) through (E14).

$$(Q_y)_A = (30)(5)(30) = 4.5(10^3) \text{ mm}^3 = 4.5(10^{-6}) \text{ m}^3 \quad (E11)$$

$$(Q_y)_B = (32.5)(5)\left(\frac{32.5}{2}\right) + (47.5)(5)(30) = 9.766(10^3) \text{ mm}^3$$

$$= 9.766(10^{-6}) \text{ m}^3 \quad (E12)$$

$$(\tau_{s_2x})_A = -\left[\frac{V_z(Q_y)_A}{I_{yy}t}\right] = -\left[\frac{(-2)(10^3)(4.5)(10^{-6})}{(1.0858)(10^{-6})(0.005)}\right] = 1.658(10^6)\ \text{N/m}^2 \quad \text{(E13)}$$

$$(\tau_{s_3x})_B = -\left[\frac{V_z(Q_y)_B}{I_{yy}t}\right] = -\left[\frac{(-2)(10^3)(9.766)(10^{-6})}{(1.0858)(10^{-6})(0.005)}\right] = 3.597(10^6)\ \text{N/m}^2 \quad \text{(E14)}$$

Noting that s_2 in Figure 6.54(b) is in the negative y direction at point A and s_3 in Figure 6.54(c) is in the negative z direction at point B, we obtain the shear stresses in Cartesian coordinates at points A and B as shown in Equation (E15).

$$(\tau_{xy})_A = -1.658\ \text{MPa} \qquad \text{and} \qquad (\tau_{xz})_B = -3.597\ \text{MPa} \qquad \text{(E15)}$$

5. **Superposition:** The total shear stresses at points A and B can be found by adding the stresses in Equations (E7), (E10), and (E15) as shown in Equations (E16) and (E17).

$$(\tau_{xy})_A = 0.958 + 4.919 - 1.658 \qquad \text{(E16)}$$

ANS. $\quad (\tau_{xy})_A = 4.22\ \text{MPa}$

$$(\tau_{xz})_B = -0.958 + 0 - 3.597 \qquad \text{(E17)}$$

ANS. $\quad (\tau_{xz})_B = -4.56\ \text{MPa}$

The results of the stresses at points A and B are as shown in Figure 6.55.

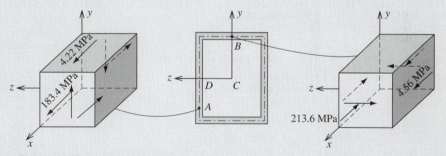

Figure 6.55 Stress cubes for Example 6.14.

COMMENTS

1. Note that the double symmetry about the y and z axes resulted in significant simplification. We did not need to find either the shear center or the starting value of the shear flow in a closed section.

2. The total shear stress (flow) is not zero at point B or D. But if we consider just one shear force V_y or V_z, then the symmetry arguments give us the zero points, which we can use in our calculations as shown in Figure 6.54.

PROBLEM SET 6.3

6.53 The cross section shown in Figure P6.53 is subjected to a torque of $T = 100$ in · kips. Calculate the maximum shear stress in the cross section and the angle of twist per unit length of the cross section. Use a shear modulus of $G = 4000$ ksi.

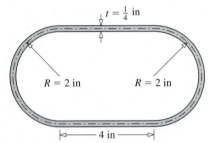

Figure P6.53

6.54 The cross section shown in Figure P6.54 is subjected to a torque of $T = 900$ N · m. Calculate the maximum shear stress in the cross section and the angle of twist per unit length of the cross section. Use a shear modulus of $G = 28$ GPa.

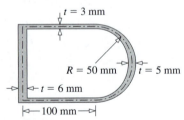

Figure P6.54

ANS. $|\tau_{max}| = 10.8$ MPa

$$\frac{d\phi}{dx} = 4.77 \text{ rad/mm}$$

6.55 The cross section shown in Figure P6.55 is subjected to a torque of $T = 15$ kN · m. Calculate the maximum shear stress in the cross section and the angle of twist per unit length of the cross section. Use a shear modulus of $G = 80$ GPa.

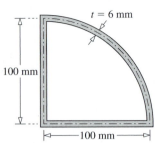

Figure P6.55

6.56 A thin tube having the cross section shown in Figure P6.56 is of uniform thickness t and has a torque T applied to it. Determine the maximum shear stress in terms of t, a, and T.

Figure P6.56

ANS. $\tau_{max} = \frac{2}{\sqrt{3}} \frac{T}{a^2 t}$

6.57 A thin tube having the cross section shown in Figure P6.57 is of uniform thickness t and has a torque T applied to it. Determine the maximum shear stress in terms of t, a, and T.

Figure P6.57

6.58 A thin tube having the cross section shown in Figure P6.58 is of uniform thickness t and has a torque T applied to it. Determine the maximum shear stress in terms of t, a, and T.

Figure P6.58

6.59 A thin tube having the cross section shown in Figure P6.59 is of uniform thickness t and has a torque T applied to it. Determine the maximum shear stress in terms of t, a, and T.

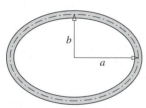

Figure P6.59

6.60 A hexagonal tube of uniform thickness is loaded as shown in Figure P6.60. Determine (a) the magnitude of the maximum torsional shear stress in the tube and (b) the relative rotation of section at D with respect to A. Use a shear modulus of $G = 28$ GPa.

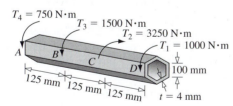

Figure P6.60

6.61 A rectangular tube is loaded as shown in Figure P6.61. Determine (a) the magnitude of the maximum torsional shear stress in the tube and (b) the relative rotation of section at D with respect to A. Use a shear modulus of $G = 4000$ ksi.

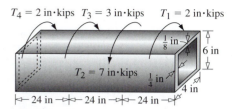

Figure P6.61

ANS. $|\tau_{max}| = 833$ psi

$\phi_D - \phi_A = 1.458(10^{-3})$ rad ccw

6.62 The three tubes shown in Problems 6.56 through 6.58 are to be compared for the maximum torque-carrying capability, assuming that all tubes have the same thickness t, that the maximum shear stress in each tube can be τ, and that the amount of material used in the cross section of each tube is A. Which shape would you use? What percentage of torque would have to be carried by the remaining two shapes to give the most efficient structural shape?

6.63 The thin cross section shown in Figure P6.63 is subjected to a torque of $T = 100$ in · kips. Calculate the maximum shear stress in the cross section and the angle of twist per unit length of the cross section. Use a shear modulus of $G = 4000$ ksi.

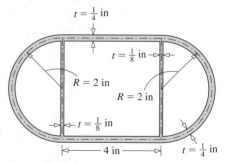

Figure P6.63

ANS. $\tau_{max} = 6.36$ ksi

$\dfrac{d\phi}{dx} = 0.61(10^{-3})$ rad/in

6.64 The cross section shown in Figure P6.64 is subjected to a torque of $T = 200$ in · kips. Calculate the maximum shear stress in the cross section and the angle of twist per unit length of the cross section. Use a shear modulus of $G = 4000$ ksi.

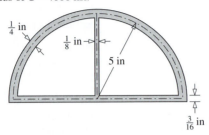

Figure P6.64

6.65 Two square tubes of uniform thickness t are attached together as shown in Figure P6.65. Determine the maximum shear stress in terms of t, a, and the internal torque T.

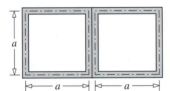

Figure P6.65

6.66 A cross section of uniform thickness t is shown in Figure P6.66. Determine the maximum shear stress in terms of t, a, and the internal torque T.

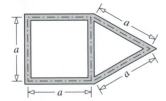

Figure P6.66

ANS. $\tau_{max} = 0.372T/(a^2 t)$

6.67 A cantilever beam is loaded as shown in Figure P6.67. The cross section has a uniform thickness of $t = 1/4$ in. Determine the normal and shear stresses at points A and B in Cartesian coordinates on a section next to the wall.

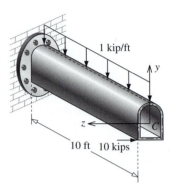

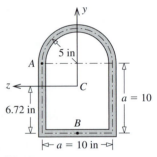

Figure P6.67

6.68 A cantilever beam is loaded as shown in Figure P6.68. Determine the normal and shear stresses at points A and B in Cartesian coordinates on a section next to the wall.

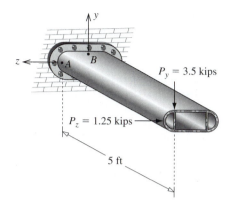

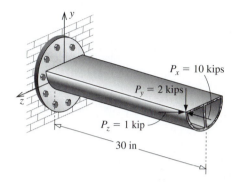

6.69 A cantilever beam is loaded as shown in Figure P6.69. Determine the normal and shear stresses at points A and B in Cartesian coordinates on a section next to the wall.

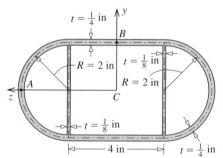

Figure P6.68

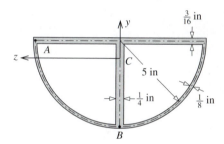

Figure P6.69

ANS. $(\sigma_{xx})_A = 7.23$ ksi (C)
$(\sigma_{xx})_B = 26.89$ ksi (T)
$(\tau_{xy})_A = -1.79$ ksi
$(\tau_{xz})_B = 0.095$ ksi

6.6 CLOSURE

In this chapter we changed the kinematics for bending problems by incorporating deflection in the y and z directions, but the material model, static equivalency, and the equilibrium equations remained unchanged. The simplest theory for unsymmetric bending was developed by making assumptions similar to those that were made in Table 3.3. These assumptions represent points at which complexities can be added, as shown for symmetric bending in Chapters 4 and 5. For the theory to be valid, the magnitude of the maximum bending normal stress must be at least an order of magnitude greater than that of the maximum bending shear stress. We saw that the applied load must be along the line passing through the centroid and the shear center of the cross section to decouple the axial, torsion, and bending problem; that the orientation of the neutral axis depended upon the geometry as well as the loading; that the displacement may not be in the direction of the applied load but is always perpendicular to the neutral axis; and that the maximum bending normal stress may be at the point furthest from the neutral axis, rather than on the top or bottom surface of the beam cross section.

In calculating the shear center for an open section, we could avoid calculating the shear flow in part of the cross section by choosing an appropriate point for torsion equivalency. To find the shear flow at the starting point in a closed section, however, the shear flow must be calculated in the entire cross section. Symmetry in geometry and loading could be used in determining the point of zero shear flow in a closed section, and in such cases the calculations were very similar to those for an open section.

When the loads did not pass through the shear center, a combined torsion and bending problem had to be solved. The shear center was used to determine the torsion and bending loads that were applied to the cross section. Once the stresses at a point had been determined, these could be used for finding principal stresses, accounting for stress concentration, determining a factor of safety by using failure theories, or accounting for cracks or fatigue, as was elaborated in Chapters 1 and 2.

7 Energy Methods

7.0 OVERVIEW

The bungee cord of the bungee jumper shown in Figure 7.1 is undeformed at the start of the jump and stores energy by deforming as the jump progresses. This stored energy due to deformation is called strain energy, and one could use energy methods to calculate the lowest point the jumper could reach safely (i.e., without hitting the ground). Energy methods are an alternative to equilibrium equations. Methods based on energy principles can be used to reduce significantly the algebraic effort needed to obtain deformation at a point or to determine reaction forces in statically indeterminate structures. Energy methods are the basis of a number of approximate methods, the most popular of which is the finite element method, discussed in Chapter 9.

The perspective of energy methods is very different from the perspective of equilibrium equations. In Sections 7.1 through 7.3 we will study the former perspective and related concepts. Virtual work, discussed in Section 7.4, is a very general concept that can be used for elastic and inelastic, linear and nonlinear material behavior. From this general concept we will derive the dummy unit load method, discussed in Section 7.5, and Castigliano's theorems, discussed in Section 7.6. These two methods can be used for finding the displacement at a point and reactions in statically indeterminate structures. In Section 7.7 we will apply the two methods for finding the deflection of curved beams. In Section 7.8 we will study minimum potential energy, which is applicable to linear or nonlinear conservative systems. In Section 7.9 we will study an approximate method called the Rayleigh-Ritz method, which is based on minimum potential energy. The Rayleigh-Ritz method was a precursor to the finite element method in structural analysis. Many of the conclusions derived by means of the Rayleigh-Ritz method will be used in the finite element method discussed in Chapter 9. We conclude the chapter with a brief discussion of functionals, a generalization that is used when energy methods are considered as part of variational calculus.

Figure 7.1 Bungee jumper. (*Photograph courtesy of James Banghort, 2006.*)

The learning objectives in this chapter are:

1. To understand the perspective and concepts in energy methods.
2. To understand the use of the dummy unit load method and Castigliano's theorems for calculating displacements in statically determinate and indeterminate structures.
3. To understand the use of the Rayleigh-Ritz method in determining approximate displacements in structures.

7.1 STRAIN ENERGY

Several definitions are associated with the change in internal energy in a material during deformation as discussed in this section.

> **Definition 1** The change in internal energy in a body during deformation is called the *strain energy*.
>
> **Definition 2** The strain energy per unit volume is called the *strain energy density* and is the area under the stress–strain curve up to the point of deformation.

Equation (7.1) shows the relationship between the strain energy and the strain energy density,

$$U = \int_V U_0 \, dV \tag{7.1}$$

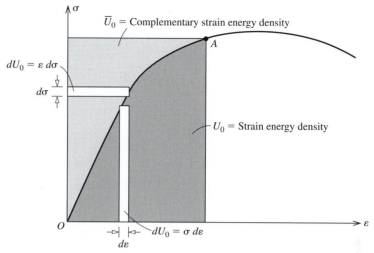

Figure 7.2 Energy densities.

where U is the strain energy, U_0 is the strain energy density, and V is the volume of the body. Noting that the strain energy density is the area under the curve shown in Figure 7.2, we obtain Equation (7.2).

$$U_0 = \int_0^\varepsilon \sigma \, d\varepsilon \qquad (7.2)$$

Equation (7.2) shows that strain energy density has the same dimensions as that of stress energy density because strain is dimensionless. But the units of strain energy density are *units of energy per unit volume,* which are different from those of stress. The units for strain energy density are newton-meters per cubic meter ($N \cdot m/m^3$), joules per cubic meter (J/m^3), inch-pounds per cubic inch ($in \cdot lb/in^3$), and foot-pounds per cubic foot ($ft \cdot lb/ft^3$).

Another related concept is the complementary strain energy density ($\overline{U}_0$), shown in Figure 7.2 and defined as follows:

$$\overline{U}_0 = \int_0^\sigma \varepsilon \, d\sigma \qquad (7.3)$$

EXAMPLE 7.1

A titanium alloy has the stress–strain curve given in Figure 7.3. Determine (a) the strain energy density at a stress level of 136 ksi and (b) the complementary strain energy density at a stress level of 136 ksi.

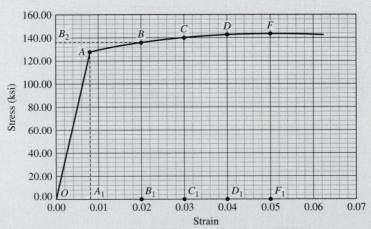

Figure 7.3 Area under the curve for Example 7.1.

PLAN

(a) On the stress–strain curve we locate the stress value of 136 ksi. The area under the curve gives us the strain energy density. (b) We can subtract the area of part (a) from the rectangle formed by the axis and the lines parallel to the axis that pass through the 136 ksi stress value to obtain the complementary strain energy density.

SOLUTION

(a) Point B in Figure 7.3 is at 136 ksi. The strain energy density at point B is the area AOA_1 plus the area AA_1BB_1. The area AA_1BB_1 is the area of a trapezoid and can be found as shown by Equation (E2).

$$AOA_1 = \frac{(128)(0.008)}{2} = 0.512 \tag{E1}$$

$$AA_1BB_1 = \frac{(128 + 136)(0.012)}{2} = 1.584 \tag{E2}$$

Thus the strain energy density at B is $U_B = 0.512 + 1.584$ or

ANS. $U_B = 2.1 \text{ in} \cdot \text{kips/in}^3$

(b) The complementary strain energy density at B can be found by subtracting U_B from the area of the rectangle OB_2BB_1. Thus, $\overline{U}_B = (136)(0.02) - 2.1$ or

ANS. $\overline{U}_B = 0.62 \text{ in} \cdot \text{kips/in}^3$

COMMENT

The approximation of the curve by straight lines for the purpose of finding areas is the same as using the trapezoidal rule of integration. We can obtain more accurate results if we use the data from which the stress–strain curve was created and do the approximation by means of a straight line between two consecutive data points. This would become tedious unless we uses a spread-sheet or a computer program and would not significantly increase the accuracy.

7.2 LINEAR STRAIN ENERGY DENSITY

Most engineering structures are designed for linear elastic materials. In the linear region, the area under the stress–strain curve is a triangle, and we obtain Equation (7.4) for the strain energy density function.

$$U_0 = \frac{1}{2}\sigma\varepsilon \tag{7.4}$$

If instead of a curve representing normal stress vs normal strain, we have a curve of shear stress vs shear strain, then we will have a similar expression for strain energy density in terms of shear stress and shear strain, as shown by Equation (7.5).

$$U_0 = \frac{1}{2}\tau\gamma \tag{7.5}$$

Strain energy, hence strain energy density, is a scalar quantity. We can add the strain energy density due to individual stress and strain components to obtain Equation (7.6) for the total linear strain energy density during deformation.

$$U_0 = \frac{1}{2}[\sigma_{xx}\varepsilon_{xx} + \sigma_{yy}\varepsilon_{yy} + \sigma_{zz}\varepsilon_{zz} + \tau_{xy}\gamma_{xy} + \tau_{yz}\gamma_{yz} + \tau_{zx}\gamma_{zx}] \tag{7.6}$$

In the sections that follow, we will obtain expressions for strain energy for axial rods, torsion of circular shafts, and symmetric beams and see how these can be extended to laminated composite structural members and the unsymmetric bending of beams.

7.2.1 Axial Strain Energy

For axial problems, we saw that all stress components except σ_{xx} were zero, $\sigma_{xx} = E\varepsilon_{xx}$, and $\varepsilon_{xx} = du/dx$. Substituting this information into Equation (7.6) and noting that the volume integral dV can be written as an integral over the length L and over the cross-sectional area A, we obtain Equation (7.7a).

$$U_A = \int_V \frac{1}{2}E\varepsilon_{xx}^2 dV = \int_L \left[\int_A \frac{1}{2}E\left(\frac{du}{dx}\right)^2 dA \right] dx \qquad (7.7a)$$

We obtain Equation (7.7b) by noting that du/dx is only a function of x and does not change across the cross section.

$$U_A = \int_L \left[\frac{1}{2}\left(\frac{du}{dx}\right)^2 \int_A E\, dA \right] dx \qquad (7.7b)$$

We assume that the material is homogeneous across the cross section—that is, E does not change in the integral across the cross section—and we obtain Equation (7.7c).

$$\int_A E\, dA = E\int_A dA = EA \qquad (7.7c)$$

For composite axial members (see Problems 7.11 and 7.12), the integral over the area A would be replaced by an integral over each material area, and the value of the integral in Equation (7.7c) would be the total sum of the axial rigidities of all the materials. Substituting Equation (7.7c) into Equation (7.7b), we obtain Equation (7.7d),

$$U_A = \int_L U_a\, dx \qquad (7.7d)$$

where U_a is the axial strain energy per unit length as given in Equation (7.8a) of Table 7.1. Noting that $du/dx = N/EA$, we can obtain the complementary strain energy $\overline{U}_A$ as given in Equation (7.7e),

$$\overline{U}_A = \int_L \overline{U}_a\, dx \qquad (7.7e)$$

where $\overline{U}_a$ is the complementary axial strain energy per unit length as given in Equation (7.8b) of Table 7.1. In Equations (7.8a) and (7.8b), the variables E, A, and N can be functions of x.

TABLE 7.1 Linear Strain Energy per Unit Length

	Strain Energy per Unit Length		Complementary Strain Energy per Unit Length	
Axial	$U_a = \frac{1}{2}EA\left(\frac{du}{dx}\right)^2$	(7.8a)	$\overline{U}_a = \frac{1}{2}\frac{N^2}{EA}$	(7.8b)
Torsion of circular shafts	$U_t = \frac{1}{2}GJ\left(\frac{d\phi}{dx}\right)^2$	(7.9a)	$\overline{U}_t = \frac{1}{2}\frac{T^2}{GJ}$	(7.9b)
Symmetric bending of beams	$U_b = \frac{1}{2}EI_{zz}\left(\frac{d^2v}{dx^2}\right)^2$	(7.10a)	$\overline{U}_b = \frac{1}{2}\frac{M_z^2}{EI_{zz}}$	(7.10b)

7.2.2 Torsional Strain Energy of Circular Shafts

In the torsion of circular shafts, we saw that all stress components except $\tau_{x\theta}$ are zero in polar coordinates. The equivalent form of Equation (7.6) in polar coordinates is Equation (7.11a).

$$U_T = \frac{1}{2}\int_V \tau_{x\theta}\gamma_{x\theta}\,dV \tag{7.11a}$$

Substituting torsional shear stress $\tau_{x\theta} = G\gamma_{x\theta}$ and torsional shear strain $\gamma_{x\theta} = \rho\,d\phi/dx$ and noting that $d\phi/dx$ is just a function of x and does not change across the cross section, we obtain Equation (7.11b).

$$U_T = \int_V \frac{1}{2}G\gamma_{x\theta}^2\,dV = \int_L\left[\int_A \frac{1}{2}G\left(\rho\frac{d\phi}{dx}\right)^2 dA\right]dx = \int_L\left[\frac{1}{2}\left(\frac{d\phi}{dx}\right)^2\int_A G\rho^2\,dA\right]dx \tag{7.11b}$$

We assume that the material is homogeneous across the cross section; that is, G does not change in the integral across the cross section, as shown in Equation (7.11c).

$$\int_A G\rho^2\,dA = G\int_A \rho^2\,dA = GJ \tag{7.11c}$$

For composite circular shafts (see Problems 7.13 and 7.14), the integral over the area A would be replaced by an integral over each material area, and the value of the integral in Equation (7.11c) would be the sum of the total torsional rigidities of all the materials. Substituting Equation (7.11c) into Equation (7.11b), we obtain Equation (7.11d),

$$U_T = \int_L U_t\,dx \tag{7.11d}$$

where U_t is the torsional strain energy per unit length as given in Equation (7.9a) of Table 7.1. Noting that $d\phi/dx = T/GJ$, we can obtain the complementary strain energy $\overline{U}_T$ as

$$\overline{U}_T = \int_L \overline{U}_t\,dx \tag{7.11e}$$

where $\overline{U}_t$ is the complementary torsional strain energy per unit length as given in Equation (7.9b) of Table 7.1. In Equations (7.9a) and (7.9b), the variables G, J, and T can be functions of x.

7.2.3 Bending Strain Energy

When symmetric sections[1] are bent about the z axis, there are two nonzero stress components, σ_{xx} and τ_{xy}. We shall consider strain energy due to each separately. Substituting $\sigma_{xx} = E\varepsilon_{xx}$ and $\varepsilon_{xx} = -y(d^2v/dx^2)$ in Equation (7.6) and noting that (d^2v/dx^2) is just a function of x and does not change across the cross section, we obtain the bending strain energy U_B as shown in Equation (7.12a).

$$U_B = \int_V \frac{1}{2}E\varepsilon_{xx}^2\,dV = \int_L\left[\int_A \frac{1}{2}E\left(y\frac{d^2v}{dx^2}\right)^2 dA\right]dx = \int_L\left[\frac{1}{2}\left(\frac{d^2v}{dx^2}\right)^2\int_A Ey^2\,dA\right]dx \tag{7.12a}$$

---

[1] See Example 7.2 and Problem 7.17 for unsymmetric bending.

Assume that the material is homogeneous across the cross section: that is, E does not change in the integral across the cross section, as shown in Equation (7.12b).

$$\int_A Ey^2 dA = E\int_A y^2 dA = EI_{zz} \tag{7.12b}$$

For composite symmetric beams (see Problems 7.15 and 7.16), the integral over the area A would be replaced by an integral over each material area, and the value of the integral in Equation (7.12b) would be the sum of the bending rigidities of all the materials. Substituting Equation (7.12b) into Equation (7.12a) we obtain Equation (7.12c),

$$U_B = \int_L U_b \, dx \tag{7.12c}$$

where U_b is the bending strain energy per unit length due to just normal stresses and strains as given in Equation (7.10a) of Table 7.1. Noting that $d^2v/dx^2 = M_z/EI_{zz}$, we can obtain the complementary strain energy $\overline{U}_B$ as shown in Equation (7.12d),

$$\overline{U}_B = \int_L \overline{U}_b \, dx \tag{7.12d}$$

where $\overline{U}_b$ is the complementary bending strain energy per unit length due to just normal stresses and strains, as given in Equation (7.10b) of Table 7.1. In Equations (7.12c) and (7.12d), the variables E, A and M_z can be functions of x.

The strain energy due to shear in bending is $U_S = \int_V \frac{1}{2}\tau_{xy}\gamma_{xy}dV$. We note that the maximum shear stress τ_{xy} and shear strain γ_{xy} are an order of magnitude smaller than the maximum normal stress σ_{xx} and the maximum normal strain ε_{xx}. Thus, U_S will be two orders of magnitude smaller than U_B and can be neglected in our comparison.

EXAMPLE 7.2

Obtain an expression for linear strain energy per unit length for a beam in unsymmetric bending in terms of modulus of elasticity E, second area moments of inertia I_{yy}, I_{zz}, and I_{yz}, and curvatures d^2v/dx^2 and d^2w/dx^2. Neglect the shear strain energy in bending.

PLAN

In unsymmetric bending, the dominant stress and strain components are σ_{xx} and ε_{xx}. From Hooke's law, the linear strain energy can be written in terms of ε_{xx}. Substituting ε_{xx} from Equation (6.5) and integrating across the cross section, we obtain the desired expression for linear strain energy density per unit length.

SOLUTION

Equation (6.5) gives an expression of strain shown in Equation (E1).

$$\varepsilon_{xx} = -y\frac{d^2v}{dx^2} - z\frac{d^2w}{dx^2} \qquad\qquad (E1)$$

We neglect the shear strain energy. The dominant stress and strain components in unsymmetric bending are σ_{xx} and ε_{xx}, which are related by Hooke's law as $\sigma_{xx} = E\varepsilon_{xx}$. The strain energy in unsymmetric bending can be written as in Equation (E2).

$$U_{UB} = \int_V \frac{1}{2}E\varepsilon_{xx}^2\,dV = \frac{1}{2}\int_L\left[\int_A E\left\{-y\frac{d^2v}{dx^2} - z\frac{d^2w}{dx^2}\right\}^2 dA\right]dx \qquad \text{or}$$

$$U_{UB} = \frac{1}{2}\int_L\left[\int_A E\left\{y^2\left(\frac{d^2v}{dx^2}\right)^2 + z^2\left(\frac{d^2w}{dx^2}\right)^2 + 2yz\left(\frac{d^2v}{dx^2}\right)\left(\frac{d^2w}{dx^2}\right)\right\}dA\right]dx \qquad (E2)$$

We note that the curvatures do not change across the cross section and are functions only of x, hence can be taken outside the integration across the cross section to obtain Equation (E3).

$$U_{UB} = \frac{1}{2}\int_L\left[\left(\frac{d^2v}{dx^2}\right)^2 \int_A Ey^2 dA + \left(\frac{d^2w}{dx^2}\right)^2 \int_A Ey^2 dA + 2\left(\frac{d^2v}{dx^2}\right)\left(\frac{d^2w}{dx^2}\right)\int_A Eyz\,dA\right]dx \qquad (E3)$$

Assuming that the material is homogeneous and noting the definitions of the second area moments of inertia, we obtain Equation (E4).

$$U_{UB} = \frac{1}{2}\int_L E\left[\left(\frac{d^2v}{dx^2}\right)^2 I_{zz} + \left(\frac{d^2w}{dx^2}\right)^2 I_{yy} + 2\left(\frac{d^2v}{dx^2}\right)\left(\frac{d^2w}{dx^2}\right)I_{yz}\right]dx \qquad (E4)$$

From Equation (E4), the linear strain energy density per unit length for unsymmetric bending can be written as

$$\textbf{ANS.} \quad U_{ub} = \frac{1}{2}E\left[\left(\frac{d^2v}{dx^2}\right)^2 I_{zz} + \left(\frac{d^2w}{dx^2}\right)^2 I_{yy} + 2\left(\frac{d^2v}{dx^2}\right)\left(\frac{d^2w}{dx^2}\right)I_{yz}\right]$$

COMMENTS

1. Though the algebra is slightly more complex than for symmetric beam bending, the basic steps in the derivation of linear strain energy per unit length for unsymmetric bending are the same, namely: substitution of the strain expression in the strain energy; noting that the curvatures do not change across the cross section; assuming material homogeneity across the cross section; and using the definition of area moment of inertia.

2. Expressions for linear strain energy per unit length can be obtained in a similar manner for laminated composite members. See Problems 7.11 through 7.16.

EXAMPLE 7.3

Show that the maximum octahedral shear stress theory of Section 2.3.2. can be obtained from the following statement.

A material will fail when the maximum distortion strain energy density is equal to the distortion strain energy density at yield point in a tension test.

"Distortion strain energy density" here is the linear strain energy density minus the strain energy density associated with the hydrostatic state of stress.

PLAN

We can write the strain energy density in terms of principal stresses and strains, using the generalized version of Hooke's law to replace the principal strains by principal stresses. The strain energy density corresponding to the hydrostatic state of stress can be found and subtracted from the total strain energy density to obtain the distortion strain energy density. The uniaxial state of stress can be substituted into the distortion strain energy density expression, and the failure criterion written and compared to the octahedral shear stress theory.

SOLUTION

The linear strain energy density in terms of principal stresses and strains can be written as Equation (E1).

$$U_0 = \frac{1}{2}(\sigma_1\varepsilon_1 + \sigma_2\varepsilon_2 + \sigma_3\varepsilon_3) \tag{E1}$$

Hooke's law in principal coordinates given by Equations (2.10a) through (2.10c) is written as Equations (E2) through (E4).

$$\varepsilon_1 = [\sigma_1 - \nu(\sigma_2 + \sigma_3)]/E \tag{E2}$$

$$\varepsilon_2 = [\sigma_2 - \nu(\sigma_3 + \sigma_1)]/E \tag{E3}$$

$$\varepsilon_3 = [\sigma_3 - \nu(\sigma_1 + \sigma_2)]/E \tag{E4}$$

Substituting Equations (E2) through (E4) into Equation (E1), we obtain Equation (E5).

$$U_0 = \frac{1}{2E}[\sigma_1\{\sigma_1 - \nu(\sigma_2 + \sigma_3)\} + \sigma_2\{\sigma_2 - \nu(\sigma_3 + \sigma_1)\} + \sigma_3\{\sigma_3 - \nu(\sigma_1 + \sigma_2)\}] \qquad \text{or}$$

$$U_0 = \frac{1}{2E}[\sigma_1^2 + \sigma_2^2 + \sigma_3^2 - 2\nu(\sigma_1\sigma_2 + \sigma_2\sigma_3 + \sigma_3\sigma_1)] \tag{E5}$$

The hydrostatic stress corresponds to $\sigma_{\text{hydro}} = (\sigma_1 + \sigma_2 + \sigma_3)/3$. Figure 7.4 shows that by subtracting and adding σ_{hydro} from each principal stress, we generate a stress cube that is subjected to a uniform stress corresponding to the hydrostatic stress. We can find the strain energy density for this state by substituting σ_{hydro} for all three principal stresses in Equation (E5) to obtain Equation (E6).

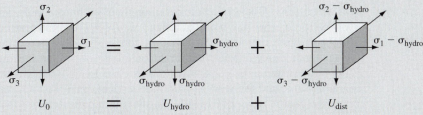

Figure 7.4 Interpretation in the calculation of distortion strain energy density.

$$U_{hydro} = \frac{3(1-2\nu)}{2E}\sigma_{hydro}^2 = \frac{3(1-2\nu)}{2E}\left(\frac{\sigma_1 + \sigma_2 + \sigma_3}{3}\right)^2 \quad \text{or}$$

$$U_{hydro} = \frac{(1-2\nu)}{6E}[\sigma_1^2 + \sigma_2^2 + \sigma_3^2 + 2(\sigma_1\sigma_2 + \sigma_2\sigma_3 + \sigma_3\sigma_1)] \quad \text{(E6)}$$

Subtracting Equation (E6) from Equation (E5), we obtain the distortion strain energy as given in Equation (E7).

$$U_{dist} = U_0 - U_{hydro} = \frac{(1+\nu)}{6E}\left[2\left(\sigma_1^2 + \sigma_2^2 + \sigma_3^2\right) - 2(\sigma_1\sigma_2 + \sigma_2\sigma_3 + \sigma_3\sigma_1)\right] \quad \text{or}$$

$$U_{dist} = \frac{(1+\nu)}{6E}\left[(\sigma_1 - \sigma_2)^2 + (\sigma_2 - \sigma_3)^2 + (\sigma_3 - \sigma_1)^2\right] \quad \text{(E7)}$$

We substitute $\sigma_1 = \sigma_{yield}$ and $\sigma_2 = \sigma_3 = 0$ in Equation (E7) to obtain the distortion strain energy at yield in a tension test, as given in Equation (E8).

$$U_{yield} = \frac{(1+\nu)}{3E}\sigma_{yield}^2 \quad \text{(E8)}$$

The failure criterion is $U_{dist} \leq U_{yield}$, which yields the result given by Equation (E9).

$$\frac{1}{\sqrt{2}}\sqrt{(\sigma_1 - \sigma_2)^2 + (\sigma_2 - \sigma_3)^2 + (\sigma_3 - \sigma_1)^2} \leq \sigma_{yield} \quad \text{(E9)}$$

Equation (E9) is the same as Equation (2.21), the statement for the maximum octahedral shear stress theory.

COMMENTS

1. This example shows that the maximum octahedral shear stress theory carries the intrinsic assumption that the hydrostatic pressure (the normal octahedral stress) does not influence the failure of ductile materials.

2. In plastic flow theories,[2] the hydrostatic pressure is subtracted from the normal stresses to produce a stress deviator matrix as described in Problem 1.39.

---

[2] See Mendelson [1968] in Appendix D.

7.3 WORK

If a force moves through a distance, then work has been done by the force. If the point at which force F is applied moves through an infinitesimal distance du in the direction of the force,[3] then the work is defined as $dW = F\,du$. Integrating the work expression along the path of movement gives the total work done by the force. If the total work done by the force depends not on the path but only on the starting and end points, then the force is said to be conservative for the following reason: if the start and end points are the same and the work is path independent, then the total work done must be zero and the energy or work expended in moving a point forward is recovered (conserved) when the point moves back to the starting point. Clearly, if friction is present, the work done to overcome friction must vary with the length of path; hence the frictional force is nonconservative. In a similar manner, if a body deforms plastically, the work done to create permanent deformation will not be recovered, and once more we have a non-conservative system. Thus, elastic deformation is a conservative deformation, while plastic deformation is nonconservative. Elastic deformation can, however, be linear or nonlinear. Work done in stretching a rubber that has a nonlinear stress–strain curve is recovered when the forces are released and the rubber returns to the undeformed position. Thus, there is a distinction between nonconservative and nonlinear systems. We record the following definition and observations:

Definition 3 Work done by a force is conservative if it is path independent.

- Nonlinear systems and nonconservative systems are two independent system descriptions.

We shall now consider work from external forces that cause infinitesimal deformation designated by the prefix (δ). Figure 7.5 shows work done in axial, torsion, and symmetric bending. The work could be for conservative (elastic) or nonconservative (plastic), linear or nonlinear material. Figure 7.5 shows that work is done by forces as well as by moments and that the deformation used in the work may be due to displacement or rotation. We introduce the following definition for generalized displacement.

Definition 4 Any variable that can be used for describing deformation is called a *generalized displacement.*

Thus, as shown in Figure 7.5, the generalized displacement can be the axial displacement $u(x)$, the torsional rotation $\phi(x)$, the bending displacement $v(x)$, or the bending rotation $\theta(x) = (dv/dx)$. Generalized displacement does not have to be a physical quantity. It can be a variable used in a mathematical approximation of displacement or rotation. For example, the deflection of a beam $v(x)$ could be written as in Equation (7.14),

$$v(x) = C_1 f_1(x) + C_2 f_2(x) + \cdots + \cdots + C_i f_i(x) + \cdots + C_n f_n(x) \tag{7.14}$$

---

[3] The more general form of work is the dot product between the force vector ($\overline{\mathbf{F}}$) and the displacement vector ($\overline{\mathbf{u}}$), that is, $dW = \overline{F} \cdot d\overline{u}$.

Loading mode

Work

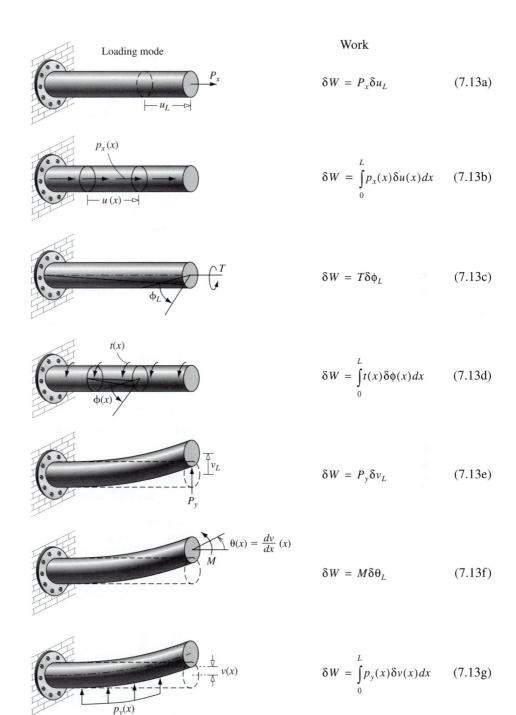

$$\delta W = P_x \delta u_L \qquad (7.13\text{a})$$

$$\delta W = \int_0^L p_x(x)\delta u(x)dx \qquad (7.13\text{b})$$

$$\delta W = T\delta\phi_L \qquad (7.13\text{c})$$

$$\delta W = \int_0^L t(x)\delta\phi(x)dx \qquad (7.13\text{d})$$

$$\delta W = P_y \delta v_L \qquad (7.13\text{e})$$

$$\delta W = M\delta\theta_L \qquad (7.13\text{f})$$

$$\delta W = \int_0^L p_y(x)\delta v(x)dx \qquad (7.13\text{g})$$

Figure 7.5 Different modes of work.

where the functions $f_i(x)$ are somehow known or approximated. If the constants C_i change, then the displacement $v(x)$ changes. Hence by Definition 4, the constants C_i in the series are the generalized displacements. We can now define generalized force, as follows.

Definition 5 Any variable that can be used for describing the cause that produces deformation is called a *generalized force*.

Thus as shown in Figure 7.5, the generalized force can be the concentrated axial force P_x, the distributed axial force $p_x(x)$, the concentrated torque T, the distributed torque $t(x)$, the concentrated bending force P_y, the concentrated bending moment M, or the distributed bending force $p_y(x)$. Generalized force does not have to be a physical quantity. It may be a variable used in a mathematical representation of a force or moment. For example, we may construct an integral of a distributed force $p(x)$ as

$$F_i = \int_0^L f_i(x)p(x)dx \tag{7.15}$$

where the functions f_i are somehow known or approximated. If $f_1 = 1$, then F_1 would represent the total force from the distributed load. If $f_2 = x$, then F_2 would represent the total moment of the distributed force at the origin. The functions f_i are called the weighting functions, and F_i is called the generalized force.

7.4 VIRTUAL WORK

Virtual work methods are applicable to linear and nonlinear systems, to conservative as well as nonconservative systems. The principle of virtual work is deceptively simple. It states:

The total virtual work done on a body at equilibrium is zero.

But what is virtual work? How can we calculate it? To answer these questions, we need to develop the concepts of virtual displacement and virtual force, which in turn require the development of the concepts of kinematically and statically admissible functions. We must then generalize the concepts of kinematically and statically admissible functions to accommodate the boundary conditions that appear in axial, torsion, and bending problems. Before embarking upon these various concepts, we write the statement of virtual work in symbolic form:

$$\delta W = 0 \tag{7.16}$$

The prefix δ will be used to designate a virtual quantity, as shown for work in Equation (7.16). In a similar manner, δu refers to virtual displacement, δF refers to virtual force, and so on.

The total virtual work can be divided into work done by external and internal virtual forces. Since the internal forces are always opposed to the external forces, the internal virtual work will always be opposite in sign to the external virtual work. We rewrite Equation (7.16) as

$$\delta W_{ext} = \delta W_{int} \tag{7.17}$$

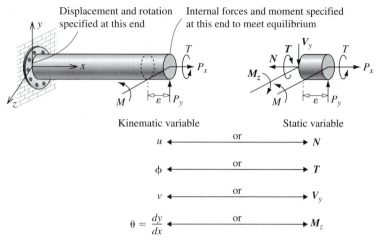

Displacement and rotation specified at this end

Internal forces and moment specified at this end to meet equilibrium

Kinematic variable   Static variable

u ← or → N

ϕ ← or → T

v ← or → V_y

$\theta = \dfrac{dy}{dx}$ ← or → M_z

Figure 7.6 Kinematic and static boundary conditions.

7.4.1 Types of Boundary Conditions

Conditions specified on displacements and rotations are restrictions on the geometric (kinematic) shapes a body can deform into. These boundary conditions are called geometric boundary conditions or *kinematic boundary conditions* and, more recently, essential boundary conditions. The variables describing displacement and rotation are called the *kinematic* or geometric variables, or primary variables.[4]

The conditions specified on the internal forces and moments at the end of a structural member are derived from the static equilibrium of a small differential element (ε) as shown in Figure 7.6. These boundary conditions are called *statical boundary conditions* or, more recently, natural boundary conditions. The internal forces and moments are called *static variables* or, more recently, secondary variables.[4]

It is the nature of boundary conditions that at a boundary point, one must specify either a condition on the kinematic variable or the associated static variable, as shown in Figure 7.6. We used this idea in statics to determine the reaction forces and moment at a support. If a point cannot move (kinematic variable specified), then a reaction force is needed to hold it in place. Since the reaction force is unknown, we cannot specify any condition on the static variable. Similarly if a point is free to move, then the kinematic variable cannot be specified; but the equilibrium of a small element at the end will give us the condition on the static variable. Note that the internal forces and moments are in opposite directions on the two surfaces of an imaginary cut and are positive or negative in accordance with our sign convention (see Table 3.3). Thus, to get the *correct sign* for

---

[4] Primary and secondary variable usage entails a greater generalization than is implied for kinematic and statical variables. For example, in heat conduction problems, temperature is the primary variable and heat is the secondary variable. In fluid mechanics, stream functions or potential functions are the primary variables and velocity is the secondary variable. Essential and natural boundary conditions are boundary conditions on the primary and secondary variables, respectively. See Reddy [1993] in Appendix D for additional details.

static (natural) boundary conditions, it is critical to *draw the free body diagram* of a small element at the boundary point.

7.4.2 Kinematically Admissible Functions

> **Definition 6** Functions that are continuous and satisfy all the kinematic boundary conditions are called *kinematically admissible functions*.

The actual displacement solution is always a kinematically admissible function. Definition 6 is less restrictive, however, because kinematically admissible functions can result in forces and moments that do not satisfy the equilibrium or static boundary conditions. To determine the kinematically admissible functions, we first list all the kinematic boundary conditions. Next we decide upon a class of functions that satisfy the kinematic boundary conditions, as demonstrated shortly in Example 7.4. Polynomials and trigonometric functions are often used for kinematically admissible functions because these two classes of functions satisfy the continuity requirement in Definition 6. But functions usually are chosen by means of educated guesses that are checked to see whether the functions satisfy the kinematic boundary conditions.

7.4.3 Statically Admissible Functions

> **Definition 7** Functions that satisfy all the static boundary conditions, satisfy the equilibrium equations at all points, and are continuous at all points except where a concentrated force or moment is applied are called *statically admissible functions*.

The actual solutions for internal forces and moments are always statically admissible functions. Definition 7 is significantly less restrictive, however, because the statically admissible functions can result in displacements that do not satisfy the kinematic boundary conditions. The statically admissible functions for internal forces and moments can be written by using the discontinuity functions or by drawing a free body diagram after making an imaginary cut at an arbitrary location x and writing the equilibrium equations. In the case of a statically determinate structure, this gives us the actual internal forces or moments. But in the case of a statically indeterminate structure, we have unknown reactions in the expressions of internal forces and moments. A statically indeterminate structure has more supports than are needed to support the external loads; that is, a statically indeterminate structure has more unknown reactions than are needed to maintain equilibrium. These extra unknowns indicate the degree of static redundancy, as defined in Equation (7.18).

$$\begin{aligned} \text{degree of static redundancy} \ = \ &\text{number of unknown reactions} \\ &- \text{number of equilibrium equations} \end{aligned} \tag{7.18}$$

For any value assigned to the extra unknown reactions, the equilibrium equation is still valid; that is, the conditions of static admissibility are still met. It is this flexibility of choosing any value for the extra unknowns that differentiates the statically admissible internal forces and moments from the actual internal forces and moments. We record this observation as follows:

- In determining statically admissible internal forces and moments, the number of reactions that can be assigned arbitrary values is equal to the degree of static redundancy.

EXAMPLE 7.4

For the beams shown in Figure 7.7, determine a class of kinematically admissible functions.

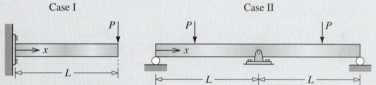

Case I   Case II

Figure 7.7 Beams for Example 7.4.

PLAN

The boundary conditions are deflection and slope. In case I they are zero at the built-in end, while in case II the deflections at $x = 0$, $x = L$ and $x = 2L$ are zero. We can examine polynomials in x that satisfy these boundary conditions.

SOLUTION

Case I: The displacement boundary conditions for this bending problem are

$$v(0) = 0 \tag{E1}$$

$$\frac{dv(0)}{dx} = 0 \tag{E2}$$

We are looking for functions that go to zero at $x = 0$, and x to any exponent would satisfy this requirement. However, a constant or linear function of x will not correspond to any beam deformation. Hence x^2 and greater exponent powers of x will satisfy all requirements. Thus the answer is a class of kinematically admissible functions in which C_i are arbitrary and undetermined constants:

$$\textbf{ANS.} \quad v(x) = C_i x^{i+1} \qquad i \geq 1$$

Case II: The displacement boundary conditions for this bending problem are

$$v(0) = 0 \tag{E3}$$

$$v(L) = 0 \tag{E4}$$

$$v(2L) = 0 \tag{E5}$$

The function

- x^i satisfies the boundary condition given by Equation (E3),
- $(x-L)^j$ satisfies the boundary condition given by Equation (E4),
- $(x-2L)^k$ satisfies the boundary condition given by Equation (E5).

The product of these functions will satisfy all three boundary conditions. Equation (E6) shows a class of kinematically admissible functions in which C_{ijk} are arbitrary and undetermined constants:

$$\text{ANS.} \quad v(x) = C_{ijk}x^i(x-L)^j(x-2L)^k \quad \text{(E6)}$$

COMMENTS

1. Another alternative is to use trigonometric functions. For case I we can use $v(x) = C_i[1 - \cos i\pi(x/L)]$, since it satisfies the two boundary conditions given by Equations (E1) and (E2). For case II we can use $v(x) = C_i\sin[2i\pi(x/L)]$, since it satisfies all three boundary conditions given by Equations (E3), (E4), and (E5).

2. Later on we shall use kinematically admissible functions for virtual work methods and approximate methods. The foregoing solutions only ensure that we have chosen kinematically admissible functions correctly; they do not imply that we have made good choices. The adjective "good" requires additional criteria,[5] which the kinematically admissible functions must satisfy.

---

[5] These additional criteria, which are related to the convergence of a series, are as follows: the set of functions should be independent and complete; a set of functions that converge rapidly are a better choice than functions that converge slowly.

EXAMPLE 7.5

An axial rod has a uniform distributed force of p_0 applied as shown. Determine a statically admissible internal axial force.

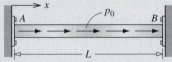

Figure 7.8 Axial rod for Example 7.5.

PLAN

We can make an imaginary cut at an arbitrary location x, and draw the free body diagram. By balancing forces in the x direction, we can obtain a statically admissible internal axial force in terms of the wall reaction. We can assign any value to the reaction because the problem has one degree of static redundancy.

SOLUTION

By equilibrium of force in the x direction on the free body diagram shown in Figure 7.9, we obtain the internal axial force as shown in Equation (E1).

$$N = R_A - p_0 x \qquad \text{(E1)}$$

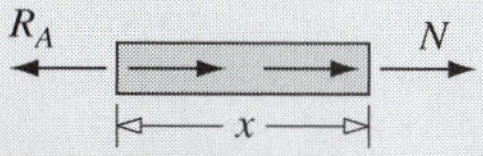

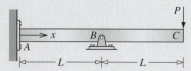

Figure 7.9 Free body diagram in Example 7.5.

We can choose any arbitrary value for the reaction R_A in Equation (E1), including zero. We obtain the statically admissible axial force given by Equation (E2).

$$\textbf{ANS.} \quad N = -p_0 x \qquad \text{(E2)}$$

COMMENTS

1. The N in Equation (E2) is not the actual internal axial force, since the displacement solution obtained by using this value will not satisfy the zero displacement conditions at both ends.

2. We could have chosen $R_A = p_0(L/2)$ by invoking symmetry. In such a case we would have had the exact solution for N.

EXAMPLE 7.6

Determine a statically admissible bending moment for the beam shown in Figure 7.10.

Figure 7.10 Beam for Example 7.6.

PLAN

The problem has one degree of static redundancy, and thus we can assign any value to one of the three reactions. The reaction force B causes the definition of "bending moment" to be represented by two functions, a function for section AB and another for section BC. By setting the reaction force at B to be zero, we can find a statically admissible bending moment that is represented by only one function for the entire beam.

SOLUTION

We make an imaginary cut at some arbitrary location x and take the right-hand part to draw the free body diagram shown in Figure 7.11, to avoid calculating the wall reactions.

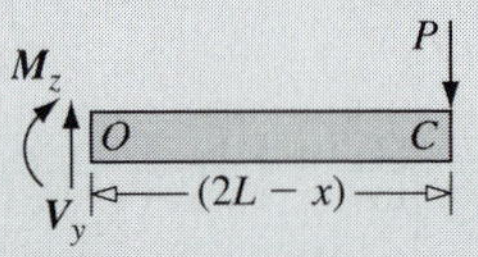

Figure 7.11 Free body diagram for Example 7.6.

By the moment equilibrium about point O in Figure 7.11, we can determine the statically admissible internal moment as shown in Equation (E1).

$$M_z = -P(2L - x) \qquad 0 \le x \le 2L \qquad \text{(E1)}$$

COMMENTS

1. The moment expression in Equation (E1) is not the actual moment, for the elastic curve obtained from the moment can satisfy at most two of the three boundary conditions.

2. Usually, but not always, a zero value for the statically redundant reactions will simplify calculations. Each problem must be examined, and statically redundant reaction values chosen with a view to reducing the algebra.

7.4.4 Virtual Displacement Method

The "virtual" part of the method of virtual displacements implies that the displacement is not real but an imaginary displacement imposed on a body.

Definition 8 Virtual displacement is an infinitesimal, imaginary, kinematically admissible displacement field imposed on a body.

The word "infinitesimal" in Definition 8 *implies that neither the internal nor the external forces or moments change* during the imposition of the virtual displacement. Figure 7.12(a) shows a simply supported beam under uniform distributed load. The kinematic condition of zero deflection at each support can be met by an infinite number of kinematically admissible functions, one of which is shown in Figure 7.12(b). The difference

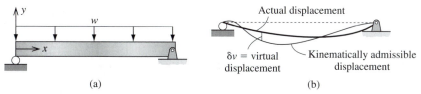

Figure 7.12 Virtual displacement.

between the actual displacement function and the kinematically admissible function is the virtual displacement. Unlike the kinematically admissible displacement functions, the actual displacement function meets not only the kinematic requirements but also the equilibrium conditions. Thus, we can look at the virtual displacement as a mathematical experimentation to find the "one function" that would meet the equilibrium conditions. Virtual work provides the mechanism by which we select the one out of the infinite functions that would meet the equilibrium conditions. We record the following definitions:

> **Definition 9** Virtual work is the work done by forces in moving through a virtual displacement.

- Of all the kinematically admissible functions, the one that satisfies the equilibrium condition is the one for which the virtual work is zero.

- In virtual displacement methods, zero virtual work implies that all equilibrium equations are met.

When the virtual work condition of Equation (7.17) is identically met, we obtain the actual displacement such as in the dummy unit load method described in Section 7.5. When the virtual work condition is approximately met, we obtain an approximate solution to the displacement function such as in the Rayleigh-Ritz method to be described in Section 7.9.

7.4.5 Virtual Force Method

The word "virtual" implies that the forces are imaginary forces that are applied to the body.

> **Definition 10** The virtual force is an infinitesimal imaginary statically admissible force field imposed on a body.

The word "infinitesimal" in Definition 10 *implies that the body does not go through additional deformation* due to the imposition of a virtual force or virtual moment. To find the statically admissible force field, we apply an imaginary small force to the structure and find the work done in moving through the actual displacement or deformations, as shown shortly in the Example 7.8. We record the following.

- Of all the static admissible forces, the one that satisfies the compatibility equations is the one for which the virtual work is zero.

- In virtual force methods, "zero virtual work" implies that all compatibility equations are met.

EXAMPLE 7.7

The roller at P shown in Figure 7.13 slides in the slot because of the force $F = 100$ kN. Member AP has a cross-sectional area $A = 100$ mm^2 and a modulus of elasticity $E = 200$ GPa. Determine the axial stress in the member AP by the virtual displacement method.

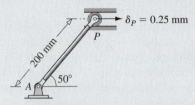

Figure 7.13 Axial member for Example 7.7.

PLAN

We can let the roller have a virtual displacement of δu_P. By drawing the approximate deformed shape, we can obtain the compatibility equation relating δu_P and δu. We can compute the virtual work done by the external force F and the internal force in AP and equate the virtual work in both cases in accordance with Equation (7.17). From the internal force, we can find the axial stress.

SOLUTION

We draw the approximate deformed shape as shown in Figure 7.14 and find a kinematically admissible displacement of the bar due to the virtual displacement of the roller, as shown in Equation (E1).

$$\delta u = \delta u_P \cos 50 \tag{E1}$$

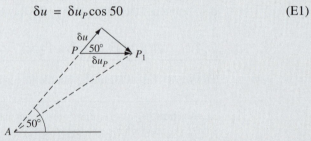

Figure 7.14 Deformed shape for Example 7.7.

We can find the internal and external virtual work by multiplying the internal and external forces by the respective displacements, as shown in Equations (E2).

$$\delta W_{\text{int}} = N \delta u \quad \text{and} \quad \delta W_{\text{ext}} = F \delta u_P \tag{E2}$$

Substituting Equations (E1) and (E2) into the statement of virtual work given by Equation (7.17), we obtain Equation (E3).

$$F \delta u_P = N \delta u_P \cos 50 \quad \text{or} \quad F = N \cos 50 \tag{E3}$$

Noting that $F = 10$ kN and $A = 100(10^{-6})$ m^2, we obtain the axial stress as follows:

$$N = \frac{10(10^3)}{\cos 50} = 15.56(10^3) \qquad \sigma_{xx} = \frac{N}{A} = \frac{15.56(10^3)}{100(10^{-6})}$$

ANS. $\sigma_{xx} = 156$ MPa (T)

COMMENTS

1. Equation (E3) is the equilibrium equation, as can be seen by equilibrium of forces in the horizontal direction in the free body diagram of the roller shown in Figure 7.15. But Equation (E3) was obtained by means of the virtual work principle, not by any force equilibrium. This once more emphasizes that the virtual displacement method enforces equilibrium, or, that the principle of virtual work is an alternative statement of equilibrium.

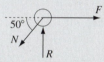

Figure 7.15 Free body diagram in Example 7.7.

2. Equation (E1) is a method of finding a kinematically admissible deformation. In kinematic analysis we are implicitly imposing conditions of continuity.

EXAMPLE 7.8

The action of force F in Example 7.7 causes the roller to move by 0.15 mm. Use the virtual force method to determine the deformation of bar AP.

PLAN

We apply a virtual force δF on the roller. From the free body diagram of the roller (Figure 7.15), we can find the virtual internal axial force δN. We compute the external virtual work and the internal virtual work and equate the two to obtain a relationship between the displacement of the roller and the deformation of the bar.

SOLUTION

We draw the free body diagram of the roller with virtual forces as shown in Figure 7.16. By equilibrium, we obtain a statically admissible internal axial force as shown in Equation (E1).

$$\delta F = \delta N \cos 50 \tag{E1}$$

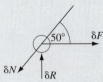

Figure 7.16 Equilibrium of virtual forces on the roller in Example 7.8.

We can find the virtual work due to the virtual forces as shown in Equations (E2).

$$\delta W_{\text{ext}} = \delta F \, \Delta u_P \quad \text{and} \quad \delta W_{\text{int}} = \delta N \, \Delta u \qquad \text{(E2)}$$

Substituting Equations (E1) and (E2) into the principle of virtual work ($\delta W_{\text{ext}} = \delta W_{\text{int}}$), we obtain Equation (E3).

$$\delta N \cos 50 (\Delta u_P) = (\delta N) \Delta u \quad \text{or} \quad \Delta u = \Delta u_P \cos 50 \qquad \text{(E3)}$$

Substituting for $\Delta u_P = 0.15$ mm into Equation (E3), we obtain the deformation of the bar as shown in Equation (E4).

$$\Delta u = 0.15 \cos 50 \qquad \textbf{ANS.} \quad \Delta u = 0.0964 \text{ mm} \qquad \text{(E4)}$$

COMMENT

Equation (E3) is the compatibility equation, but we obtained it from virtual work, not from the geometry of the deformed shape as in Example 7.7. This once more emphasizes that the virtual force method enforces the compatibility equations.

PROBLEM SET 7.1

7.1 Determine a class of kinematically admissible displacement functions for the beam shown in Figure P7.1.

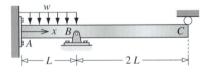

Figure P7.1

7.2 For the beam and loading shown in Figure P7.1, determine a statically admissible bending moment.

7.3 Determine a class of kinematically admissible displacement functions for the beam shown in Figure P7.3.

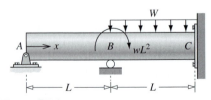

Figure P7.3

7.4 For the beam and loading shown in Figure P7.3, determine a statically admissible bending moment.

7.5 The roller at P shown in Figure P7.5 slides in the slot as a result of the force $F = 20$ kN. Member AP has a cross-sectional area $A = 100$ mm^2 and a modulus of elasticity $E = 200$ GPa. Determine the axial stress in the member AP by the virtual displacement method.

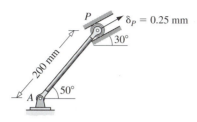

Figure P7.5

ANS. $\sigma = 212.8$ MPa (T)

7.6 The action of force F in Figure P7.5 causes the roller to move by 0.25 mm. Use the virtual force method to determine the axial strain in bar AP.

ANS. $\varepsilon = 1174.6$ μmm/mm

7.7 The roller at P shown in Figure P7.7 slides in the slot as a result of the force $F = 20$ kN. Both bars have a cross-sectional area $A = 100$ mm^2 and a modulus of elasticity $E = 200$ GPa. Bars AP and BP have lengths of $L_{AP} = 200$ mm and $L_{BP} = 250$ mm, respectively. Determine the axial stress in the member AP by the virtual displacement method.

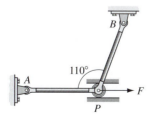

Figure P7.7

7.8 A force $F = 20$ kN is applied to pin P shown in Figure P7.8. Both bars have a

cross-sectional area $A = 100$ mm^2 and a modulus of elasticity $E = 200$ GPa. Bars AP and BP have lengths of $L_{AP} = 200$ mm and $L_{BP} = 250$ mm, respectively. Use the virtual force method to determine the movement of the pin in the direction of the force F.

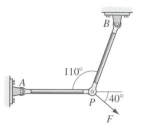

Figure P7.8

7.9 The roller at P shown in Figure P7.9 slides in the slot as a result of the force $F = 20$ kN. Both bars have a cross-sectional area $A = 100$ mm^2 and a modulus of elasticity $E = 200$ GPa. Bars AP and BP have lengths of $L_{AP} = 200$ mm and $L_{BP} = 250$ mm, respectively. Use the virtual displacement method to determine the axial stress in the member AP.

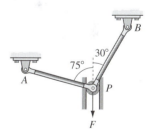

Figure P7.9

ANS. $\sigma_{AP} = 77.7$ MPa (T)

7.10 A force $F = 20$ kN is applied to pin P shown in Figure P7.10. Both bars have a cross-sectional area $A = 100$ mm^2 and a modulus of elasticity $E = 200$ GPa. Bar, AP and BP have lengths of $L_{AP} = 200$ mm and $L_{BP} = 250$ mm, respectively. Use the virtual force method to determine the movement of the pin in the direction of the force F.

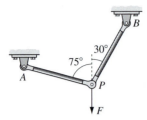

Figure P7.10

ANS. $\Delta u_p = 0.305$ mm

7.11 For a laminated axial member having n materials, obtain an expression for the linear strain energy per unit length in terms of moduli of elasticity $E_1, E_2, \ldots, E_n$, cross-sectional areas $A_1, A_2, \ldots, A_n$, and rate of axial deformation du/dx.

7.12 For a laminated axial member having n materials, obtain an expression for the linear complementary strain energy per unit length in terms of moduli of elasticity $E_1, E_2, \ldots, E_n$, cross-sectional areas $A_1, A_2, \ldots, A_n$, and internal axial force N.

7.13 For a laminated circular shaft having n materials, obtain an expression for the linear strain energy per unit length in terms of shear moduli of elasticity $G_1, G_2, \ldots, G_n$, polar moments of inertia of cross-sectional areas $J_1, J_2, \ldots, J_n$, and rate of twist $d\phi/dx$.

7.14 For a laminated circular shaft having n materials, obtain an expression for the complementary linear strain energy per unit length in terms of shear moduli of elasticity $G_1, G_2, \ldots, G_n$, polar moments of inertia on cross-sectional areas $J_1, J_2, \ldots, J_n$, and internal torque T.

7.15 For a laminated symmetric beam having n materials bending about the z axis, obtain an expression for the linear strain energy per unit length in terms of moduli of elasticity $E_1, E_2, \ldots, E_n$, second area moments of inertia $(I_{zz})_1, (I_{zz})_2, \ldots, (I_{zz})_n$, and curvature d^2v/dx^2. Neglect the shear strain energy in bending.

7.16 For a laminated symmetric beam having n materials bending about the z axis, obtain an expression for the linear strain energy per unit length in terms of moduli of elasticity $E_1, E_2, \ldots, E_n$, second area moments of inertia $(I_{zz})_1, (I_{zz})_2, \ldots, (I_{zz})_n$, and internal bending moment M_z. Neglect the shear strain energy in bending.

7.17 Obtain an expression for the complementary linear strain energy per unit length for a beam in unsymmetric bending in terms of modulus of elasticity E, second area moments of inertia I_{yy}, I_{zz}, and I_{yz}, and internal moments M_y and M_z. Neglect the shear strain energy in bending.

7.5   DUMMY UNIT LOAD METHOD

In Example 7.8, we saw that a virtual force (δF) imposed on a roller dropped out of the calculation. Hence, had we started with $\delta F = 1$, we would have conducted exactly the same calculations without writing δF and δN in each equation. This is the basic idea that is formalized in the dummy unit load method.

In the dummy unit load method, we impose a force or moment of unit magnitude at the point at which we want to find the displacement or the slope. We next compute statically admissible force and moment fields, and then compute the internal and external virtual work. By invoking the virtual work principle, we obtain a formula relating the displacement (or slope) to the virtual forces (or moments).

7.5.1 Axial Members

We call the actual axial member rod 1. We represent the actual internal axial force by $N_1(x)$ and the axial displacement by $u_1(x)$. Our next step is to draw another rod with exactly the same supports as rod 1 but with only a unit load placed at point x_p, at which we want to calculate the displacement. For rod 2, find a statically admissible axial force. Note that to find a statically admissible axial force, we can assign arbitrary values to as many reactions as there are degrees of static redundancy. Let $N_2(x)$ be the statically admissible axial force and $u_2(x)$ be the kinematically admissible displacement for rod 2. It should be noted that the displacement u_2 and axial force N_2 are not related because they are independently chosen to satisfy the kinematic and static admissibility requirements, respectively.

The internal virtual work for rod 2 is the internal axial force multiplied by the displacement at a point that is integrated over the entire rod, as shown in Equation (7.19a).

$$\delta W_{int} = \int_0^L N_2(x)du_2 = \int_0^L N_2(x)\frac{du_2}{dx}dx \tag{7.19a}$$

The supports for rods 2 and 1 are the same; hence the kinematically admissible displacement field u_2 is also the kinematically admissible displacement field for rod 1. The actual displacement u_1 is always kinematically admissible, hence can be used in place of u_2. Thus the internal virtual work can be written as shown in Equation (7.19b).

$$\delta W_{int} = \int_0^L N_2(x)\frac{du_1}{dx}dx \tag{7.19b}$$

We now note that the actual displacement u_1 is related to the actual internal force N_1 through the relation $du_1/dx = N_1/EA$. We obtain Equation (7.19c).

$$\delta W_{int} = \int_0^L \frac{N_2(x)N_1(x)}{EA}dx \tag{7.19c}$$

For finding displacement at x_p, we apply a virtual force of value $\delta F = 1$ at x_p. The corresponding external virtual work for rod 2 is given by Equation (7.19d).

$$\delta W_{ext} = (\delta F = 1)u_2(x_P) = (\delta F = 1)u_1(x_P) \tag{7.19d}$$

Equating the external virtual work to the internal virtual work, we obtain Equation (7.20).

$$\boxed{(\delta F = 1)u_1(x_P) = \int_0^L \frac{N_1(x)N_2(x)}{EA}dx} \tag{7.20}$$

The axial displacement obtained from Equation (7.20) is considered to be positive in the direction of the applied unit force on rod 2. This is because the work done by the unit force can be positive only if the force and the displacement are in the same direction.

EXAMPLE 7.9

The column shown in Figure 7.17 has a modulus of elasticity of $E = 4000$ ksi and a cross-sectional area of $A = 100$ in^2. Use the dummy unit load method to determine (a) the reaction force at A and (b) the movement of the rigid plate at C.

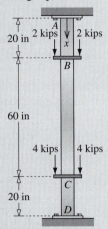

Figure 7.17 Column for Example 7.9.

PLAN

The support at A can be replaced with a reaction force, and the discontinuity function can be used to write the axial force N_1 in terms of the reaction force. (a) First we find N_2 by applying a unit force at A and no other force. We can use Equation (7.20) to evaluate the displacement at A and equate it to zero to determine the reaction at A. (b) We then rewrite N_2 by applying a unit force at C with no other force. The displacement at C can be found from Equation (7.20).

SOLUTION

Figure 7.18(a) shows rod 1 with the support at A replaced by a reaction force. By using the discontinuity functions, the internal axial force can be written as Equation (E1).

$$N_1 = [R_A - 4\langle x - 20 \rangle^0 - 8 \langle x - 80 \rangle^0]. \text{ kips} \qquad \text{(E1)}$$

(a) Figure 7.18(b) shows rod 2 with a unit force applied at A. The axial force N_2 can be written as Equation (E2).

$$N_2 = 1 \qquad \text{(E2)}$$

Substituting Equations (E1) and (E2) into Equation (7.20), we obtain the displacement at A, which we equate to zero as shown in Equation (E3).

$$u_A = \int_0^{100} \frac{[1][R_A - 4\langle x - 20 \rangle^0 - 8 \langle x - 80 \rangle^0]}{EA} dx = 0 \qquad \text{or} \qquad \text{(E3)}$$

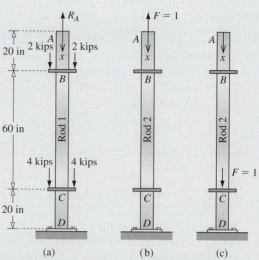

Figure 7.18 Calculations for Example 7.9: (a) N_1 calculations, (b) N_2 calculations for reaction, and (c) N_2 calculations for displacement at C.

$$\frac{[R_A x - 4\langle x - 20\rangle^1 - 8\langle x - 80\rangle^1]}{EA}\bigg|_0^{100} = \frac{1}{EA}[R_A(100) - 4(80) - 8(20)] = 0 \quad \text{or}$$

ANS. $R_A = 4.8$ kips

(b) Figure 7.18(c) shows rod 2 with a unit force applied at C. The axial force N_2 can be written as shown in Equation (E4).

$$N_2 = -(1)\langle x - 80\rangle^0 \tag{E4}$$

Substituting Equations (E1) and (E4) into Equation (7.20), we obtain the displacement at C as shown in Equation (E5).

$$u_C = \frac{1}{EA}\int_0^{100} [-\langle x - 80\rangle^0][R_A - 4\langle x - 20\rangle^0 - 8\langle x - 80\rangle^0]dx \tag{E5}$$

Noting that $\langle x - 80\rangle^0 = 0$ for $x < 80$, the integration limit can be changed from 80 to 100. For $x > 80$, $\langle x - 80\rangle^0 = 1$, and we obtain Equation (E6).

$$u_C = \frac{1}{EA}\int_{80}^{100} [-1][R_A - 4(1) - 8(1)]dx =$$

$$-\left(\frac{1}{EA}\right)[4.8 - 4(1) - 8(1)]x\bigg|_{80}^{100} = \frac{(7.2)(20)}{(4000)(100)} \tag{E6}$$

ANS. $u_C = 0.00036$ inch downward

COMMENTS

1. The degree of static redundancy in this problem is one. Thus in calculating a statically admissible axial force N_2, we can set any value to one reaction force in rod 2. We set the reaction at support A in rod 2 in Figure 7.18(b) and (c) to zero, thus simplifying our calculations.

2. The use of the discontinuity functions simplified the writing of N_1 and N_2. It also simplified the evaluation of the integrals in Equations (E3) and (E5).

7.5.2 Torsion of Circular Shafts

The process for obtaining an expression for the shaft rotation at a section is similar to that for obtaining the axial displacement in a rod. The actual shaft is labeled shaft 1. We represent the actual internal torque by $T_1(x)$ and the rotation by $\phi_1(x)$. We draw another shaft with exactly the same supports as shaft 1 but with *only* unit torque ($\delta T = 1$) placed at section at x_p at which we want to calculate the rotation. For shaft 2, we find a statically admissible torque. Note that in finding a statically admissible torque in shaft 2, we can assign arbitrary values to as many reaction torques as there are degrees of static redundancy, which for a shaft on a single axis is at most one. Let $T_2(x)$ be the statically admissible torque and $\phi_2(x)$ be the kinematically admissible rotation for shaft 2. It should be noted that the rotation ϕ_2 and torque T_2 are independently chosen to satisfy the kinematic and static admissibility requirements and hence are not related. We calculate the internal and external virtual work, recognize that $\phi_1(x)$ is kinematically admissible, recognize the relationship $d\phi_1/dx = T_1/GJ$, and equate the internal and external virtual work to obtain Equation (7.21).

$$(\delta T = 1)\phi_1(x_P) = \int_0^L \frac{T_2(x)T_1(x)}{GJ}dx \qquad (7.21)$$

The rotation obtained from Equation (7.21) is considered to be positive in the direction of the applied unit torque on shaft 2. This is because the work done by the unit torque can be positive only if the torque and the rotation are in the same direction.

7.5.3 Symmetric Bending of Beams

The process for finding the deflection and slope in beam-bending problems is similar to that for finding the axial displacement and torsional rotation. However there are some differences that need to be accounted for in the derivation.

Let $M_1(x)$ and $v_1(x)$ be the actual internal moment and displacement in a beam. Let the actual beam be called beam 1. Draw another beam with exactly the same supports as beam 1 but with only unit load ($\delta F = 1$), placing it at point x_p, at which we want to calculate the displacement or slope. For beam 2, find a statically admissible bending moment. Note that in finding a statically admissible bending moment for beam 2, we can assign arbitrary values to as many reactions as there are degrees of static redundancy. Let $M_2(x)$ be the statically admissible bending moment and $v_2(x)$ be the kinematically admissible displacement for beam 2. It should be noted that the displacement v_2 and the moment M_2 are independently chosen to satisfy the kinematic and static admissibility requirements and hence are not related.

The internal virtual work for beam 2 is the internal moment multiplied by the change in slope at a point that is integrated over the entire beam, as shown in Equation (7.22a).

$$\delta W_{\text{int}} = \int_0^L M_2(x)d\theta_2 = \int_0^L M_2(x)\frac{d\theta_2}{dx}\,dx = \int_0^L M_2(x)\frac{d}{dx}\left(\frac{dv_2}{dx}\right)dx$$

$$= \int_0^L M_2(x)\frac{d^2v_2}{dx^2}dx \tag{7.22a}$$

The supports for beam 2 and beam 1 are the same. Hence the kinematically admissible displacement field v_2 is also the kinematically admissible displacement field for beam 1. The actual displacement v_1 is always kinematically admissible, hence can be used in place of v_2. Thus the internal virtual work can be written as Equation (7.22b).

$$\delta W_{\text{int}} = \int_0^L M_2(x)\frac{d^2v_1}{dx^2}\,dx \tag{7.22b}$$

We now note that the actual displacement v_1 is related to the actual moment M_1 through the moment curvature relation, and we can replace the curvature by $d^2v_1/dx^2 = M_1/EI$ to obtain Equation (7.22c).

$$\delta W_{\text{int}} = \int_0^L \frac{M_2(x)M_1(x)}{EI}dx \tag{7.22c}$$

For finding the displacement at x_p, we apply a virtual force of value 1 at x_p. The corresponding external virtual work for beam 2 is as shown in Equation (7.22d).

$$\delta W_{\text{ext}} = (\delta F = 1)v_2(x_P) = (\delta F = 1)v_1(x_P) \tag{7.22d}$$

Equating the external virtual work to the internal virtual work, we obtain Equation (7.23).

$$\boxed{(\delta F = 1)v_1(x_P) = \int_0^L \frac{M_2(x)M_1(x)}{EI}\,dx} \tag{7.23}$$

- A positive sign for v_1 implies that the deflection is in the same direction as the applied dummy unit force.[6]

---

[6] The statement is valid because a positive work quantity implies that force and displacement are in the same direction.

For finding the slope at x_p, we apply a virtual unit moment ($\delta M = 1$) at x_p and the corresponding external virtual work for beam 2 is as shown in Equation (7.24).

$$\delta W_{\text{ext}} = (\delta M = 1)\frac{dv_2}{dx}(x_P) = (\delta M = 1)\frac{dv_1}{dx}(x_P) \tag{7.24}$$

Equating the external virtual work to the internal virtual work, we obtain Equation (7.25).

$$\boxed{(\delta M = 1)\frac{dv_1}{dx}(x_P) = \int_0^L \frac{M_2(x)M_1(x)}{EI}dx} \tag{7.25}$$

- A positive sign for dv_1/dx implies that the slope is in the same direction as the applied dummy unit moment.
- The slope and the deflection are equal to the same integrals in Equations (7.23) and (7.25), because the left- and right-hand sides in both equations have units of work.

Example 7.10 demonstrates the application of the dummy unit method for a statically determinate problem, and Example 7.11 demonstrates the method for statically indeterminate beams.

EXAMPLE 7.10

For the beam and loading shown in Figure 7.19, determine by dummy unit load method (a) the deflection at $x = L/2$ and (b) the slope at $x = L/2$. Assume that EI is constant for the beam.

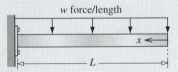

Figure 7.19 Beam for Example 7.10.

PLAN

The internal moment M_1 as a function of x can be found by drawing the free body diagram. (a) A cantilever beam with a force of 1 unit applied at the center can be used to find the internal moment M_2 for deflection calculations. The deflection can be found by using Equation (7.23). (b) A cantilever beam with a moment of 1 unit applied at the center can be used to find the internal moment M_2 for the slope calculation. The slope can be found by using Equation (7.25).

SOLUTION

We make an imaginary cut in the given beam and take the right-hand part for drawing the free body diagram as shown in Figure 7.20. After replacing the distributed load by an equivalent load, we can take the moment about point O and calculate M_1 as shown in Equation (E1).

$$M_1 = -\left(\frac{wx^2}{2}\right) \tag{E1}$$

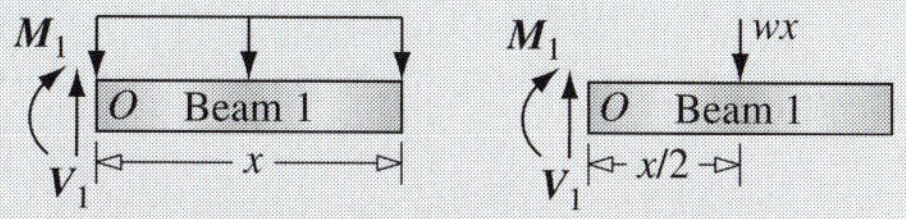

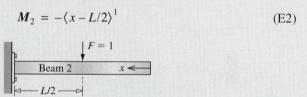

Figure 7.20 Finding M_1 for Example 7.10.

(a) We apply a unit force at $x = L/2$ on a cantilever beam as shown in Figure 7.21. The moment M_2 can be written as shown in Equation (E2).

$$M_2 = -\langle x - L/2 \rangle^1 \tag{E2}$$

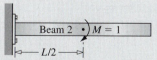

Figure 7.21 Finding M_2 for in part (a) of Example 7.10.

Substituting Equations (E1) and (E2) into Equation (7.23) and noting that the discontinuity function is zero between $x = 0$ and $x = L/2$, we obtain Equation (E3).

$$v_1\left(\frac{L}{2}\right) = \int_0^L \frac{\left[-\langle x - \frac{L}{2} \rangle^1\right]\left[-\left(\frac{wx^2}{2}\right)\right]}{EI} dx = \frac{1}{EI}\int_{L/2}^L \left[-\left(x - \frac{L}{2}\right)\right]\left[-\left(\frac{wx^2}{2}\right)\right]dx \quad \text{or}$$

$$v_1\left(\frac{L}{2}\right) = \frac{w}{2EI}\int_{L/2}^L \left(x^3 - \frac{Lx^2}{2}\right)dx = \frac{w}{2EI}\left(\frac{x^4}{4} - \frac{Lx^3}{6}\right)\Big|_{L/2}^L = \frac{w}{2EI}\left[\left(\frac{L^4}{4} - \frac{L^4}{6}\right) - \left(\frac{L^4}{64} - \frac{L^4}{48}\right)\right] \tag{E3}$$

ANS. $v_1\left(\frac{L}{2}\right) = \frac{17wL^4}{384EI}$ downward

(b) We apply a unit moment at $x = L/2$ on a cantilever beam as shown in Figure 7.22. The moment M_2 can be written as Equation (E4).

$$M_2 = -\langle x - L/2 \rangle^0 \tag{E4}$$

Figure 7.22 Finding M_2 for part (b) of Example 7.10.

Substituting Equations (E1) and (E4) into Equation (7.23) and noting that the discontinuity function is zero between $x = 0$ and $x = L/2$ and 1 between $x = L/2$ and $x = L$, we obtain Equation (E5).

$$\frac{dv_1}{dx}\left(\frac{L}{2}\right) = \int_0^L \frac{\left[-\langle x - \frac{L}{2}\rangle^0\right]\left[-\left(\frac{wx^2}{2}\right)\right]}{EI} dx$$

$$= \frac{1}{EI}\int_{L/2}^L [-1]\left[-\left(\frac{wx^2}{2}\right)\right]dx = \frac{w}{2EI}\left(\frac{x^3}{3}\right)\Bigg|_{L/2}^L \qquad \text{or} \qquad \text{(E5)}$$

$$\textbf{ANS.} \quad \frac{dv_1}{dx}\left(\frac{L}{2}\right) = \frac{7wL^4}{48EI} \text{ cw}$$

COMMENTS

1. The deflection in Equation (E3) results in a positive value. Hence, it is in the same direction as the dummy unit force in Figure 7.21 (i.e., downward, as shown in the answer).

2. The slope in Equation (E5) results in a positive value. Hence, it is in the same direction as the dummy unit moment in Figure 7.22 (i.e., clockwise, as shown in the answer).

EXAMPLE 7.11

Determine the slope at $x = L/2$ for the beam and loading shown in Figure 7.23. Assume that EI is constant for the beam.

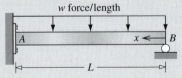

w force/length

Figure 7.23 Beam for Example 7.11.

PLAN

We can replace the right-hand support with an unknown reaction R_B and find the moment M_1 in terms of R_B. We find M_2 by placing a unit force at B of a cantilever beam. We can use Equation (7.23) to find the displacement at B and equate it to zero to find the reaction force. We can then proceed to find the slope as in Example 7.10.

SOLUTION

Let the reaction force at the right-hand support be R_B. We make a cut at an arbitrary location x and take the right-hand part for calculating M_1 as shown in Figure 7.24. By equilibrium of the moment, we obtain

$$M_1 = -\left(\frac{wx^2}{2}\right) + R_B x \qquad \text{(E1)}$$

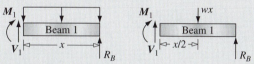

Figure 7.24 Calculating M_1 in Example 7.11.

We apply a unit force at $x = L$ on the cantilever beam as shown in Figure 7.25. We note that the degree of static redundancy is 1, and hence we can assign an arbitrary value to one reaction and obtain a statically admissible moment as

$$M_2 = -x \qquad \text{(E2)}$$

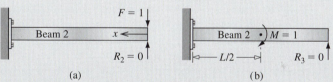

Figure 7.25 Calculating M_2 in Example 7.11 (a) for reaction and (b) for slope.

Substituting Equations (E1) and (E2) in Equation (7.23) and setting the resulting displacement to zero, we obtain

$$v_1(L) = \int_0^L \frac{[-x]\left[-\left(\frac{wx^2}{2}\right) + R_B x\right]}{EI}\, dx = \frac{1}{EI}\int_0^L \left(\frac{wx^3}{2} - R_B x^2\right)dx = \frac{1}{EI}\left(\frac{wx^4}{8} - \frac{R_B x^3}{3}\right)\bigg|_0^L \quad \text{or} \quad \text{(E3)}$$

$$v_1(L) = \frac{1}{EI}\left(\frac{wL^4}{8} - \frac{R_B L^3}{3}\right) = 0 \quad \text{or} \quad R_B = \frac{3wL}{8} \qquad \text{(E4)}$$

Substituting Equation (E4) in (E1), we obtain

$$M_1 = -\left(\frac{wx^2}{2}\right) + \left(\frac{3wL}{8}\right)x \qquad \text{(E5)}$$

To find the slope at $x = L/2$, we apply a unit moment at $x = L/2$ as shown in Figure 7.25(b) and write the moment M_2 as

$$M_2 = -\langle x - L/2\rangle^0 \qquad \text{(E6)}$$

Substituting Equations (E5) and (E6) into Equation (7.25), and noting the properties of the discontinuity functions, we obtain

$$\frac{dv_1}{dx}\left(\frac{L}{2}\right) = \frac{1}{EI}\int_{L/2}^{L}\left[-\langle x - \frac{L}{2}\rangle^0\right]\left[-\left(\frac{wx^2}{2}\right) + \left(\frac{3wL}{8}\right)x\right]dx$$

$$= \frac{1}{EI}\int_{L/2}^{L}[-1]\left[-\left(\frac{wx^2}{2}\right) + \left(\frac{3wL}{8}\right)x\right]dx \qquad \text{or}$$

$$\frac{dv_1}{dx}\left(\frac{L}{2}\right) = \frac{w}{EI}\left(\frac{x^3}{6} - \frac{3Lx^2}{16}\right)\Big|_{L/2}^{L} = \frac{w}{EI}\left[\left(\frac{L^3}{6} - \frac{3L^3}{16}\right) - \left(\frac{L^3}{48} - \frac{3L^3}{64}\right)\right] \qquad \text{or} \quad \text{(E7)}$$

ANS. $\quad \dfrac{dv_1}{dx}\left(\dfrac{L}{2}\right) = \dfrac{wL^3}{192EI} \text{ cw}$

COMMENTS

1. The sign of the slope in Equation (E7) is positive, which means that it is in the direction of the dummy unit moment shown in Figure 7.25(b) (i.e., clockwise as reported in the answer).

2. If there were more than one unknown reaction, then for each reaction a dummy unit load would have to be introduced at the support and the corresponding deflection or slope set to zero. This would lead to a very tedious process. Fortunately there is an easier method for finding reactions in statically indeterminate structure, as discussed in the next section.

7.6 CASTIGLIANO'S THEOREMS

The method for statically indeterminate structures that we study in this section is simpler to implement than the dummy unit load method. The formulas we will derive here are specific cases of Castigliano's general theorems.[7] We will derive the formulas for beam bending and then generalize for axial members and torsion of circular shafts.

In Equation (7.23), M_1 is the actual moment in the beam and M_2 is a statically admissible bending moment obtained by placing a unit force at point x_p, at which we want to calculate the deflection. If instead of a unit force we consider a force F applied at x_p, then the corresponding statically admissible moment ($\tilde{M}_2$) would be F multiplied by M_2 (i.e., $\tilde{M}_2 = FM_2$). If we take the derivative with respect to F, we obtain $M_2 = \partial \tilde{M}_2/\partial F$, and from Equation (7.23) we obtain Equation (7.26a)

$$v_1(x_P) = \int_0^L \frac{1}{EI}\left(\frac{\partial \tilde{M}_2}{\partial F} M_1(x)\right)dx \qquad (7.26a)$$

---

[7] The theorems were first derived by the Italian engineer C. A. Castigliano in 1879. See Timoshenko [1983] in Appendix D for additional details.

Noting that the actual moment is a statically admissible moment, and hence we can substitute $\tilde{\boldsymbol{M}}_2 = \boldsymbol{M}_1$ in Equation (7.26a), we obtain Equation (7.26b).

$$v_1(x_P) = \int_0^L \frac{1}{EI}\left(\frac{\partial \boldsymbol{M}}{\partial F}^1 \boldsymbol{M}_1(x)\right) dx = \int_0^L \frac{1}{2EI}\left(\frac{\partial \boldsymbol{M}_1^2}{\partial F}\right) dx = \frac{\partial}{\partial F}\left[\int_0^L \frac{\boldsymbol{M}_1^2}{2EI} dx\right] \quad (7.26b)$$

The last term in the brackets is the complementary strain energy in bending. Substituting Equation (7.12d) into Equation (7.26b), we obtain Equation (7.27a).

$$\boxed{v_1(x_P) = \frac{\partial \overline{U}_B}{\partial F}} \quad (7.27a)$$

In exactly the same manner, we can start with Equation (7.25) and instead of a unit moment place a moment M at point x_p and obtain Equation (7.27b).

$$\boxed{\frac{dv_1}{dx}(x_P) = \frac{\partial \overline{U}_B}{\partial M}} \quad (7.27b)$$

In linear systems, the complementary strain energy[8] is same as the strain energy. Hence $\overline{U}_B$ could be replaced by U_B in Equations (7.27a) and (7.27b). The equations were derived for beams, but similar form of equations are valid for any elastic structure.

For finding the axial displacement, we can replace v with u and $\overline{U}_B$ with $\overline{U}_A$ in Equation (7.27a) to obtain Equation (7.27c).

$$\boxed{u_1(x_P) = \frac{\partial \overline{U}_A}{\partial F}} \quad (7.27c)$$

For finding torsional rotation we can replace dv_1/dx with ϕ_1 and $\overline{U}_B$ with $\overline{U}_T$ in Equation (7.27b) to obtain Equation (7.27d),

$$\boxed{\phi_1(x_P) = \frac{\partial \overline{U}_T}{\partial T}} \quad (7.27d)$$

which is the general form of Equations (7.27a) through (7.27d) that is referred to as Castigliano's theorem.

> The partial derivative of complementary strain energy of a structure with respect to a force is equal to the displacement at the point of application of the force, and the partial derivative of complementary strain energy with respect to a moment is equal to the rotation at the point of application of the moment.

In the dummy unit load method, a positive value for the displacement (or rotation) implied that the displacement (or rotation) was in the direction of the force (or moment). This is also true for Equations (7.27a) through (7.27d).

---

[8] The complementary strain energy is not equal to the strain energy for nonlinear materials. However, Castigliano's theorems are applicable to nonlinear materials provided the material is elastic.

In the calculations of deflection and slopes, the derivative with regard to the force and moment should be performed before the integration; otherwise the algebra will be significantly more complex. If there is no force (or moment) at a point at which we wish to find the displacement (or slope), we initially introduce a force (or moment) and calculate the internal force and moment in Equations (7.27a) through (7.27d). After performing the derivatives, we substitute a zero value for the force (or moment) and obtain the displacement (or rotation).

We record the following observations.

- Positive values for displacement and rotation by Castigliano's theorems imply that the displacement and rotation are in the direction of the force and moment with which the derivation is being performed.

- Differentiation with respect to the force or moment before integration will save algebraic effort in applying Castigliano's theorems.

EXAMPLE 7.12

Use Castigliano's theorems to determine the reaction force at A and the displacement at C in the column shown in Figure 7.17 of Example 7.9.

PLAN

We can use the axial force N_1 in Example 7.9. We can find the derivative of N_1 with respect to the reaction R_A. Substituting N_1 and the derivative of N_1 into Equation (7.27c) and equating the displacement at A to zero, we can determine the reaction force at A. We can now place a force F at C and find a new N_1. Substituting the new N_1 and its derivative with respect to F evaluated at $F = 0$, we obtain the displacement at C from Equation (7.27c).

SOLUTION

From Example 7.9 we have the internal axial force shown in Equation (E1).

$$N_1 = \left[R_A - 4\langle x - 20 \rangle^0 - 8\langle x - 80 \rangle^0 \right] \text{ kips} \qquad \text{(E1)}$$

We can write the derivative of N_1 with respect to R_A as shown in Equation (E2).

$$\frac{\partial N_1}{\partial R_A} = 1 \qquad \text{(E2)}$$

Performing the derivative with respect to R_A in Equation (7.27a) and using Equations (E1) and (E2), we obtain the displacement at A, which we equate to zero to determine R_A as shown in Equation (E3).

$$u_A = \frac{\partial \overline{U}_A}{\partial R_A} = \frac{\partial}{\partial R_A} \left[\int_0^L \frac{N_1^2}{2EA} dx \right] = \frac{1}{EA} \int_0^L N_1 \left(\frac{\partial N_1}{\partial R_B} \right) dx \qquad \text{or}$$

$$u_A = \frac{1}{EA} \int_0^L \left[R_A - 4\langle x - 20 \rangle^0 - 8\langle x - 80 \rangle^0 \right](1)dx = 0 \qquad \text{or}$$

$$u_A = \frac{1}{EA} \left[R_A x - 4\langle x - 20 \rangle^1 - 8\langle x - 80 \rangle^1 \right]\Bigg|_0^{100}$$

$$= \frac{1}{EA} \left[R_A(100) - 4(80) - 8(20) \right] = 0 \tag{E3}$$

ANS. $R_A = 4.8$ kips

Figure 7.26 shows a dummy force applied at C for calculating the displacement at C. The axial force N_1 can be written as shown in Equation (E4).

$$N_1 = \left[4.8 - 4\langle x - 20 \rangle^0 - 8\langle x - 80 \rangle^0 - F\langle x - 80 \rangle^0 \right] \text{kips} \tag{E4}$$

The derivative of N_1 with respect to F is as shown in Equation (E5).

$$\frac{\partial N_1}{\partial F} = -\langle x - 80 \rangle^0 \tag{E5}$$

Performing the derivative with respect to F in Equation (7.27a) and evaluating it at $F = 0$, we obtain Equation (E6).

$$u_C = \frac{\partial \bar{U}_A}{\partial F}\bigg|_{F=0} = \frac{\partial}{\partial F}\left[\int_0^L \frac{N_1^2}{2EA} dx \right]\Bigg|_{F=0} = \frac{1}{EA}\int_0^L \left[N_1\left(\frac{\partial N_1}{\partial F}\right) \right]\Bigg|_{F=0} dx \tag{E6}$$

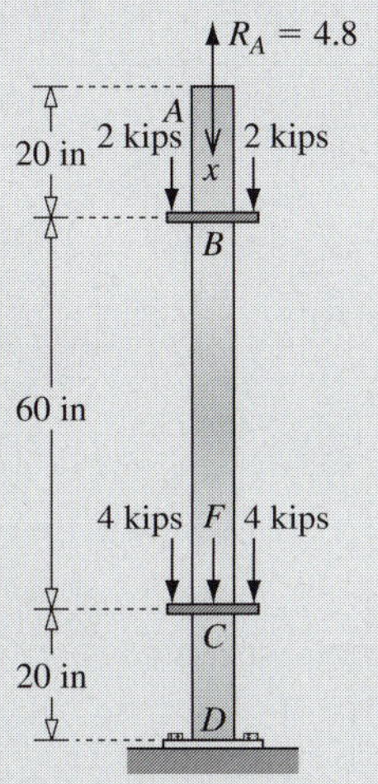

Figure 7.26 Dummy force for calculating the displacement at C in Example 7.12.

Substituting Equations (E4) and (E5) into Equation (E6), we obtain the displacement at C as shown in Equation (E7).

$$u_C = \frac{1}{EA} \int_0^{100} [-\langle x - 80 \rangle^0][4.8 - 4\langle x - 20 \rangle^0 - 8\langle x - 80 \rangle^0]dx \qquad (E7)$$

Evaluating this integral as in Example 7.9, we obtain the displacement at C.

ANS. $u_C = 0.00036$ inch downward

COMMENTS

1. The integrals in Equation (E3) in this example and Equation (E3) of Example 7.9 are the same, but arriving at them is simpler by Castigliano's theorems because we do not need to find the statically admissible force N_2.

2. Equation (E7) in this example is same as Equation (E5) of Example 7.9, but it is more tedious because a force must be first introduced and then set equal to zero.

EXAMPLE 7.13

Each steel bar ($E = 30,000$ ksi) shown in Figure 7.27 has a cross-sectional area of $1\,\text{in}^2$ and a length of 5 inches. Determine the displacement of point D.

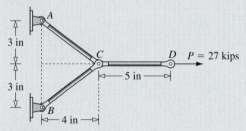

Figure 7.27 Geometry for Example 7.13.

PLAN

The total complementary strain energy for the structure is the sum of axial complementary strain energies of all the members. The displacement of point D can be written by taking the derivative of the total complementary strain energy in accordance with Equation (7.27c), which will result in an expression containing derivatives of the axial forces with respect to P for each member. By drawing free body diagrams, the internal axial force in each member can be written in terms of the applied force P and substituted to obtain the displacement at D.

SOLUTION

The total complementary strain energy for the structure can be written as in Equation (E1).

$$\bar{U}_A = \sum_{i=1}^{3} \frac{1}{2} \int_{0}^{L_i} \frac{N_i^2}{E_i A_i} dx = \sum_{i=1}^{3} \frac{1}{2} \frac{N_i^2 L_i}{E_i A_i} = \frac{1}{2} \left[\frac{N_{AC}^2 L_{AC}}{E_{AC} A_{AC}} + \frac{N_{BC}^2 L_{BC}}{E_{BC} A_{BC}} + \frac{N_{CD}^2 L_{CD}}{E_{CD} A_{AC}} \right] \quad \text{(E1)}$$

We take the derivative of Equation (E1) with respect to P as in Equation (7.27c) to obtain a displacement expression for point D as shown in Equation (E2).

$$U_D = \frac{\partial \bar{U}_A}{\partial P} = \left[\left(\frac{N_{AC} L_{AC}}{E_{AC} A_{AC}} \right) \frac{\partial N_{AC}}{\partial P} + \left(\frac{N_{BC} L_{BC}}{E_{BC} A_{BC}} \right) \frac{\partial N_{BC}}{\partial P} + \left(\frac{N_{CD} L_{CD}}{E_{CD} A_{AC}} \right) \frac{\partial N_{CD}}{\partial P} \right] \quad \text{(E2)}$$

From the free body diagram in Figure 7.28(a), we obtain Equation (E3).

$$N_{CD} = P \quad \text{(E3)}$$

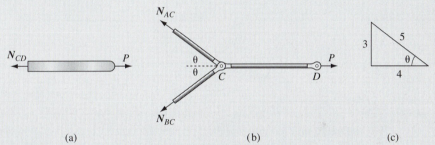

(a)   (b)   (c)

Figure 7.28 (a, b) Free body diagrams and (c) geometry for Example 7.13

From the free body diagram in Figure 7.28(b) and geometry in Figure 7.28(c), we obtain Equations (E4) and (E5).

$$N_{AC} \sin \theta - N_{BC} \sin \theta = 0 \quad \text{or} \quad N_{AC} = N_{BC} \quad \text{(E4)}$$

$$-N_{AC} \cos \theta - N_{BC} \cos \theta + P = 0 \quad \text{or} \quad 2N_{AC} \left(\frac{4}{5} \right) = P \quad \text{or} \quad N_{AC} = 0.625P \quad \text{(E5)}$$

Using Equations (E3), (E4), and (E5), we obtain the derivatives with respect to P as shown in (E6).

$$\frac{\partial N_{CD}}{\partial P} = 1 \qquad \frac{\partial N_{AC}}{\partial P} = 0.625 \qquad \frac{\partial N_{BC}}{\partial P} = 0.625 \quad \text{(E6)}$$

Substituting Equations (E3), (E4), (E5), and (E6) into Equation (E2), we obtain Equation (E7).

$$u_D = \left[\left(\frac{0.625 P L_{AC}}{E_{AC} A_{AC}} \right) (0.625) + \left(\frac{0.625 P L_{BC}}{E_{BC} A_{BC}} \right) (0.625) + \left(\frac{P L_{CD}}{E_{CD} A_{AC}} \right) (1) \right] \quad \text{(E7)}$$

In Equation (E7) we substitute the following for all members: the length as 5 in, the modulus of elasticity as 30,000 ksi, the cross-sectional area as 1 in², and $P = 27$ kips. Then we obtain the displacement at D as shown in Equation (E8).

$$u_D = \left[\frac{(0.625)(27)(5)}{(30,000)(1)}(0.625) + \frac{(0.625)(27)(5)}{(30,000)(1)}(0.625) + \frac{(27)(5)}{(30,000)(1)}\right]$$

$$= 8.016(10^{-3}) \text{ in} \qquad\qquad\qquad\qquad \text{(E8)}$$

ANS. $u_D = 0.008$ in

COMMENT

This example demonstrates the use of Castigliano's theorems for structures. The scalar nature of complementary strain energy permits one not only to add the strain energy of axial members but to work with a structure made up of members of various types. Example 7.16 will show this for a structure consisting of beams and axial members.

EXAMPLE 7.14

For the beam and loading shown in Figure 7.23, use Castigliano's theorems to calculate the slope at $x = L/2$.

PLAN

We can use the moment M_1 in Example 7.11. We can find the derivative of M_1 with respect to the reaction at B. Substituting M_1 and the derivative of M_1 into Equation (7.27a) and equating the displacement at B to zero, we can determine the reaction at B. We can now place a moment M at $x = L/2$ and find the new M_1. Substituting the new M_1 and its derivative with respect to M evaluated at $M = 0$, we obtain the slope from Equation (7.27b).

SOLUTION

From Example 7.11 we have the internal bending moment shown in Equation (E1).

$$M_1 = -\left(\frac{wx^2}{2}\right) + R_B x \qquad\qquad\qquad\qquad \text{(E1)}$$

We can write the derivative of M_1 with respect to R_B as shown in Equation (E2).

$$\frac{\partial M_1}{\partial R_B} = x \qquad\qquad\qquad\qquad \text{(E2)}$$

Performing the derivative with respect to R_B in Equation (7.27c) and using Equations (E1) and (E2), we obtain the displacement at B as shown in Equation (E3).

$$v_B = \frac{\partial \overline{U}_B}{\partial R_B} = \frac{\partial}{\partial R_B}\left[\int_0^L \frac{M_1^2}{2EI}dx\right] = \frac{1}{EI}\int_0^L M_1\left(\frac{\partial M_1}{\partial R_B}\right)dx = \frac{1}{EI}\int_0^L \left[-\left(\frac{wx^2}{2}\right) + R_Bx\right](x)dx \quad \text{(E3)}$$

Equating the displacement at B to zero and solving for the reaction as in Example 7.11, we obtain Equation (E4).

$$R_B = \frac{3wL}{8} \tag{E4}$$

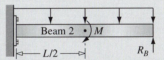

Figure 7.29 Calculation of M_1 for finding the slope in Example 7.14.

Figure 7.29 shows the original beam and loading with a moment applied at $x = L/2$. By using the discontinuity functions, we can write M_1 as shown in Equation (E5).

$$M_1 = -\left(\frac{wx^2}{2}\right) + R_Bx - M\left\langle x - \frac{L}{2}\right\rangle^0 \tag{E5}$$

Taking the derivative with respect to M, we obtain Equations (E6).

$$\frac{\partial M_1}{\partial M} = -\left\langle x - \frac{L}{2}\right\rangle^0 \tag{E6}$$

Performing the derivation with respect to M in Equation (7.27b), then using Equations (E5) and (E6) after substituting $M = 0$, we obtain Equation (E7).

$$\frac{dv_1}{dx}\left(\frac{L}{2}\right) = \frac{\partial \overline{U}_B}{\partial M}\bigg|_{M=0} = \frac{1}{EI}\int_0^L \left[M_1\left(\frac{\partial M_1}{\partial M}\right)\right]\bigg|_{M=0} dx$$

$$= \frac{1}{EI}\int_0^L \left[-\left(\frac{wx^2}{2}\right) + R_Bx\right]\left[-\left\langle x - \frac{L}{2}\right\rangle^0\right]dx \tag{E7}$$

Evaluating this integral as in Example 7.11, we obtain:

ANS. $\quad \dfrac{dv_1}{dx}\left(\dfrac{L}{2}\right) = \dfrac{wL^3}{192EI}$ cw

COMMENTS

1. To find reaction R_B we needed only moment M_1 in Equation (E1), whereas in Example 7.11 we also needed to find M_2, corresponding to a unit force applied at B. However, the resulting integrals in Equation (E3) in this example and Equation (E3) of Example 7.11 differ only in sign which is unimportant as both integrals are equaled to zero. This example and Example 7.12 show that finding reactions is simpler by Castigliano's theorems than with the dummy unit load method. As the number of reactions increases, the advantages of Castigliano's theorems become more pronounced.

2. The integral in Equation (E7) in this example is the same as the integral in Equation (E7) in Example 7.11, but it is more tedious because a moment must be first introduced and then set equal to zero. This example and Equation 7.12 show that if there is no concentrated force or moment present in the original loading at the point at which we seek displacement or slope, then the dummy unit load method is simpler than Castigliano's theorems.

3. Based on Comments 1 and 2, we can record the following observations:
   - It is easier to use Castigliano's theorems than the dummy unit load method to find reactions in statically indeterminate structures.
   - It is easier to use the dummy unit load method than Castigliano's theorems to find displacements and slopes at points at which *no* concentrated forces or moments are present in the original loading.

EXAMPLE 7.15

For the beam and loading shown in Figure 7.30, determine the reaction forces at A and B in terms of w, L, E, and I.

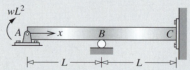

Figure 7.30 Beam for Example 7.15.

PLAN

We can replace the supports at A and B by reactions forces R_A and R_B, respectively. We can then use discontinuity functions to write the moment M_1. The partial derivatives of M_1 with respect to R_A and R_B can be found. We can use Equation (7.27a) to find the displacements at A and B in terms of R_A and R_B and set them equal to zero. The two equations can be solved simultaneously to obtain the reactions R_A and R_B.

SOLUTION

Figure 7.31 shows the beam in which the supports at A and B are replaced by reaction forces. From the templates shown in Figure 7.34, we can write the moment M_1 by using the discontinuity functions as in Equation (E1).

$$M_1 = wL^2 - R_A x + R_B \langle x - L \rangle^1 \qquad \text{(E1)}$$

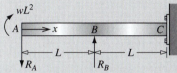

Figure 7.31 Beam without supports for Example 7.15.

The partial derivatives of M_1 in Equation (E1) with respect to R_A and R_B are shown in Equations (E2).

$$\frac{\partial M_1}{\partial R_A} = -x \qquad \text{and} \qquad \frac{\partial M_1}{\partial R_B} = \langle x - L \rangle^1 \qquad \text{(E2)}$$

Deflections at A can be written by using Equation (7.27a) as shown in Equation (E3).

$$v_A = \frac{\partial \overline{U}_B}{\partial R_A} = \frac{1}{EI} \int_0^{2L} M_1 \left(\frac{\partial M_1}{\partial R_A} \right) dx = \frac{1}{EI} \int_0^{2L} [wL^2 - R_A x + R_B \langle x - L \rangle^1](-x) dx \qquad \text{(E3)}$$

We note that the discontinuity function $\langle x - L \rangle^1$ is zero for $x < L$ and equal to $(x - L)$ for $x > L$. The deflection in Equation (E3) can be evaluated and equated to zero as shown in Equation (E4).

$$v_A = \frac{1}{EI} \left[\int_0^{2L} [wL^2 - R_A x](-x) dx + \int_0^L [R_B(0)](x) dx + \int_L^{2L} [R_B(x - L)](-x) dx \right] \qquad \text{or}$$

$$v_A = \frac{1}{EI} \left[\left(-wL^2 \left(\frac{x^2}{2} \right) + R_A \frac{x^3}{3} \right) \Big|_0^{2L} - R_B \left(\frac{x^3}{3} - L \frac{x^2}{2} \right) \Big|_L^{2L} \right] \qquad \text{or}$$

$$v_A = \frac{1}{EI} \left[\left(-2wL^4 + R_A \frac{8L^3}{3} \right) - R_B \left\{ \left(\frac{8L^3}{3} - 2L^3 \right) - \left(\frac{L^3}{3} - \frac{L^3}{2} \right) \right\} \right] \qquad \text{or}$$

$$v_A = \frac{1}{EI} \left[-2wL^4 + R_A \frac{8L^3}{3} - R_B \frac{5L^3}{6} \right] = 0 \qquad \text{or} \qquad 16R_A - 5R_B = 12wL \qquad \text{(E4)}$$

The deflections at B can be written by using Equation (7.27a) as shown in Equation (E5).

$$v_B = \frac{\partial \overline{U}_B}{\partial R_B} = \frac{1}{EI} \int_0^{2L} M_1 \left(\frac{\partial M_1}{\partial R_B} \right) dx$$

$$= \frac{1}{EI} \int_0^{2L} [wL^2 - R_A x + R_B \langle x - L \rangle^1] \langle x - L \rangle^1 dx \qquad \text{(E5)}$$

Once more using the property of the discontinuity function, Equation (E5) can be written as shown in Equation (E6).

$$v_B = \frac{1}{EI} \int_L^{2L} [wL^2 - R_A x + R_B (x - L)](x - L) dx \qquad \text{(E6)}$$

We substitute $\zeta = (x - L)$ in Equation (E6) and integrate to obtain the deflection at B, which we equate to zero as shown in Equation (E7).

$$v_B = \frac{1}{EI} \int_0^L [wL^2 - R_A(\zeta + L) + R_B \zeta](\zeta) dx$$

$$= \frac{1}{EI} \left[(wL^2)\frac{\zeta^2}{2} - R_A \left(\frac{\zeta^3}{3} + L\frac{\zeta^2}{2} \right) + R_B \frac{\zeta^3}{3} \right]_0^L \qquad \text{or}$$

$$v_B = \frac{1}{EI} \left[\left(\frac{wL^4}{2} \right) - R_A \left(\frac{L^3}{3} + \frac{L^3}{2} \right) + R_B \frac{L^3}{3} \right] = 0 \qquad \text{or} \qquad 5R_A - 2R_B = 3wL \qquad \text{(E7)}$$

Solving Equations (E4) and (E7), we obtain the reactions R_A and R_B.

ANS. $R_A = \frac{9}{7}wL \qquad R_B = \frac{12}{7}wL$

COMMENTS

1. Castigliano's theorems offer an elegant method of formulating a set of algebraic equations for unknown reactions in statically indeterminate structures as shown by Equations (E4) and (E7).

2. The use of the discontinuity functions simplifies the writing of moment equations and the generation of the required integrals.

EXAMPLE 7.16

A rectangular wooden beam having a cross section of 60 mm × 180 mm is supported at the right end by an aluminum circular rod of 8 mm diameter as shown in Figure 7.32. The moduli of elasticity for wood and aluminum are $E_{wood} = 12.6$ GPa and $E_{Al} = 70$ GPa. Determine the axial stress in the aluminum rod AC.

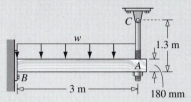

Figure 7.32 Structure for Example 7.16.

PLAN

The structure's complementary strain energy is the sum of the complementary strain energies of the beam AB and the axial rod AC. The structure has one degree of static redundancy. The internal axial force and the internal bending moment can be written in terms of the reaction at support C. By using Castigliano's theorems, we can find the displacement at C and equate it to zero to find the reaction at C. The internal axial force in AC is equal to the reaction force at C, from which the axial stress in AC can be found.

SOLUTION

The complementary strain energy of the structure can be written as Equation (E1).

$$\overline{U} = \overline{U}_A + \overline{U}_B = \frac{1}{2}\int_0^{L_{AC}} \frac{N_{AC}^2}{E_{AC}A_{AC}}dx + = \frac{1}{2}\left[\frac{N_{AC}^2 L_{AC}}{E_{AC}A_{AC}}\right] + \frac{1}{2}\int_0^{L_{AB}} \frac{M_z^2}{E_{AB}I_{zz}}dx \tag{E1}$$

The displacement at C can be found by taking the derivative of Equation (E1) with respect to the reaction force at C to obtain Equation (E2).

$$u_C = \frac{\partial \overline{U}}{\partial R_C} = \left(\frac{N_{AC}L_{AC}}{E_{AC}A_{AC}}\right)\frac{\partial N_{AC}}{\partial R_C} + \frac{1}{E_{AB}I_{zz}}\int_0^{L_{AB}} M_z\frac{\partial M_z}{\partial R_C}dx \tag{E2}$$

We can make an imaginary cut in the beam AB and draw the free body diagram as shown in Figure 7.33. By inspection, we obtain the axial force in AC as shown in Equation (E3) and by moment equilibrium about point O we obtain the internal bending moment, as shown in Equation (E4).

$$N_{AC} = R_C \tag{E3}$$

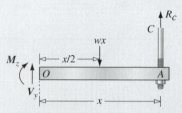

Figure 7.33 Free body diagram for Example 7.16.

$$M_z = R_C x - \frac{wx^2}{2} \tag{E4}$$

Taking the derivatives of Equations (E3) and (E4) with respect to R_C, we obtain Equations (E5).

$$\frac{\partial N_{AC}}{\partial R_C} = 1 \quad \text{and} \quad \frac{\partial M_z}{\partial R_C} = x \tag{E5}$$

Substituting Equations (E3), (E4), and (E5) into Equation (E2) and integrating, we obtain Equation (E6).

$$u_C = \left(\frac{R_C L_{AC}}{E_{AC} A_{AC}}\right)(1) + \frac{1}{E_{AB} I_{zz}} \int_0^{L_{AB}} \left(R_C x - \frac{wx^2}{2}\right)(x)\,dx$$

$$= \left(\frac{R_C L_{AC}}{E_{AC} A_{AC}}\right) + \frac{R_C \dfrac{L_{AB}^3}{3} - \dfrac{wL_{AB}^4}{8}}{E_{AB} I_{zz}} \tag{E6}$$

The various variables in Equation (E6) are calculated, as shown in Equations (E7) and (E8).

$$L_{AC} = 1.3 \text{ m} \qquad E_{AC} = 70(10^9) \text{ N/m}^2$$
$$A_{AC} = \frac{\pi}{4}(0.008^2) = 50.265(10^{-6}) \text{ m}^2 \tag{E7}$$

$$L_{AB} = 3.0 \text{ m} \qquad E_{AB} = 12.6(10^9) \text{ N/m}^2$$
$$I_{zz} = \frac{1}{12}(0.06)(0.18^3) = 29.16(10^{-6}) \text{ m}^4 \tag{E8}$$

Substituting Equations (E7) and (E8) into Equation (E6) and equating to zero, we can obtain the reaction force R_C as shown in Equation (E9).

$$u_C = \frac{R_C(1.3)}{(70)(10^9)(50.265)(10^{-6})} + \frac{R_C \dfrac{3^3}{3} - \dfrac{(4)(10^3)(3^4)}{8}}{12.6(10^9)[29.16(10^{-6})]} = 0 \quad \text{or}$$

$$0.3695(10^{-6})R_C + 24.495(10^{-6})R_C - 110.229(10^{-3}) = 0 \quad \text{or}$$

$$R_C = 4.433(10^3) \text{ N} \tag{E9}$$

The axial stress in AC can be found as shown in Equation (E10).

$$\sigma_{AC} = \frac{N_{AC}}{A_{AC}} = \frac{4.433(10^3)}{50.265(10^{-6})} = 88.197(10^6) \text{ N/m}^2 \qquad \text{(E10)}$$

ANS. $\sigma_{AC} = 88.2$ MPa (T)

COMMENT

The maximum bending normal stress in the beam can be found by first using Equation (E4) to find the maximum bending moment. Similarly, the maximum shear force can be found by force equilibrium of the free body diagram in Figure 7.33. Thus, the maximum stresses in the structure can be found. This emphasizes that Castigliano's theorems could be used to solve the static indeterminacy, clearing the way to perform any aspect of analysis or design.

7.7 DEFLECTION OF CURVED BEAMS

In Section 3.5, we divided curved beams into two categories:

1. Beams with very large radii of curvature (i.e., $r_c/h > 5$), where r_c is the radius of curvature of the line joining the centroids of the cross sections and h is the depth of the beam cross section (see Figure 3.47),

2. Thick beams with $r_c/h < 5$.

The stresses and strains in beams with $r_c/h > 5$ could be calculated as for straight beams, and hence the complementary strain energies for such curved beams would be similar to those for straight beams. In this section we will use energy methods to find the deflection of beams for which $r_c/h > 5$, which would be the difficult by equilibrium methods.

Figure 7.34(a) shows a curved beam with in-plane loads P_x and P_y. Figure 7.34(b) shows a free body diagram after an imaginary cut has been made. From Figure 7.34(b) we conclude that the total complementary strain energy is the sum of the complementary strain energies due to the axial force, shear force, and bending moment. The complementary strain energy due to shear is neglected in comparison to that due to moment because the shear stresses in bending are an order of magnitude less than the normal stresses if the approximation of the plane section before deformation remains plane and perpendicular to the centerline during deformation. Generally speaking, the axial normal stress is an order of magnitude smaller than the maximum bending normal stress at a cross section when both stresses are produced from the same force. Thus, in analysis we consider only the complementary strain energy for a curved beam ($\overline{U}_{IB}$) under an in-plane load due to the moment, as given by Equation (7.28),

$$\boxed{\overline{U}_{IB} = \int_L \frac{1}{2} \frac{M_z^2}{EI_{zz}} ds} \qquad \text{(7.28)}$$

where s is the tangential coordinate along the line joining the centroids of the curved beam.

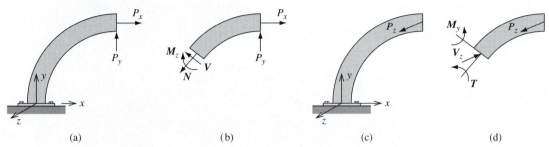

Figure 7.34 Curved beam under (a) in-plane and (c) out-of-plane loading, with associated free body diagrams (b, d).

Figure 7.34(c) shows a curved beam an with out-of-plane load P_z. Figure 7.34(d) shows the free body diagram after an imaginary cut has been made. From Figure 7.34(d) we conclude that the total complementary strain energy is the sum of the complementary strain energies due to the shear force, bending moment about the y axis, and torsion. The complementary strain energy due to shear is once more neglected. We will restrict ourselves to *circular cross sections,* for which we have a well-developed theory of torsion. The total complementary strain energy for a curved beam due to out-of-plane load ($\overline{U}_{OB}$) due to bending moment and torsion is given by Equation (7.29).

$$\overline{U}_{OB} = \int_L \frac{1}{2} \frac{M_y^2}{EI_{yy}} ds + \int_L \frac{1}{2} \frac{T^2}{GJ} ds \qquad (7.29)$$

EXAMPLE 7.17

A circular beam having a rectangular cross section is shown in Figure 7.35. For the loading and geometry shown, find the ratio of the complementary axial strain energy to the complementary bending strain energy, for the two cases (a) $\alpha = 0$ and (b) $\alpha = \pi/2$.

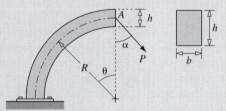

Figure 7.35 Circular beam for Example 7.17.

PLAN

By drawing a free body diagram after making a cut at an angle θ, we can find the internal axial force N and the internal bending moment M_z in terms of P and the other variables in the problem. These values can be substituted in the complementary strain energy expressions and integrated, and the ratio found in terms of the angle α. The two values of α can then be substituted and the result obtained.

SOLUTION

Figure 7.36 shows the free body diagram after an imaginary cut has been made at an angle θ. Line AB is drawn parallel to line OC, and the angle between the line AB and the applied force P can be determined as shown in Figure 7.36. Since the axial force is perpendicular to line AB, by the equilibrium of forces in the axial direction, we obtain Equation (E1).

$$N = P\sin(\alpha - \theta) \qquad (E1)$$

By equilibrium of the moment about point O in Figure 7.36, we obtain Equation (E2).

$$M_z + (P\cos\alpha)(R\sin\theta) + (P\sin\alpha)R(1 - \cos\theta) = 0 \qquad \text{or}$$

$$M_z + PR(\cos\alpha\sin\theta - \sin\alpha\cos\theta) + (PR\sin\alpha) = 0 \qquad \text{or}$$

$$M_z = PR[\sin(\alpha - \theta) - \sin\alpha] \qquad (E2)$$

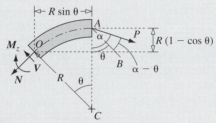

Figure 7.36 Free body diagram for Example 7.17.

Complementary axial strain energy: The complementary axial strain energy is given by Equation (E3).

$$\overline{U}_A = \int_L \frac{1}{2}\frac{N^2}{EA}ds \qquad (E3)$$

We note that the cross-sectional area $A = bh$, and $ds = R\,d\theta$. Substituting Equation (E1) into Equation (E3) and integrating, we obtain Equation (E4).

$$\overline{U}_A = \frac{1}{2EA}\int_0^{\pi/2} P^2\sin^2(\alpha - \theta)R\,d\theta = \frac{P^2R}{2Ebh}\int_0^{\pi/2} [1 - \cos 2(\alpha - \theta)]d\theta \qquad \text{or}$$

$$\overline{U}_A = \frac{P^2R}{2Ebh}\left[\theta + \frac{\sin 2(\alpha - \theta)}{2}\right]\Bigg|_0^{\pi/2} = \frac{P^2R}{2Ebh}\left[\frac{\pi}{2} + \frac{\sin(2\alpha - \pi) - \sin 2\alpha}{2}\right]$$

$$= \frac{P^2R}{4Ebh}[\pi - 2\sin 2\alpha] \qquad (E4)$$

Complementary bending strain energy: We note that $I_{zz} = bh^3/12$. Substituting Equation (E2) into Equation (7.28) and integrating, we obtain Equation (E5).

$$\overline{U}_{IB} = \int_L \frac{1}{2}\frac{M_z^2}{EI_{zz}}ds = \frac{1}{2(EI_{zz})}\int_0^{\pi/2} P^2R^2[\sin(\alpha - \theta) - \sin\alpha]^2 R\,d\theta$$

$$\overline{U}_{IB} = \frac{6P^2R^3}{Ebh^3} \int_0^{\pi/2} [\sin^2(\alpha - \theta) + \sin^2\alpha - 2\sin(\alpha - \theta)\sin\alpha] \, d\theta$$

$$\overline{U}_{IB} = \frac{6P^2R^3}{Ebh^3} \int_0^{\pi/2} \left[\frac{1 - \cos 2(\alpha - \theta)}{2} + \sin^2\alpha - 2\sin(\alpha - \theta)\sin\alpha\right] d\theta$$

$$\overline{U}_{IB} = \frac{3P^2R^3}{Ebh^3} \left[\left(\theta + \frac{\sin 2(\alpha - \theta)}{2}\right) + 2\theta \, \sin^2\alpha - 4 \, \cos(\alpha - \theta)\sin\alpha\right]_0^{\pi/2}$$

$$\overline{U}_{IB} = \frac{3P^2R^3}{Ebh^3} \left[\frac{\pi}{2} + \frac{\sin(2\alpha - \pi) - \sin 2\alpha}{2} + \pi\sin^2\alpha - 4\sin\alpha\left\{\cos\left(\alpha - \frac{\pi}{2}\right) - \cos\alpha\right\}\right]$$

$$\overline{U}_{IB} = \frac{3P^2R^3}{Ebh^3} \left[\frac{\pi}{2} + \sin 2\alpha + (\pi - 4)\sin^2\alpha\right] \tag{E5}$$

We find the ratio of the two complementary strain energies by dividing Equation (E4) by Equation (E5) to obtain Equation (E6).

$$\frac{\overline{U}_A}{\overline{U}_{IB}} = \frac{\dfrac{P^2R}{4Ebh}[\pi - 2 \, \sin 2\alpha]}{\dfrac{3P^2R^3}{Ebh^3}\left[\dfrac{\pi}{2} + \sin 2\alpha + (\pi - 4)\sin^2\alpha\right]} = \frac{h^2}{12R^2}\left[\frac{\pi - 2 \, \sin 2\alpha}{\dfrac{\pi}{2} + \sin 2\alpha + (\pi - 4)\sin^2\alpha}\right] \tag{E6}$$

(a) Substituting $\alpha = 0$ in Equation (E6), we obtain Equation (E7)

$$\frac{\overline{U}_A}{\overline{U}_{IB}} = \frac{h^2}{12R^2}\left[\frac{\pi}{\pi/2}\right] \tag{E7}$$

ANS. $\dfrac{\overline{U}_A}{\overline{U}_{IB}} = 0.167\left(\dfrac{h}{R}\right)^2$

(b) Substituting $\alpha = \pi/2$ in Equation (E6), we obtain Equation (E8).

$$\frac{\overline{U}_A}{\overline{U}_{IB}} = \frac{h^2}{12R^2}\left[\frac{\pi}{\pi/2 + (\pi - 4)}\right] \tag{E8}$$

ANS. $\dfrac{\overline{U}_A}{\overline{U}_{IB}} = 0.367\left(\dfrac{h}{R}\right)^2$

COMMENT

Note that our theory is valid for $R/h > 5$. For $R/h = 5$, the complementary axial strain energy is less than that for bending by two orders of magnitudes for the two extreme cases. Thus, it is reasonable to neglect the complementary axial strain energy instead of that for bending when the two strain energies are produced from the same force.

EXAMPLE 7.18

For the circular beam and loading shown in Figure 7.35, determine the displacement at point A in the direction of the applied force for two cases: (a) $\alpha = 0$ and (b) $\alpha = \pi/2$.

PLAN

The complementary bending strain energy for the circular beam in Figure 7.35 was found in Example 7.17. The derivative of this energy with respect to P will give the displacement in the direction of P.

SOLUTION

The complementary bending strain energy for the circular beam in Figure 7.35 was found in Example 7.17 as Equation (E5), which is repeated as Equation (E1) for convenience.

$$\overline{U}_{IB} = \frac{3P^2R^3}{Ebh^3}\left[\frac{\pi}{2} + \sin 2\alpha + (\pi - 4)\sin^2\alpha\right] \tag{E1}$$

The displacement in the direction of force P can be obtained by taking the derivative of Equation (E1) with respect to P as shown in Equation (E2).

$$v_P = \frac{\partial \overline{U}_{IB}}{\partial P} = \frac{6PR^3}{Ebh^3}\left[\frac{\pi}{2} + \sin 2\alpha + (\pi - 4)\sin^2\alpha\right] \tag{E2}$$

(a) Substituting $\alpha = 0$ in Equation (E2), we obtain Equation (E3).

$$v_P = \frac{6PR^3}{Ebh^3}\left[\frac{\pi}{2}\right] \qquad \textbf{ANS.} \quad v_P = \frac{9.425P}{Eb}\left(\frac{R}{h}\right)^3 \tag{E3}$$

(b) Substituting $\alpha = \pi/2$ in Equation (E2), we obtain Equation (E4).

$$v_P = \frac{6PR^3}{Ebh^3}\left[\frac{\pi}{2} + (\pi - 4)\right] \qquad \textbf{ANS.} \quad v_P = \frac{4.274P}{Eb}\left(\frac{R}{h}\right)^3 \tag{E4}$$

COMMENT

The results show that for the same amount of force, a circular beam deflects more in the radial direction ($\alpha = 0$) than in the tangential ($\alpha = \pi/2$) direction. Alternatively, we can say that a curved beam is stiffer in the tangential direction than in the radial direction.

EXAMPLE 7.19

For the circular beam and loading shown in Figure 7.35, determine the displacement at point A in the horizontal direction (tangential direction) for $\alpha = 0$.

PLAN

We can apply a dummy unit force in the horizontal direction and calculate a statically admissible moment M_2. The actual moment M_1 is known from Example 7.17. We can then use Equation (7.23) to find the displacement in the horizontal direction.

SOLUTION

Figure 7.37 shows the free body diagram for calculating the statically admissible bending moment due to a dummy unit force applied at point A in the horizontal direction. By equilibrium of the moment about point O, we obtain Equation (E1).

$$M_2 = -(1)[R(1 - \cos \theta)] \tag{E1}$$

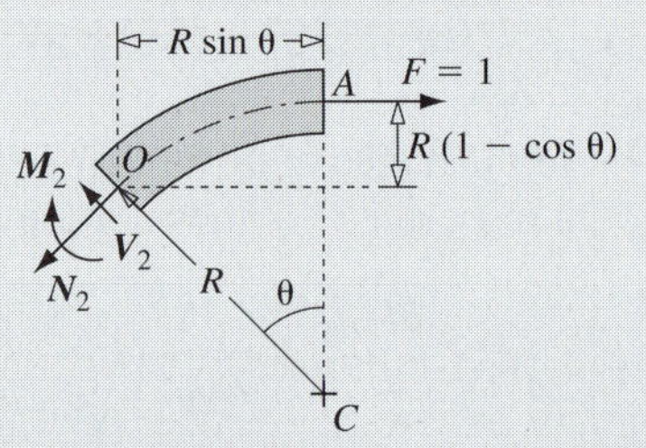

Figure 7.37 Free body diagram for Example 7.19.

Substituting $\alpha = 0$ in Equation (E2) in Example 7.17, we obtain the actual moment M_1 as shown in Equation (E2).

$$M_1 = PR[\sin(\alpha - \theta) - \sin \alpha]\big|_{\alpha=0} = -PR \sin \theta \tag{E2}$$

Substituting Equations (E1) and (E2) into Equation (7.23) and integrating over the curved beam, we obtain the horizontal displacement v_{Ax} as shown in Equation (E3).

$$v_{Ax} = \int_0^L \frac{M_2(s)M_1(s)}{EI} ds = \frac{1}{EI} \int_0^{\pi/2} \{-(1)[R(1 - \cos \theta)]\}\{-PR \sin \theta\} R d\theta$$

or

$$v_{Ax} = \frac{PR^3}{EI} \int_0^{\pi/2} \left[\sin \theta - \frac{\sin 2\theta}{2} \right] d\theta = \frac{12PR^3}{Ebh^3} \left[-\cos \theta + \frac{1}{4}\cos 2\theta \right]\Bigg|_0^{\pi/2}$$

$$= \frac{12PR^3}{Ebh^3} \left[\frac{1}{4}\cos \pi + 1 - \frac{1}{4} \right] \tag{E3}$$

ANS. $v_{Ax} = \dfrac{6P}{Eb}\left(\dfrac{R}{h}\right)^3$

COMMENT

In this example for a circular beam, the force is applied in the radial direction and the deflection is obtained in the tangential direction. In straight beams we would not have obtained any deflection perpendicular to the load for small-deformation theory, while for curved beams, even for small-deformation theory, we obtain deflection perpendicular to the load.

EXAMPLE 7.20

A circular beam is loaded by an out-of-plane load P as shown in Figure 7.38. Determine the deflection at point A in the direction of the load.

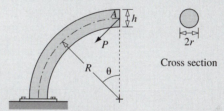

Figure 7.38 Circular beam for Example 7.20.

PLAN

By drawing a free body diagram after a cut has been made at an angle θ, we can find the internal torque T and the internal bending moment M_y in terms of P and the other variables in the problem. These values can be substituted in the complementary strain energy expressions of Equation (7.29). By taking the derivative with respect to P, we can obtain an integral expression for the displacement that can be integrated to give the desired result.

SOLUTION

Figure 7.39(a) shows a free body diagram after an imaginary cut has been made at an angle θ. The moments M_{OB} and M_{OC} about the axes OB and OC can be obtained as shown in Figure 7.39(b). By taking components of the moments M_{OB} and M_{OC}, we can obtain the torque T and the moment M_y as shown in Equations (E1) and (E2).

$$M_y = -M_{OB} \sin \theta - M_{OC} \cos \theta = -[PR(1 - \cos \theta)] \sin \theta - [PR \sin \theta] \cos \theta$$

$$= -PR \sin \theta \tag{E1}$$

$$T = -M_{OB}\cos\theta + M_{OC}\sin\theta = -[PR(1-\cos\theta)]\cos\theta + [PR\sin\theta]\sin\theta$$

$$= PR(1-\cos\theta) \tag{E2}$$

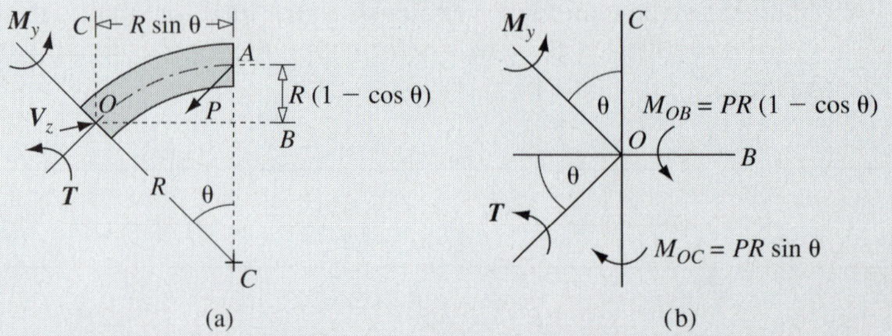

Figure 7.39 (a) Free body diagram and (b) moments OC and OB for Example 7.20.

Taking the derivative of Equation (7.29) with respect to P, we obtain the displacement in the direction of P as shown in Equation (E3).

$$w_P = \frac{\partial \overline{U}_{OB}}{\partial P} = \frac{\partial}{\partial P}\left[\int_L \frac{1}{2}\frac{M_y^2}{EI_{yy}}ds + \int_L \frac{1}{2}\frac{T^2}{GJ}ds\right]$$

$$= \frac{1}{EI_{yy}}\int_0^{\pi/2} M_y\frac{\partial M_y}{\partial P}Rd\theta + \frac{1}{GJ}\int_0^{\pi/2} T\frac{\partial T}{\partial P}Rd\theta \tag{E3}$$

Substituting Equations (E1) and (E2) into Equation (E3) and integrating, we obtain Equation (E4).

$$w_P = \frac{1}{EI_{yy}}\int_0^{\pi/2}[-PR\sin\theta][-R\sin\theta]Rd\theta +$$

$$\frac{1}{GJ}\int_0^{\pi/2}[PR(1-\cos\theta)][R(1-\cos\theta)]Rd\theta \qquad \text{or}$$

$$w_P = \frac{PR^3}{EI_{yy}}\int_0^{\pi/2}\sin^2\theta\, d\theta + \frac{PR^3}{GJ}\int_0^{\pi/2}[1-2\cos\theta+\cos^2\theta]Rd\theta$$

$$w_P = \frac{PR^3}{2EI_{yy}}\left[\theta-\frac{1}{2}\sin 2\theta\right]\Big|_0^{\pi/2} + \frac{PR^3}{GJ}\left[\theta-2\sin\theta+\frac{1}{2}\left(\theta+\frac{1}{2}\sin 2\theta\right)\right]\Big|_0^{\pi/2}$$

$$= \frac{\pi PR^3}{4EI_{yy}} + \left(\frac{3\pi}{4}-2\right)\frac{PR^3}{GJ} \tag{E4}$$

For a circular cross section, $I_{yy} = (\pi/4)r^4$ and $J = (\pi/2)r^4$. Substituting these values into Equation (E4) and simplifying, we obtain Equation (E5).

$$w_P = \frac{\pi P R^3}{4E\left(\frac{\pi}{4}r^4\right)} + \left(\frac{3\pi}{4} - 2\right)\frac{PR^3}{G\left(\frac{\pi}{2}r^4\right)} \tag{E5}$$

ANS. $\quad w_P = \dfrac{PR^3}{r^4}\left[\dfrac{1}{E} + \dfrac{1}{G}\left(\dfrac{3\pi - 8}{2\pi}\right)\right]$

COMMENT

The example emphasizes that the out-of-plane deflection due to a force is a result of bending as well as torsion. This coupling of bending and torsion would not be seen in a straight beam under a bending force.

PROBLEM SET 7.2

7.18 Use the dummy unit load method to find the deflection and slope at the free end of the beam shown in Figure P7.18. Assume that EI is constant.

Figure P7.18

ANS. $\quad v(L) = PL^3/(3EI)$

downward $\quad \dfrac{dv}{dx} = (PL^2)/(2EI)$

7.19 Use the dummy unit load method to find the deflection at the midpoint for the beam and loading shown in Figure P7.19. Assume that EI is constant.

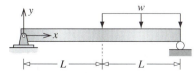

Figure P7.19

ANS. $\quad v(L) = 5wL^4/(48EI)$ downward

7.20 Use the dummy unit load method to find the deflection at the midpoint for the beam and loading shown in Figure P7.20. Assume that EI is constant.

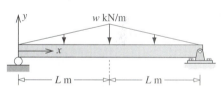

Figure P7.20

ANS. $\quad v(L) = 2wL^4/(15EI)$ downward

7.21 Use the dummy unit load method to determine the reaction force at A and the deflection and slope at B in terms of P, E, I, and L for the beam shown in Figure P7.21.

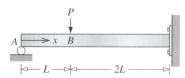

Figure P7.21

7.22 Use the dummy unit load method to find the reaction force at B and the deflection at A in terms of w, E, I, and L for the beam shown in Figure P7.22.

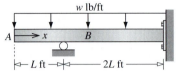

w lb/ft

A ⟶ x B

⟵ L ft ⟶⟵ $2L$ ft ⟶

Figure P7.22

ANS. $R_B = 17wL/8$

$v(0) = 5wL^4/(24EI)$ downward

7.23 A force $F = 20$ kN is applied to a pin as shown in Figure P7.23. Both bars have a cross-sectional area $A = 100$ mm^2 and a modulus of elasticity $E = 200$ GPa. Bars AP and BP have lengths of $L_{AP} = 200$ mm and $L_{BP} = 250$ mm, respectively. Use Castigliano's theorems to determine the displacement of point P in the direction of F.

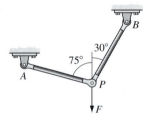

Figure P7.23

7.24 A force $F = 20$ kN is applied to a pin as shown in Figure P7.24. Bars AP and BP have a cross-sectional area $A = 100$ mm^2 and a modulus of elasticity $E = 200$ GPa. The lengths are $L_{AP} = 200$ mm and $L_{BP} = 250$ mm. Use Castigliano's theorems to determine the displacement of point P in the direction of F.

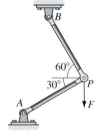

Figure P7.24

7.25 A steel ($G_{st} = 12,000$ ksi) shaft and a bronze ($G_{Cu/Sn} = 5600$ ksi) shaft are securely connected at B as shown in Figure P7.25. Use Castigliano's theorems to determine the reaction torque at A.

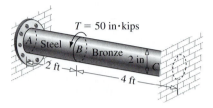

$T = 50$ in·kips

A Steel B Bronze 2 in C

2 ft — 4 ft

Figure P7.25

ANS. $T_A = 40.54$ in · kips

7.26 A solid circular steel ($G_{st} = 12,000$ ksi, $E_{st} = 30,000$ ksi) shaft of 4-inch diameter is loaded as shown in Figure P7.26. Use Castigliano's theorems to determine the reaction torque at A.

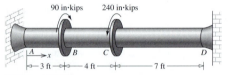

90 in·kips   240 in·kips

A ⟶ x B C D

⟵ 3 ft ⟶⟵ 4 ft ⟶⟵ 7 ft ⟶

Figure P7.26

7.27 Use Castigliano's theorems to find the reaction force at A and the deflection at B in terms of P, E, I, and L for the beam shown in Figure P7.21.

7.28 Use Castigliano's theorems to find the reaction force at B and the deflection at A in terms of w, E, I, and L for the beam shown in Figure P7.22.

7.29 Determine the deflection at the midpoint on the beam shown in Figure P7.29 in terms of w, E, I, and L.

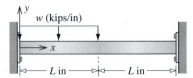

y

w (kips/in)

⟶ x

⟵ L in ⟶⟵ L in ⟶

Figure P7.29

ANS. $v(L) = wL^4/(48EI)$ downward

7.30 A semicircular beam is loaded as shown in Figure P7.30. Determine the deflection of point B in the direction of the applied load P in terms of E, I, P, and R.

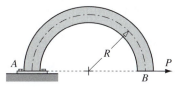

Figure P7.30

ANS. $v_B = \pi P R^3/(2EI)$

7.31 For the semicircular beam shown in Figure P7.30, determine the deflection in terms of E, I, P, and R at point B in a direction $45°$ counterclockwise to the applied load P.

ANS. $v_{45} = (\pi + 2)PR^3/(2\sqrt{2}EI)$

7.32 A semicircular beam is loaded as shown in Figure P7.32. Determine the deflection of point B in the direction of the applied load P in terms of E, I, P, and R.

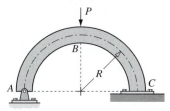

Figure P7.32

7.33 For the semicircular beam shown in Figure P7.32, determine the deflection of point C along the smooth surface in terms of E, I, P, and R.

7.34 A curved beam is loaded as shown in Figure P7.34. Determine the deflection of point B in the direction of the applied load P in terms of E, I, P, and R.

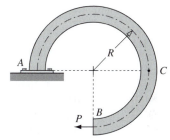

Figure P7.34

7.35 For the curved beam shown in Figure P7.34, determine the horizontal deflection of point C in terms of E, I, P, and R.

ANS. $v_C = (\pi + 4)PR^3/(2EI)$ leftward

7.36 For the curved beam shown in Figure P7.34, determine the vertical deflection of point B in terms of E, I, P, and R.

7.8 MINIMUM POTENTIAL ENERGY

We start by defining the potential energy function Ω as

$$\Omega = U - W \qquad (7.30)$$

where U is the strain energy and W is the work potential of a force. Unlike work, which is a concept associated with any force (conservative or nonconservative) that moves, the "work potential of a force" is associated with conservative forces only and implies that there is a potential function from which such a force can be obtained. Gravitational forces and electromagnetic forces[9] are two examples of conservative forces that can be obtained from a

---

[9] Conservative forces can be written as $F_i = -(\partial W/\partial x_i)$, where W is the work potential of the force F_i. If we integrate the equation between two points, we obtain $W_2 - W_1 = \int_{x_1}^{x_2} F_i dx_i$, irrespective of the path of integration between x_1 and x_2, which matches the definiton of conservative forces in definition 3, given earlier.

potential function. If we assume that the external forces are acting on a *elastic* (linear or nonlinear) structure and are conservative, then the "work potential of a force" is calculated as the work term in Figure 7.5 without the symbol for virtual displacement. Thus, from our perspective, the work potential and the work done by the force are calculated similarly, provided the external forces are conservative and are applied to elastic systems.

To obtain the statement of the theorem of minimum potential energy, we consider the statement of virtual work as given by Equation (7.17). Restricting ourselves to conservative elastic systems, we can state that the internal virtual work is the variation in elastic strain energy during deformation (i.e., $\delta W_{int} = \delta U$) and the external virtual work is the variation in the work potential of the force (i.e., $\delta W_{ext} = \delta W$). Thus, from Equation (7.17) we obtain

$$\delta W_{int} - \delta W_{ext} = \delta U - \delta W = \delta \Omega = 0 \qquad (7.31)$$

Equation (7.31) implies that at equilibrium, the virtual variation in the potential energy function is zero—which occurs where the slope of the potential energy function with respect to the parameters defining the potential function is zero. The parameters defining the potential function are the generalized displacements. The zero slope condition can occur where the potential function is at maximum, at a saddle point, or at minimum. The maximum of the potential function occurs at points of unstable equilibrium, and the saddle points occur at points of neutral stability, neither of which we shall consider here. The minimum in potential function occurs at the stable equilibrium point, which is the only equilibrium we will consider here. The theorem of minimum potential energy can be stated as follows.

> Of all the kinematically admissible displacement functions, the actual displacement function is the one that minimizes the potential energy function at stable equilibrium.

In Section 7.4.2 it was noted that there are many kinematically admissible displacement functions, and there is no requirement that these functions satisfy the equilibrium equations or the static boundary conditions.[10] The actual displacement is kinematically admissible and satisfies all the equilibrium conditions and the static boundary conditions. Thus, if we choose an arbitrary kinematically admissible function and calculate the potential energy function, the value so obtained will always be greater than the value of the potential energy function at equilibrium. In other words, we approach the potential energy function value at equilibrium from above. Thus, if we have two approximations for the displacement functions, the one that gives a lower potential energy value is the better approximation. A corollary to the preceding statement is that if we add another term (increase the degrees of freedom) in an approximation, the potential energy value can only decrease (i.e., improve the accuracy of the approximation). We record the following observations.

- The better approximation of displacement function is the one that yields the lower potential energy.
- The greater the degrees of freedom, the lower will be the potential energy for a given set of kinematically admissible functions.[11]

---

[10] The virtual displacement method in Section 7.4.4 is one way of finding the kinematic functions that satisfy the equilibrium equations.

[11] This is the basis of improving accuracy by making a finer mesh (increasing degrees of freedom) in the finite element methods discussed in Section 9.3.6.

7.9   RAYLEIGH–RITZ METHOD

The Rayleigh–Ritz method is an approximate method of finding displacements that is based on the theorem of minimum potential energy. The method is restricted to conservative systems, which may be linear or nonlinear.

Let the displacement $u(x)$ be represented by a series of kinematically admissible functions $f_i(x)$ as shown in Equation (7.32a),

$$u(x) = \sum_{i=1}^{n} C_i f_i(x) \tag{7.32a}$$

where C_i are undetermined constants (i.e., generalized displacements as in Definition 4). By substituting Equation (7.32a) into the strain energy U and the work potential W, we can obtain the potential energy function Ω in terms of the constants C_i. The constants C_i are to be determined such that Ω is minimized. The necessary condition at the minimum value of Ω is given by Equation (7.32b).

$$\frac{\partial \Omega}{\partial C_i} = 0 \qquad i = 1, n \tag{7.32b}$$

The preceding statement represents a set of n algebraic equations in the constants C_i, which can be solved simultaneously to obtain the values of C_i. Once C_i are known, the stresses can be found from the displacement function in Equation (7.32a). We will discuss the use of the Rayleigh-Ritz method in the context of one-dimensional linear structural elements, although it is equally applicable to two and three dimensions, and for linear and nonlinear material behavior, provided the system is conservative.

7.9.1   Axial Members

The linear elastic strain energy density for axial members was given by Equation (7.8a). Taking the derivative of Equation (7.32a), we obtain Equation (7.33a).

$$\frac{du}{dx} = \sum_{j=1}^{n} C_j \frac{df_j}{dx} \tag{7.33a}$$

The axial strain energy density can be written as shown in Equation (7.33b).

$$U_a = \frac{1}{2} EA \left(\frac{du}{dx}\right)\left(\frac{du}{dx}\right) = \frac{1}{2} EA \left(\sum_{j=1}^{n} C_j \frac{df_j}{dx} \right)\left(\sum_{k=1}^{n} C_k \frac{df_k}{dx} \right)$$

$$= \frac{1}{2} EA \sum_{j=1}^{n} \sum_{k=1}^{n} C_j C_k \left(\frac{df_j}{dx}\right)\left(\frac{df_k}{dx}\right) \tag{7.33b}$$

The axial strain energy is as shown in Equation (7.33c),

$$U_A = \int_0^L U_a dx = \frac{1}{2} \sum_{j=1}^{n} \sum_{k=1}^{n} C_j C_k \left[\int_0^L EA\left(\frac{df_j}{dx}\right)\left(\frac{df_k}{dx}\right) dx \right] = \frac{1}{2} \sum_{j=1}^{n} \sum_{k=1}^{n} C_j C_k K_{jk} \tag{7.33c}$$

where the matrix K_{jk} is called the stiffness matrix and is given by Equation (7.33d).

$$K_{jk} = \int_0^L EA\left(\frac{df_j}{dx}\right)\left(\frac{df_k}{dx}\right)dx \tag{7.33d}$$

In Equation (7.33d), EA can be a function of x. If the axial member is tapered, then A as a function of x can be found and used in Equation (7.33d). If the axial member is made up segments of different materials, or if the cross section changes for each segment, then the integration over the length of the axial member can be written as the sum of the integrals over the segments, and the changing properties of E and A can be accounted for in Equation (7.33d).

We assume an axial distributed force $p_x(x)$ and concentrated axial forces applied at m points with coordinates x_q. The work potential of these forces can be written as follows:

$$W_A = \int_0^L p_x(x)u(x)dx + \sum_{q=1}^m F_q u(x_q) \tag{7.33e}$$

Substituting Equation (7.32a) in Equation (7.33e), we obtain Equation (7.33f),

$$W_A = \sum_{j=1}^n C_j \int_0^L p_x(x)f_j(x)dx + \sum_{q=1}^m \sum_{j=1}^n F_q C_j f_j(x_q) = \sum_{j=1}^n C_j R_j \tag{7.33f}$$

where the vector R_j is as given in Equation (7.33g).

$$R_j = \int_0^L p_x(x)f_j(x)dx + \sum_{q=1}^m F_q f_j(x_q) \tag{7.33g}$$

The potential energy function for axial members can be written as shown in Equation (7.34a).

$$\Omega_A = U_A - W_A = \frac{1}{2}\sum_{j=1}^n \sum_{k=1}^n C_j C_k K_{jk} - \sum_{j=1}^n C_j R_j \tag{7.34a}$$

In accordance with Equation (7.32b), we minimize the potential energy function by taking the derivative of Equation (7.34a) with respect to C_i to obtain Equation (7.34b).

$$\frac{\partial \Omega_A}{\partial C_i} = \frac{1}{2}\sum_{j=1}^n \sum_{k=1}^n \left[\frac{\partial C_j}{\partial C_i}C_k + C_j\frac{\partial C_k}{\partial C_i}\right]K_{jk} - \sum_{j=1}^n \frac{\partial C_j}{\partial C_i}R_j = 0 \qquad i = 1, n \tag{7.34b}$$

We note that the C's are generalized displacements that are independent of each other. Thus, the property of the partial derivatives of the C's is as shown in Equation (7.34c).

$$\frac{\partial C_j}{\partial C_i} = \begin{cases} 1 & i = j \\ 0 & i \neq j \end{cases} \tag{7.34c}$$

Substituting Equation (7.34c) into Equation (7.34b), we obtain Equation (7.34d).

$$
\frac{\partial \Omega_A}{\partial C_i} = \frac{1}{2}\left(\sum_{k=1}^{n} C_k K_{ik} + \sum_{j=1}^{n} C_j K_{ji} \right) - R_i
$$

$$
= \frac{1}{2}\left(\sum_{j=1}^{n} C_j (K_{ij} + K_{ji}) \right) - R_i = 0 \qquad i = 1, n \tag{7.34d}
$$

Equation (7.33d) shows that the matrix K is symmetric, as shown in Equation (7.34e).

$$
\boxed{K_{ji} = K_{ij}} \tag{7.34e}
$$

Substituting Equation (7.34e) into Equation (7.34d), we obtain the following set of algebraic equations:

$$
\boxed{\sum_{j=1}^{n} K_{ij} C_j = R_i \qquad i = 1, n} \tag{7.35a}
$$

By using matrix notation, Equation (7.35a) can be written as follows:

$$
\boxed{[K]\{C\} = \{R\}} \tag{7.35b}
$$

The set of n algebraic equations in Equation (7.35a) can be solved to obtain the values of C_j. The displacement function in Equation (7.32a) now is completely known. With axial displacement known, axial stresses, axial strains, the internal axial force on a section, and the potential energy of the axial member can be found. Let the solution of Equation (7.35a) be represented by C_j^*. The expression for potential energy in Equation (7.34a) can be simplified by using Equation (7.35a) as shown in Equation (7.35c).

$$
\Omega_A = \frac{1}{2}\sum_{j=1}^{n} C_j^*\left(\sum_{k=1}^{n} K_{jk} C_k^* \right) - \sum_{j=1}^{n} C_j^* R_j = \frac{1}{2}\sum_{j=1}^{n} C_j^* R_j - \sum_{j=1}^{n} C_j^* R_j
$$

$$
= -\left(\frac{1}{2}\right)\sum_{j=1}^{n} C_j^* R_j \tag{7.35c}
$$

It should be emphasized that Equation (7.35c) is valid only at equilibrium. Equation (7.35c) also highlights that at equilibrium, the strain energy is half the work potential of the forces, as shown in Equation (7.36).

$$
\boxed{U_A = \frac{W_A}{2} = -\Omega_A = \frac{1}{2}\sum_{j=1}^{n} C_j^* R_j \qquad \text{at equilibrium}} \tag{7.36}
$$

The strain energy is always positive. Thus at equilibrium, the potential energy is negative for the elastic systems we will study. Thus, improved accuracies from approximations will make the potential energy more negative.

Clearly for the large values of n (i.e., number of unknowns) in Equation (7.35a), we will need numerical methods to solve the set of algebraic equations. But we can demonstrate the Rayleigh-Ritz method by using small values of n, as will be illustrated after the development of the Rayleigh-Ritz method for the torsion of circular shafts and the symmetric bending of beams (see Example 7.21).

7.9.2 Torsion of Circular Shafts

The derivation of the algebraic equations for the torsion of circular shafts is similar to that of axial members and hence only the key equations are described here. The torsional rotation of a section in terms of a kinematically admissible function $f_i(x)$ can be written as follows:

$$\phi(x) = \sum_{i=1}^{n} C_i f_i(x) \tag{7.37a}$$

The torsional strain energy is given by Equation (7.9a). We assume that there is a distributed torque $t(x)$ and m concentrated torques (T_q) applied at points with coordinates x_q. The work potential of these forces can be written as Equation (7.37b).

$$W_T = \int_0^L t(x)\phi(x)dx + \sum_{q=1}^{m} T_q\phi(x_q) \tag{7.37b}$$

It can be shown (see later: Problem 7.37) that the matrix K_{jk} and the right-hand-side vector R_j in Equation (7.35a) can be written as Equations (7.37c) and (7.37d).

$$K_{jk} = \int_0^L GJ\left(\frac{df_j}{dx}\right)\left(\frac{df_k}{dx}\right)dx \tag{7.37c}$$

$$R_j = \int_0^L t(x)f_j(x)dx + \sum_{q=1}^{m} T_q f_j(x_q) \tag{7.37d}$$

In Equation (7.37a), GJ can be function of x. If the shaft is tapered, then J as a function of x can be found and used in Equation (7.37a). If the shaft is made up of segments of different materials, or if the cross section changes for each segment, then the integration over the length of the shaft can be written as the sum of the integrals over the shaft segments, and the changing properties of G and J can be accounted for in Equation (7.37a).

Once more, Equation (7.35a) can be solved for the constants C_i, and thus rotation at all points is now known from Equation (7.37a). With rotation known, the torsional shear strain, torsional shear stress, internal torque on a section, and potential energy in the shaft can be found.

7.9.3 Symmetric Bending of Beams

The derivation of the algebraic equations for the symmetric bending of beams is similar to that for axial members and torsion of circular shafts. Once more, only the key equations are described here. The bending displacement in terms of a kinematically admissible function $f_i(x)$ can be written as Equation (7.38a).

$$v(x) = \sum_{i=1}^{n} C_i f_i(x) \tag{7.38a}$$

The bending strain energy was given by Equation (7.10a). We assume that there is a distributed load $p_y(x)$ and m_1 concentrated forces (F_q) and m_2 concentrated moments (M_q). The work potential of these forces and moments can be written as Equation (7.38b).

$$W_B = \int_0^L p_y(x)v(x)dx + \sum_{q=1}^{m_1} F_q v(x_q) + \sum_{q=1}^{m_2} M_q \frac{dv}{dx}(x_q) \tag{7.38b}$$

It can be shown (see Problem 7.38) that the matrix K_{jk} and the right-hand-side vector R_j in Equation (7.35a) can be written as Equations (7.38c) and (7.38d).

$$\boxed{K_{jk} = \int_0^L (EI_{zz})\left(\frac{d^2 f_j}{dx^2}\right)\left(\frac{d^2 f_k}{dx^2}\right)dx} \tag{7.38c}$$

$$\boxed{R_j = \int_0^L p_y(x)f_j(x)dx + \sum_{q=1}^{m_1} F_q f_j(x_q) + \sum_{q=1}^{m_1} M_q \frac{df_j}{dx}(x_q)} \tag{7.38d}$$

In Equation (7.38a), EI_{zz} can be function of x. If the beam is tapered, then I_{zz} as a function of x can be found and used in Equation (7.38a). If the beam is made up of segments of different materials, or if the cross section changes for each segment, then the integration over the length of the beam can be written as the sum of the integrals over the beam segments, and the changing properties of E and I_{zz} can be accounted for in Equation (7.38a).

Once more, Equation (7.35a) can be solved for the constants C_i, and thus the deflection at all points is now known from Equation (7.38a). With the displacement known, the bending normal strain, bending normal stress, internal bending moment on a section, and potential energy in the beam can be found.

EXAMPLE 7.21

A beam and its loading are shown in Figure 7.40. Use the Rayleigh-Ritz method with one and two parameters to determine the deflection at $x = 0.25L$, $x = 0.5L$, $x = 0.75L$, and $x = L$, and the potential energy function. Compare your results with the analytical solution. Assume that EI is constant for the beam.

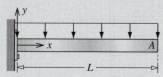

Figure 7.40 Beam for Example 7.21.

PLAN

The kinematic boundary conditions for the deflection and slope are zero at $x = 0$, as in Example 7.4. We can use the set of kinematically admissible functions obtained in Example 7.4, even though the loading is different. The matrix elements and the elements of the right-hand-side vector can be found from Equations (7.38c) and (7.38d). The algebraic equations can be solved for the constants C_i for each case (number of parameters) and the required quantities calculated.

SOLUTION

From Example 7.4 we have the following kinematically admissible displacement field:

$$v(x) = \sum_{i=1}^{n} C_i x^{i+1} \tag{E1}$$

The first and second derivatives of the functions f_i can be written as shown in Equation (E2).

$$f_i(x) = x^{i+1} \qquad \frac{df_i}{dx} = (i+1)x^i \qquad \frac{d^2 f_i}{dx^2} = (i+1)(i)x^{i-1} \tag{E2}$$

We note that the only loading is a distributed load, hence $p_y = -w$, $F_q = 0$, and $M_q = 0$.

One parameter ($n = 1$): From Equations (7.38c) and (7.38d) we obtain Equations (E3) and (E4).

$$K_{11} = (EI)\int_0^L \left(\frac{d^2 f_1}{dx^2}\right)\left(\frac{d^2 f_1}{dx^2}\right)dx = (EI)\int_0^L (2)(2)dx = 4(EI)L \tag{E3}$$

$$R_1 = \int_0^L (-w)f_1(x)dx = -w\int_0^L x^2 dx = -\left(\frac{wL^3}{3}\right) \tag{E4}$$

From Equation (7.35a) we obtain Equation (E5).

$$4(EI)LC_1 = -\left(\frac{wL^3}{3}\right) \qquad \text{or} \qquad C_1 = -\left(\frac{wL^2}{12EI}\right) \tag{E5}$$

The potential energy can be found from Equation (7.35c) as shown in Equation (E6).

$$\Omega = -\left(\frac{1}{2}\right)C_1 R_1 = -\frac{1}{2}\left(\frac{wL^2}{12EI}\right)\left(\frac{wL^3}{3}\right) = -\left(\frac{w^2 L^5}{72EI}\right) = -0.0139\left(\frac{w^2 L^5}{EI}\right) \tag{E6}$$

The constant C_1 can be substituted into Equation (E1) and the deflection evaluated at $x = 0.25L$, $x = 0.5L$, $x = 0.75L$, and $x = L$. The results are shown in Table 8.1.

Two parameter (n = 2): We calculate K_{11} and R_1 as for the one-parameter solution. The rest of the quantities can be found from Equations (7.38c) and (7.38d), as shown in Equations (E7), (E8), and (E9).

$$K_{12} = (EI)\int_0^L \left(\frac{d^2 f_1}{dx^2}\right)\left(\frac{d^2 f_2}{dx^2}\right)dx = (EI)\int_0^L (2)(6x)dx = 6(EI)L^2 \tag{E7}$$

$$K_{22} = (EI)\int_0^L \left(\frac{d^2 f_2}{dx^2}\right)\left(\frac{d^2 f_2}{dx^2}\right)dx = (EI)\int_0^L (6x)(6x)dx = 12(EI)L^3 \tag{E8}$$

$$R_2 = \int_0^L (-w) f_2(x)dx = -w\int_0^L x^3 dx = -\left(\frac{wL^4}{4}\right) \tag{E9}$$

Noting that $K_{21} = K_{12}$, we obtain Equation (E10) from Equation (7.35a):

$$\begin{bmatrix} 4(EI)L & 6(EI)L^2 \\ 6(EI)L^2 & 12(EI)L^3 \end{bmatrix} \begin{Bmatrix} C_1 \\ C_2 \end{Bmatrix} = \begin{Bmatrix} -w(L^3/3) \\ -(wL^4)/4 \end{Bmatrix} \tag{E10}$$

Solving Equation (E10), we obtain the values of C_1 and C_2 as shown in Equations (E11).

$$C_1 = -\left(\frac{5wL^2}{24EI}\right) \quad \text{and} \quad C_2 = \left(\frac{wL}{12EI}\right) \tag{E11}$$

The potential energy can be found from Equation (7.35c) as shown in Equation (E12).

$$\Omega = -\left(\frac{1}{2}\right)(C_1 R_1 + C_2 R_2) = -\frac{1}{2}\left[\left(\frac{5wL^2}{24EI}\right)\left(\frac{wL^3}{3}\right) - \left(\frac{wL}{12EI}\right)\left(\frac{wL^4}{4}\right)\right] = -\left(\frac{7wL^5}{288EI}\right) \tag{E12}$$

The constants C_1 and C_2 can be substituted in Equation (E1) and the deflection evaluated at $x = 0.25L$, $x = 0.5L$, $x = 0.75L$, and $x = L$. The results are given in Table 8.1.

Analytical solution: The analytical solution for the deflection can be obtained by integration as follows:

$$v(x) = -\left(\frac{w}{24EI}\right)(x^4 - 3Lx^3 + 3L^2 x^2) \tag{E13}$$

Equation (E13) can be used to evaluate the deflection at $x = 0.25L$, $x = 0.5L$, $x = 0.75L$, and $x = L$. The results are given in Table 8.1.

The strain energy in bending can be found from Equation (7.10a), as shown in Equation (E14).

$$U_B = \frac{1}{2}EI\int_0^L \left(\frac{d^2 v}{dx^2}\right)^2 dx = \frac{w^2}{8EI}\int_0^L (x - L)^4 dx = \frac{w^2 L^5}{40EI} \tag{E14}$$

The work potential can be found from Equation (7.38b), as shown in Equation (E15).

$$W = \int_0^L (-w)\left[-\left(\frac{w}{24EI}\right)(x^4 - 4Lx^3 + 6L^2x^2)\right]$$

$$= \frac{w^2}{24EI}\left(\frac{x^5}{5} - 4L\frac{x^4}{4} + 6L^2\frac{x^3}{3}\right)\Bigg|_0^L = \frac{w^2L^5}{20EI} \tag{E15}$$

The potential energy can be found from Equation (7.30), as shown in Equation (E16).

$$\Omega = U_B - W = \frac{w^2L^5}{40EI} - \frac{w^2L^5}{20EI} = -\frac{w^2L^5}{40EI} = -0.025\left(\frac{w^2L^5}{EI}\right) \tag{E16}$$

The results of the deflection and potential energy are given in Table 8.1. The percentage difference is calculated by using Equation (E17).

$$\% \, \text{diff} = \left|\frac{\text{analytical value} - \text{calculated value}}{\text{analytical value}}\right| \times 100 \tag{E17}$$

TABLE 8.1 Results for Example 7.21

	Deflection $v / \left(\dfrac{wL^4}{EI}\right)$								Potential Energy $\Omega / \left(\dfrac{w^2L^5}{EI}\right)$	
	$v\left(\dfrac{L}{4}\right)$	% diff	$v\left(\dfrac{L}{2}\right)$	% diff	$v\left(\dfrac{3L}{4}\right)$	% diff	$v(L)$	% diff	Value	% diff
$n = 1$	−0.0052	60.5	−0.0208	52.94	−0.0469	43.86	−0.0833	33.3	−0.0139	44.44
$n = 2$	−0.0117	11.1	−0.0417	5.88	−0.0820	1.75	−0.125	0	−0.0243	2.78
Analytical	−0.0132		−0.0443		−0.0835		−0.125		−0.025	

COMMENTS

1. The improvement of results with two parameters over the solution by one parameter may be less dramatic for more complex problems than those shown in Table 8.1.

2. Notice that there is no difference between the analytical results and two-parameter results at $x = L$, but at $x = L/4$ there is an 11.1% difference. Great care must be taken in drawing conclusions of improvement from point values of displacements and stresses.

3. Table 8.1 shows that the potential energy decreases with an increase in parameters. With the addition of parameters, the potential energy either decreases or remains the same, but it will never increase. Thus, the decrease in potential energy is a surer measure of improvement in accuracy than the use of values at a point.

4. For three parameters, we would get the analytical results of this problem because the three kinematic functions x^2, x^3, and x^4 in the approximation are three terms in the analytical solution of Equation (E13).

5. Analytical results are not same as exact results because an analytical solution also starts with a model that is an approximation of reality. Beam theory is an approximate theory.

7.10 FUNCTIONALS

Strain energy, the potential of work, and potential energy are all functions of the displacements. The displacements are functions of the position x. Thus, strain energy, the potential of work, and potential energy are functions of a function. Such functions of a function are called *functionals*. *Variational calculus*[12] is the branch of mathematics dealing with finding the maximum and minimum values of functionals. The minimum potential energy is finding a minimum of a functional. Another area of application of variational calculus is optimization methods. Designing structures for minimum weight is a problem in optimization in which the weight is the functional that is minimized. Industrial engineering contains many applications of variational calculus involving minimizing cost, time, and so on. There are properties of functionals that can be effectively used in reducing tedious algebra, as shall be shown in discussion that follows.

A functional $l(u)$ is said to be linear if and only if it satisfies the relationship in Equation (7.39a),

$$l(\alpha_1 u + \alpha_2 v) = \alpha_1 l(u) + \alpha_2 l(v) \tag{7.39a}$$

where α_1 and α_2 are any scalars. In Equation (7.33e) the work potential is a linear functional of the displacement u; in Equation (7.37b) it is a linear functional of rotation ϕ; in Equation (7.38b) it is a linear functional of the displacement v.

A functional $B(u, v)$ is said to be a bilinear functional if it is a linear functional in each of its arguments of u and v, as shown in Equation (7.39b).

$$B(\alpha_1 u_1 + \alpha_2 u_2, v) = \alpha_1 B(u_1, v) + \alpha_2 B(u_2, v)$$
$$B(u, \alpha_1 v_1 + \alpha_2 v_2) = \alpha_1 B(u, v_1) + \alpha_2 B(u, v_2) \tag{7.39b}$$

Strain energy and potential of work can be written as a bilinear functional and as linear functionals for axial, torsion, and bending problems, as shown in Equations (7.40a) and (7.40b). By substituting $u_1 = u_2 = u$, $\phi_1 = \phi_2 = \phi$, and $v_1 = v_2 = v$ into Equations (7.40a) and (7.40b), we obtain the equations for the individual structural members.

$$U_A = \frac{1}{2} \int_L \left[EA \frac{du_1}{dx} \frac{du_2}{dx} \right] dx = \frac{1}{2} B(u_1, u_2)$$

$$U_T = \frac{1}{2} \int_L \left[GJ \frac{d\phi_1}{dx} \frac{d\phi_2}{dx} \right] dx = \frac{1}{2} B(\phi_1, \phi_2) \tag{7.40a}$$

$$U_B = \frac{1}{2} \int_L \left[EI_{zz} \frac{d^2 v_1}{dx^2} \frac{d^2 v_2}{dx^2} \right] dx = \frac{1}{2} B(v_1, v_2)$$

---

[12] See Lanczos [1986] in Appendix D for additional details.

$$W_A = \int_0^L p_x(x)u_1(x)dx + \sum_{q=1}^{m} F_q u_1(x_q) = l(u_1)$$

$$W_T = \int_0^L t(x)\phi_1(x)dx + \sum_{q=1}^{m} T_q \phi_1(x_q) = l(\phi_1) \tag{7.40b}$$

$$W_B = \int_0^L p_y(x)v_1(x)dx + \sum_{q=1}^{m_1} F_q v_1(x_q) + \sum_{q=1}^{m_2} M_q \frac{dv_1}{dx}(x_q) = l(v_1)$$

Equation (7.40a) also shows that the bilinear functionals representing strain energy are symmetric in their arguments, as shown in Equation (7.41).

$$B(u_1, u_2) = B(u_2, u_1) \qquad B(\phi_1, \phi_2) = B(\phi_2, \phi_1) \qquad B(v_1, v_2) = B(v_2, v_1) \tag{7.41}$$

The potential energy for axial, torsion, and bending problems thus has the generic form given by Equation (7.42).

$$\Omega = \frac{1}{2}B(u, u) - l(u) \tag{7.42}$$

By representing u by a series of kinematically admissible functions and using the properties of linear and bilinear functionals, we can obtain the stiffness matrix and the right-hand-side vectors as shown in Equation (7.43). This will be demonstrated for axial problems in Example 7.22 (see also Problems 7.44 and 7.45).

$$K_{jk} = B(f_j, f_k) \qquad \text{and} \qquad R_j = l(f_j) \tag{7.43}$$

In Equation (7.43) the symmetric property of the bilinear functional results in a symmetric stiffness matrix.

EXAMPLE 7.22

Representing u by a series of kinematically admissible functions [Equation (7.32a)] and using the properties of linear and bilinear functionals, show that the stiffness matrix and the right-hand-side vectors are given by Equations (7.33d) and (7.33g), respectively.

PLAN

Equation (7.32a) will be substituted for u_1 in the bilinear and linear functionals and by using the properties of Equations (7.39a) and (7.39b), the series will be obtained. Then Equation (7.32a) will be substituted for u_2 and another summation obtained. Equations (7.40a) and (7.40b) will provide the desired results.

SOLUTION

We can write Equation (7.32a) as Equation (E1).

$$u_1(x) = \sum_{j=1}^{n} C_j f_j(x) \quad \text{and} \quad u_2(x) = \sum_{k=1}^{n} C_k f_k(x) \tag{E1}$$

Substituting u_1 of Equation (E1) into the bilinear and linear functionals and using Equations (7.39a) and (7.39b), we obtain Equations (E2) and (E3).

$$l(u_1) = l\left(\sum_{j=1}^{n} C_j f_j(x)\right) = \sum_{j=1}^{n} C_j l(f_j(x)) \tag{E2}$$

$$B(u_1, u_2) = B\left(\sum_{j=1}^{n} C_j f_j(x), u_2\right) = \sum_{j=1}^{n} C_j B(f_j(x), u_2) \tag{E3}$$

Substituting u_1 from Equation (E1) into Equation (E3) and using Equation (7.39b), we obtain Equation (E4).

$$B(u_1, u_2) = \sum_{j=1}^{n} C_j B\left(f_j(x), \sum_{k=1}^{n} C_k f_k(x)\right) = \sum_{j=1}^{n} C_j \sum_{k=1}^{n} C_k B(f_j(x), f_k(x))$$

$$B(u_1, u_2) = \sum_{j=1}^{n} \sum_{k=1}^{n} C_j C_k B(f_j(x), f_k(x)) \tag{E4}$$

Substituting f_j for u_1 and f_k for u_2 in the axial terms of Equations (7.40a) and (7.40b) and noting the definitions in Equations (7.33d) and (7.33g), we obtain Equations (E5) and (E6).

ANS. $\quad B(f_j(x), f_k(x)) = \int_{0}^{L} EA\left(\frac{df_j}{dx}\right)\left(\frac{df_k}{dx}\right) dx = K_{jk} \tag{E5}$

ANS. $\quad l(f_j(x)) = \int_{0}^{L} p_x(x) f_j(x) dx + \sum_{q=1}^{m} F_q f_j(x_q) = R_j \tag{E6}$

COMMENTS

1. The potential energy can be written in the form of Equation (E7).

$$\Omega_A = \frac{1}{2} \sum_{j=1}^{n} \sum_{k=1}^{n} C_j C_k B(f_j(x), f_k(x)) - \sum_{j=1}^{n} C_j l(f_j(x)) \tag{E7}$$

By minimizing the potential energy with respect to C_i, we can obtain the algebraic equations represented by Equation (7.35b).

2. The definitions of bilinear and linear functionals make it possible to write equations in compact form, and the properties given by Equations (7.39a) and (7.39b) reduce the algebra significantly.

PROBLEM SET 7.3

7.37 Starting with the torsional rotation approximation of Equation (7.37a), show that minimization of potential energy results in the shaft stiffness matrix given by Equation (7.37c) and the load vector given by Equation (7.37d).

7.38 Starting with the bending displacement approximation of Equation (7.38a), show that minimization of potential energy results in the beam stiffness matrix given by Equation (7.38c) and the load vector given by Equation (7.38d).

7.39 For the beam shown in Figure P7.39, determine the two-parameter Rayleigh-Ritz solution using the approximation for the bending displacement given. Is this solution better than the one in Example 7.21? Justify your answer by comparing potential energies.

$$v(x) = \sum_{i=1}^{n} C_i\left(1 - \cos\frac{i\pi x}{L}\right)$$

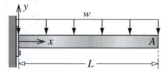

Figure P7.39

7.40 For the beam shown in Figure P7.40, determine one- and two-parameter Rayleigh-Ritz solutions, using the approximation for the bending displacement given. Calculate the potential energy function for the one- and two-parameter problems.

$$v(x) = \sum_{i=1}^{n} C_i\left(1 - \cos\frac{2i\pi x}{L}\right)$$

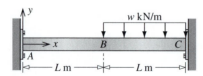

Figure P7.40

ANS.

$$C_1 = -\left(\frac{wL^4}{\pi^4 EI}\right) \qquad C_2 = -\left(\frac{wL^4}{16\pi^4 EI}\right)$$

$$\Omega_1 = -\left(\frac{w^2 L^5}{2\pi^4 EI}\right) \qquad \Omega_2 = -\left(\frac{17 w^2 L^5}{32\pi^4 EI}\right)$$

7.41 For the beam shown in Figure P7.41, determine the one- and two-parameter Rayleigh-Ritz solutions, using the approximation for the bending displacement given. Calculate the potential energy function for the one- and two-parameter problems. Assume that EI is constant.

$$v(x) = \sum_{i=1}^{n} C_i x^i (x - 2L)$$

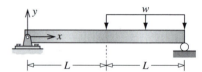

Figure P7.41

7.42 For the beam shown in Figure P7.41, determine the one- and two-parameter Rayleigh-Ritz displacement solutions, using the approximation for the bending displacement given. Calculate the potential energy function for the one- and two-parameter problems.

$$v(x) = \sum_{i=1}^{n} C_i \sin\frac{i\pi x}{2L}$$

7.43 For the beam shown in Figure P7.43, determine a two-parameter Rayleigh-Ritz displacement solution, using a polynomial approximation for the bending displacement.

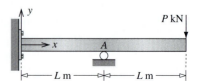

Figure P7.43

7.44 Representing ϕ by means of a series of kinematically admissible functions [Equation (7.37a)] and using the properties of linear and bilinear functionals, show that the stiffness matrix and the right-hand-side vectors are given by Equations (7.37c) and (7.37d), respectively.

7.45 Representing v by means of a series of kinematically admissible functions [Equation (7.38a)] and using the properties of linear and bilinear functionals, show that the stiffness matrix and the right-hand-side vectors are given by Equations (7.38c) and (7.38d), respectively.

7.11   CLOSURE

In this chapter we saw three methods of calculating displacements and rotations by using energy methods that were obtained from the theorem of virtual work. The three methods were the dummy unit load method, Castigliano's theorems, and the Rayleigh-Ritz method. These three methods utilized the concepts of strain energy, complementary strain energy, work, kinematically admissible functions, statically admissible functions, generalized displacements, generalized forces, virtual force, and virtual displacements. Castigliano's theorems are elegant and require the least computational work for finding reaction forces and moments in statically indeterminate structures. They are also effective if there is a point force (or point moment) already present at the point for which we wish to find displacement (or rotation). If there is no point force (or point moment) present in the actual loading, the dummy unit load method is easier to implement in determining the displacement (or rotation). The Rayleigh-Ritz method, an elegant formalism for implementing the minimum potential energy theorem to find the approximate displacement or slope, can be used for linear and nonlinear systems provided the systems are conservative. Effective use of the three methods intrinsically depends on an understanding of the underlying concepts utilized by each method.

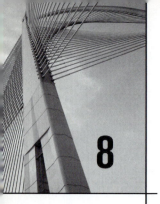

8 | Elasticity and the Mechanics of Materials

8.0 OVERVIEW

Stress and stiffness analysis and design of three-dimensional (and many two-dimensional) machine components cannot be achieved by means of a mechanics of materials approach. Numerical methods are often used, but even these call for an understanding of equations of elasticity. Figure 8.1 shows a bracket that was modeled by using three-dimensional tetrahedron elements in the finite element method. To obtain the properties (element stiffness matrix) of a tetrahedron element, we need equations of elasticity. In "elasticity," we study how the variables of mechanics (displacements, strains, stresses, internal forces, and moments) vary from point to point on a elastic body. This is in contrast to a mechanics of materials approach, in which we average the variables of mechanics by approximating their variation across the cross section or thickness[1] of an elastic body. The field of elasticity is very large and beyond the scope of a single course, let alone a chapter. Here we simply introduce some basic concepts and equations of elasticity, show the relationship of these equations to mechanics of materials equations and results, and use the equations of elasticity to derive formulas for such applications as torsion of noncircular shafts, thick cylinders, and rotating disks.

The basic equations of elasticity are the strain–displacement equations (studied in Chapter 1), the generalized version of Hooke's law relating stresses and strains (studied in Chapter 2), the equilibrium equations on stresses, and the compatibility equations for strains, which ensure single-valued solutions for displacements. We will see that the equilibrium equations on internal forces and moments in the mechanics of materials can be obtained from the equilibrium equations of elasticity.

---

[1] In case of plates and shells, we average the variables across the thickness and study the variation of the variables across the plan form of an elastic body.

Figure 8.1 Finite element mesh of a bracket. (*Courtesy of Professor C.R. Vilmann.*)

An important technique for solving an elasticity problem is to use a function from which stresses can be derived that satisfy the equilibrium equations. Such a function is called an Airy stress function. We will see how the Airy stress functions can be used to obtain the results in mechanics of materials problems.

The learning objectives in this chapter are:

1. To become familiar with the equations of elasticity and the Airy stress function.

2. To understand the application of the equations of elasticity to thick cylinders, rotating disks, and the torsion of noncircular bars.

8.1 ELASTICITY EQUATIONS

The basic equations of linear elasticity are the strain–displacement equations, the compatibility equations, which are derived from the strain–displacement equations, the generalized version of Hooke's law, and the equilibrium equations.[2] These are discussed in this section.

8.1.1 Compatibility Equations

The strain–displacement relationships, given earlier as Equations (1.35a) through (1.35f), are repeated as Equations (8.1a) through (8.1f) for convenience.

$$\boxed{\varepsilon_{xx} = \frac{\partial u}{\partial x}} \qquad (8.1a)$$

$$\boxed{\varepsilon_{yy} = \frac{\partial v}{\partial y}} \qquad (8.1b)$$

---

[2] See Barber [1992], Boresi and Chong [1987], and Fung [1965] in Appendix D for additional details on elasticity.

$$\varepsilon_{zz} = \frac{\partial w}{\partial z} \qquad (8.1c)$$

$$\gamma_{xy} = \gamma_{yx} = \frac{\partial u}{\partial y} + \frac{\partial v}{\partial x} \qquad (8.1d)$$

$$\gamma_{yz} = \gamma_{zy} = \frac{\partial v}{\partial z} + \frac{\partial w}{\partial y} \qquad (8.1e)$$

$$\gamma_{zx} = \gamma_{xz} = \frac{\partial w}{\partial x} + \frac{\partial u}{\partial z} \qquad (8.1f)$$

Six compatibility equations ensure that the strain fields are such that the displacement field so obtained is single valued. We derive one of the six compatibility equations by taking the derivatives of Equation (8.1d) and interchanging their order of as shown in Equation (8.2a).

$$\frac{\partial^2 \gamma_{xy}}{\partial x \partial y} = \frac{\partial^2}{\partial x \partial y}\left(\frac{\partial u}{\partial y} + \frac{\partial v}{\partial x}\right) = \frac{\partial^3 u}{\partial x \partial y^2} + \frac{\partial^3 v}{\partial x^2 \partial y} = \frac{\partial^2}{\partial y^2}\left(\frac{\partial u}{\partial x}\right) + \frac{\partial^2}{\partial x^2}\left(\frac{\partial v}{\partial y}\right) \qquad (8.2a)$$

Substituting Equations (8.1a) and (8.1b) in the right-hand side of Equation (8.2a), we obtain Equation (8.2b).

$$\frac{\partial^2 \varepsilon_{xx}}{\partial y^2} + \frac{\partial^2 \varepsilon_{yy}}{\partial x^2} = \frac{\partial^2 \gamma_{xy}}{\partial x \partial y} \qquad (8.2b)$$

The other five compatibility equations can be derived (see Problems 8.1 and 8.2) in a similar manner to Equation (8.2b). We can also obtain the strain–displacement relationship in polar coordinates, which we will use in later sections. For two-dimensional problems, the relationships are given by Equation (1.48) and are shown for convenience, as Equations (8.3a) through (8.3c),

$$\varepsilon_{rr} = \frac{\partial u_r}{\partial r} \qquad (8.3a)$$

$$\varepsilon_{\theta\theta} = \frac{u_r}{r} + \frac{1}{r}\frac{\partial v_\theta}{\partial \theta} \qquad (8.3b)$$

$$\gamma_{x\theta} = \frac{1}{r}\frac{\partial u_r}{\partial \theta} + \frac{\partial v_\theta}{\partial r} - \frac{v_\theta}{r} \qquad (8.3c)$$

where u_r and v_θ are the displacements in the r and θ direction, respectively.

8.1.2   Plane Stress and Plane Strain

The two-dimensional problems of plane stress and plane strain were discussed in Section 2.2. The equations of plane stress and plane strain can be derived from the equations of the generalized Hooke's law [see Equations (2.12) through (2.15)] and are stated in slightly different form for convenience.

For *plane stress* we have

$$\boxed{E\varepsilon_{xx} = \sigma_{xx} - \nu\sigma_{yy} \qquad E\varepsilon_{yy} = \sigma_{yy} - \nu\sigma_{xx} \qquad G\gamma_{xy} = \tau_{xy}}$$ (8.4a)

Alternatively, Equation (8.4a) can be written as

$$\boxed{\sigma_{xx} = E[\varepsilon_{xx} + \nu\varepsilon_{yy}]/(1 - \nu^2) \qquad \sigma_{yy} = E[\varepsilon_{yy} + \nu\varepsilon_{xx}]/(1 - \nu^2) \qquad \tau_{xy} = G\gamma_{xy}}$$ (8.4b)

For *plane strain* we have

$$\boxed{2G\varepsilon_{xx} = (1 - \nu)\sigma_{xx} - \nu\sigma_{yy} \qquad 2G\varepsilon_{yy} = (1 - \nu)\sigma_{yy} - \nu\sigma_{xx} \qquad G\gamma_{xy} = \tau_{xy}}$$ (8.5a)

Alternatively, Equation (8.5a) can be written as:

$$\boxed{\sigma_{xx} = \frac{2G}{(1 - 2\nu)}[(1 - \nu)\varepsilon_{xx} + \nu\varepsilon_{yy}] \qquad \sigma_{yy} = \frac{2G}{(1 - 2\nu)}[(1 - \nu)\varepsilon_{yy} + \nu\varepsilon_{xx}] \qquad \tau_{xy} = G\gamma_{xy}}$$

(8.5b)

Equations (8.4) and (8.5) are valid for any orthogonal system. By replacing x with r and y with θ, we can obtain the relationship between stresses and strains in polar coordinates.

The equations for plane stress and plane strain are similar in form. It can be verified that we can obtain the equations of plane stress by substituting $\nu/(1 + \nu)$ in place of ν and keeping G unchanged in plane strain equations. This device can be used in obtaining a solution for plane stress from the plane strain solution,[3] as we shall describe in Section 8.5. We record this observation as Equation (8.6) for future use.

$$\boxed{\text{Plane strain} \qquad \begin{array}{c} \nu \to \nu/(1 + \nu) \\ G \to G \end{array} \qquad \text{Plane stress}}$$ (8.6)

8.1.3 Equilibrium Equations

In the two-dimensional differential element in plane stress (or plane strain) shown in Figure 8.2(a), the body forces F_x and F_y are acting at the point and have the dimensions of force per unit volume. The stresses can be converted to forces by multiplying by the areas of the plane on which they act. Then, by multiplying the body force by the differential volume, we obtain the free body diagram shown in Figure 8.2(b).

By taking the force equilibrium in the x direction in Figure 8.2(b), we obtain Equation (8.7a).

$$\left(\sigma_{xx} + \frac{\partial\sigma_{xx}}{\partial x}dx\right)(dy\,dz) - (\sigma_{xx})(dy\,dz) + \left(\tau_{yx} + \frac{\partial\tau_{yx}}{\partial y}dy\right)(dx\,dz)$$

$$- (\tau_{yx})(dx\,dz) + (F_x)(dx\,dy\,dz) = 0 \qquad \text{or}$$

---

[3] See Problems 8.3 and 8.4 for a compact form of writing equations for plane stress and plane strain.

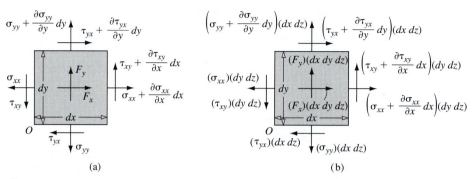

Figure 8.2 (a) A two-dimensional differential element and (b) its free body diagram.

$$\frac{\partial \sigma_{xx}}{\partial x}(dx\,dy\,dz) + \frac{\partial \tau_{yx}}{\partial y}(dx\,dy\,dz) + (F_x)(dx\,dy\,dz) = 0 \qquad \text{or}$$

$$\boxed{\frac{\partial \sigma_{xx}}{\partial x} + \frac{\partial \tau_{yx}}{\partial y} + F_x = 0} \tag{8.7a}$$

By taking the force equilibrium in the y direction, we obtain Equation (8.7b).

$$\left(\sigma_{yy} + \frac{\partial \sigma_{yy}}{\partial y}dy\right)dx\,dz - (\sigma_{yy})(dx\,dz) + \left(\tau_{xy} + \frac{\partial \tau_{xy}}{\partial x}dx\right)(dy\,dz)$$

$$- (\tau_{xy})(dy\,dz) + (F_y)(dx\,dy\,dz) = 0 \qquad \text{or}$$

$$\frac{\partial \sigma_{yy}}{\partial y}(dx\,dy\,dz) + \frac{\partial \tau_{xy}}{\partial x}(dx\,dy\,dz) + (F_y)(dx\,dy\,dz) = 0 \qquad \text{or}$$

$$\boxed{\frac{\partial \tau_{xy}}{\partial x} + \frac{\partial \sigma_{yy}}{\partial y} + F_y = 0} \tag{8.7b}$$

We now consider moment equilibrium about a point about the center of the differential element in Figure 8.2(b). The moment at the center will be from shear stresses only. We can neglect the terms containing the product of four differentials, since these terms tend to zero faster in the limit than the terms with the products of three differentials. As a result, we obtain Equation (8.7c).

$$\left(\tau_{xy} + \frac{\partial \tau_{xy}}{\partial x}dx\right)(dy\,dz)\left(\frac{dx}{2}\right) + (\tau_{xy})(dy\,dz)\left(\frac{dx}{2}\right) - \left(\tau_{yx} + \frac{\partial \tau_{yx}}{\partial y}dy\right)(dx\,dz)\left(\frac{dy}{2}\right)$$

$$- (\tau_{yx})(dx\,dz)\left(\frac{dy}{2}\right) = 0 \qquad \text{or}$$

$$(\tau_{xy})(dy\,dz)(dx) - (\tau_{yx})(dx\,dz)(dy) = 0 \qquad \text{or}$$

$$\boxed{\tau_{xy} = \tau_{yx}} \tag{8.7c}$$

The equilibrium equations in three dimensions can be obtained as shown earlier by considering the equilibrium of a differential element (cube) in three dimensions.[4] In addition to equations of symmetric of shear stresses we have Equations (8.8a), (8.8b), and (8.8c).

$$\frac{\partial \sigma_{xx}}{\partial x} + \frac{\partial \tau_{yx}}{\partial y} + \frac{\partial \tau_{zx}}{\partial z} + F_x = 0 \qquad (8.8a)$$

$$\frac{\partial \tau_{xy}}{\partial x} + \frac{\partial \sigma_{yy}}{\partial y} + \frac{\partial \tau_{zy}}{\partial z} + F_y = 0 \qquad (8.8b)$$

$$\frac{\partial \tau_{xz}}{\partial x} + \frac{\partial \tau_{yz}}{\partial y} + \frac{\partial \sigma_{zz}}{\partial z} + F_z = 0 \qquad (8.8c)$$

Equations (8.7a), (8.7b), and (8.7c) are the equilibrium equations at a point of a two-dimensional elastic body in Cartesian coordinates. By considering a differential element in polar coordinates, we can obtain the equilibrium equations shown in Equations (8.9a), (8.9b), and (8.9c).

$$\frac{\partial \sigma_{rr}}{\partial r} + \frac{1}{r}\frac{\partial \tau_{r\theta}}{\partial \theta} + \frac{\sigma_{rr} - \sigma_{\theta\theta}}{r} + F_r = 0 \qquad (8.9a)$$

$$\frac{1}{r}\frac{\partial \sigma_{\theta\theta}}{\partial \theta} + \frac{\partial \tau_{r\theta}}{\partial r} + \frac{2\tau_{r\theta}}{r} + F_\theta = 0 \qquad (8.9b)$$

$$\tau_{r\theta} = \tau_{\theta r} \qquad (8.9c)$$

8.1.4 Boundary Conditions

To solve the elasticity equations, we need boundary conditions. At a given point on the surface of a body, we must specify either displacement or traction in a given direction. The traction stress vector was given by Equation (1.12). In long form, the boundary conditions at a point on a surface are given by Equations (8.10a), (8.10b), and (8.10c).

$$u = u_0 \qquad \text{or} \qquad \sigma_{xx}n_x + \tau_{xy}n_y + \tau_{xz}n_z = t_x \qquad (8.10a)$$

$$v = v_0 \qquad \text{or} \qquad \tau_{yx}n_x + \sigma_{yy}n_y + \tau_{yz}n_z = t_y \qquad (8.10b)$$

$$w = w_0 \qquad \text{or} \qquad \tau_{zx}n_x + \tau_{zy}n_y + \sigma_{zz}n_z = t_z \qquad (8.10c)$$

where u_0, v_0, and w_0 represent the specified displacement in the x, y, and z directions, respectively;

t_x, t_y, and t_z represent the specified traction in the x, y, and z direction, respectively; and n_x, n_y, and n_z represent the direction cosines of the unit normal to the surface in the x, y, and z direction, respectively.

---

[4] A compact notation for the equilibrium is $\sum_j (\partial \sigma_{ji}/\partial x_j) + F_i = 0$, where i and j are x, y, and z.

EXAMPLE 8.1

Starting from the equilibrium equations of elasticity, obtain the equilibrium equations for the mechanics of materials for (a) axial members and (b) symmetric bending of beams.[5] Assume that the cross-sectional area is constant.

PLAN

In the mechanics of materials, the equilibrium equations are on the internal forces and moments that act on a cross section. The equilibrium equations of elasticity are on stress components. By appropriate integration of the elasticity equations across the cross sections, interchanging the order of integration and differentiation, and using the equivalency of internal forces and moments to stresses, we can obtain the equilibrium equations of the mechanics of materials.

SOLUTION

Axial: The only nonzero stress component in axial members is σ_{xx}. From Equation (8.8a) we obtain Equation (E1).

$$\frac{\partial \sigma_{xx}}{\partial x} + F_x = 0 \qquad (E1)$$

Multiplying Equation (E1) by dA and integrating across the cross section, we obtain Equation (E2).

$$\int_A \frac{\partial \sigma_{xx}}{\partial x} dA + \int_A F_x dA = 0 \qquad \text{or} \qquad \frac{\partial}{\partial x}\int_A \sigma_{xx} dA + \int_A F_x dA = 0 \qquad (E2)$$

We note that the integral of the stress over the area A is the internal axial force N, and the integral of the body force per unit volume F_x is p_x, the axial force per unit length. We also note that the stress σ_{xx} is a function of x only, hence N is a function of x only. Substituting, we obtain the equilibrium equation for axial members as given by Equation (3.11-A), and repeated here:

ANS. $\quad \dfrac{dN}{dx} + p_x = 0$

Symmetric bending about the z axis: The nonzero stress components in symmetric bending are σ_{xx} and τ_{xy}. The only distributed force is in the y direction. From Equations (8.8a) and (8.8b), we obtain Equations (E3) and (E4).

$$\frac{\partial \sigma_{xx}}{\partial x} + \frac{\partial \tau_{yx}}{\partial y} = 0 \qquad (E3)$$

$$\frac{\partial \tau_{xy}}{\partial x} + F_y = 0 \qquad (E4)$$

---

[5] For torsion of circular shafts, see Problem 8.5.

We multiply Equation (E3) by $y\,dA$ and integrate over the cross-sectional area to obtain Equation (E5).

$$\int_A y\frac{\partial\sigma_{xx}}{\partial x}dA + \int_A y\frac{\partial\tau_{yx}}{\partial y}dA = 0 \quad\text{or}\quad \frac{\partial}{\partial x}\int_A y\sigma_{xx}dA + \int_A y\frac{\partial\tau_{yx}}{\partial y}dA = 0 \qquad\text{(E5)}$$

We note that $dA = dy\,dz$. Integrating the second integral by parts and making use of the definition of M_z, we obtain Equation (E6).

$$\frac{\partial}{\partial x}[-M_z] + (y\tau_{yx})\big|_{\text{surface}} - \int_A \tau_{yx}dA = 0 \qquad\text{(E6)}$$

The shear stress τ_{yx} is zero on the surface of the beam (free surface), and hence the second term in Equation (E6) is zero. The last term by definition is the shear force V_y. We also note that M_z is a function only of x. Substituting this, we obtain the equilibrium equation for the bending of beams, as given by Equation (3.11b-B) and repeated here in slightly different form.

ANS. $\dfrac{dM_z}{dx} + V_y = 0$

We multiply Equation (E4) by dA and integrate over the cross-sectional area to obtain Equation (E7).

$$\int_A \frac{\partial\tau_{xy}}{\partial x}dA + \int_A F_y dA = 0 \quad\text{or}\quad \frac{\partial}{\partial x}\int_A \tau_{xy}dA + \int_A F_y dA = 0 \qquad\text{(E7)}$$

Once more we note that the integration of the shear stresses over the area results in the shear force V_y, and integration of body forces per unit volume F_y results in the distributed force per unit length p_y. Substituting, we obtain the second equilibrium equation for the bending of beams [given earlier by Equation (3.11a-B)] as follows:

ANS. $\dfrac{dV_y}{dx} + p_y(x) = 0$

We multiply Equation (E4) by dA and integrate over the area A_s, which is between the free surface and some point y as shown in Equation (E8).

$$\int_{A_s} \frac{\partial\sigma_{xx}}{\partial x}dA + \int_{A_s} \frac{\partial\tau_{yx}}{\partial y}dA = 0 \quad\text{or}\quad \frac{\partial}{\partial x}\int_{A_s} \sigma_{xx}dA + \int_{A_s} \frac{\partial\tau_{yx}}{\partial y}dA = 0 \qquad\text{(E8)}$$

We note that the integral of σ_{xx} over the area A_s results in axial force N_s (see Section 3.2.3) in Equation (E8). Assuming thin beams, we can write $dA = t\,dy$, where t is the thickness. Since we are integrating from a free surface where τ_{yx} is zero to any point on the cross section, we obtain Equation (E9).

$$\frac{\partial N_s}{\partial x} + \tau_{yx}t = 0 \qquad\text{(E9)}$$

Noting that N_s is only a function of x in Equation (E9), we obtain the equilibrium given as follows [see also Equation (3.15-B)]:

ANS. $\dfrac{dN_s}{dx} + \tau_{yx}t = 0$

COMMENT

To obtain the equilibrium equations appropriate for the mechanics of materials from the equilibrium equations of elasticity, we had to assume that the cross-sectional area did not change with x. The process would not work for tapered cross sections. In the mechanics of materials, we first replace the stresses over the cross section by internal forces and moments; then we consider the variation of these over a differential element dx. In so doing we are implicitly making the assumption that both the stresses and the cross section can change with x. In obtaining the equilibrium equations of elasticity, when we take the differential element we permit only the stresses to vary over the element.

8.2 AXISYMMETRIC PROBLEMS

Axisymmetric problems are those in which the loading, geometry, and material properties are all independent of angular location (θ). Under the conditions of axisymmetry, the results of displacements, strains, and stresses should also be independent of θ. Thus, all derivatives with respect to θ will be zero, and all variables are functions of the radial coordinate r only; partial derivatives of r can be written as ordinary derivatives of r. The strain–displacement Equations (8.3a) and (8.3b) can be written as Equations (8.11a) and (8.11b).

$$\varepsilon_{rr} = \frac{\partial u_r}{\partial r} = \frac{du_r}{dr} \tag{8.11a}$$

$$\varepsilon_{\theta\theta} = \frac{u_r}{r} \tag{8.11b}$$

The equilibrium Equation (8.9a) can be written as Equation (8.12).

$$\frac{d\sigma_{rr}}{dr} + \frac{\sigma_{rr} - \sigma_{\theta\theta}}{r} + F_r = 0 \tag{8.12}$$

8.2.1 Axisymmetric Plane Strain

For plane strain, Equation (8.5b) can be written as Equations (8.13a) and (8.13b).

$$\sigma_{rr} = \frac{2G}{(1-2v)}[(1-v)\varepsilon_{rr} + v\varepsilon_{\theta\theta}] = \frac{2G}{(1-2v)}\left[(1-v)\frac{du_r}{dr} + v\frac{u_r}{r}\right] \tag{8.13a}$$

$$\sigma_{\theta\theta} = \frac{2G}{(1-2v)}[(1-v)\varepsilon_{\theta\theta} + v\varepsilon_{rr}] = \frac{2G}{(1-2v)}\left[(1-v)\frac{u_r}{r} + v\frac{du_r}{dr}\right] \tag{8.13b}$$

Substituting Equations (8.13a) and (8.13b) into Equation (8.12) and simplifying, we obtain a differential equation:

$$\frac{d^2u_r}{dr^2} + \frac{1}{r}\frac{du_r}{dr} - \frac{u_r}{r^2} + \frac{(1-2v)}{(1-v)G}F_r = 0 \quad \text{or} \quad \frac{d}{dr}\left[\frac{1}{r}\frac{d}{dr}(ru_r)\right] + \frac{(1-2v)}{(1-v)G}F_r = 0 \tag{8.14}$$

The solution to Equation (8.14) can be written as the solution to the homogeneous part $(u_r)_{\text{homo}}$ corresponding to $F_r = 0$ and a particular solution $(u_r)_{\text{part}}$ that will depend upon the value of F_r. The homogeneous solution can be obtained in terms of the constants C_1 and C_2, which must be determined from the boundary conditions of a particular application. The homogeneous solution is as shown in Equation (8.15).

$$u_r = (u_r)_{\text{homo}} + (u_r)_{\text{part}} \qquad (u_r)_{\text{homo}} = C_1 r + \frac{C_2}{r} \tag{8.15}$$

8.2.2 Axisymmetric Plane Stress

For plane stress, Equation (8.4b) can be written as Equations (8.16a) and (8.16b).

$$\sigma_{rr} = \frac{E}{(1-v^2)}[\varepsilon_{rr} + v\varepsilon_{\theta\theta}] = \frac{E}{(1-v^2)}\left[\frac{du_r}{dr} + v\frac{u_r}{r}\right] \tag{8.16a}$$

$$\sigma_{\theta\theta} = \frac{E}{(1-v^2)}[\varepsilon_{\theta\theta} + v\varepsilon_{rr}] = \frac{E}{(1-v^2)}\left[\frac{u_r}{r} + v\frac{du_r}{dr}\right] \tag{8.16b}$$

Substituting the Equations (8.16a) and (8.16b) into Equation (8.12) and simplifying, we obtain a differential equation:

$$\frac{d^2 u_r}{dr^2} + \frac{1}{r}\frac{du_r}{dr} - \frac{u_r}{r^2} + \frac{(1-v^2)}{E}F_r = 0 \qquad \text{or} \qquad \frac{d}{dr}\left[\frac{1}{r}\frac{d}{dr}(ru_r)\right] + \frac{(1-v^2)}{E}F_r = 0 \tag{8.17}$$

The solution to Equation (8.17) is given by Equation (8.15), with the particular part of the solution dependent upon the value of F_r.

8.3 ROTATING DISKS

A grinding wheel or a disk brake can be modeled as a rotating disk. We assume that the disk is thin (plane stress) and is rotating at a constant angular speed ω. This would subject a point of the disk that is at a radial distance r to an acceleration of $\omega^2 r$ in the radial direction. If the mass density of the disk material is ρ, then the radial force will be $F_r = \rho\omega^2 r$. Substituting this radial force into Equation (8.17), we can obtain Equation (8.18).

$$\frac{d}{dr}\left[\frac{1}{r}\frac{d}{dr}(ru_r)\right] + \frac{(1-v^2)}{E}(\rho\omega^2 r) = 0 \tag{8.18}$$

The particular solution to Equation (8.18) is

$$(u_r)_{\text{part}} = -\frac{(1-v^2)(\rho\omega^2)(r^3)}{E}\frac{1}{8}$$

The total solution given by Equation (8.15) can now be written as shown in Equation (8.19a).

$$u_r = C_1 r + \frac{C_2}{r} - \frac{(1-v^2)(\rho\omega^2 r^3)}{8E} \tag{8.19a}$$

Substituting Equation (8.19a) into Equations (8.16a) and (8.16b), we obtain the radial and tangential normal stresses as shown in Equations (8.19b) and (8.19c).

$$\sigma_{rr} = \frac{E}{(1-v^2)}\left[C_1(1+v) - \frac{C_2(1-v)}{r^2}\right] - \frac{(3+v)}{8}\rho\omega^2 r^2 \tag{8.19b}$$

$$\sigma_{\theta\theta} = \frac{E}{(1-v^2)}\left[C_1(1+v) + \frac{C_2(1-v)}{r^2}\right] - \frac{(1+3v)}{8}\rho\omega^2 r^2 \tag{8.19c}$$

The constants C_1 and C_2 in Equations (8.19a), (8.19b) and (8.19c) can now be determined from the boundary conditions for a particular application. The following typical boundary conditions are to be used in our examples and problems.

1. *A solid rotating disk* (see Problem 8.6): The outer boundary is stress free, thus $\sigma_{rr}(r=R_o) = 0$. For a solution to be finite at the center ($r=0$) of a solid disk requires $C_2 = 0$.

2. *A rotating disk with a hole* (see Problem 8.7): The inner boundary ($r=R_i$) and outer boundary ($r=R_o$) are stress free. Thus the boundary conditions are $\sigma_{rr}(r=R_i) = 0$ and $\sigma_{rr}(r=R_o) = 0$.

3. *A rotating disk bonded on a rigid shaft:* The outer boundary is stress free, and the point on the inner boundary cannot be displaced. Thus the boundary conditions are $u_r(r=R_i) = 0$ and $\sigma_{rr}(r=R_o) = 0$.

EXAMPLE 8.2

The maximum rotational speed at which a grinding wheel can operate is called the "bursting speed," since if this speed is exceeded, maximum tensile stress will cause the wheel to burst. Consider a grinding wheel with inner radius a, outer radius $2a$, Poisson ratio $v = 1/3$, and modulus of elasticity E. Use the maximum tensile stress failure theory to obtain a relationship between the burst speed ω_{max} and the allowable stress σ_{allow} in terms r, a, and E. Assume that the grinding wheel is mounted (bonded) on a rigid shaft.

PLAN

The boundary conditions are $u_r(r=a) = 0$ and $\sigma_{rr}(r=2a) = 0$. The constants C_1 and C_2 in Equations (8.19a), (8.19b), and (8.19c) can be determined by using these boundary conditions. The normal stresses can now be determined from Equations (8.19b) and (8.19c) and the maximum normal stress then equated to σ_{allow} to obtain the required relationship.

SOLUTION

Substituting $v = 1/3$ into Equations (8.19a), (8.19b), and (8.19c), we obtain Equations (E1), (E2), and (E3).

$$u_r = C_1 r + \frac{C_2}{r} - \frac{(1-(1/3)^2)(\rho\omega^2 r^3)}{8E} = C_1 r + \frac{C_2}{r} - \frac{\rho\omega^2 r^3}{9E} \tag{E1}$$

$$\sigma_{rr} = \frac{E}{1-(1/3)^2}\left[C_1(1+1/3) - \frac{C_2(1-1/3)}{r^2}\right] - \frac{(3+1/3)}{8}\rho\omega^2 r^2$$

$$= \frac{3E}{8}\left[4C_1 - \frac{2C_2}{r^2}\right] - \frac{5}{12}\rho\omega^2 r^2 \qquad \text{(E2)}$$

$$\sigma_{\theta\theta} = \frac{E}{1-(1/3)^2}\left[C_1(1+1/3) + \frac{C_2(1-1/3)}{r^2}\right] - \frac{(1+3(1+1/3))}{8}\rho\omega^2 r^2 \qquad \text{or}$$

$$\sigma_{\theta\theta} = \frac{3E}{8}\left[4C_1 + \frac{2C_2}{r^2}\right] - \frac{1}{4}\rho\omega^2 r^2 \qquad \text{(E3)}$$

The point on the inner boundary cannot be displaced. Substituting $r = a$ into Equation (E1) and equating it to zero, we obtain Equation (E4).

$$C_1 a + \frac{C_2}{a} - \frac{\rho\omega^2 a^3}{8E} = 0 \qquad \text{or} \qquad C_1 + \frac{C_2}{a^2} = \frac{\rho\omega^2 a^2}{8E} \qquad \text{(E4)}$$

The outer surface of the wheel is stress free. Substituting $r = 2a$ into Equation (E2) and equating it to zero, we obtain Equation (E5).

$$\frac{3E}{8}\left[4C_1 - \frac{2C_2}{(2a)^2}\right] - \frac{5}{12}\rho\omega^2(2a)^2 = 0 \qquad \text{or} \qquad C_1 - \frac{C_2}{8a^2} = \frac{10}{9E}\rho\omega^2 a^2 \qquad \text{(E5)}$$

Solving Equations (E4) and (E5), we obtain Equations (E6).

$$C_1 = 1.0015\frac{\rho\omega^2 a^2}{E} \qquad \text{and} \qquad C_2 = -0.8765\frac{\rho\omega^2 a^4}{E} \qquad \text{(E6)}$$

Substituting Equations (E6) into Equations (E2) and (E3), we obtain the normal stresses shown in Equations (E7) and (E8).

$$\sigma_{rr} = \left[\frac{3}{8}\left\{4(1.0015a^2) - \frac{2(-0.8765a^4)}{r^2}\right\} - \frac{5}{12}r^2\right](\rho\omega^2)$$

$$= \rho\omega^2 a^2\left[1.502 + 0.6574\frac{a^2}{r^2} - 0.4167\frac{r^2}{a^2}\right] \qquad \text{or}$$

$$\sigma_{rr} = \left[1.502 + 0.6574\frac{a^2}{r^2} - 0.4167\frac{r^2}{a^2}\right]\rho\omega^2 a^2 \qquad \text{(E7)}$$

$$\sigma_{\theta\theta} = \left[\frac{3}{8}\left\{4(1.0015a^2) + \frac{2(-0.8765a^4)}{r^2}\right\} - \frac{1}{4}r^2\right](\rho\omega^2)$$

$$= \rho\omega^2 a^2\left[1.502 - 0.6574\frac{a^2}{r^2} - 0.25\frac{r^2}{a^2}\right] \qquad \text{or}$$

$$\sigma_{\theta\theta} = \left[1.502 - 0.6574\frac{a^2}{r^2} - 0.25\frac{r^2}{a^2}\right]\rho\omega^2 a^2 \qquad \text{(E8)}$$

The maximum normal stress σ_{rr} will be at the inner radius. Substituting $r = a$ into Equation (E7) and noting that the stress should be less than the allowable stress, we obtain Equation (E9).

$$\sigma_{rr} = [1.502 + 0.6574 - 0.4167]\rho\omega^2 a^2 = 1.7427\rho\omega^2 a^2 \le \sigma_{allow} \qquad \text{or}$$

$$\omega \le 0.7575 \sqrt{\frac{\sigma_{allow}}{\rho a^2}} \qquad (E9)$$

From Equation (E9), the maximum rotational speed (i.e., burst speed is):

ANS. $\quad \omega_{max} = 0.7575 \sqrt{\dfrac{\sigma_{allow}}{\rho a^2}}$

COMMENTS

1. Materials for grinding wheels are brittle, and thus the use of the maximum tensile stress theory is appropriate.

2. Heat will be generated during a grinding operation, originating at the point of contact. This will generate thermal stresses, which in general will not be axisymmetric.

8.4 THICK HOLLOW CYLINDERS

Thick hollow cylinders that are very long (theoretically infinite) relative to the outer radius can be modeled as axisymmetric problems in plane strain.

The axisymmetric cross section shown in Figure 8.3 could be a thick cylinder with an inside radius R_i and an outside radius R_o. We suppose that an internal pressure p_i acts

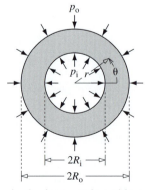

Figure 8.3 A pressurized axisymmetric problem.

on the inner surface and a pressure p_o acts on the outer surface. We shall assume that there are no body forces (i.e., $F_r = 0$).

With $F_r = 0$, the particular solution is zero. Substituting $(u_r)_{part} = 0$ into Equation (8.15), we obtain Equation (8.20),

$$u_r = C_1 r + \frac{C_2}{r} \tag{8.20}$$

where C_1 and C_2 are constants to be determined from the boundary conditions. We note that the pressure will produce compressive stress in the radial direction on the inner and outer surfaces. The boundary conditions can be written as Equations (8.21a) and (8.21b).

$$\sigma_{rr}(r = R_i) = -p_i \tag{8.21a}$$

$$\sigma_{rr}(r = R_o) = -p_o \tag{8.21b}$$

Substituting Equation (8.20) into Equations (8.13a) and (8.13b) and simplifying, we obtain Equations (8.22a) and (8.22b).

$$\sigma_{rr} = \frac{2G}{(1-2v)}\left[C_1 - \frac{(1-2v)}{r^2}C_2\right] \tag{8.22a}$$

$$\sigma_{\theta\theta} = \frac{2G}{(1-2v)}\left[C_1 + \frac{(1-2v)}{r^2}C_2\right] \tag{8.22b}$$

Substituting Equation (8.22a) into Equations (8.21a) and (8.21b), we obtain Equations (8.23a) and (8.23b).

$$\frac{2G}{(1-2v)}\left[C_1 - \frac{(1-2v)}{R_i^2}C_2\right] = -p_i \tag{8.23a}$$

$$\frac{2G}{(1-2v)}\left[C_1 - \frac{(1-2v)}{R_o^2}C_2\right] = -p_o \tag{8.23b}$$

Solving Equations (8.23a) and (8.23b), we obtain Equations (8.24a) and (8.24b).

$$C_1 = \frac{-(1-2v)}{2G}\left(\frac{p_o R_o^2 - p_i R_i^2}{R_o^2 - R_i^2}\right) \tag{8.24a}$$

$$C_2 = \left(\frac{R_i^2 R_o^2}{2G}\right)\frac{p_i - p_o}{R_o^2 - R_i^2} \tag{8.24b}$$

Substituting Equations (8.24a) and (8.24b) into Equations (8.22a), and (8.22b), we obtain Equations (8.25a) and (8.25b).

$$\boxed{\sigma_{rr} = \frac{1}{R_o^2 - R_i^2}\left[-(p_o R_o^2 - p_i R_i^2) - \frac{R_i^2 R_o^2 (p_i - p_o)}{r^2}\right]} \tag{8.25a}$$

$$\boxed{\sigma_{\theta\theta} = \frac{1}{R_o^2 - R_i^2}\left[-(p_o R_o^2 - p_i R_i^2) + \frac{R_i^2 R_o^2 (p_i - p_o)}{r^2}\right]} \tag{8.25b}$$

Substituting Equations (8.24a) and (8.24b) into Equation (8.20), we obtain Equation (8.25c).

$$u_r = \frac{1}{2G(R_o^2 - R_i^2)}\left[-(1-2\nu)(p_oR_o^2 - p_iR_i^2)r + \frac{R_i^2R_o^2(p_i - p_o)}{r}\right] \tag{8.25c}$$

It should be noted that the axisymmetric nature of the problem results in zero displacement in the θ direction. Thus, all points move along the radial lines by an amount u_r. Thus a circle of radius r becomes a circle with radius $(r + u_r)$ owing to the applied pressure. There are no changes of angle between the r and θ directions; hence the shear strain $\gamma_{r\theta}$ is zero at all points, and the corresponding shear stress $\tau_{r\theta}$ is zero at all points. The equilibrium equation (8.9b) is implicitly satisfied.

We now consider simplified results corresponding to pressure applied either just on the interior or just on the exterior surface.

8.4.1 Internal Pressure Only

Substituting $p_o = 0$ in Equations (8.25a), (8.25b), and (8.25c), we obtain Equations (8.26a), (8.26b), and (8.26c).

$$\sigma_{rr} = \left[\frac{p_iR_i^2}{R_o^2 - R_i^2}\right]\left[1 - \frac{R_o^2}{r^2}\right] = \frac{p_i}{(R_o/R_i)^2 - 1}\left[1 - \left(\frac{R_o}{r}\right)^2\right] \tag{8.26a}$$

$$\sigma_{\theta\theta} = \left[\frac{p_iR_i^2}{R_o^2 - R_i^2}\right]\left[1 + \frac{R_o^2}{r^2}\right] = \frac{p_i}{(R_o/R_i)^2 - 1}\left[1 + \left(\frac{R_o}{r}\right)^2\right] \tag{8.26b}$$

$$u_r = \frac{p_iR_i^2}{2G(R_o^2 - R_i^2)}\left[(1-2\nu)r + \frac{R_o^2}{r}\right] = \frac{p_ir}{2G[(R_o/R_i)^2 - 1]}\left[(1-2\nu) + \left(\frac{R_o}{r}\right)^2\right] \tag{8.26c}$$

In Equation (8.26a) we note that $r < R_o$; hence the radial normal stress σ_{rr} is compressive at all points, with a maximum compressive value of p_i at the inner surface. The tangential normal stress $\sigma_{\theta\theta}$ is always tensile, and its maximum value is also the inner surface. Equation (8.26c) shows that all points move in the positive radial direction. That is, a circle of radius r will enlarge to a radius of $(r + u_r)$ owing to the internal pressure—consistent with our intuitive expectation.

8.4.2 External Pressure Only

We substitute $p_i = 0$ into Equations (8.25a), (8.25b), and (8.25c) to obtain Equations (8.27a), (8.27b), and (8.27c).

$$\sigma_{rr} = -\left[\frac{p_oR_o^2}{R_o^2 - R_i^2}\right]\left[1 - \frac{R_i^2}{r^2}\right] = -\left[\frac{p_o}{1 - (R_i/R_o)^2}\right]\left[1 - \frac{R_i^2}{r^2}\right] \tag{8.27a}$$

$$\sigma_{\theta\theta} = -\left[\frac{p_oR_o^2}{R_o^2 - R_i^2}\right]\left[1 + \frac{R_i^2}{r^2}\right] = -\left[\frac{p_o}{1 - (R_i/R_o)^2}\right]\left[1 + \frac{R_i^2}{r^2}\right] \tag{8.27b}$$

$$u_r = -\left[\frac{p_o R_o^2}{2G[R_o^2 - R_i^2]}\right]\left[(1-2v)r + \frac{R_i^2}{r}\right]$$

$$= -\left[\frac{p_o r}{2G[1-(R_i/R_o)^2]}\right]\left[(1-2v) + \left(\frac{R_i}{r}\right)^2\right] \qquad (8.27c)$$

In Equation (8.27a) we note that $r > R_i$; hence the radial normal stress σ_{rr} is compressive at all points, with a maximum compressive value of p_o at the outer surface. The tangential normal stress $\sigma_{\theta\theta}$ is always compressive, and its maximum value is at the inner surface. Equation (8.27c) shows that all points move in the negative radial direction. That is, consistent with our intuitive expectation, the external pressure will cause the radius of the circle to shrink from r to $r - u_r$.

8.5 THIN DISKS

Thin-disk problems are axisymmetric problems in plane stress. Another alternative for plane stress may comprise cylinders of finite length in which there are no axial forces or stresses. Both these types of plane stress problem have the same equations relating displacements and stresses to applied pressures.

The axisymmetric cross section shown earlier in Figure 8.3 could be of a thin circular sheet of radius of R_o with a hole radius of R_i. If the front and back surfaces of the sheet are stress free, we can make the assumption of plane stress.

The equations for stress in plane strain problems were independent of the material constants. Thus, the stress expressions derived for plane strain are applicable to plane stress problems also. To obtain the displacement expressions, we make use of the observation recorded in Equation (8.6) and substitute it in Equation (8.25c) to obtain Equation (8.28).

$$u_r = \frac{1}{2G(R_o^2 - R_i^2)}\left[-\left(1 - 2\frac{v}{1+v}\right)(p_o R_o^2 - p_i R_i^2)r + \frac{R_i^2 R_o^2 (p_i - p_o)}{r}\right] \qquad \text{or}$$

$$u_r = \frac{1}{2G(1+v)(R_o^2 - R_i^2)}\left[-(1-v)(p_o R_o^2 - p_i R_i^2)r + (1+v)\frac{R_i^2 R_o^2 (p_i - p_o)}{r}\right] \qquad \text{or}$$

$$u_r = \frac{1}{E(R_o^2 - R_i^2)}\left[-(1-v)(p_o R_o^2 - p_i R_i^2)r + (1+v)\frac{R_i^2 R_o^2 (p_i - p_o)}{r}\right] \qquad (8.28)$$

The displacement for the two specific cases of just external or just internal pressure can now be obtained.

8.5.1 Internal Pressure Only

The stress expressions are given by Equations (8.26a) and (8.26b). Substituting $p_o = 0$ into Equation (8.28), we obtain the radial displacement shown in Equation (8.29).

$$u_r = \left[\frac{p_i R_i^2}{E(R_o^2 - R_i^2)}\right]\left[(1 - \nu)r + (1 + \nu)\frac{R_o^2}{r}\right]$$

$$= \left[\frac{p_i r}{E[(R_o/R_i)^2 - 1]}\right]\left[(1 - \nu) + (1 + \nu)\left(\frac{R_o}{r}\right)^2\right] \tag{8.29}$$

8.5.2 External Pressure Only

The stress expressions are given by Equations (8.27a) and (8.27b). Substituting $p_i = 0$ into Equation (8.28), we obtain the radial displacement shown in Equation (8.30).

$$u_r = -\left[\frac{p_o R_o^2}{E(R_o^2 - R_i^2)}\right]\left[(1 - \nu)r + (1 + \nu)\frac{R_i^2}{r}\right]$$

$$= -\left[\frac{p_o r}{E[(R_o/R_i)^2 - 1]}\right]\left[(1 - \nu) + (1 + \nu)\left(\frac{R_o}{r}\right)^2\right] \tag{8.30}$$

8.5.3 Shrink Fitting

A cylinder with a slightly smaller inner radius (called the jacket) is shrink-fitted onto another cylinder (called the tube) with a slightly bigger outer radius. This procedure, called shrink fitting, introduces residual stresses in opposite directions to the stresses that will be seen in the assembled two-cylinder unit during service, thus increasing the load-carrying capacity of the assembly. Shrink fitting is applied in the fabrication of gun barrels, hydraulic cylinders, and laminated pressure vessels, among other items.

To fit the jacket (outer cylinder) onto a tube (inner cylinder), we may heat the jacket to expand it, then slip it onto the tube and let it cool. When the assembly reaches ambient temperature, the inner cylinder will be subjected to an external contact pressure p_c and the outer cylinder will be subjected to the same contact pressure p_c on the inside, as shown in Figure 8.4. We assume that the initial difference between the outer

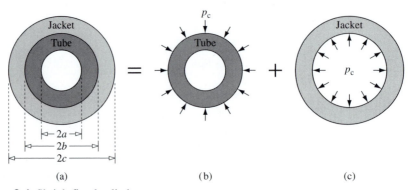

Figure 8.4 Shrink-fitted cylinders.

diameter of the tube and inner diameter of the jacket is Δ. This difference of diameters is called *interference*. We assume that the interference is small compared with the diameter $2b$ in Figure 8.4(a). After the assembly, the outer surface of the tube will move radially inward, the inner surface of the jacket tube will displace radially outward, and the total movement of the points on the interface should be equal to the half the interference, that is,

$$-(u_r)_{\mathrm{T}} + (u_r)_{\mathrm{J}} = \Delta/2 \qquad (8.31)$$

where $(u_r)_{\mathrm{T}}$ and $(u_r)_{\mathrm{J}}$ are the radial displacements of the tube and the jacket, respectively, at radius $r = b$. Note that u_r represents the change in radius, while Δ is the difference in diameter, these two observations are the reason for the factor of 1/2 in Equation (8.31).

Equation (8.31) is applicable to both plane strain and plane stress problems. We shall now assume that there is *no axial force*, hence no axial stress, and the problem is thus a plane stress problem. For the tube we substitute $R_{\mathrm{i}} = a$, $R_{\mathrm{o}} = b$, and $p_{\mathrm{o}} = p_{\mathrm{c}}$ into Equation (8.30) to obtain Equation (8.32a).

$$
\begin{aligned}
(u_r)_{\mathrm{T}} &= -\frac{p_c b^2}{E(b^2 - a^2)}\left[(1-\nu)r + (1+\nu)\frac{a^2}{r}\right]\Bigg|_{r=b} \\
&= -\frac{p_c b}{E(b^2 - a^2)}[(1-\nu)b^2 + (1+\nu)a^2]
\end{aligned}
\qquad (8.32a)
$$

For the jacket we substitute $R_{\mathrm{i}} = b$, $R_{\mathrm{o}} = c$, and $p_{\mathrm{i}} = p_{\mathrm{c}}$ into Equation (8.29) to obtain Equation (8.32b).

$$
\begin{aligned}
(u_r)_{\mathrm{J}} &= \frac{p_c b^2}{E(c^2 - b^2)}\left[(1-\nu)(r) + (1+\nu)\frac{c^2}{r}\right]\Bigg|_{r=b} \\
&= \frac{p_c b}{E(c^2 - b^2)}[(1-\nu)(b^2) + (1+\nu)c^2]
\end{aligned}
\qquad (8.32b)
$$

Substituting Equations (8.32a) and (8.32b) into (8.31), we obtain Equation (8.33).

$$-\frac{p_c b}{E(b^2 - a^2)}[(1-\nu)b^2 + (1+\nu)a^2] + \frac{p_c b}{E(c^2 - b^2)}[(1-\nu)(b^2) + (1+\nu)c^2] = \frac{\Delta}{2} \qquad \text{or}$$

$$\Delta = \frac{4b^3 p_c(c^2 - a^2)}{E(b^2 - a^2)(c^2 - b^2)} = \frac{4b p_c[(c/b)^2 - (a/b)^2]}{E[1 - (a/b)^2][(c/b)^2 - 1]} \qquad (8.33)$$

From Equation (8.33) we can find the contact pressure p_{c} for a given interference. Once p_{c} is known, the stresses in the tube and in the jacket can be found. Equation (8.33) is valid for plane stress problems; for plane strain shrink fitting, see Problem 8.14.

In our derivation, we assumed that the materials for the tube and the jacket were the same. This is not a requirement for solving shrink-fit problems. If the materials are different, the algebra would be more complicated, but Equation (8.31) would remain valid, as would the expressions for finding the radial displacements for plane stress and plane strain.

EXAMPLE 8.3

A small steel cylinder with no axial forces has an inside diameter of 100 mm and an outside diameter of 300 mm. The steel has a modulus of elasticity of $E = 200$ GPa, a Poisson ratio of 0.3, and a yield stress of $\sigma_{yield} = 200$ MPa. Use the octahedral shear stress theory to determine the maximum internal pressure if yielding is to be avoided.

PLAN

With no axial forces, the problem is treated as one of plane stress. From Equations (8.26a) and (8.26b) we can obtain the radial and tangential normal stresses in terms of the internal pressure p_i at the inner radius, where these stresses are maximum. The radial and tangential normal stresses are principal stresses 1 and 2, and the third principal stress is zero. The von Mises stress can be found in terms of p_i by using Equation (2.22) and equated to yield stress to obtain the internal pressure.

SOLUTION

Substituting $R_i = 0.05$ m, $R_o = 0.15$ m, and $r = R_i$ in Equations (8.26a) and (8.26b), we obtain Equations (E1) and (E2).

$$\sigma_{rr} = -p_i \tag{E1}$$

$$\sigma_{\theta\theta} = \frac{p_i}{(R_o/R_i)^2 - 1}\left[1 + \left(\frac{R_o}{R_i}\right)^2\right] = \frac{p_i}{(3)^2 - 1}[1 + (3)^2] = 1.25\,p_i \tag{E2}$$

Noting that the shear strain $\gamma_{r\theta}$ is zero, we see that the principal stresses are the radial and the tangential normal stresses as given in Equations (E1) and (E2). The third principal stress for plane stress is zero, as shown in Equation (E3).

$$\sigma_1 = \sigma_{\theta\theta} = 1.25\,p_i \qquad \sigma_2 = \sigma_{rr} = -p_i \qquad \sigma_3 = 0 \tag{E3}$$

From Equation (2.22) we can find von Mises stress as shown in Equation (E4).

$$\sigma_{von} = \frac{1}{\sqrt{2}}\sqrt{(1.25\,p_i + p_i)^2 + (-p_i)^2 + (-1.25\,p_i)^2} = 1.9526\,p_i \tag{E4}$$

The von Mises stress should be less than or equal to the yield stress of $200\,(10^6)$ N/m^2. From this information, we obtain the internal pressure as shown in Equation (E5).

$$\sigma_{von} = 1.9526\,p_i \leq 200(10^6) \qquad \text{or} \qquad p_i \leq 102.4(10^6)\text{ N/m}^2 \tag{E5}$$

The maximum internal pressure is $\qquad\qquad\qquad$ **ANS.** $(p_i)_{max} = 102$ MPa

COMMENT

If the problem was considered to be one of plane strain, we would have used the three principal stresses given by Equation (E6).

$$\sigma_1 = \sigma_{\theta\theta} = 1.25\,p_i \qquad \sigma_2 = \sigma_{rr} = -p_i \qquad \sigma_3 = \nu(\sigma_1 + \sigma_2) = 0.075\,p_i \tag{E6}$$

From Equation (2.22) we can find the von Mises stress as shown in Equation (E7).

$$\sigma_{\text{von}} = \frac{1}{\sqrt{2}}\sqrt{(1.25p_i + p_i)^2 + (-p_i - 0.075p_i)^2 + (0.075p_i - 1.25p_i)^2} = 1.9492p_i \quad (E7)$$

Since the von Mises stresses for plane stress and plane strain are nearly the same, the maximum internal pressure will be the same.

EXAMPLE 8.4

A small steel cylinder with no axial forces is constructed by shrink fitting a steel jacket having an outside diameter of 300 mm onto a steel tube (outside diameter, 200 mm; inside diameter, 100 mm). Steel has a modulus of elasticity of $E = 200$ GPa, a Poisson ratio of 0.3, and a yield stress of $\sigma_{\text{yield}} = 200$ MPa. (a) Determine the maximum interference between the tube and the jacket if yielding is to be avoided. Use the octahedral shear stress theory. (b) If the assembled cylinder is subjected to the internal pressure calculated in Example 8.3, determine the von Mises stress on the inner surface.

PLAN

With no axial forces, the problem is treated as one of plane stress. (a) We can use Equations (8.27a) and (8.27b) to find the maximum external pressure p_c to which the tube can be subjected without exceeding the yield stress. We can also use Equations (8.26a) and (8.26b) to find the maximum internal pressure p_c to which the jacket can be subjected without exceeding the yield stress. The lower of the two values of p_c is the contact pressure that can be developed by interference between the tube and the jacket. The interference can be found from Equation (8.33). (b) The tube will be subjected to external pressure p_c calculated in part (a) and internal pressure p_i of Example 8.3. The total normal stresses on the inner surface in the radial and tangential directions can be obtained by superposition of the stresses due to p_c calculated in this example, and the stresses calculated in Example 8.3. From the total stresses, the von Mises stress can be found.

SOLUTION

(a) The tube is subjected to external pressure $p_o = p_c$. Substituting $R_i = 0.05$ m, $R_o = 0.1$ m, and $r = R_i$ in Equations (8.27a) and (8.27b), we obtain Equations (E1) and (E2).

$$\sigma_{rr} = 0 \quad (E1)$$

$$\sigma_{\theta\theta} = -\left[\frac{p_c}{1 - (R_i/R_o)^2}\right]\left[1 + \frac{R_i^2}{R_i^2}\right] = -\left(\frac{p_c}{1 - 0.5^2}\right)[1 + 1] = -2.667p_c \quad (E2)$$

We note that the radial and tangential stresses in Equations (E1) and (E2) are principal stresses and that for plane stress, the third principal stress is zero. From Equation (2.22) we can find the von Mises stress as shown in Equation (E3).

$$\sigma_{\text{von}} = \frac{1}{\sqrt{2}}\sqrt{(2.667 p_c)^2 + (2.667 p_c)^2} = 2.667 p_c \leq 200(10^6) \qquad \text{or}$$

$$p_c \leq 75(10^6) \text{ N/m}^2 \tag{E3}$$

The jacket is subjected to internal pressure $p_i = p_c$. Substituting $R_i = 0.1$ m, $R_o = 0.15$ m, and $r = R_i$ in Equations (8.26a) and (8.26b), we obtain Equations (E4) and (E5).

$$\sigma_{rr} = -p_c \tag{E4}$$

$$\sigma_{\theta\theta} = \frac{p_c}{(R_o/R_i)^2 - 1}\left[1 + \left(\frac{R_o}{R_i}\right)^2\right] = \frac{p_c}{(1.5)^2 - 1}[1 + (1.5)^2] = 2.6 p_c \tag{E5}$$

From Equation (2.22) we can find the von Mises stress as shown in Equation (E6).

$$\sigma_{\text{von}} = \frac{1}{\sqrt{2}}\sqrt{(2.6 p_c + p_c)^2 + (-p_c)^2 + (-2.6 p_c)^2} = 3.2187 p_c \leq 200(10^6) \quad \text{or}$$

$$p_c \leq 62.137(10^6) \text{ N/m}^2 \tag{E6}$$

Comparing Equations (E3) and (E6), we obtain the maximum contact pressure as:

ANS. $p_c = 62.1$ MPa

Substituting $a = 0.05$ m, $b = 0.1$ m, $c = 0.15$ m, $E = 200(10^9)$ N/m^2, and $p_c = 62.1(10^6)$ N/m^2 into Equation (8.33), we obtain the interference shown in Equation (E7).

$$\Delta = \frac{4bp_c[(c/b)^2 - (a/b)^2]}{E[1 - (a/b)^2][(c/b)^2 - 1]} = \frac{4(0.1)(62.1)(10^6)[(1.5)^2 - (0.5)^2]}{(200)(10^9)[1 - (0.5)^2][(1.5)^2 - 1]}$$

$$= 0.265(10^{-3}) \text{ m} \tag{E7}$$

The maximum interference without exceeding yield stress is:

ANS. $\Delta = 0.265$ mm

(b) Substituting $p_c = 62.1$ MPa into Equation (E2), we obtain the stresses given by Equation (E8) due to contact pressure alone.

$$(\sigma_{rr})_o = 0 \qquad (\sigma_{\theta\theta})_o = -2.667(62.1) = -165.6 \text{ MPa} \tag{E8}$$

Without interference, the stresses in the cylinder were calculated in Equations (E1) and (E2) of Example 8.3 for internal pressure $p_i = 102$ MPa and are as follows:

$$(\sigma_{rr})_i = -p_i = -102 \text{ MPa} \tag{E9}$$

$$(\sigma_{\theta\theta})_i = 1.25 p_i = 127.5 \text{ MPa} \tag{E10}$$

The total stresses acting on the tube are the superposition of stresses in Equations (E8), (E9), and (E10) as shown in Equations (E11) and (E12).

$$\sigma_{rr} = (\sigma_{rr})_o + (\sigma_{rr})_i = -102 \text{ MPa} \tag{E11}$$

$$\sigma_{\theta\theta} = (\sigma_{\theta\theta})_o + (\sigma_{\theta\theta})_i = -165.6 + 127.5 = -38.1 \text{ MPa} \tag{E12}$$

From Equation (2.22) we can find the von Mises stress as shown in Equation (E13).

$$\sigma_{\text{von}} = \frac{1}{\sqrt{2}} \sqrt{(-102 + 38.1)^2 + (-102)^2 + (-38.1)^2} \tag{E13}$$

ANS. $\sigma_{\text{von}} = 89.3 \text{ MPa}$

COMMENTS

1. In Example 8.3 the von Mises stress on the inner surface of the cylinder was at yield stress (i.e., at 200 MPa). By assembling the cylinder by shrink fitting, this stress was reduced to 89.3 MPa as shown by Equation (E13).

2. The mechanism of the reduction in stress in the shrink-fitted cylinder is seen in Equation (E12). The residual normal stress in the tangential direction is compressive, while the tangential normal stress due to the internal pressure is tensile, and the superposition results in a reduction in the total tangential normal stress.

8.6   AIRY STRESS FUNCTION

Many elasticity problems can be solved with significant ease by using the Airy stress function. The function is chosen such that the equilibrium equations in the absence of body forces are implicitly satisfied by the stresses in two dimensions. It may be verified by substitution that if stresses are defined as in Equations (8.34a), (8.34b), and (8.34c), then the Equations (8.7a) and (8.7b) are identically met if $F_x = 0$ and $F_y = 0$.

$$\sigma_{xx} = \frac{\partial^2 \psi}{\partial y^2} \tag{8.34a}$$

$$\sigma_{yy} = \frac{\partial^2 \psi}{\partial x^2} \tag{8.34b}$$

$$\tau_{xy} = -\left(\frac{\partial^2 \psi}{\partial x \partial y}\right) \tag{8.34c}$$

The function ψ is called the Airy stress function. To determine ψ, we need to satisfy the compatibility condition of Equation (8.2b) and all the boundary conditions. We can obtain the compatibility equation on stresses by substituting the strains in terms of stresses. Substituting the plane stress Equation (8.4a) into Equation (8.2b), we obtain Equation (8.35).

$$\frac{\partial^2}{\partial y^2}\left[\frac{\sigma_{xx} - \nu\sigma_{yy}}{E}\right] + \frac{\partial^2}{\partial x^2}\left[\frac{\sigma_{yy} - \nu\sigma_{xx}}{E}\right] = \frac{\partial^2}{\partial x \, \partial y}\left[\frac{\tau_{xy}}{G}\right] = \frac{\partial^2}{\partial x \, \partial y}\left[\frac{2\tau_{xy}(1 + \nu)}{E}\right] \qquad \text{or}$$

$$\frac{\partial^2}{\partial y^2}([\sigma_{xx} - \nu\sigma_{yy}]) + \frac{\partial^2}{\partial x^2}([\sigma_{yy} - \nu\sigma_{xx}]) = 2(1 + \nu)\frac{\partial^2 \tau_{xy}}{\partial x \, \partial y} \qquad (8.35)$$

Substituting Equations (8.34a) through (8.34c) into the Equation (8.35), we obtain Equation (8.36a).

$$\frac{\partial^2}{\partial y^2}\left[\frac{\partial^2 \psi}{\partial y^2} - \nu\frac{\partial^2 \psi}{\partial x^2}\right] + \frac{\partial^2}{\partial x^2}\left[\frac{\partial^2 \psi}{\partial x^2} - \nu\frac{\partial^2 \psi}{\partial y^2}\right] = -2(1 + \nu)\frac{\partial^2}{\partial x \, \partial y}\frac{\partial^2 \psi}{\partial x \, \partial y} \qquad \text{or}$$

$$\left[\frac{\partial^4 \psi}{\partial y^4} - \nu\frac{\partial^2}{\partial y^2}\left(\frac{\partial^2 \psi}{\partial x^2}\right)\right] + \left[\frac{\partial^4 \psi}{\partial x^4} - \nu\frac{\partial^2}{\partial x^2}\left(\frac{\partial^2 \psi}{\partial y^2}\right)\right] = -2(1 + \nu)\frac{\partial^4 \psi}{\partial x^2 \, \partial y^2} \qquad \text{or}$$

$$\boxed{\frac{\partial^4 \psi}{\partial x^4} + 2\frac{\partial^4 \psi}{\partial x^2 \, \partial y^2} + \frac{\partial^4 \psi}{\partial y^4} = 0} \qquad (8.36a)$$

Alternatively, Equation (8.36a) can be written as Equation (8.36b).

$$\boxed{\nabla^4 \psi = 0} \qquad (8.36b)$$

The operator ∇^4 is called the biharmonic operator because it can be written as $\nabla^4 = \nabla^2\nabla^2$, and $\nabla^2 = \partial^2/\partial x^2 + \partial^2/\partial y^2$ is called the harmonic operator. Thus, if we can find an Airy stress function that satisfies the biharmonic operator of Equation (8.36a), we will have satisfied the equilibrium and compatibility equations. If in the Airy stress function there are some parameters that can be chosen such that the boundary conditions can be met, then we have the complete solution to the boundary value problem. For an arbitrary boundary geometry with arbitrary conditions, the problem is still difficult,[6] we will consider some simple cases for which we can find some solutions.

Equation (8.36a) was derived by using plane stress equations. Since, however, there are no material constants in the equation, the equation is also valid for plane strain.

8.7   SOLUTION BY POLYNOMIALS

We can find solutions to Equation (8.36a) by substituting polynomials of different order and determining the coefficients of the polynomials. Some polynomials are considered next.

---

[6] The complex variable method and the Fourier transform method are two general methods that are used in finding general elasticity solutions. See Muskhelishvili [1963] and Sneddon [1980] in Appendix D for additional details.

8.7.1 Quadratic Polynomials

For stresses to be nonzero, the lowest order of polynomial is quadratic. A quadratic polynomial will implicitly satisfy the biharmonic equation. We start with the Airy stress function defined as in Equation (8.37a).

$$\psi = a_2\frac{x^2}{2} + b_2 xy + c_2\frac{y^2}{2} \tag{8.37a}$$

The stresses can be obtained by substituting Equation (8.37a) into Equations (8.34a) through (8.34c) as follows:

$$\sigma_{xx} = c_2 \qquad \sigma_{yy} = a_2 \qquad \tau_{xy} = -b_2 \tag{8.37b}$$

All three stress components are constant throughout the body, irrespective of the shape of the body. For this state of stress to be the solution, the conditions on tractions on the boundary, in accordance with Equations (1.13a) and (1.13b), must be as shown in Equation (8.37c),

$$S_x = c_2 n_x - b_2 n_y \qquad S_y = -b_2 n_x + a_2 n_y \tag{8.37c}$$

where n_x and n_y are the direction cosines of the unit normal to the boundary at a point. For a rectangular body, the state of stress is shown in Figure 8.5.

Several cases of interest in this constant stress state are briefly described as follows:

- If $b_2 = a_2 = 0$, and $c_2 = \sigma$, we have uniaxial tension.
- If $b_2 = \tau$ and $a_2 = c_2 = 0$, we have a state of pure shear.
- If $b_2 = 0$ and $a_2 = c_2 = \sigma$, we have a hydrostatic state of stress (i.e., the normal stress σ is the same in all directions).
- If $b_2 = 0$ and $a_2 = -c_2 = \sigma$, we have a state of pure shear in a coordinate system that is 45° to the xy coordinate system.

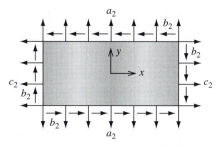

Figure 8.5 Constant stress state.

8.7.2 Cubic Polynomials

A cubic polynomial will implicitly satisfy the biharmonic equation. We start with the Airy stress function defined as in Equation (8.38a).

$$\psi = \frac{a_3}{6}x^3 + \frac{b_3}{2}x^2 y + \frac{c_3}{2}xy^2 + d_3\frac{y^3}{6} \tag{8.38a}$$

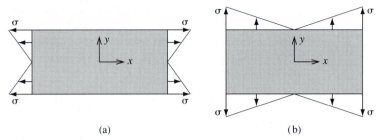

(a)   (b)

Figure 8.6 Two states of pure bending with cubic polynomials.

The stresses can be obtained from Equation (8.34a) through (8.34c) as follows:

$$\sigma_{xx} = (c_3 x + d_3 y) \qquad \sigma_{yy} = (a_3 x + b_3 y) \qquad \tau_{xy} = -(b_3 x + c_3 y) \tag{8.38b}$$

All three stress components are linear in x and y irrespective of the shape of the body. Once more, we consider several subcases for this stress state.

- If $a_3 = b_3 = c_3 = 0$, and $d_3 = \sigma$, we have pure bending of a rectangular cross section as shown in Figure 8.6(a).

- If $b_3 = c_3 = a_3 = 0$ and $a_3 = \sigma$, we have pure bending of a rectangular cross section as shown in Figure 8.6(b).

8.7.3 Fourth-Order Polynomials

A fourth-order polynomial will not implicitly satisfy the biharmonic equation, and the relationship between the constants must be determined. We start with the following Airy stress function:

$$\psi = \frac{a_4}{12}x^4 + \frac{b_4}{6}x^3 y + \frac{c_4}{2}x^2 y^2 + \frac{d_4}{6}xy^3 + \frac{e_4}{12}y^4 \tag{8.39a}$$

Substituting Equation (8.39a) into Equation (8.36a), we obtain the following condition on the constants:

$$a_4 + 2c_4 + e_4 = 0 \tag{8.39b}$$

Substituting Equations (8.39a) and (8.39b) into Equations (8.34a) through (8.34c), we obtain the stresses as follows:

$$\begin{aligned}
\sigma_{xx} &= c_4 x^2 + d_4 xy - (a_4 + 2c_4)y^2 \\
\sigma_{yy} &= a_4 x^2 + b_4 xy + c_4 y^2 \\
\tau_{xy} &= -\left(\frac{b_4}{2}x^2 + 2c_4 xy + \frac{d_4}{2}y^2\right)
\end{aligned} \tag{8.39c}$$

EXAMPLE 8.5

Figure 8.7 shows a cantilever beam with a rectangular cross section. Equation (E1) gives an Airy stress function that could be used.[7] Determine the stress components σ_{xx}, σ_{yy}, and τ_{xy} in terms of P_1, P_2, b, h, x, y, and L.

$$\psi = a_1\left[3\frac{y}{h} - \left(\frac{y}{h}\right)^3\right](x - L) + a_2\left(\frac{y}{h}\right)^2 \tag{E1}$$

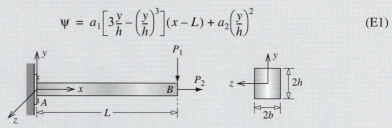

Figure 8.7 Cantilever beam in Example 8.7.

PLAN

The stress components σ_{xx}, σ_{yy}, and τ_{xy} can be obtained in terms of the constants a_1 and a_2 by using Equations (8.34a) through (8.34c). The constants in the stress functions can be determined from the boundary conditions on the axial and shear forces at end B and the required stress expression obtained.

SOLUTION

Substituting Equation (E1) into Equations (8.34a) through (8.34c), we obtain Equations (E2) through (E4).

$$\sigma_{xx} = \frac{\partial^2 \psi}{\partial y^2} = -6a_1\frac{y(x - L)}{h^3} + 2\frac{a_2}{h^2} \tag{E2}$$

ANS. $$\sigma_{yy} = \frac{\partial^2 \psi}{\partial x^2} = 0 \tag{E3}$$

$$\tau_{xy} = -\left(\frac{\partial^2 \psi}{\partial x\,\partial y}\right) = -\left(\frac{3a_1}{h}\right)\left[1 - \left(\frac{y}{h}\right)^2\right] \tag{E4}$$

To obtain the boundary conditions at the end of beam, we make an imaginary cut an infinitesimal distance from the end and draw the internal forces and moments in accordance with our sign conventions in Table 3.3 to obtain Figure 8.8. By the equilibrium of forces and moments, we obtain the conditions given by Equations (E5) through (E7).

$$N(x = L) = P_2 \tag{E5}$$

---

[7] The function is an outcome of a method proposed by C.Y. Neou. See Boresi and Chong [1987] in Appendix D for additional details.

$$V_y(x = L) = -P_1 \qquad \text{(E6)}$$

$$M_z(x = L) = 0 \qquad \text{(E7)}$$

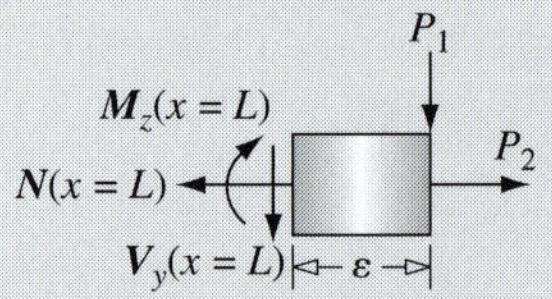

Figure 8.8 Free body diagram of an infinitesimal element at the end of a beam.

Substituting Equations (E2) and (E4) into Equations (3.4a-A) and (3.4c-B) and evaluating the result at $x = L$, we obtain Equations (E8) and (E9).

$$N(x = L) = \int_A \sigma_{xx} dA = \int_{-h}^{h} \left(2\frac{a_2}{h^2}\right)(2b\ dy) = 4\frac{ba_2}{h^2} y \Big|_{-h}^{h} = P_2 \qquad \text{or} \qquad a_2 = \frac{h}{8b}P_2 \quad \text{(E8)}$$

$$V_y = \int_A \tau_{xy} dA = -\left(\frac{3a_1}{h}\right)\int_{-h}^{h}\left[1 - \left(\frac{y}{h}\right)^2\right](2b\ dy) = -\left(\frac{6ba_1}{h}\right)\left[y - \frac{y^3}{3h^2}\right]\Big|_{-h}^{h} = -P_1 \qquad \text{or}$$

$$a_1 = \frac{P_1}{8b} \qquad \text{(E9)}$$

Substituting Equations (E8) and (E9) into Equations (E2) and (E4), we obtain the following stresses:

ANS. $\quad \sigma_{xx} = -\left(\frac{3P_1}{4bh^3}\right)y(x - L) + \frac{P_2}{4bh} \qquad \text{(E10)}$

ANS. $\quad \tau_{xy} = -\left(\frac{3P_1}{8bh}\right)\left[1 - \left(\frac{y}{h}\right)^2\right] \qquad \text{(E11)}$

COMMENTS

1. The condition on the moment in Equation (E7) is implicitly met because the stress σ_{xx} is uniform across the cross section at $x = L$.

2. The boundary condition of zero traction on top and bottom is implicitly met because the stress components σ_{yy} and τ_{xy} are zero at $y = \pm h$.

3. The boundary condition of zero traction at all points except where axial force is applied is not met at $x = L$ because the stress components σ_{xx} and τ_{xy} are not zero. We satisfied the boundary condition at $x = L$ in an average sense across the cross section when we used the conditions for equivalent internal forces and moments.

4. Note that $I_{zz} = (2bh^3)/3$, $A = 4bh$, $M_z = P_1(x - L)$, and $N = P_2$. Substituting these values for normal stress under combined loading will result in the expression given by Equation (E10). Similarly, if the bending shear stress is calculated at any point, Equation (E11) will result. This emphasizes that the stress function of Equation (E1) incorporates the approximations of beam theory implicitly. Other stress functions could give other results.

8.8   TORSION OF NONCIRCULAR SHAFTS

In the introductory course on the mechanics of materials, we saw the theory for the torsion of circular shafts. In Section 6.4 we saw the theory for the torsion of thin tubes of arbitrary cross section. In this section we develop a more general theory in which there is no limitation on cross-sectional shape or on shaft thickness. Saint-Venant was the first to develop a theory for the torsion for noncircular shafts. Prandtl later developed an alternative based on Airy's stress function approach. We will use both theories. We will use Prandtl's approach for obtaining the stresses and Saint-Venant's to obtain the deformation.

8.8.1   Saint-Venant's Method

Saint-Venant observed that a displacement field[8] for a cross section of a noncircular shaft under torsion, as in Figure 8.9, should account for the following:

(i) The cross section would warp under torsion.

(ii) The only nonzero stress components would be τ_{xy} and τ_{xz}, which for isotropic materials implies that the only nonzero strain components would be γ_{xy} and γ_{xz}.

Saint-Venant proposed a displacement field that included a warping function χ and a form that would result in shear strains γ_{xy} and γ_{xz} only, as given by Equation (8.40),

$$u = \chi(y,z)\frac{d\phi}{dx} \qquad v = -xz\frac{d\phi}{dx} \qquad w = xy\frac{d\phi}{dx} \tag{8.40}$$

where u, v, and w are the displacements in the x, y, and z direction, respectively; $d\phi/dx$ is considered to be a *constant* representing the rate of twist per unit length; and the warping function $\chi(y,z)$ describes the movement of points out of the cross-sectional plane.

With the displacement field known, the logic schematized in Figure 3.2 can be used to obtain the required formulas, as shown in the discussion that follows.

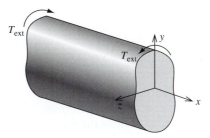

Figure 8.9 Torsion of a noncircular shaft.

---

[8] Saint-Venant described his method as *semi-inverse*, since the validation of his displacement field was that it could satisfy all equations of elasticity. See Timoshenko [1983] in Appendix D for additional details.

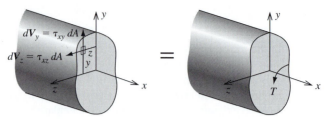

Figure 8.10 Static equivalency for noncircular shafts.

Substituting Equation (8.40) into the strain–displacement relations given by Equations (8.1a) through (8.1f), we obtain the nonzero strain components given by Equations (8.41a) and (8.41b).

$$\gamma_{xy} = \frac{\partial}{\partial y}\left[\chi(y,z)\frac{d\phi}{dx}\right] + \frac{\partial}{\partial x}\left[-xz\frac{d\phi}{dx}\right] = \frac{d\phi}{dx}\left[\frac{\partial\chi}{\partial y} - z\right] \tag{8.41a}$$

$$\gamma_{xz} = \frac{\partial}{\partial x}\left[xy\frac{d\phi}{dx}\right] + \frac{\partial}{\partial z}\left[\chi(y,z)\frac{d\phi}{dx}\right] = \frac{d\phi}{dx}\left[y + \frac{\partial\chi}{\partial z}\right] \tag{8.41b}$$

From Hooke's law for isotropic materials, the shear stresses can be written as shown in Equations (8.42a) and (8.42b).

$$\tau_{xy} = G\left(\frac{\partial\chi}{\partial y} - z\right)\frac{d\phi}{dx} \tag{8.42a}$$

$$\tau_{xz} = G\left(\frac{\partial\chi}{\partial z} + y\right)\frac{d\phi}{dx} \tag{8.42b}$$

The shear stresses can be replaced by an equivalent internal torque as shown in Figure 8.10. By equating the moment about the origin, we obtain Equation (8.43).

$$T = \int_A (y\tau_{xz} - z\tau_{xy})dA \tag{8.43}$$

We could substitute Equations (8.42a) and (8.42b) into Equation (8.43) and obtain other formulas, but at this stage we will switch to Prandtl's method, which results in a similar form of equations.

8.8.2   Prandtl's Method

The nonzero stress components τ_{xy} and τ_{xz} that will be generated to resist the external torque must satisfy the equilibrium equations (8.8a) through (8.8c), which in the absence of body forces can be written as shown by Equations (8.44a) through (8.44c).

$$\frac{\partial\tau_{yx}}{\partial y} + \frac{\partial\tau_{zx}}{\partial z} = 0 \tag{8.44a}$$

$$\frac{\partial\tau_{xy}}{\partial x} = 0 \tag{8.44b}$$

$$\frac{\partial \tau_{xz}}{\partial x} = 0 \tag{8.44c}$$

Equations (8.44b) and (8.44c) imply that shear stresses cannot be functions of x, which is possible as long as there is no distributed torque on the shaft. To satisfy Equation (8.44a), we define a stress function as in Equations (8.45a) and (8.45b).

$$\boxed{\tau_{yx} = \tau_{xy} = \frac{\partial \psi}{\partial z}} \tag{8.45a}$$

$$\boxed{\tau_{zx} = \tau_{xz} = -\left(\frac{\partial \psi}{\partial y}\right)} \tag{8.45b}$$

The boundary condition on the surface of the noncircular shaft is zero traction. Substituting $p_x = 0$, $\sigma_{xx} = 0$, and Equations (8.45a) and (8.45b) into Equation (8.10a), we obtain Equation (8.46).

$$\frac{\partial \psi}{\partial z} n_y - \frac{\partial \psi}{\partial y} n_z = 0 \tag{8.46}$$

The outward normal to the surface of the noncircular shaft will lie in the y-z plane. Figure 8.11 shows the geometry by which we can relate the direction cosines of the unit normal to the surface geometry. Noting that the tangent to the surface is 90° to the unit normal (i.e., $\lambda_z = 90 + \theta_z$), we obtain Equations (8.47a) and (8.47b).

$$t_z = \frac{dz}{ds} = \cos \lambda_z = \cos(90 + \theta_z) = -\sin \theta_z = -n_y \qquad \text{or} \qquad n_y = -\left(\frac{dz}{ds}\right) \tag{8.47a}$$

$$t_y = \frac{dy}{ds} = \sin \lambda_z = \sin(90 + \theta_z) = \cos \theta_z = n_z \qquad \text{or} \qquad n_z = \frac{dy}{ds} \tag{8.47b}$$

Substituting Equations (8.47a) and (8.47b) into Equation (8.46), we obtain Equation (8.48).

$$-\left(\frac{\partial \psi}{\partial z}\right)\left(\frac{dz}{ds}\right) - \frac{\partial \psi}{\partial y}\left(\frac{dy}{ds}\right) = 0 \qquad \text{or} \qquad -\left(\frac{d\psi}{ds}\right) = 0 \tag{8.48}$$

Equation (8.48) implies that ψ is a constant on the shaft surface (i.e., on the boundary of the cross section). This conclusion plays a critical role in selection of the stress function, and we record it for future use.

> The stress function ψ must be constant on the boundary of the cross section of the noncircular shaft $\tag{8.49}$

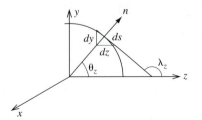

Figure 8.11 Direction cosines of a unit normal.

Substituting Equations (8.45a) and (8.45b) into Equation (8.43) and integrating, we obtain Equation (8.50),

$$T = \iint_A \left(y\left[-\left(\frac{\partial \psi}{\partial y}\right)\right] - z\left[\frac{\partial \psi}{\partial z}\right]\right) dy \ dz = -\oint [y\psi] dz - \oint [z\psi] dy + \iint_A [\psi + \psi] dy \ dz \quad (8.50)$$

where $\oint$ represents integration on the closed boundary defining the cross section. From Equation (8.49), ψ is constant on the boundary of the cross section. The contour integrals can be written as shown in Equation (8.51).

$$-\oint [y\psi] dz - \oint [z\psi] dy = -\psi \oint [y \ dz + z \ dy] = -\psi \oint d(yz) = 0 \quad (8.51)$$

The integral in Equation (8.51) is zero because if we start at any point on the boundary, go around it, and return to the starting point, the y and z coordinates do not change. Substituting Equation (8.51) into Equation (8.50), we obtain the final result for the internal torque as shown in Equation (8.52).

$$\boxed{T = 2\iint_A \psi dy \ dz} \quad (8.52)$$

To obtain the displacements from the stresses in Equations (8.45a) and (8.45b), we could first obtain strains and then integrate the partial derivatives. We choose an easier alternative, as described in the discussion that follows.

We substitute Equations (8.42a) and (8.42b) into Equations (8.45a) and (8.45b) to obtain Equations (8.53a) and (8.53b).

$$\frac{\partial \psi}{\partial z} = G\left(\frac{\partial \chi}{\partial y} - z\right)\frac{d\phi}{dx} \quad (8.53a)$$

$$\left(\frac{\partial \psi}{\partial y}\right) = -G\left(\frac{\partial \chi}{\partial z} + y\right)\frac{d\phi}{dx} \quad (8.53b)$$

Taking the partial derivative of Equation (8.53a) with respect to z and the partial derivative of Equation (8.53b) with respect to y and adding, we obtain Equation (8.54).

$$\boxed{\frac{\partial^2 \psi}{\partial y^2} + \frac{\partial^2 \psi}{\partial z^2} = -2G\frac{d\phi}{dx}} \quad (8.54)$$

In obtaining Equation (8.54), we made use of the fact that $d\phi/dx$ and G are constants. We now have all the equations we need to solve problems of torsion of noncircular shafts in Section 8.8.3 gives the procedure.

8.8.3 Procedure for Solving Problems of Torsion of Noncircular Shafts

Step 1. For a given cross-sectional shape, obtain a stress function ψ that is constant on the boundary of the cross section in accordance with Equation (8.49).

Step 2. Determine any constant in the stress function in terms of the internal torque T by integrating over the cross section in accordance with Equation (8.52).

Step 3. Determine the shear stresses in accordance with Equations (8.45a) and (8.45b).

Step 4. Determine the rate of twist $d\phi/dx$ in accordance with Equation (8.54).

EXAMPLE 8.6

A shaft has an elliptical cross section as shown in Figure 8.12. Determine the equations for maximum shear stress τ_{max} at a cross section and the relative rotation $(\phi_2 - \phi_1)$ of the cross sections at points x_1 and x_2 along the length of the shaft in terms of internal torque T, shear modulus G, a, b, x_1, and x_2.

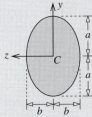

Figure 8.12 Elliptical cross section for Example 8.6.

PLAN

The equation of an ellipse, which is known, can be written so that all terms are on the left and there is a zero on the right. If the stress function ψ is chosen equal to the terms on the left-hand side of the ellipse equation and multiplied by any arbitrary constant, the stress function ψ will give a zero value at any point on the boundary. If this stress function is used, the required quantities can be determined by means of the steps outlined in Section 8.8.3.

SOLUTION

Step 1. The stress function ψ, which will have a constant value on the boundary, can be written as shown by Equation (E1),

$$\psi = K\left(\frac{y^2}{a^2} + \frac{z^2}{b^2} - 1\right) \tag{E1}$$

where K is a constant to be determined.

Step 2. Substituting Equation (E1) into Equation (8.52) and integrating, we obtain Equation (E2),

$$T = 2\iint_A K\left(\frac{y^2}{a^2} + \frac{z^2}{b^2} - 1\right)dy\,dz = 2K\left[\frac{I_{zz}}{a^2} + \frac{I_{yy}}{b^2} - A\right] \tag{E2}$$

where I_{yy} and I_{zz} are the area moments of inertia of an ellipse and A is the cross-sectional area of an ellipse. Substituting $I_{yy} = (\pi a b^3)/4$, $I_{zz} = (\pi b a^3)/4$, and $A = \pi a b$ into Equation (E2), we obtain Equation (E3).

$$T = 2K\left[\frac{\pi a b^3}{4a^2} + \frac{\pi a b^3}{4b^2} - \pi a b\right] = -K\pi a b \qquad \text{or} \qquad K = -\frac{T}{\pi a b} \qquad \text{(E3)}$$

Substituting Equation (E3) into Equation (E1), we obtain Equation (E4).

$$\psi = \left(-\frac{T}{\pi a b}\right)\left(\frac{y^2}{a^2} + \frac{z^2}{b^2} - 1\right) \qquad \text{(E4)}$$

Step 3. The shear stress components can be obtained by substituting Equation (E4) into Equations (8.45a) and (8.45b) to obtain Equations (E5) and (E6).

$$\tau_{yx} = \tau_{xy} = -\frac{2Tz}{\pi a b^3} \qquad \text{(E5)}$$

$$\tau_{zx} = \tau_{xz} = \frac{2Ty}{\pi a^3 b} \qquad \text{(E6)}$$

Since $a > b$, the magnitude of the maximum shear stress will be given by Equation (E5) when $z = b$ as shown by Equation (E7).

ANS. $\quad \tau_{\max} = \dfrac{2T}{\pi a b^2} \qquad \text{(E7)}$

Step 4. The rate of rotation can be found by substituting Equation (E4) into Equation (8.54) as shown in Equation (E8).

$$-2G\frac{d\phi}{dx} = \left(-\frac{T}{\pi a b}\right)\left(\frac{2}{a^2} + \frac{2}{b^2}\right) \qquad \text{or} \qquad \frac{d\phi}{dx} = \frac{T(a^2 + b^2)}{G\pi a^3 b^3} \qquad \text{(E8)}$$

We note that because the rate of rotation is a constant, it can be written as $d\phi/dx = (\phi_2 - \phi_1)/(x_2 - x_1)$ to obtain Equation (E9).

ANS. $\quad \phi_2 - \phi_1 = \dfrac{T(a^2 + b^2)(x_2 - x_1)}{G\pi a^3 b^3} \qquad \text{(E9)}$

COMMENTS

1. The torque in Equations (E7) and (E9) is an internal torque. It can be related to the external torque by making an imaginary cut and drawing a free body diagram as demonstrated next, in Example 8.7.

2. For a circular shaft $a = b$. Substituting $a = b$ into Equations (E7) and (E9), we obtain Equations (E10).

$$\tau_{max} = \frac{2T}{\pi a^3} \qquad \phi_2 - \phi_1 = \frac{2T(x_2 - x_1)}{G\pi a^4} \qquad \text{(E10)}$$

The values of τ_{max} and $\phi_2 - \phi_1$ are the same as those that would be obtained from the theory of the torsion of circular shafts.

3. If we compare Equation (E8) with Equation (3.8-T), we can say that the torsional rigidity of the elliptical shaft is $G\pi a^3 b^3/(a^2 + b^2)$.

EXAMPLE 8.7

An aluminum ($G = 4000$ ksi) elliptical shaft is loaded as shown in Figure 8.13. Determine the maximum shear stress in the shaft and the rotation of section D with respect to the rotation of section A.

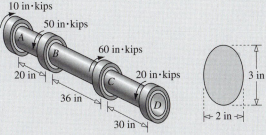

Figure 8.13 Elliptical shaft for Example 8.7.

PLAN

The internal torque in each segment of the shaft can be found by making an imaginary cut and drawing free body diagrams. The equations of the maximum shear stress and relative rotation of two sections were developed in Example 8.6 in terms of the internal torque. By substituting the internal torque for each segment, the maximum shear stress and relative rotation of segment ends can be found, and the desired results can be calculated from these values.

SOLUTION

The maximum shear stress and relative rotation for an elliptical shaft were determined in Example 8.6 as shown in Equations (E1) and (E2).

$$\tau_{max} = \frac{2T}{\pi ab^2} \qquad \text{(E1)}$$

$$\phi_2 - \phi_1 = \frac{T(a^2 + b^2)(x_2 - x_1)}{G\pi a^3 b^3} \tag{E2}$$

After making imaginary cuts in segments AB, BC, and CD of Figure 8.13, we can draw the free body diagrams shown in Figure 8.14. The internal torque is drawn in accordance with our sign convention (Table 3.3).

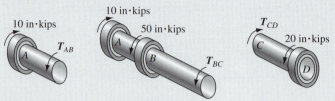

Figure 8.14 Free body diagrams for Example 8.7.

By the equilibrium of moment about the shaft axis in Figure 8.14, we obtain the following internal torques:

$$T_{AB} = 10 \text{ in} \cdot \text{kips} \qquad T_{BC} = -40 \text{ in} \cdot \text{kips} \qquad T_{CD} = 20 \text{ in} \cdot \text{kips} \tag{E3}$$

With a and b constant for the entire shaft, the maximum shear stress will be in segment BC, where the internal torque has the largest magnitude. Substituting $a = 1.5$ in, $b = 1$ in, and T_{BC} from Equation (E3) into Equation (E1), we obtain the magnitude of maximum shear stress as shown in Equation (E4).

$$\tau_{\max} = \left| \frac{2T_{BC}}{\pi a b^2} \right| = \frac{2(40)}{\pi(1.5)(1)^2} \qquad \textbf{ANS.} \quad \tau_{\max} = 16.98 \text{ ksi} \tag{E4}$$

Substituting $G = 4000$ ksi, $a = 1.5$ in, $b = 1$ in, and the internal torques in Equation (E3) into Equation (E2), we obtain the relative rotations of each segment, as shown in Equations (E5), (E6), and (E7).

$$\phi_B - \phi_A = \frac{T_{AB}(a^2 + b^2)(x_B - x_A)}{G\pi a^3 b^3} = \frac{(10)(1.5^2 + 1^2)(20)}{(4000)\pi(1.5)^3 1^3} = 15.326(10^{-3}) \text{ rad} \tag{E5}$$

$$\phi_C - \phi_B = \frac{T_{BC}(a^2 + b^2)(x_C - x_B)}{G\pi a^3 b^3} = \frac{(-40)(1.5^2 + 1^2)(36)}{(4000)\pi(1.5)^3 1^3} = -110.347(10^{-3}) \text{ rad} \tag{E6}$$

$$\phi_D - \phi_C = \frac{T_{CD}(a^2 + b^2)(x_D - x_C)}{G\pi a^3 b^3} = \frac{(20)(1.5^2 + 1^2)(30)}{(4000)\pi(1.5)^3 1^3} = 45.978(10^{-3}) \text{ rad} \tag{E7}$$

Adding Equations (E5), (E6), and (E7), we obtain the relative rotation of section D with respect to section A as shown in Equation (E8).

$$\phi_D - \phi_A = [15.326 - 110.347 + 45.978](10^{-3}) = -49.04(10^{-3}) \text{ rad} \tag{E8}$$

$$\textbf{ANS.} \quad \phi_D - \phi_A = 0.049 \text{ rad cw}$$

COMMENT

The problem once more emphasizes that the equilibrium equations relating internal forces and moments to external forces and moments are independent of kinematics. The change in kinematics impacted the relationship of rotation and maximum stress to internal torque.

PROBLEM SET 8.1

8.1 Starting with $\partial^2 \gamma_{yz}/\partial y\, \partial z$, derive the compatibility expression of Equation (8.55a).

$$\frac{\partial^2 \varepsilon_{yy}}{\partial z^2} + \frac{\partial^2 \varepsilon_{zz}}{\partial y^2} = \frac{\partial^2 \gamma_{yz}}{\partial y\, \partial z} \qquad (8.55a)$$

8.2 Starting with $\partial^2 \varepsilon_{zz}/\partial x\, \partial y$, derive the expression of Equation (8.55b).

$$\frac{\partial^2 \varepsilon_{zz}}{\partial x\, \partial y} = \frac{1}{2}\frac{\partial}{\partial z}\left[\frac{\partial \gamma_{yz}}{\partial x} + \frac{\partial \gamma_{zx}}{\partial y} - \frac{\partial \gamma_{xy}}{\partial z}\right] \qquad (8.55b)$$

8.3 Show that the Equation (8.4a) of plane stress and Equation (8.5a) of plane strain can be written in the compact form given by Equations (8.56a) through (8.56c),

$$8G\varepsilon_{xx} = (\kappa + 1)\sigma_{xx} - (3 - \kappa)\sigma_{yy} \quad (8.56a)$$

$$8G\varepsilon_{yy} = (\kappa + 1)\sigma_{yy} - (3 - \kappa)\sigma_{xx} \quad (8.56b)$$

where

$$\kappa = \begin{cases} \dfrac{(3 - v)}{(1 + v)} & \text{for plane stress} \\[2mm] 3 - 4v & \text{for plane strain} \end{cases} \qquad (8.56c)$$

8.4 Show that Equations (8.56a) and (8.56b) can be written as Equations (8.56d) and (8.56e),

$$(\kappa - 1)\sigma_{xx} = G[(\kappa + 1)\varepsilon_{xx}$$
$$+ (3 - \kappa)\varepsilon_{yy}] \qquad (8.56d)$$

$$(\kappa - 1)\sigma_{yy} = G[(\kappa + 1)\varepsilon_{yy}$$
$$+ (3 - \kappa)\varepsilon_{xx}] \qquad (8.56e)$$

where κ is as defined in Equation (8.56c).

8.5 The equilibrium equations in the three-dimensional polar coordinates r, θ, and x are as follows:

$$\frac{\partial \sigma_{rr}}{\partial r} + \frac{1}{r}\frac{\partial \tau_{r\theta}}{\partial \theta} + \frac{\sigma_{rr} - \sigma_{\theta\theta}}{r} + \frac{\partial \tau_{xr}}{\partial x} + F_r = 0$$

$$\frac{1}{r}\frac{\partial \sigma_{\theta\theta}}{\partial \theta} + \frac{\partial \tau_{r\theta}}{\partial r} + \frac{2\tau_{r\theta}}{r} + \frac{\partial \tau_{x\theta}}{\partial x} + F_\theta = 0 \quad (8.57)$$

$$\frac{\partial \tau_{rx}}{\partial r} + \frac{1}{r}\frac{\partial \tau_{\theta r}}{\partial \theta} + \frac{\partial \sigma_{xx}}{\partial x} + \frac{\tau_{rx}}{r} + F_x = 0$$

In these equations, F_r, F_θ, and F_x are the body force per unit volume in the r, θ, and x directions, respectively. Use Equations (8.57) to obtain the equilibrium equation (3.11-T) for a circular shaft in torsion. Assume a constant cross-sectional area.

8.6 A thin solid disk of radius a, modulus of elasticity E, Poisson ratio v, and density ρ is rotating at an angular speed of ω. Determine the radial and tangential normal stresses in terms of a, E, v, ρ, ω, and r.

ANS. $\sigma_{rr}(r) = \dfrac{(3 + v)}{8}\rho\omega^2(a^2 - r^2)$

$$\sigma_{\theta\theta}(r) = \frac{\rho\omega^2}{8}[(3 + v)a^2 - (1 + 3v)r^2]$$

8.7 A thin disk with of radius $2a$ has a hole of radius a, modulus of elasticity E, Poisson ratio v, and density ρ; it is rotating at an angular speed of ω. Assume that the inner and outer surfaces are stress free. Determine the radial and tangential normal stresses in terms of a, E, ν, ρ, ω, and r.

8.8 A grinding wheel has an allowable stress of 12 ksi, a modulus of elasticity of 1500 ksi, a Poisson ratio of 1/3, and a specific weight of 0.1 lb/in³. The inner radius of the wheel is 3 in, and the outer radius is 6 in. Determine the maximum rotational speed (burst speed) of the grinding wheel.

ANS. $\omega_{max} = 87.5$ rad/s

8.9 A steel cylinder with an inside diameter of 8 inches and an outside diameter of 12 inches is subjected to an internal pressure of 12 ksi. Determine (a) the maximum tensile stress in the cylinder and (b) the radial and tangential stresses in the middle (i.e., at $r = 5$ in).

8.10 A thick-walled cylinder having an inner radius of 6 inches is to be subjected to an internal pressure of 15 ksi. The maximum allowable tensile stress is not to exceed 25 ksi. Determine the thickness of the cylinder.

ANS. $t = 6$ in

8.11 A small laminated steel cylinder with no axial forces is constructed by shrink fitting a steel jacket (outside diameter 300 mm) onto a steel tube (outside diameter 200 mm, inside diameter 100 mm). The steel has a modulus of elasticity of $E = 200$ GPa. The interference between the tube and the jacket is 0.12 mm. Determine (a) the contact pressure between the tube and the jacket and (b) the maximum tensile and compressive normal stresses in the laminated cylinder for an internal pressure of 100 MPa.

8.12 A steel cylinder with an inside diameter of 8 inches and an outside diameter of 12 inches has a yield stress of 30 ksi. Use the octahedral shear stress theory to determine the maximum internal pressure that the cylinder can hold without exceeding its yield stress.

8.13 A small laminated steel cylinder with no axial forces is constructed by shrink fitting a steel jacket (outside diameter 12 in) onto a

steel tube (outside diameter 10 in, inside diameter 8 in). Steel has a modulus of elasticity of $E = 30,000$ ksi, a Poisson ratio of 0.3, and a yield stress of $\sigma_{yield} = 30$ ksi. (a) Determine the maximum interference between the tube and the jacket if yielding is to be avoided. Use the octahedral shear stress theory. (b) Determine the von Mises stress on the inner surface if the assembled cylinder is subjected to an internal pressure of 10 ksi.

8.14 Consider a laminated cylinder that is assembled by shrink fitting a jacket onto a tube. Assuming that this is a plane strain problem, obtain a relationship between contact pressure p_c and interference Δ.

8.15 Starting with the fifth-order polynomial given, determine the relationship between the constants for the biharmonic function to be satisfied.

$$\psi = \frac{a_5}{20}x^5 + \frac{b_5}{12}x^4y + \frac{c_5}{6}x^3y^2$$
$$+ \frac{d_5}{6}x^2y^3 + \frac{e_5}{12}xy^4 + \frac{f_5}{20}y^5$$

8.16 Two solid elliptical steel ($G_{st} = 80$ GPa) shafts and a solid elliptical bronze ($G_{Cu/Sn} = 40$ GPa) shaft are securely connected by a coupling at C. The major diameter of the elliptical cross section is 100 mm, and the minor diameter is 80 mm. A torque of $T = 10$ kN · m is applied to the rigid wheel B as shown in Figure P8.16. The coupling plates cannot rotate relative to each other. Determine the angle of rotation of the wheel B due to the applied torque and the maximum shear stress in the shaft. Use the results of Example 8.6 to solve the problem.

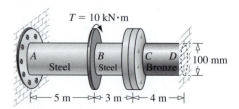

Figure P8.16

ANS. $\phi_B = 0.07$ rad cw

$\tau_{max} = 54.7$ MPa

8.17 The stress function for the cross section shown in Figure P8.17, shaped as an equilateral triangle, is given by the equation

$$\psi = K[y + z\sqrt{3} - 2h/3]$$
$$\times [y - z\sqrt{3} - 2h/3][y + h/3] \quad (8.58)$$

where K is a constant to be determined. Determine the maximum shear stress and the rate of twist in terms of internal torque T, shear modulus G, and h.

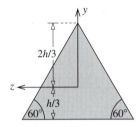

Figure P8.17

ANS. $d\phi/dx = 15\sqrt{3}T/Gh^4$
$\tau_{max} = 15\sqrt{3}T/2h^3$

8.18 The shaft shown in Figure P8.18 has the cross section of an equilateral triangle. Each side of the triangle is 200 mm, and the shear modulus of elasticity is 70 GPa. Use the results of Problem 8.17 to determine the maximum shear stress in the shaft and the rotation of section D with respect to section A.

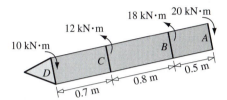

Figure P8.18

8.9 CLOSURE

The kinematic equations relating strains and displacements, the compatibility equations on strains, the generalized version of Hooke's law, the equilibrium equations, and the boundary conditions on tractions or displacements are the equations of elasticity that were introduced in this chapter. From arguments of axisymmetry, the equations of elasticity can be used to obtain results for plane strain problems of thick long cylinders and for plane stress problems of thin disks or thick short cylinders with no axial load. Equations of elasticity can also be used to obtain matrix equations for the finite element method in two and three dimensions, as briefly described in the next chapter (see Section 9.6). Equilibrium equations on stresses are intrinsically satisfied by the Airy stress function, a powerful concept whose one application is for the torsion of noncircular shafts.

9 Finite Element Method

9.0 OVERVIEW

The finite element method is a versatile numerical method that is ubiquitous in stress analysis and in the design of machines and structures. Figure 9.1 shows the frame of a building representing an assembly of beams, columns, and axial members. Associated with each structural element of the building frame is a stiffness matrix, and all these matrices together can be assembled into a global stiffness matrix to represent the structure. This process of assembly is methodically done in the finite element method, as will be seen in this chapter.

The finite element method began as a matrix method of analysis. There are two versions of it: the "stiffness method," in which the displacements of points on the structure are unknown, and the "flexibility method," in which the internal forces in the structural members are unknown. Commercial computer programs are usually based on the stiffness method, which is described in this chapter. Such simple elements as axial rods, circular shafts, and symmetric beams will be the primary focus in this introduction to the finite element method. Our treatment of two-dimensional elements (Section 9.6) will be brief.

The set of equations used in the stiffness method are the equilibrium equations relating the displacement of points. The Rayleigh-Ritz method, which predates the finite element method, is a formal procedure for deriving equilibrium equations in matrix form, as was seen in Section 7.9. From a theoretical viewpoint, the primary difference between the Rayleigh-Ritz method and the finite element method is that the kinematically admissible functions used in finding the approximate solutions are defined over the entire structural member in the Rayleigh-Ritz method, while in the finite element method the functions are piecewise kinematically admissible. The small theoretical difference, however, results in a dramatically different perspective of solving problems by the finite element method.

Many of the conclusions and equations of the Rayleigh-Ritz method are applicable to the finite element method. We will briefly study "Lagrange polynomials," which

Figure 9.1 Discrete elements of a building frame.

are used pervasively in the finite element method as the piecewise kinematically admissible displacement functions. The key steps of the finite element method will be elaborated for simple structures made from axial rods, circular shafts, and symmetric beams.

The learning objectives in this chapter are:

1. To understand the perspective, the key issues, and the terminology of the finite element method.

2. To understand the procedural steps of solving problems by the finite element method.

9.1 TERMINOLOGY

The finite element method (FEM) originated in structures as a matrix method. Each row of the matrix represents an equilibrium equation. In the matrix method, the equilibrium equations were obtained by force[1] and moment balance. In this chapter we will obtain the equilibrium equations by minimizing the potential energy.[2] Thus, the finite element method formulation presented here is very similar to the Rayleigh-Ritz method with one important difference: the kinematically admissible displacement

---

[1] Visualize a truss in which the displacements of pins are unknown. If we now use the "method of joints" and write the equilibrium of forces at each joint, we will obtain a set of algebraic equations in which the unknowns are the pin displacements and the right-hand-side quantities are the external forces acting on the pin.

[2] Today the finite element method is used extensively in solving engineering problems. It is viewed as a numerical method for solving partial differential equations, and algebraic equations are obtained from an approach called the "weak form." See Reddy [1993] in Appendix D for additional details.

functions in the finite element method are defined piecewise continuously over small (finite) domains; these are the "elements." The boundary points of the elements are called "nodes," although nodes can also be points inside the element. The constants multiplying the piecewise kinematically admissible functions are the displacements of the nodes, and the kinematically admissible functions are called interpolation functions because they can be used to interpolate the values of displacements between the nodes. The representation of a structure by elements and nodes is called a mesh. A mesh with boundary conditions, applied loads, and material property is called a model. A model is a finite element representation of a real-life problem, and the accuracy of the model's predictions is determined by the assumptions and limitations that are made in constructing the finite element model and the errors introduced in solving the model by numerical methods.

The use of piecewise kinematically admissible functions changes the perspective with which we view and solve a problem by means of the finite element method. To elaborate this perspective, consider the statement of minimum potential energy in Equation (7.31). A structure could be made up of axial members, circular shafts, symmetric beams, and other members such as curved beams, plates, and shells. Potential energy is a scalar quantity and can be written as the sum of the potential energies of all the structural members ($\delta\Omega^{(i)}$). Equation (7.31) can be written as:

$$\delta\Omega = \sum_{i=1}^{n} \delta\Omega^{(i)} \tag{9.1}$$

Equation (9.1) is valid for structural members of all types, irrespective of orientation. We could thus develop the potential energy in matrix form for each member separately. That is, we could develop matrices at the element level in a local coordinate system without regard to how a member is used in the structure. The individual local matrices, called element stiffness[3] matrices, could be assembled by using Equation (9.1) to form the global (stiffness) matrix of the entire structure. This perspective of reducing the *complexity of analyzing large structures to the analysis of simple individual members (elements)* is what makes the finite element method such a versatile and popular tool in structural (and engineering application) analysis.

We will develop element stiffness matrices for axial members, circular shafts, and symmetric beams. We will use these matrices to analyze simple structures to elaborate the principles of assembly represented by Equation (9.1). The analysis of complex structure requires the use of computers. The FEM is now a very mature technique, and there are many commercially available software packages that can be used for solving engineering problems.

The foregoing description introduced terms extensively used in the FEM. We formally define the terms as follows.

---

[3] In a less-used version of the finite element method called the flexibility method, the nodal forces are the unknowns. The flexibility matrix is derived from the complementary potential energy.

Definition 1 Nodes are points on the structure at which displacements and rotations are to be found or prescribed.

Definition 2 An element is a small domain on which we can solve a boundary value problem in terms of the displacements and forces of the nodes on the element. It can also be defined as a small domain on which the displacement is represented by a kinematically admissible polynomial function.

Definition 3 An interpolation function[4] is a kinematically admissible displacement function that is defined on an element and can be used for interpolating displacement values between the nodes.

Definition 4 A discrete representation of structural geometry by elements and nodes is called a mesh.

Definition 5 The process of creating a mesh (comprising discrete entities) is called discretization.

Definition 6 The mesh, boundary conditions, loads, and material properties representing the actual structure comprise what is called a model.

Definition 7 An element stiffness matrix relates the displacements to the forces at the element nodes.

Definition 8 A global stiffness matrix is an assembly of element stiffness matrices that relates the displacements of the nodes on the mesh to applied external forces.

9.2 LAGRANGE POLYNOMIALS

Lagrange polynomials were discovered independently of the finite element method. These functions are used for interpolations of many quantities. The polynomials are introduced by using the axial members to provide the motivation and practical relevance of these functions in the FEM.

Figure 9.2(a) shows an axial member (element) with two nodes. The axial displacements u_1 and u_2 are the degrees of freedom (generalized displacements) in terms of which we plan to derive the element stiffness matrix. The element has 2 degrees of freedom, and

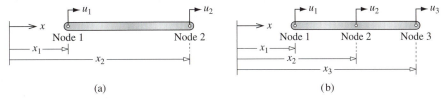

(a)   (b)

Figure 9.2 Axial (a) linear and (b) quadratic elements.

---

[4]In two and three dimensions, the interpolation functions are also used for approximating the geometry and are called *shape functions*. When the same interpolation functions are used for approximating both geometry and displacement, the element is called *isoparametric*. Isoparametric elements are very popular, and many FEM practitioners use the terms "interpolation function" and "shape function" interchangeably.

so we choose a linear function $u(x) = C_1 + C_2 x$ with two unknown parameters. We note that at $x = x_1$ the displacement $u(x_1) = u_1$, and at $x = x_2$ the displacement $u(x_2) = u_2$. The constants C_1 and C_2 can be solved in terms of u_1 and u_2 and substituted into the linear function representation to obtain Equation (9.2a),

$$u(x) = u_1\left(\frac{x - x_2}{x_1 - x_2}\right) + u_2\left(\frac{x - x_1}{x_2 - x_1}\right) = u_1\mathcal{L}_1(x) + u_2\mathcal{L}_2(x) = \sum_{i=1}^{2} u_i\mathcal{L}_i(x) \qquad (9.2a)$$

where

$$\mathcal{L}_1(x) = \left(\frac{x - x_2}{x_1 - x_2}\right) \qquad \text{and} \qquad \mathcal{L}_2(x) = \left(\frac{x - x_1}{x_2 - x_1}\right) \qquad (9.2b)$$

A linear representation of displacement is sufficient if the forces are applied only at the element end and only if the cross-sectional area does not change across the element. If the axial member has a distributed load, or if the member is tapered, then the axial displacement is no longer linear inside the element. A quadratic or higher-order polynomial may converge to the actual solution faster than a linear element. Figure 9.2(b) shows an element with three nodes. With 3 degrees of freedom, we can start with a quadratic displacement function $u(x) = C_1 + C_2 x + C_3 x^2$, solve the constant in terms of the nodal displacement, and obtain an equation analogous to Equation (9.2a). This process would be tedious for higher-order polynomials. So we use an alternative approach. We represent the displacement in the element by Equation (9.3a),

$$u(x) = \sum_{i=1}^{n} u_i\mathcal{L}_i(x) \qquad (9.3a)$$

where n is the degrees of freedom (number of nodes, in this case) of the element that can be used for representing the $(n-1)$ order of polynomials. Now at the jth node, the displacement $u(x_j) = u_j$, and we obtain Equation (9.3b).

$$u(x_j) = \sum_{i=1}^{n} u_i\mathcal{L}_i(x_j) = u_j \qquad (9.3b)$$

For Equation (9.3b) to be true, the property given in Equation (9.3c) must hold.

$$\mathcal{L}_i(x_j) = \begin{cases} 1 & i = j \\ 0 & i \neq j \end{cases} \qquad (9.3c)$$

Equation (9.3c) implies that the polynomials $\mathcal{L}_i(x)$ are such that the value is 1 on its own ith node and zero at other nodes. Figure 9.3 shows the approximate plots for linear and quadratic $\mathcal{L}_i(x)$ that meet this requirement.

Consider now $\mathcal{L}_1$ for the quadratic. If we represent $\mathcal{L}_1(x) = a_1(x - x_2)(x - x_3)$, its value is zero at nodes 2 and 3. We can now determine the constant a_1 such that $\mathcal{L}_1$ at node 1 is equal to one, and we obtain Equation (9.4a).

$$\mathcal{L}_1(x) = \left(\frac{x - x_2}{x_1 - x_2}\right)\left(\frac{x - x_3}{x_1 - x_3}\right) \qquad (9.4a)$$

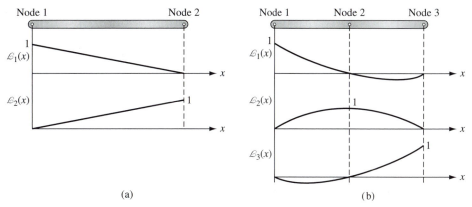

Figure 9.3 (a) Linear and (b) quadratic Lagrange polynomials.

In a similar manner, we can start with $\mathcal{L}_2(x) = a_2(x - x_3)(x - x_1)$ and $\mathcal{L}_3(x) = a_3(x - x_1)(x - x_2)$ and determine the value of a_2 and a_3 such that $\mathcal{L}_2$ and $\mathcal{L}_3$ at nodes 2 and 3, respectively, have a value of one to obtain Equations (9.4b).

$$\mathcal{L}_2(x) = \left(\frac{x - x_1}{x_2 - x_1}\right)\left(\frac{x - x_3}{x_2 - x_3}\right) \quad \text{and} \quad \mathcal{L}_3(x) = \left(\frac{x - x_1}{x_3 - x_1}\right)\left(\frac{x - x_2}{x_3 - x_2}\right) \tag{9.4b}$$

The process we used to obtain the polynomials for the quadratic can now be generalized to obtain Equation (9.5),

$$\mathcal{L}_i(x) = \prod_{\substack{j=1 \\ i \neq j}}^{n}\left[\frac{(x - x_j)}{(x_i - x_j)}\right] \tag{9.5}$$

where $\prod_{i=1}^{n}[\cdots]$ represents the product of the terms in square brackets. The functions defined by Equation (9.5) are called *Lagrange polynomials*.

Functions represented by Lagrange polynomials will be continuous at the element ends. Inside the elements, all orders of derivatives are defined as polynomials are continuous. However, the continuity of the derivative of the function cannot be ensured at the element end irrespective of the order of polynomials when Lagrange polynomials are used for representing the function. Figure 9.4 shows a possible variation of a displacement field represented by Lagrange polynomials over two adjoining elements. Displacements at nodes are independent parameters that can have any value; hence the

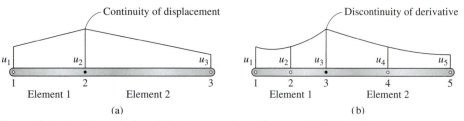

Figure 9.4 Possible variation of displacement for (a) linear and (b) quadratic elements.

variation shown in Figure 9.4 is a possibility. As can be seen from Figure 9.4, the continuity of the displacement is maintained, but its first derivative is not continuous for either the linear or the quadratic element at the element end node.

9.3   AXIAL ELEMENTS

The strains, stresses, and internal forces on any element e can be obtained once the nodal displacements in Equation (9.3a) have been determined. To differentiate quantities at the element and global levels, we will use a superscript to designate quantities at the element level and no superscript at the global level. Thus $u_3^{(1)}$ and $u_3^{(2)}$ will refer to the displacement of node 3 for element 1 and 2, respectively, while u_3 refers to the displacement of node 3 on the actual structure. The strains, stresses, and internal force can be written as follows:

$$\varepsilon_{xx}^{(e)} = \frac{du^{(e)}}{dx} = \sum_{i=1}^{n} u_i^{(e)} \frac{d\mathcal{L}_i}{dx} \tag{9.6a}$$

$$\sigma_{xx}^{(e)} = E^{(e)} \varepsilon_{xx}^{(e)} = E^{(e)} \sum_{i=1}^{n} u_i^{(e)} \frac{d\mathcal{L}_i}{dx} \tag{9.6b}$$

$$N^{(e)} = A^{(e)} \sigma_{xx}^{(e)} = E^{(e)} A^{(e)} \sum_{i=1}^{n} u_i^{(e)} \frac{d\mathcal{L}_i}{dx} \tag{9.6c}$$

The element stiffness matrix and the element load vector can be obtained by comparing Equation (9.3a) and Equation (7.32a) of Rayleigh-Ritz method. We note the generalized displacements $C_i = u_i^{(e)}$ and the kinematically admissible functions $f_i(x) = \mathcal{L}_i(x)$.

9.3.1   Element Stiffness Matrix

Substituting $f_i(x) = \mathcal{L}_i(x)$ into Equation (7.33d), we obtain the element stiffness matrix as shown in Equation (9.6d).

$$K_{jk}^{(e)} = \int_0^L E^{(e)} A^{(e)} \left(\frac{d\mathcal{L}_j}{dx} \right) \left(\frac{d\mathcal{L}_k}{dx} \right) dx \tag{9.6d}$$

9.3.2   Element Load Vector

To obtain the element load vector, we assume that point forces can be applied only at the element end nodes. This requirement is easily met during mesh creation, where we create an element such that the point forces are at the end. From this requirement and from Equation (7.33g), we obtain Equation (9.6e).

$$R_j^{(e)} = \int_0^L p_x(x)\mathcal{L}_j(x)dx + F_1^{(e)}\mathcal{L}_j(x_1) + F_n^{(e)}\mathcal{L}_j(x_n) \tag{9.6e}$$

We know from Equation (9.3c) that $\mathcal{L}_j(x_1)$ is zero except when $j = 1$, and $\mathcal{L}_j(x_n)$ is zero except when $j = n$. Thus, if p_x is zero,[5] only the forces at the end point are nonzero, in accordance with our requirement that external forces be applied at the element end. For $p_x = 0$:

$$R_1^{(e)} = F_1^{(e)} \qquad R_n^{(e)} = F_n^{(e)} \qquad R_j^{(e)} = 0 \qquad j = 2 \text{ to } (n-1) \tag{9.6f}$$

It should be noted that the force $F_j^{(e)}$ is positive in the positive direction $u_j^{(e)}$. This property will be important during assembly.

9.3.3 Assembly of Global Matrix and Global Load Vector

To assemble a global matrix, we must use Equation (9.1). The assembly process is primarily a careful bookkeeping effort to ensure that the matrix components at the element level add to the correct components in the global matrix. The governing criterion is that the displacement function at the node where two elements meet be continuous. The assembly process is elaborated by using two quadratic axial elements to model simple structure, as shown in Figure 9.5. We will assume that there is no distributed force (i.e., $p_x = 0$).

The virtual variation in potential energy due to the virtual displacement in an element can be written as in Equation (9.7a).

$$\delta\Omega^{(m)} = \sum_{i=1}^{n} \frac{\partial\Omega^{(m)}}{\partial u_i^{(m)}} \delta u_i^{(m)} \tag{9.7a}$$

In Equation (7.34a) we evaluated the derivative of potential energy with respect to C_i. Following steps similar to those in Equations (7.34a) through (7.34e), we obtain the derivative of the element's potential energy, which we substitute in Equation (9.7a) to obtain Equation (9.7b).

$$\delta\Omega^{(m)} = \sum_{i=1}^{n} \delta u_i^{(m)} \sum_{j=1}^{n} (K_{ij}^{(m)} u_j^{(m)} - R_i^{(m)}) \tag{9.7b}$$

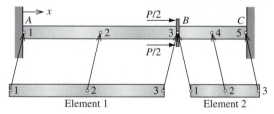

Figure 9.5 Assembly of two quadratic axial elements.

---

[5] Recall that the distributed load from gravity is $p_x = \pm\gamma A$, where γ is the specific weight of the material, A is the cross-sectional area; the plus sign is used if x is in the direction of gravity and the minus sign if it is opposite. To avoid evaluating the integral, we sometimes take the weight of the element and distribute it equally over the element nodes. For a quadratic element, this means that f_j is equal to a third of the element weight.

Equation (9.7b) can be written in matrix form for the two elements shown in Figure 9.5 as shown in Equations (9.8a) and (9.8b).

$$
\delta\Omega^{(1)} = \left\{ \begin{matrix} \delta u_1^{(1)} \\ \delta u_2^{(1)} \\ \delta u_3^{(1)} \end{matrix} \right\}^T \left(\begin{bmatrix} K_{11}^{(1)} & K_{12}^{(1)} & K_{13}^{(1)} \\ K_{21}^{(1)} & K_{22}^{(1)} & K_{23}^{(1)} \\ K_{31}^{(1)} & K_{32}^{(1)} & K_{33}^{(1)} \end{bmatrix} \left\{ \begin{matrix} u_1^{(1)} \\ u_2^{(1)} \\ u_3^{(1)} \end{matrix} \right\} - \left\{ \begin{matrix} F_1^{(1)} \\ 0 \\ F_3^{(1)} \end{matrix} \right\} \right)
\tag{9.8a}
$$

$$
\delta\Omega^{(2)} = \left\{ \begin{matrix} \delta u_1^{(2)} \\ \delta u_2^{(2)} \\ \delta u_3^{(2)} \end{matrix} \right\}^T \left(\begin{bmatrix} K_{11}^{(2)} & K_{12}^{(2)} & K_{13}^{(2)} \\ K_{21}^{(2)} & K_{22}^{(2)} & K_{23}^{(2)} \\ K_{31}^{(2)} & K_{32}^{(2)} & K_{33}^{(2)} \end{bmatrix} \left\{ \begin{matrix} u_1^{(2)} \\ u_2^{(2)} \\ u_3^{(2)} \end{matrix} \right\} - \left\{ \begin{matrix} F_1^{(2)} \\ 0 \\ F_3^{(2)} \end{matrix} \right\} \right)
\tag{9.8b}
$$

From Figure 9.5 we note the following relationship between the displacements at the element nodes to the displacements on the original structure.

$$
\begin{matrix} u_1^{(1)} = u_1 & u_2^{(1)} = u_2 & u_3^{(1)} = u_3 \\ u_1^{(2)} = u_3 & u_2^{(2)} = u_4 & u_3^{(2)} = u_5 \end{matrix}
\tag{9.9}
$$

We note that there are five nodes on the actual structure. Thus the stiffness matrix for the potential energy of the entire structure will have 5 rows and 5 columns. We can use Equation (9.9) to write Equations (9.8a) and (9.8b) as Equations (9.10a) and (9.10b).

$$
\delta\Omega^{(1)} = \left\{ \begin{matrix} \delta u_1 \\ \delta u_2 \\ \delta u_3 \\ \delta u_4 \\ \delta u_5 \end{matrix} \right\}^T \left(\begin{bmatrix} K_{11}^{(1)} & K_{12}^{(1)} & K_{13}^{(1)} & 0 & 0 \\ K_{21}^{(1)} & K_{22}^{(1)} & K_{23}^{(1)} & 0 & 0 \\ K_{31}^{(1)} & K_{32}^{(1)} & K_{33}^{(1)} & 0 & 0 \\ 0 & 0 & 0 & 0 & 0 \\ 0 & 0 & 0 & 0 & 0 \end{bmatrix} \left\{ \begin{matrix} u_1 \\ u_2 \\ u_3 \\ u_4 \\ u_5 \end{matrix} \right\} - \left\{ \begin{matrix} F_1^{(1)} \\ 0 \\ F_3^{(1)} \\ 0 \\ 0 \end{matrix} \right\} \right)
\tag{9.10a}
$$

$$
\delta\Omega^{(2)} = \left\{ \begin{matrix} \delta u_1 \\ \delta u_2 \\ \delta u_3 \\ \delta u_4 \\ \delta u_5 \end{matrix} \right\}^T \left(\begin{bmatrix} 0 & 0 & 0 & 0 & 0 \\ 0 & 0 & 0 & 0 & 0 \\ 0 & 0 & K_{11}^{(2)} & K_{12}^{(2)} & K_{13}^{(2)} \\ 0 & 0 & K_{21}^{(2)} & K_{22}^{(2)} & K_{23}^{(2)} \\ 0 & 0 & K_{31}^{(2)} & K_{32}^{(2)} & K_{33}^{(2)} \end{bmatrix} \left\{ \begin{matrix} u_1 \\ u_2 \\ u_3 \\ u_4 \\ u_5 \end{matrix} \right\} - \left\{ \begin{matrix} 0 \\ 0 \\ F_1^{(2)} \\ 0 \\ F_3^{(2)} \end{matrix} \right\} \right)
\tag{9.10b}
$$

From Equation (9.1) we know that the total potential energy of the structure is the sum of potential energies of all the elements (i.e., $\delta\Omega = \delta\Omega^{(1)} + \delta\Omega^{(2)}$). We thus obtain Equation (9.11).

$$\delta\Omega = \begin{Bmatrix} \delta u_1 \\ \delta u_2 \\ \delta u_3 \\ \delta u_4 \\ \delta u_5 \end{Bmatrix}^T \left(\begin{bmatrix} K_{11}^{(1)} & K_{12}^{(1)} & K_{13}^{(1)} & 0 & 0 \\ K_{21}^{(1)} & K_{22}^{(1)} & K_{23}^{(1)} & 0 & 0 \\ K_{31}^{(1)} & K_{32}^{(1)} & (K_{33}^{(1)}+K_{11}^{(2)}) & K_{12}^{(2)} & K_{13}^{(2)} \\ 0 & 0 & K_{21}^{(2)} & K_{22}^{(2)} & K_{23}^{(2)} \\ 0 & 0 & K_{31}^{(2)} & K_{32}^{(2)} & K_{33}^{(2)} \end{bmatrix} \begin{Bmatrix} u_1 \\ u_2 \\ u_3 \\ u_4 \\ u_5 \end{Bmatrix} - \begin{Bmatrix} F_1^{(1)} \\ 0 \\ F_3^{(1)}+F_1^{(2)} \\ 0 \\ F_3^{(2)} \end{Bmatrix} \right) \tag{9.11}$$

Equation (9.11) highlights that the element stiffness matrix and the element load components that are added correspond to the degrees of freedom associated with the shared node of the elements.

9.3.4 Incorporating the External Concentrated Forces

From Figure 9.5 we see that at nodes 1 and 5 there will be reaction forces at supports A and C, which we label R_A and R_C. Force $F_3^{(1)}$ is in the direction of $u_3^{(1)}$, force $F_1^{(2)}$ is in the direction of $u_1^{(2)}$, and the applied force is in the direction of u_3. Since $u_3^{(1)}$, $u_1^{(2)}$, and u_3 are equal, we see that the applied force P is equal to the sum of the two forces applied at the element nodes. We thus have the equivalence relationship of Equation (9.12).

$$F_1^{(1)} = R_A \qquad F_3^{(1)} + F_1^{(2)} = P \qquad F_3^{(2)} = R_C \tag{9.12}$$

Substituting Equation (9.12) into Equation (9.11), we obtain Equation (9.13).

$$\delta\Omega = \begin{Bmatrix} \delta u_1 \\ \delta u_2 \\ \delta u_3 \\ \delta u_4 \\ \delta u_5 \end{Bmatrix}^T \left(\begin{bmatrix} K_{11}^{(1)} & K_{12}^{(1)} & K_{13}^{(1)} & 0 & 0 \\ K_{21}^{(1)} & K_{22}^{(1)} & K_{23}^{(1)} & 0 & 0 \\ K_{31}^{(1)} & K_{32}^{(1)} & (K_{33}^{(1)}+K_{11}^{(2)}) & K_{12}^{(2)} & K_{13}^{(2)} \\ 0 & 0 & K_{21}^{(2)} & K_{22}^{(2)} & K_{23}^{(2)} \\ 0 & 0 & K_{31}^{(2)} & K_{32}^{(2)} & K_{33}^{(2)} \end{bmatrix} \begin{Bmatrix} u_1 \\ u_2 \\ u_3 \\ u_4 \\ u_5 \end{Bmatrix} - \begin{Bmatrix} R_A \\ 0 \\ P \\ 0 \\ R_C \end{Bmatrix} \right) \tag{9.13}$$

9.3.5 Incorporating the Boundary Conditions on Displacements

The stiffness matrix in Equation (9.13) is singular (i.e., its determinant is zero). The singular nature of the matrix reflects the fact that the two element structures can move as a rigid body. To eliminate this rigid body mode, we impose the boundary conditions of zero displacement at nodes 1 and 5 shown in Equation (9.14).

$$u_1 = 0 \qquad \delta u_1 = 0 \qquad u_5 = 0 \qquad \delta u_5 = 0 \tag{9.14}$$

Substituting Equation (9.14) in Equation (9.13), we obtain Equation (9.15). Note that the zero values of displacement caused all the components of the matrix in the corresponding row to be multiplied by zero, thus eliminating rows 1 and 5. Also, the zero value in the variation of the displacement caused all the components of the matrix in the corresponding columns to be multiplied by zero, thus eliminating columns 1 and 5. By comparing Equations (9.13) and (9.15), we observe that rows and column corresponding to nodes 1 and 5 are indeed eliminated.

$$
\delta\Omega = \left\{ \begin{matrix} \delta u_2 \\ \delta u_3 \\ \delta u_4 \end{matrix} \right\}^T \left(\begin{bmatrix} K_{22}^{(1)} & K_{23}^{(1)} & 0 \\ K_{32}^{(1)} & (K_{33}^{(1)} + K_{11}^{(2)}) & K_{12}^{(2)} \\ 0 & K_{21}^{(2)} & K_{22}^{(2)} \end{bmatrix} \left\{ \begin{matrix} u_2 \\ u_3 \\ u_4 \end{matrix} \right\} - \left\{ \begin{matrix} 0 \\ P \\ 0 \end{matrix} \right\} \right)
\tag{9.15}
$$

By the principle of minimum potential energy, the virtual variation of the potential energy due to virtual displacement must be zero (i.e., $\delta\Omega = 0$). Since the virtual displacement cannot be zero, the remaining terms in the bracket must be zero. We obtain the set of algebraic equations shown in Equation (9.16).

$$
\begin{bmatrix} K_{22}^{(1)} & K_{23}^{(1)} & 0 \\ K_{32}^{(1)} & K_{33}^{(1)} + K_{11}^{(2)} & K_{12}^{(2)} \\ 0 & K_{21}^{(2)} & K_{22}^{(2)} \end{bmatrix} \left\{ \begin{matrix} u_2 \\ u_3 \\ u_4 \end{matrix} \right\} = \left\{ \begin{matrix} 0 \\ P \\ 0 \end{matrix} \right\}
\tag{9.16}
$$

The set of equations in Equation (9.16) can be solved to obtain the displacements u_2, u_3, and u_4. Thus, we now know the displacements at all nodes. The strains, stresses, and internal forces can be found in each element by using Equations (9.6a), (9.6b), and (9.6c). The reaction forces R_A and R_C can be found from the equations corresponding to first and fifth rows in Equation (9.13). These equations can be written as shown in Equations (9.17).

$$
R_A = K_{12}^{(1)} u_2 + K_{13}^{(1)} u_3 \qquad \text{and} \qquad R_C = K_{31}^{(2)} u_3 + K_{32}^{(2)} u_4
\tag{9.17}
$$

9.3.6   Element Strain Energy

Equation (7.36) showed that the strain energy is half the work potential of the forces at equilibrium. We note that the C_j in Equation (7.36) are the element nodal displacements. The element strain energy can be written as Equation (9.18a).

$$
U_A^{(e)} = \frac{1}{2} \sum_{j=1}^{n} u_j^{(e)} R_j^{(e)}
\tag{9.18a}
$$

In the absence of distributed forces, we can substitute Equation (9.6e) into Equation (9.18a) to obtain Equation (9.18b).

$$
U_A^{(e)} = \frac{1}{2}(u_1^{(e)} F_1^{(e)} + u_n^{(e)} F_n^{(e)})
\tag{9.18b}
$$

Figure 9.6 shows relationship between the external and internal nodal forces. We note that $F_1^{(e)} = -N_1^{(e)}$ and $F_n^{(e)} = N_n^{(e)}$. In the absence of distributed forces, the internal force in the element is constant: $N_1^{(e)} = N_n^{(e)} = N^{(e)}$. Substituting these relationship into Equation (9.18b), we obtain Equation (9.18c).

$$
U_A^{(e)} = \frac{1}{2}(u_n^{(e)} N_n^{(e)} - u_1^{(e)} N_1^{(e)}) = \frac{1}{2}(u_n^{(e)} - u_1^{(e)}) N^{(e)}
\tag{9.18c}
$$

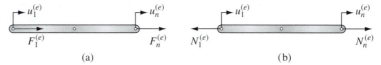

(a)   (b)

Figure 9.6 (a) External and (b) internal nodal forces.

The strain energy values are used for deciding whether the FEM model needs to be improved. In regions of very large stress gradients, the elements will be highly stressed, and hence elements in these regions will have large strain energies. We would like to refine the mesh so that we can get good resolution of the stress gradients. Elements with high strain energy identify the region of the body in which the mesh should be refined.

There are three major techniques of mesh refinement based on the observation made in footnote 11 on page 460 that the higher the number of degrees of freedom, the lower the potential energy, and the better the solution. The *h-method* of mesh refinement reduces the size of the element. The *p-method* of mesh refinement increases the order of polynomials in an element. The *r-method* of mesh refinement relocates the position of a node. There are also combinations such as the hr- and hp-methods of mesh refinement. These mesh refinement methods of reducing the element size, increasing the polynomial order, and reallocating the nodes are used in regions containing elements with large strain energies.

9.3.7 Transformation Matrix

In the preceding discussion, the orientation of the coordinate system was the same at the element and global levels. In general, axial members are at various orientations in a truss. Thus the element stiffness matrix must be transformed before assembly to correspond to the degrees of freedom in the global coordinate system.

Figure 9.7 shows the global coordinate and the local coordinate, which is along the axial direction of the member. The local coordinate is oriented at an angle θ to the global coordinate. In global coordinates a node point displaces in two directions. The displacement vector of node 1 can be written as $\overline{\mathbf{D}}_1 = u_{G1}^{(1)}\,\overline{\mathbf{i}} + v_{G1}^{(1)}\,\overline{\mathbf{j}}$, where $\overline{\mathbf{i}}$ and $\overline{\mathbf{j}}$ are the unit vectors in the global x and y directions. The unit vector along the local coordinate can be written as: $\overline{\mathbf{e}}_L = \cos\theta\,\overline{\mathbf{i}} + \sin\theta\,\overline{\mathbf{j}}$. The displacement u_1 in the local coordinate is the component of the vector $\overline{\mathbf{D}}_1$ in the local direction and can be obtained by a dot product as shown in Equation (9.19).

$$u_1^{(1)} = \overline{\mathbf{D}}_1 \cdot \overline{\mathbf{e}}_L = u_{G1}^{(1)} \cos\theta + v_{G1}^{(1)} \sin\theta \qquad (9.19)$$

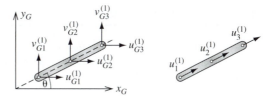

Figure 9.7 Coordinate transformation.

The displacements at other nodes also transform as in Equation (9.19). The equations transforming the local displacement into global displacements are written in matrix form as shown in Equation (9.20).

$$
\left\{ \begin{array}{c} u_1^{(1)} \\ u_2^{(1)} \\ u_3^{(1)} \end{array} \right\} = \begin{bmatrix} \cos\theta & \sin\theta & 0 & 0 & 0 & 0 \\ 0 & 0 & \cos\theta & \sin\theta & 0 & 0 \\ 0 & 0 & 0 & 0 & \cos\theta & \sin\theta \end{bmatrix} \left\{ \begin{array}{c} u_{G1}^{(1)} \\ v_{G1}^{(1)} \\ u_{G2}^{(1)} \\ v_{G2}^{(1)} \\ u_{G3}^{(1)} \\ v_{G3}^{(1)} \end{array} \right\} = [T] \left\{ \begin{array}{c} u_{G1}^{(1)} \\ v_{G1}^{(1)} \\ u_{G2}^{(1)} \\ v_{G2}^{(1)} \\ u_{G3}^{(1)} \\ v_{G3}^{(1)} \end{array} \right\} \tag{9.20}
$$

The $[T]$ matrix in Equation (9.20) is 3×6 matrix relating the local and the global coordinate systems. If we had n nodes on the element, then the size of the matrix $[T]$ would be $n \times 2n$. We now rewrite Equation (9.20) in compact matrix form:

$$
\{u^{(1)}\} = [T]\{u_G^{(1)}\} \tag{9.21}
$$

We can also write Equation (9.8a) in compact matrix form, as follows:

$$
\delta\Omega^{(1)} = \{\delta u^{(1)}\}^T ([K^{(1)}]\{u^{(1)}\} - \{F_1^{(1)}\}) \tag{9.22a}
$$

Substituting Equation (9.21) into Equation (9.22a), we obtain Equation (9.22b).

$$
\delta\Omega^{(1)} = \{\delta u_G^{(1)}\}^T [T]^T ([K^{(1)}][T]\{u_G^{(1)}\} - \{F_1^{(1)}\}) \qquad \text{or}
$$

$$
\delta\Omega^{(1)} = \{\delta u_G^{(1)}\}^T ([T]^T [K^{(1)}][T]\{u_G^{(1)}\} - [T]^T \{F_1^{(1)}\}) \tag{9.22b}
$$

We define the quantities in Equations (9.22c).

$$
[K_G^{(1)}] = [T]^T [K^{(1)}][T] \qquad \{F_{G1}^{(1)}\} = [T]^T \{F_1^{(1)}\} \tag{9.22c}
$$

Equations (9.22c) use the transformation matrix to transform the stiffness and the load vector. Substituting these equations into Equation (9.22b), we obtain Equation (9.22d).

$$
\delta\Omega^{(1)} = \{\delta u_G^{(1)}\}^T ([K_G^{(1)}]\{u_G^{(1)}\} - \{F_{G1}^{(1)}\}) \tag{9.22d}
$$

Equation (9.22d) has the same form as Equation (9.22a), but now the nodal displacements and forces are in the global coordinate system. Equation (9.22d) can now be used as before in the assembly process.

9.3.8   Linear and Quadratic Axial Elements

In writing the element stiffness matrix and the element load vector, we will assume the following to account for the possibility of varying distributed loads, cross-sectional areas, and moduli of elasticity.

- The distributed load p_x is evaluated at the midpoint of the element and has a uniform value p_0.

- The cross-sectional area A is evaluated at the midpoint of the element.
- The modulus of elasticity E is constant over the element.

The stiffness matrix and the load vector can be calculated by using Equations (9.6d) and (9.6e) and are as given in Equation (9.23).

$$\{u^{(e)}\} = \begin{Bmatrix} u_1^{(e)} \\ u_2^{(e)} \end{Bmatrix} \quad [K^{(e)}] = \frac{E^{(e)}A^{(e)}}{L^{(e)}} \begin{bmatrix} 1 & -1 \\ -1 & 1 \end{bmatrix} \quad \{R^{(e)}\} = \frac{p_0^{(e)}L^{(e)}}{2} \begin{Bmatrix} 1 \\ 1 \end{Bmatrix} + \begin{Bmatrix} F_1^{(e)} \\ F_2^{(e)} \end{Bmatrix} \quad (9.23)$$

We assume that the three nodes of the quadratic element are equally spaced. The stiffness matrix and the load vector, which can be calculated by using Equations (9.6d) and (9.6e), are given in Equations (9.24).

$$\{u^{(e)}\} = \begin{Bmatrix} u_1^{(e)} \\ u_2^{(e)} \\ u_3^{(e)} \end{Bmatrix} \quad [K^{(e)}] = \frac{EA}{3L} \begin{bmatrix} 7 & -8 & 1 \\ -8 & 16 & -8 \\ 1 & -8 & 7 \end{bmatrix} \quad \{R\} = \frac{p_0 L}{6} \begin{Bmatrix} 1 \\ 4 \\ 1 \end{Bmatrix} + \begin{Bmatrix} F_1^{(e)} \\ 0 \\ F_3^{(e)} \end{Bmatrix} \quad (9.24)$$

The potential energy of the element in compact matrix form can be written as in Equation (9.25).

$$\delta\Omega^{(e)} = \{\delta u^{(e)}\}^T ([K^{(e)}]\{u^{(e)}\} - \{R^{(e)}\}) \quad (9.25)$$

9.3.9 Procedural Steps in the Finite Element Method

The analysis of a structure by the FEM proceeds in steps that are generic and can be used for structural elements other than axial rods. However we use axial rods to elaborate the steps.

Step 1. Obtain the element stiffness matrices and the element load vectors.

From the geometry, material properties, and distributed loads, the element stiffness matrices and the element load vectors can be written by using forms equivalent to Equation (9.23) [or Equation (9.24)].

Step 2. Transform from local orientation to global orientation.

Equation (9.22c) can be used to transform the element stiffness and the element load vector to the orientation of the global coordinate system.

Step 3. Assemble the global stiffness matrix and load vector.

The element nodal generalized displacements are related to the nodal generalized displacements on the structure to ensure continuity of the generalized displacements. For axial elements, the generalized displacements are just the displacement. As shall be seen, however, for beam elements the generalized displacement also include slopes (i.e., derivatives of displacements).

Step 4. Incorporate the external loads.

The sum of the nodal forces at the element end are replaced by the equivalent external forces applied to the structure.

Step 5. Incorporate the boundary conditions.

Any zero values of the generalized displacement are incorporated by eliminating the corresponding row and column from the global stiffness matrix and load vector.

Step 6. Solve the algebraic equations for the nodal displacements.

Step 7. Obtain the reaction forces, stresses, internal forces, and strain energy.

Step 8. Interpret and check the results.

Step 9. Refine the mesh if necessary, and repeat steps 1 through 8.

EXAMPLE 9.1

A rectangular tapered aluminum bar (E_{Al} = 10,000 ksi, v = 0.25) is shown in Figure 9.8. The depth in the tapered section varies as $h(x) = 4 - 0.04x$. Use the following finite element models to solve the problem.

Model 1: two linear elements AB and BC

Model 2: two equal length linear elements in BC and a linear element in AB

For the two models, find the stress at point B, the displacement at point C, and the strain energy in each element. Compare the results with analytical values and comment.

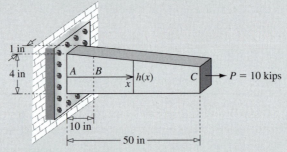

Figure 9.8 Axial member for Example 9.1.

PLAN

We can determine the cross-sectional area as a function of x, and we can determine the cross-sectional area in the middle of each element for determining the element stiffness matrix. For each model, we will follow the steps outlined in Section 9.3.9.

SOLUTION

The cross-sectional area varies as shown in Equation (E1).

$$A(x) = (1)h(x) = 4 - 0.04x \tag{E1}$$

Figure 9.9 shows the two FEM models. In each model, the cross-sectional area is evaluated from Equation (E1) at the midpoint of the element.

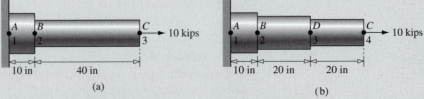

Figure 9.9 The FEM models for Example 9.1. (a) model 1 and (b) model 2.

MODEL 1

Step 1. The cross-sectional areas at the midpoints for the two elements shown in Figure 9.9(a) are as follows:

$$A_1 = A(5) = 3.8 \text{ in}^2 \quad \text{and} \quad A_2 = A(30) = 2.8 \text{ in}^2 \tag{E2}$$

Noting that $L_1 = 10$ and $L_2 = 40$, we obtain Equations (E3)

$$E_1 A_1 / L_1 = 3800 \quad \text{and} \quad E_2 A_2 / L_2 = 700 \tag{E3}$$

The element stiffness matrix and the element load vector for the two elements can be written by using Equation (9.23) as shown in Equations (E4) and (E5).

$$[K^{(1)}] = \begin{bmatrix} 3800 & -3800 \\ -3800 & 3800 \end{bmatrix} \quad \{R^{(1)}\} = \begin{Bmatrix} F_1^{(1)} \\ F_2^{(1)} \end{Bmatrix} \tag{E4}$$

$$[K^{(2)}] = \begin{bmatrix} 700 & -700 \\ -700 & 700 \end{bmatrix} \quad \{R^{(2)}\} = \begin{Bmatrix} F_1^{(2)} \\ F_2^{(2)} \end{Bmatrix} \tag{E5}$$

Step 2. The global and local orientations of the elements are the same, so no transformation of the element stiffness matrices and the load vectors is needed.

Step 3. There are three nodes at the global level. The global stiffness matrix and load vector before incorporation of boundary conditions and loads can be written as in Equation (E6).

$$[K_G] = \begin{bmatrix} 3800 & -3800 & 0 \\ -3800 & 4500 & -700 \\ 0 & -700 & 700 \end{bmatrix} \quad \{R_G\} = \begin{Bmatrix} F_1^{(1)} \\ F_2^{(1)} + F_1^{(2)} \\ F_2^{(2)} \end{Bmatrix} \tag{E6}$$

Step 4. We note that there is no concentrated force at node B, as recognized in Equation (E7).

$$F_2^{(1)} + F_1^{(2)} = 0 \tag{E7}$$

The force at both the element and global levels is in the direction of displacement at point C. Hence we can write Equation (E8).

$$F_2^{(2)} = P = 10 \text{ kips} \tag{E8}$$

Step 5. The displacement at A is zero, which corresponds to the first degree of freedom. We eliminate the first row and column to obtain the algebraic equations in the matrix form shown Equation (E9).

$$\begin{bmatrix} 4500 & -700 \\ -700 & 700 \end{bmatrix} \begin{Bmatrix} u_2 \\ u_3 \end{Bmatrix} = \begin{Bmatrix} 0 \\ 10 \end{Bmatrix} \tag{E9}$$

Step 6. Equation (E9) can be solved to obtain the results shown in Equation (E10).

$$u_2 = 2.6316(10^{-3}) \text{ in} \quad \text{and} \quad u_3 = 16.9173(10^{-3}) \text{ in} \qquad \text{(E10)}$$

The displacement at C is the displacement of node 3. **ANS.** $u_C = 0.00169$ in

Step 7. The stress at point B in each element can be found by using Equation (9.6d) as shown in Equations (E11) and (E12).

$$\sigma_B^{(1)} = E\left[u_1^{(1)}\frac{d\mathcal{L}_1}{dx} + u_2^{(1)}\frac{d\mathcal{L}_2}{dx}\right]\bigg|_{x=10} = E\left[u_1\left(\frac{1}{-L_1}\right) + u_2\left(\frac{1}{L_1}\right)\right]$$

$$= 10{,}000\left(\frac{2.6316(10^{-3})}{10}\right) \qquad \text{(E11)}$$

The axial stress at point B in element 1 is **ANS.** $\sigma_B^{(1)} = 2.63$ ksi (T)

$$\sigma_B^{(2)} = E\left[u_1^{(2)}\frac{d\mathcal{L}_1}{dx} + u_2^{(2)}\frac{d\mathcal{L}_2}{dx}\right]\bigg|_{x=10} = E\left[u_2\left(\frac{1}{-L_2}\right) + u_3\left(\frac{1}{L_2}\right)\right]$$

$$= 10{,}000\left(\frac{14.2857(10^{-3})}{40}\right) \qquad \text{(E12)}$$

The axial stress at point B in element 2 is **ANS.** $\sigma_B^{(2)} = 3.57$ ksi (T)

The strain energy in each element can be calculated as described in Section 9.3.6. The internal axial force in each element can be found from Equation (9.6c), as shown in Equations (E13) and (E14).

$$N^{(1)} = A^{(1)}\sigma_B^{(1)} = (3.8)(2.6316) = 10.0 \qquad \text{(E13)}$$

$$N^{(2)} = A^{(2)}\sigma_B^{(2)} = (2.8)(3.5714) = 10.0 \qquad \text{(E14)}$$

The strain energy in each of the elements can be found from Equation (9.18c) as shown in Equations (E15) and (E16).

$$U_A^{(1)} = \frac{1}{2}\left[u_2^{(1)} - u_1^{(1)}\right]N^{(1)} = \frac{1}{2}u_2 N^{(1)} = 0.01316 \text{ in} \cdot \text{kips} \qquad \text{(E15)}$$

$$U_A^{(2)} = \frac{1}{2}\left[u_2^{(2)} - u_1^{(2)}\right]N^{(2)} = \frac{1}{2}[u_3 - u_2] = 0.07143 \text{ in} \cdot \text{kips} \qquad \text{(E16)}$$

MODEL 2

Step 1. From Equation (E1), the cross-sectional areas at the midpoints for the three elements shown in Figure 9.9(b) can be written as in Equation (E17).

$$A_1 = A(5) = 3.8 \text{ in}^2 \qquad A_2 = A(20) = 3.2 \text{ in}^2$$

$$A_3 = A(40) = 2.4 \text{ in}^2 \tag{E17}$$

Noting that $L_1 = 10$, $L_2 = 20$, and $L_3 = 20$, we obtain Equation (E18).

$$E_1 A_1 / L_1 = 3800 \qquad E_2 A_2 / L_2 = 1600 \qquad E_3 A_3 / L_3 = 1200 \tag{E18}$$

The element stiffness matrix and element load vector for element 1 are same as in Equation (E4). The element stiffness matrix and element load vector for the remaining two elements can be written by using Equation (9.23), as shown in Equations (E19) and (E20).

$$[K^{(2)}] = \begin{bmatrix} 1600 & -1600 \\ -1600 & 1600 \end{bmatrix} \qquad \{R^{(2)}\} = \begin{Bmatrix} F_1^{(2)} \\ F_2^{(2)} \end{Bmatrix} \tag{E19}$$

$$[K^{(3)}] = \begin{bmatrix} 1200 & -1200 \\ -1200 & 1200 \end{bmatrix} \qquad \{R^{(3)}\} = \begin{Bmatrix} F_1^{(3)} \\ F_2^{(3)} \end{Bmatrix} \tag{E20}$$

Step 2. The global and local orientations of the elements are same, so no transformation of the element stiffness matrices and load vectors is needed.

Step 3. There are four nodes at the global level. The global stiffness matrix and load vector can be written as in Equation (E21).

$$[K_G] = \begin{bmatrix} 3800 & -3800 & 0 & 0 \\ -3800 & 5400 & -1600 & 0 \\ 0 & -1600 & 2800 & -1200 \\ 0 & 0 & -1200 & 1200 \end{bmatrix} \qquad \{R_G\} = \begin{Bmatrix} F_1^{(1)} \\ F_2^{(1)} + F_1^{(2)} \\ F_2^{(2)} + F_1^{(3)} \\ F_2^{(3)} \end{Bmatrix} \tag{E21}$$

Step 4. We note that there is no concentrated force at nodes B and D in Figure 9.9(b). Hence by force equivalence, we obtain Equations (E22) and (E23).

$$F_2^{(1)} + F_1^{(2)} = 0 \tag{E22}$$

$$F_2^{(2)} + F_1^{(3)} = 0 \tag{E23}$$

The force at both the element and global levels is in the direction of displacement at point C. Hence we obtain Equation (E24).

$$F_2^{(3)} = P = 10 \text{ kips} \tag{E24}$$

Step 5. The displacement at A is zero, which corresponds to the first degree of freedom. We eliminate the first row and column to obtain the algebraic equations in matrix form, as follow:

$$\begin{bmatrix} 5400 & -1600 & 0 \\ -1600 & 2800 & -1200 \\ 0 & -1200 & 1200 \end{bmatrix} \begin{Bmatrix} u_2 \\ u_3 \\ u_4 \end{Bmatrix} = \begin{Bmatrix} 0 \\ 0 \\ 10 \end{Bmatrix} \tag{E25}$$

Step 6. The displacements in Equation (E25) are solved in Equation (E26).

$$u_2 = 2.6316(10^{-3}) \text{ in} \qquad u_3 = 8.8816(10^{-3}) \text{ in} \qquad u_4 = 17.2149(10^{-3}) \text{ in} \tag{E26}$$

The displacement at point C is the displacement of node 4.

ANS. $u_C = 0.00172$ in

Step 7. The stress at point B in element 1 is as given by Equation (E11) because the displacement at node 2 did not change in this model. The stress in element 2 can be found by using Equation (9.6b) as shown in Equation (E27).

$$\sigma_B^{(2)} = E\left[u_1^{(2)}\frac{d\mathscr{L}_1}{dx} + u_2^{(2)}\frac{d\mathscr{L}_2}{dx}\right]\bigg|_{x=10} = E\left[u_2\left(\frac{1}{-L_2}\right) + u_3\left(\frac{1}{L_2}\right)\right]$$

$$= 10{,}000\left(\frac{6.25(10^{-3})}{20}\right) \tag{E27}$$

The axial stress at point B in element 2 is **ANS.** $\sigma_B^{(2)} = 3.13$ ksi (T)

The strain energy in element 1 is as given by Equation (E15) because the node 2 displacement of this model is the same as for model 1. The internal forces in the remaining two elements are as shown in Equations (E28) and (E29).

$$N^{(2)} = A^{(2)}\sigma_B^{(2)} = 10.0 \tag{E28}$$

$$N^{(3)} = A^{(3)}\sigma_B^{(3)} = 10.0 \tag{E29}$$

The strain energy in each element can be found from Equation (9.18c), as shown in Equations (E30) and (E31).

$$U_A^{(2)} = \frac{1}{2}[u_2^{(2)} - u_1^{(2)}]N^{(2)} = \frac{1}{2}[u_3 - u_2]N^{(2)} = 0.03125 \text{ in} \cdot \text{kip} \qquad \text{(E30)}$$

$$U_A^{(3)} = \frac{1}{2}[u_2^{(3)} - u_1^{(3)}]N^{(3)} = \frac{1}{2}[u_4 - u_3]N^{(3)} = 0.041667 \text{ in} \cdot \text{kip} \qquad \text{(E31)}$$

Step 8. The analytical results for displacement and stress for the tapered axial rod shown in Figure 9.9(b) are given in Equations (E32).

$$\sigma_{xx}(x) = \frac{2.5}{(1 - 0.01x)} \text{ ksi} \quad \text{and} \quad u(x) = -0.025 \ln(1 - 0.01x) \text{ in} \qquad \text{(E32)}$$

The strain energy of the entire rod can be found by using the fact that the internal force is $N = 10$ kN, as shown in Equation (E33).

$$U_A = \frac{1}{2}\int_0^{50}\left(EA\frac{du}{dx}(x)\right)\frac{du}{dx}(x)dx = \frac{1}{2}\int_0^{50}N(x)\frac{du}{dx}(x)dx = \frac{1}{2}\int_0^{50}(10)\frac{du}{dx}(x)dx \quad \text{or}$$

$$U_A = 5[u(50) - u(0)] = -0.086643 \text{ in} \cdot \text{kip} \qquad \text{(E33)}$$

The potential energy is twice the negative value of the strain energy at equilibrium, as shown in Equation (7.36). The analytical results and the results using the two FEM models are shown in Table 9.1.

TABLE 9.1 Results of Example 9.1

	u_C (inches) (10^{-3})	Stress at B: σ_B (ksi)			Strain Energy: U_A			Total Potential energy Ω
		Left of B	Right of B	Average	Element 1	Element 2	Element 3	
Model 1	16.917	2.632	3.571	3.101	0.01316	0.07143		−0.16917
Model 2	17.215	2.632	3.125	2.878	0.01316	0.03125	0.04167	−0.17215
Analytical	17.329	2.778	2.778	2.778				−0.17329

COMMENTS

1. The total potential energy of model 2 is less than that of model 1, thus we expect the results of model 2 to be better than that of model 1 and this expectation is validated by the results in every category of Table 9.1. However, commercial codes do not calculate the total potential energy of a structure. Also the total energy cannot tell us if results in a particular area of interest will improve or not. There are several other indicators that can be used to make decisions to refine a mesh or not in a given area.

2. For model 1 the strain energy of element 2 is significantly larger than element 1. The more evenly the strain energy is distributed over the elements in a mesh the better will be the results. A uniform value of strain energy over the elements will produce the best results for a given number of degrees of freedom. Thus, regions containing elements with high-strain energy should be refined.

3. There is a stress discontinuity at point B in both models. This should not be the case as B is not an interface of two materials nor is there a concentrated force to cause the stress to jump. Thus, the discontinuity is an artifact of the finite element model. The discontinuity in model 2 however is smaller than the discontinuity in model 1. This is another indicator that the regions (elements) showing large stress discontinuity must be refined.

4. The average stress value at B in both models is closer to the analytical solution than stress value in either elements. Thus, for purpose of using FEM results for design and analysis the average values should be used.

5. The reaction force at A can be found noting that $F_1^{(1)} = R_A$ in the first row of Equation (E6) as shown in Equation (E34).

$$R_A = u_1^{(1)}(3800) + u_2^{(1)}(-3800) = u_1(3800) + u_2(-3800)$$

$$= 2.6316(10^{-3})(-3800) = -10 \text{ kN} \tag{E34}$$

The minus sign indicates that the force R_A acts opposite to the positive direction $u_1^{(1)}$. In other words, the reaction is to the left, which we can see intuitively is the correct direction. In interpreting results from FEM models, one must note carefully the local and global coordinate directions.

We record the following observations.

- A good FEM mesh should not show large differences in element strain energy.

- A good FEM mesh should not show large discontinuities in stress values across element boundaries unless the element boundary is an interface of two materials or a concentrated force is applied at the element end.

- Average values of stresses at the element boundary should be used for purposes of design and analysis.

- Care must be taken in interpreting FEM results: note the orientation of the local and global coordinates and whether the variable reported is in the local or global coordinates.

EXAMPLE 9.2

A force of $F = 20$ kN is applied to a roller that slides inside a slot as shown in Figure 9.10. Both bars have a cross-sectional area of $A = 100$ mm^2 and a modulus of elasticity $E = 200$ GPa. Bars AP and BP have lengths of $L_{AP} = 200$ mm and $L_{BP} = 250$ mm, respectively. Determine the displacement of the roller and the reaction force on the roller, using linear elements to represent each bar.

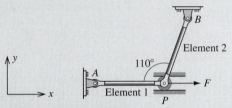

Figure 9.10 Two-bar structure for Example 9.2.

PLAN

We will model each pin as able to displace in the x as well as the y direction to account for the fact that element 2 is at an angle to the x axis. We will follow the steps outlined in Section 9.3.9 to obtain the results.

SOLUTION

Step 1. Equations (E1) and (E2) can be written from the given information.

$$\frac{E_1 A_1}{L_1} = \frac{(200)(10^9)(100)(10^{-6})}{(200)(10^{-3})} = (100)(10^6) \qquad \text{(E1)}$$

$$\frac{E_2 A_2}{L_2} = \frac{(200)(10^9)(100)(10^{-6})}{(250)(10^{-3})} = (80)(10^6) \qquad \text{(E2)}$$

We can use Equation, (9.23) to write the element stiffness matrix and load vector in the local coordinates for the two elements, as shown in Equations (E3) and (E4).

$$[K^{(1)}] = \begin{bmatrix} 100 & -100 \\ -100 & 100 \end{bmatrix}(10^6) \qquad \{R^{(1)}\} = \begin{Bmatrix} F_1^{(1)} \\ F_2^{(1)} \end{Bmatrix} \qquad \text{(E3)}$$

$$[K^{(2)}] = \begin{bmatrix} 80 & -80 \\ -80 & 80 \end{bmatrix}(10^6) \qquad \{R^{(2)}\} = \begin{Bmatrix} F_1^{(2)} \\ F_2^{(2)} \end{Bmatrix} \qquad \text{(E4)}$$

Step 2. Element 1 is in the x direction, but in the global coordinate each node has 2 degrees of freedom (u and v), and thus the element stiffness matrix and load vector in the global coordinate system can be written as in Equation (E5).

$$[K_G^{(1)}] = \begin{bmatrix} 100 & 0 & -100 & 0 \\ 0 & 0 & 0 & 0 \\ -100 & 0 & 100 & 0 \\ 0 & 0 & 0 & 0 \end{bmatrix}(10^6) \qquad \{R^{(1)}\} = \begin{Bmatrix} F_1^{(1)} \\ 0 \\ F_2^{(1)} \\ 0 \end{Bmatrix} \tag{E5}$$

We note that the element 2 makes an angle of 70° with the x axis. The transformation matrix can be written as in Equation (E6).

$$[T] = \begin{bmatrix} \cos 70 & \sin 70 & 0 & 0 \\ 0 & 0 & \cos 70 & \sin 70 \end{bmatrix}$$

$$= \begin{bmatrix} 0.3420 & 0.9397 & 0 & 0 \\ 0 & 0 & 0.3420 & 0.9397 \end{bmatrix} \tag{E6}$$

The element stiffness matrix for element 2 in the global coordinate system can be calculated as shown in Equation (E7).

$$[K_G^{(2)}] = [T]^T[K^{(2)}][T]$$

$$= [T]^T \begin{bmatrix} 80 & -80 \\ -80 & 80 \end{bmatrix} \begin{bmatrix} 0.3420 & 0.9397 & 0 & 0 \\ 0 & 0 & 0.3420 & 0.9397 \end{bmatrix}(10^6) \qquad \text{or}$$

$$[K_G^{(2)}] = \begin{bmatrix} 0.3420 & 0 \\ 0.9397 & 0 \\ 0 & 0.3420 \\ 0 & 0.9397 \end{bmatrix} \begin{bmatrix} 27.362 & 75.175 & -27.362 & -75.175 \\ -27.362 & -75.175 & 27.362 & 75.175 \end{bmatrix}(10^6)$$

$$[K_G^{(2)}] = \begin{bmatrix} 9.3582 & 25.711 & -9.3582 & 25.711 \\ 25.711 & 70.642 & -25.711 & -70.642 \\ -9.3582 & -25.711 & 9.3582 & 25.711 \\ 25.711 & -70.642 & 25.711 & 70.642 \end{bmatrix}(10^6) \tag{E7}$$

The element load vector for element 2 can be calculated as shown in Equation (E8).

$$\{R_G^{(2)}\} = [T]^T\{R^{(2)}\} = \begin{bmatrix} 0.3420 & 0 \\ 0.9397 & 0 \\ 0 & 0.3420 \\ 0 & 0.9397 \end{bmatrix} \begin{Bmatrix} F_1^{(2)} \\ F_2^{(2)} \end{Bmatrix} = \begin{Bmatrix} 0.3420 F_1^{(2)} \\ 0.9397 F_1^{(2)} \\ 0.3420 F_2^{(2)} \\ 0.9397 F_2^{(2)} \end{Bmatrix} = \begin{Bmatrix} F_{1x}^{(2)} \\ F_{1y}^{(2)} \\ F_{2x}^{(2)} \\ F_{2y}^{(2)} \end{Bmatrix} \tag{E8}$$

Step 3. With three nodes there are 6 degrees of freedom. The potential energy for each element can be written as shown in Equations (E9) and (E10).

$$
\delta\Omega^{(1)} = \{\delta u_G^{(1)}\}^T \begin{bmatrix} 100 & 0 & -100 & 0 & 0 & 0 \\ 0 & 0 & 0 & 0 & 0 & 0 \\ -100 & 0 & 100 & 0 & 0 & 0 \\ 0 & 0 & 0 & 0 & 0 & 0 \\ 0 & 0 & 0 & 0 & 0 & 0 \\ 0 & 0 & 0 & 0 & 0 & 0 \end{bmatrix} (10^6)\{u_G^{(1)}\} - \begin{Bmatrix} F_1^{(1)} \\ 0 \\ F_2^{(1)} \\ 0 \\ 0 \\ 0 \end{Bmatrix} \tag{E9}
$$

$$
\delta\Omega^{(2)} = \{\delta u_G^{(2)}\}^T \begin{bmatrix} 0 & 0 & 0 & 0 & 0 & 0 \\ 0 & 0 & 0 & 0 & 0 & 0 \\ 0 & 0 & 9.3582 & 25.711 & -9.3582 & 25.711 \\ 0 & 0 & 25.711 & 70.642 & -25.711 & -70.642 \\ 0 & 0 & -9.3582 & -25.711 & 9.3582 & 25.711 \\ 0 & 0 & 25.711 & -70.642 & 25.711 & 70.642 \end{bmatrix} (10^6)\{u_G^{(2)}\} - \begin{Bmatrix} 0 \\ 0 \\ F_{1x}^{(2)} \\ F_{1y}^{(2)} \\ F_{2x}^{(2)} \\ F_{2y}^{(2)} \end{Bmatrix} \tag{E10}
$$

The total potential energy of the system can be obtained by adding the potential energies in Equations (E9) and (E10) to obtain Equation (E11).

$$
\delta\Omega = \begin{Bmatrix} \delta u_1 \\ \delta v_1 \\ \delta u_2 \\ \delta v_2 \\ \delta u_3 \\ \delta v_3 \end{Bmatrix}^T \begin{bmatrix} 100 & 0 & -100 & 0 & 0 & 0 \\ 0 & 0 & 0 & 0 & 0 & 0 \\ -100 & 0 & 109.3582 & 25.711 & -9.3582 & 25.711 \\ 0 & 0 & 25.711 & 70.642 & -25.711 & -70.642 \\ 0 & 0 & -9.3582 & -25.711 & 9.3582 & 25.711 \\ 0 & 0 & 25.711 & -70.642 & 25.711 & 70.642 \end{bmatrix} (10^6) \begin{Bmatrix} u_1 \\ v_1 \\ u_2 \\ v_2 \\ u_3 \\ v_3 \end{Bmatrix} - \begin{Bmatrix} F_1^{(1)} \\ 0 \\ F_2^{(1)} + F_{1x}^{(2)} \\ F_{1y}^{(2)} \\ F_{2x}^{(2)} \\ F_{2y}^{(2)} \end{Bmatrix} \tag{E11}
$$

Step 4. Noting that the applied load at the roller is in the x direction and that there are reaction forces at points A and B, we obtain the load equivalences shown in Equation (E12),

$$
F_1^{(1)} = R_A \qquad F_2^{(1)} + F_{1x}^{(2)} = 20(10^3) \text{ N} \qquad F_{1y}^{(2)} = R_P \qquad F_{2x}^{(2)} = R_{Bx} \qquad F_{2y}^{(2)} = R_{By} \tag{E12}
$$

where R_A and R_P are the reaction forces at A and P, and R_{Bx} and R_{By} are the reaction forces at B in the x and y directions. Substituting Equation (E12) into Equation (E11), we obtain Equation (E13).

$$\delta\Omega = \begin{Bmatrix} \delta u_1 \\ \delta v_1 \\ \delta u_2 \\ \delta v_2 \\ \delta u_3 \\ \delta v_3 \end{Bmatrix}^T \left(\begin{bmatrix} 100 & 0 & -100 & 0 & 0 & 0 \\ 0 & 0 & 0 & 0 & 0 & 0 \\ -100 & 0 & 109.3582 & 25.711 & -9.3582 & 25.711 \\ 0 & 0 & 25.711 & 70.642 & -25.711 & -70.642 \\ 0 & 0 & -9.3582 & -25.711 & 9.3582 & 25.711 \\ 0 & 0 & 25.711 & -70.642 & 25.711 & 70.642 \end{bmatrix} (10^6) \begin{Bmatrix} u_1 \\ v_1 \\ u_2 \\ v_2 \\ u_3 \\ v_3 \end{Bmatrix} - \begin{Bmatrix} R_A \\ 0 \\ 20(10^3) \\ R_P \\ R_{Bx} \\ R_{By} \end{Bmatrix} \right) \quad (E13)$$

Step 5. The boundary conditions on displacements are given in Equation (E14).

$$u_1 = 0 \qquad v_1 = 0 \qquad v_2 = 0 \qquad u_3 = 0 \qquad v_3 = 0$$
$$\delta u_1 = 0 \qquad \delta v_1 = 0 \qquad \delta v_2 = 0 \qquad \delta u_3 = 0 \qquad \delta v_3 = 0 \qquad (E14)$$

Substituting Equation (E14) into Equation (E13) and equating $\delta\Omega = 0$, we obtain Equation (E15).

$$109.3582(10^6)u_2 = 20(10^3) \qquad \text{or} \qquad u_2 = 0.1829(10^{-3}) \text{ m} \qquad (E15)$$

The displacement of the roller is   **ANS.** $u_B = 0.1829$ mm

The reaction force can be calculated from the fourth row of Equation (E13) as shown in Equation (E16).

$$25.711(10^6)u_2 = R_P \qquad \text{or} \qquad R_P = [25.711(10^6)][0.1829(10^{-3})]$$

$$= 4.702(10^3) \text{ N} \qquad (E16)$$

ANS. $R_P = 4.7$ kN

COMMENT

If the roller were not in the slot, it would be free to displace in the y direction, and v_2 would not be zero. In such a case, Equation (E13) would yield two equations in the two unknowns u_2 and v_2, which could be solved for the displacement of the roller. See Problem 9.8.

9.4 CIRCULAR SHAFT ELEMENTS

It was seen in the Raleigh-Ritz method that the derivation of a stiffness matrix and load vectors for torsion in circular shafts is similar to that for axial members. The stiffness matrix and load vectors can be obtained by replacing (a) the axial rigidity EA by torsional rigidity GJ, (b) the distributed axial force $p_x(x)$ by the distributed torque $t(x)$, and (c) the concentrated axial forces F_i by the concentrated torques T_i. The assembly and solution procedure is same as for axial members.

9.5   SYMMETRIC BEAM ELEMENTS

The deflection v and its derivative dv/dx must be continuous at all points on the beam, including the element ends. Lagrange polynomials cannot be used for representing v because as elaborated in Section 9.2, this would result in a slope that was discontinuous at the element end irrespective of the order of polynomial used. To overcome this problem of continuity of slope, we must define v and dv/dx as degrees of freedom at the element ends. With 2 degrees of freedom at each end, we have a total of 4 degrees of freedom in the element. A cubic polynomial has four unknown constants that can be solved in terms of the 4 degrees of freedom. We write $v(x)$ and the four conditions as shown in Equations (9.26a) and (9.26b).

$$v(x) = C_1 + C_2 x + C_3 x^2 + C_4 x^3 \tag{9.26a}$$

$$v(x_1) = v_1^{(e)} \qquad \frac{dv}{dx}(x_1) = \theta_1^{(e)} \qquad v(x_2) = v_2^{(e)} \qquad \frac{dv}{dx}(x_2) = \theta_2^{(e)} \tag{9.26b}$$

Substituting Equation (9.26a) into the four conditions in Equation (9.26b), solving the constants C's, and substituting the result back into Equation (9.26a), we obtain Equations (9.27a) and (9.27b).

$$v(x) = f_1(x)v_1^{(e)} + f_2(x)\theta_1^{(e)} + f_3(x)v_2^{(e)} + f_4(x)\theta_2^{(e)} \tag{9.27a}$$

$$f_1(x) = 1 - 3\left(\frac{x-x_1}{L}\right)^2 + 2\left(\frac{x-x_1}{L}\right)^3$$

$$f_2(x) = L\left[\left(\frac{x-x_1}{L}\right) - 2\left(\frac{x-x_1}{L}\right)^2 + \left(\frac{x-x_1}{L}\right)^3\right]$$

$$f_3(x) = 3\left(\frac{x-x_1}{L}\right)^2 - 2\left(\frac{x-x_1}{L}\right)^3$$

$$f_4(x) = L\left[-\left(\frac{x-x_1}{L}\right)^2 + \left(\frac{x-x_1}{L}\right)^3\right] \tag{9.27b}$$

The interpolation functions in Equation (9.27b) belong to a class called *Hermite polynomials*. Figure 9.11 shows plots of Hermite polynomials. Note the zero value of the function and the zero value of the slopes. These values are a consequence of the fact that Equation (9.27a) must satisfy the conditions in Equation (9.26b).

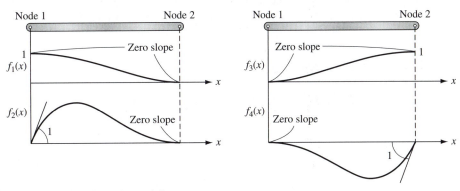

Figure 9.11 Hermite polynomials.

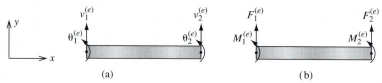

Figure 9.12 Positive directions on a beam element for (a) deflection and slope and (b) forces and moments.

The stiffness matrix was given by Equation (7.38c). To obtain the element load vector, we assume that point forces and moments can be applied only at element end nodes. This requirement is easily met during mesh creation, where we create an element such that the point forces and moments are at the end. From this requirement and from Equation (7.38d), we obtain the right-hand-side vector as shown in Equation (9.28).

$$R_j^{(e)} = \int_0^L p_y(x) f_j(x)\, dx + F_1^{(e)} f_j(x_1) + F_2^{(e)} f_j(x_2) + M_1^{(e)} \frac{df_j}{dx}(x_1) + M_2^{(e)} \frac{df_j}{dx}(x_2) \quad (9.28)$$

From Figure 9.11 we have that $f_j(x_1)$ is zero except when $j = 1$, and $f_j(x_2)$ is zero except when $j = 2$. Similarly $(df_j/dx)(x_1)$ is zero except when $j = 2$, and $(df_j/dx)(x_2)$ is zero except when $j = 4$.

Figure 9.12 shows the local coordinate system for the beam element and the positive directions for deflection and slopes. For the work potential to be positive, the nodal forces and moments must be positive in the same direction as the deflection and slope.

In writing the element stiffness matrix and element load vectors, we will assume the following to account for the possibility of varying distributed loads, cross-sectional areas, and moduli of elasticity.

- The distributed load p_y is evaluated at the midpoint of the element and has a uniform value p_0.

- The area moment of inertia I is evaluated for the cross section at the midpoint of the element.

- The modulus of elasticity E is constant over the element.

Substituting Equation (9.27b) into Equations (7.38c) and (9.28), we obtain the element stiffness matrix and load vectors shown in Equations (9.29a) and (9.29b). We also show the displacement vector with 4 degrees of freedom.

$$\{v^{(e)}\} = \begin{Bmatrix} v_1^{(e)} \\ \theta_1^{(e)} \\ v_2^{(e)} \\ \theta_2^{(e)} \end{Bmatrix} \qquad [K^{(e)}] = \frac{2E^{(e)}I^{(e)}}{(L^{(e)})^3} \begin{bmatrix} 6 & 3L^{(e)} & -6 & 3L^{(e)} \\ 3L^{(e)} & 2(L^{(e)})^2 & -3L^{(e)} & (L^{(e)})^2 \\ -6 & -3L^{(e)} & 6 & -3L^{(e)} \\ 3L^{(e)} & (L^{(e)})^2 & -3L^{(e)} & 2L^{(e)2} \end{bmatrix} \quad (9.29a)$$

$$\{R^{(e)}\} = \frac{p_0^{(e)} L^{(e)}}{12} \begin{Bmatrix} 6 \\ L^{(e)} \\ 6 \\ -L^{(e)} \end{Bmatrix} + \begin{Bmatrix} F_1^{(e)} \\ M_1^{(e)} \\ F_2^{(e)} \\ M_2^{(e)} \end{Bmatrix} \tag{9.29b}$$

The virtual variation of potential energy for the element can be written in compact matrix form as shown in Equation (9.30).

$$\delta \Omega_B^{(e)} = \{\delta v^{(e)}\}^T ([K^{(e)}]\{v^{(e)}\} - \{R^{(e)}\}) \tag{9.30}$$

The assembly and solution process, which is as described Section 9.3 for axial members, is elaborated in Example 9.3.

EXAMPLE 9.3

For the beam and loading shown in Figure 9.13, use two equal elements to determine the deflection and slope at B and the reaction force and moment at A. Assume that EI is constant for the beam.

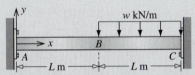

Figure 9.13 Beam and loading for Example 9.3.

PLAN

There are three nodes on the beam, resulting in a total of 6 degrees of freedom before boundary conditions are imposed. We follow the procedure outlined in Section 9.3.9.

SOLUTION

Step 1. Using Equation (9.29a) and noting that $p_0 = -w$, we write the load vectors for the two elements as shown in Equations (E1).

$$\{R^{(1)}\} = \begin{Bmatrix} F_1^{(1)} \\ M_1^{(1)} \\ F_2^{(1)} \\ M_2^{(1)} \end{Bmatrix} \quad \text{and} \quad \{R^{(2)}\} = -\left(\frac{wL}{12}\right) \begin{Bmatrix} 6 \\ L \\ 6 \\ -L \end{Bmatrix} + \begin{Bmatrix} F_1^{(2)} \\ M_1^{(2)} \\ F_2^{(2)} \\ M_2^{(2)} \end{Bmatrix} \tag{E1}$$

The element stiffness matrix is the same for both elements. From Equation (9.29a), the element stiffness matrix can be written as shown in Equation (E2).

$$[K^{(1)}] = [K^{(2)}] = \frac{2EI}{L^3}\begin{bmatrix} 6 & 3L & -6 & 3L \\ 3L & 2L^2 & -3L & L^2 \\ -6 & -3L & 6 & -3L \\ 3L & L^2 & -3L & 2L^2 \end{bmatrix} \tag{E2}$$

Step 2. We note that the relationships between the local displacements and slopes and the global displacements and slopes are as given by Equations (E3).

$$v_1^{(1)} = v_A \qquad v_2^{(1)} = v_B \qquad v_1^{(2)} = v_B \qquad v_2^{(2)} = v_C$$

$$\theta_1^{(1)} = \theta_A \qquad \theta_2^{(1)} = \theta_B \qquad \theta_1^{(2)} = \theta_B \qquad \theta_2^{(2)} = \theta_C \tag{E3}$$

Step 3. The variation of potential energy function for the two elements can be written as shown in Equations (E4) and (E5).

$$\delta\Omega^{(1)} = \begin{Bmatrix} \delta v_1^{(1)} \\ \delta\theta_1^{(1)} \\ \delta v_2^{(1)} \\ \delta\theta_2^{(1)} \end{Bmatrix}^T \left(\frac{2EI}{L^3}\begin{bmatrix} 6 & 3L & -6 & 3L \\ 3L & 2L^2 & -3L & L^2 \\ -6 & -3L & 6 & -3L \\ 3L & L^2 & -3L & 2L^2 \end{bmatrix}\begin{Bmatrix} v_1^{(1)} \\ \theta_1^{(1)} \\ v_2^{(1)} \\ \theta_2^{(1)} \end{Bmatrix} - \begin{Bmatrix} F_1^{(1)} \\ M_1^{(1)} \\ F_2^{(1)} \\ M_2^{(1)} \end{Bmatrix} \right) \quad \text{or}$$

$$\delta\Omega^{(1)} = \begin{Bmatrix} \delta v_A \\ \delta\theta_A \\ \delta v_B \\ \delta\theta_B \\ \delta v_C \\ \delta\theta_C \end{Bmatrix}^T \left(\frac{2EI}{L^3}\begin{bmatrix} 6 & 3L & -6 & 3L & 0 & 0 \\ 3L & 2L^2 & -3L & L^2 & 0 & 0 \\ -6 & -3L & 6 & -3L & 0 & 0 \\ 3L & L^2 & -3L & 2L^2 & 0 & 0 \\ 0 & 0 & 0 & 0 & 0 & 0 \\ 0 & 0 & 0 & 0 & 0 & 0 \end{bmatrix}\begin{Bmatrix} v_A \\ \theta_A \\ v_B \\ \theta_B \\ v_C \\ \theta_C \end{Bmatrix} - \begin{Bmatrix} F_1^{(1)} \\ M_1^{(1)} \\ F_2^{(1)} \\ M_2^{(1)} \\ 0 \\ 0 \end{Bmatrix} \right) \tag{E4}$$

$$\delta\Omega^{(2)} = \begin{Bmatrix} \delta v_1^{(2)} \\ \delta\theta_1^{(2)} \\ \delta v_2^{(2)} \\ \delta\theta_2^{(2)} \end{Bmatrix}^T \left(\frac{2EI}{L^3}\begin{bmatrix} 6 & 3L & -6 & 3L \\ 3L & 2L^2 & -3L & L^2 \\ -6 & -3L & 6 & -3L \\ 3L & L^2 & -3L & 2L^2 \end{bmatrix}\begin{Bmatrix} v_1^{(2)} \\ \theta_1^{(2)} \\ v_2^{(2)} \\ \theta_2^{(2)} \end{Bmatrix} - \begin{Bmatrix} F_1^{(2)} \\ M_1^{(2)} \\ F_2^{(2)} \\ M_2^{(2)} \end{Bmatrix} + \frac{wL}{12}\begin{Bmatrix} 6 \\ L \\ 6 \\ -L \end{Bmatrix} \right) \quad \text{or}$$

$$\delta\Omega^{(2)} = \begin{Bmatrix} \delta v_A \\ \delta\theta_A \\ \delta v_B \\ \delta\theta_B \\ \delta v_C \\ \delta\theta_C \end{Bmatrix}^T \left(\frac{2EI}{L^3} \begin{bmatrix} 0 & 0 & 0 & 0 & 0 & 0 \\ 0 & 0 & 0 & 0 & 0 & 0 \\ 0 & 0 & 6 & 3L & -6 & 3L \\ 0 & 0 & 3L & 2L^2 & -3L & L^2 \\ 0 & 0 & -6 & -3L & 6 & -3L \\ 0 & 0 & 3L & L^2 & -3L & 2L^2 \end{bmatrix} \begin{Bmatrix} v_A \\ \theta_A \\ v_B \\ \theta_B \\ v_C \\ \theta_C \end{Bmatrix} - \begin{Bmatrix} 0 \\ 0 \\ F_1^{(2)} \\ M_1^{(2)} \\ F_2^{(2)} \\ M_2^{(2)} \end{Bmatrix} + \frac{wL}{12} \begin{Bmatrix} 0 \\ 0 \\ 6 \\ L \\ 6 \\ -L \end{Bmatrix} \right) \quad \text{(E5)}$$

The total potential energy of the beam, $\delta\Omega = \delta\Omega^{(1)} + \delta\Omega^{(2)}$, can be written as in Equation (E6).

$$\delta\Omega = \begin{Bmatrix} \delta v_A \\ \delta\theta_A \\ \delta v_B \\ \delta\theta_B \\ \delta v_C \\ \delta\theta_C \end{Bmatrix}^T \left(\frac{2EI}{L^3} \begin{bmatrix} 6 & 3L & -6 & 3L & 0 & 0 \\ 3L & 2L^2 & -3L & L^2 & 0 & 0 \\ -6 & -3L & 12 & 0 & -6 & 3L \\ 3L & L^2 & 0 & 4L^2 & -3L & L^2 \\ 0 & 0 & -6 & -3L & 6 & -3L \\ 0 & 0 & 3L & L^2 & -3L & 2L^2 \end{bmatrix} \begin{Bmatrix} v_A \\ \theta_A \\ v_B \\ \theta_B \\ v_C \\ \theta_C \end{Bmatrix} - \begin{Bmatrix} F_1^{(1)} \\ M_1^{(1)} \\ F_2^{(1)} + F_1^{(2)} \\ M_2^{(1)} + M_1^{(2)} \\ F_2^{(2)} \\ M_2^{(2)} \end{Bmatrix} + \frac{wL}{12} \begin{Bmatrix} 0 \\ 0 \\ 6 \\ L \\ 6 \\ -L \end{Bmatrix} \right) \quad \text{(E6)}$$

Step 4. We note that there are no external forces or moments at B and that the relationships between the element nodal forces and moments and the reaction forces and moments at A and C can be written as in Equations (E7).

$$F_1^{(1)} = R_A \qquad F_2^{(1)} + F_1^{(2)} = 0 \qquad F_2^{(2)} = R_C$$
$$M_1^{(1)} = M_A \qquad M_2^{(1)} + M_1^{(2)} = 0 \qquad M_2^{(2)} = M_C \qquad \text{(E7)}$$

Substituting Equations (E7) into Equation (E6), we obtain the potential energy function as shown in Equation (E8).

$$\delta\Omega = \begin{Bmatrix} \delta v_A \\ \delta\theta_A \\ \delta v_B \\ \delta\theta_B \\ \delta v_C \\ \delta\theta_C \end{Bmatrix}^T \left(\frac{2EI}{L^3} \begin{bmatrix} 6 & 3L & -6 & 3L & 0 & 0 \\ 3L & 2L^2 & -3L & L^2 & 0 & 0 \\ -6 & -3L & 12 & 0 & -6 & 3L \\ 3L & L^2 & 0 & 4L^2 & -3L & L^2 \\ 0 & 0 & -6 & -3L & 6 & -3L \\ 0 & 0 & 3L & L^2 & -3L & 2L^2 \end{bmatrix} \begin{Bmatrix} v_A \\ \theta_A \\ v_B \\ \theta_B \\ v_C \\ \theta_C \end{Bmatrix} - \begin{Bmatrix} R_A \\ M_A \\ 0 \\ 0 \\ R_C \\ M_C \end{Bmatrix} + \frac{wL}{12} \begin{Bmatrix} 0 \\ 0 \\ 6 \\ L \\ 6 \\ -L \end{Bmatrix} \right) \quad \text{(E8)}$$

The displacement and slopes at A and C are zero as shown in Equations (E9).

$$v_A = 0 \qquad \theta_A = 0 \qquad v_C = 0 \qquad \theta_C = 0$$
$$\delta v_A = 0 \qquad \delta\theta_A = 0 \qquad \delta v_C = 0 \qquad \delta\theta_C = 0 \tag{E9}$$

Substituting Equations (E9) into Equation (E8), we can write the potential energy function as Equation (E10).

$$\delta\Omega = \begin{Bmatrix} \delta v_B \\ \delta\theta_B \end{Bmatrix}^T \left(\frac{2EI}{L^3} \begin{bmatrix} 12 & 0 \\ 0 & 4L^2 \end{bmatrix} \begin{Bmatrix} v_B \\ \theta_B \end{Bmatrix} + \frac{wL}{12} \begin{Bmatrix} 6 \\ L \end{Bmatrix} \right) \tag{E10}$$

By the principle of minimum potential energy, the virtual variation of potential energy due to virtual displacement must be zero (i.e., $\delta\Omega = 0$). Since the virtual displacement cannot be zero, the remaining terms in the brackets in Equation (E10) must be zero, and we obtain the set of algebraic equations shown in Equation (E11).

$$\frac{2EI}{L^3} \begin{bmatrix} 12 & 0 \\ 0 & 4L^2 \end{bmatrix} \begin{Bmatrix} v_B \\ \theta_B \end{Bmatrix} = -\left(\frac{wL}{12}\right) \begin{Bmatrix} 6 \\ L \end{Bmatrix} \tag{E11}$$

Solving the above two equations shown in Equation (E11), we obtain the results shown in Equation (E12).

$$\textbf{ANS.} \quad v_B = -\left(\frac{wL^4}{48EI}\right) \qquad \theta_B = -\left(\frac{wL^3}{96EI}\right) \tag{E12}$$

The reaction force and moments at A can be found from the first two rows of Equation (E8) as shown in Equations (E13) and (E14).

$$R_A = \frac{2EI}{L^3}[-6v_B + 3L\theta_B] = \frac{2EI}{L^3}\left[6\left(\frac{wL^4}{48EI}\right) - 3L\left(\frac{wL^3}{96EI}\right)\right] \tag{E13}$$

$$\textbf{ANS.} \quad R_A = \frac{3}{16}wL$$

$$M_A = \frac{2EI}{L^3}[-3Lv_B + L^2\theta_B] = \frac{2EI}{L^3}\left[3L\left(\frac{wL^4}{48EI}\right) - L^2\left(\frac{wL^3}{96EI}\right)\right] \tag{E14}$$

$$\textbf{ANS.} \quad M_A = \frac{5}{48}wL^2$$

COMMENTS

1. The primary difference in the solution procedure in this example and in the axial rod problem of Example 9.1 is that two rows and two columns are added instead of one at the shared node of the two elements. This is not surprising, since the shared node has 2 degrees of freedom.

2. In accordance with Figure 9.12 the negative sign on deflection at B implies that it is downward, and a negative sign on slope implies clockwise rotation, which makes intuitive sense if we visualize the deformed shape.

3. In accordance with Figure 9.12, the positive signs for reactions implies that force is upward and the moment is counterclockwise.

9.6   FINITE ELEMENT EQUATIONS IN TWO-DIMENSIONS

In this section the equations for the finite element method in two dimensions are presented. The strain energy density in two dimensions can be written in matrix form as shown in Equations (9.31a) and (9.31b).

$$U_0 = \frac{1}{2}[\sigma_{xx}\varepsilon_{xx} + \sigma_{yy}\varepsilon_{yy} + \tau_{xy}\gamma_{xy}] = \frac{1}{2}\{\tilde{\sigma}\}^T\{\tilde{\varepsilon}\} \tag{9.31a}$$

$$\{\tilde{\sigma}\} = \begin{Bmatrix} \sigma_{xx} \\ \sigma_{yy} \\ \tau_{xy} \end{Bmatrix} \qquad \{\tilde{\varepsilon}\} = \begin{Bmatrix} \varepsilon_{xx} \\ \varepsilon_{yy} \\ \gamma_{xy} \end{Bmatrix} \tag{9.31b}$$

Hooke's law for plane stress [Equation (8.4a)] and plane strain [Equation (8.5a)] can be written in matrix form as shown in Equations (9.32a) through (9.32c).

$$\{\tilde{\sigma}\} = [\tilde{E}]\{\tilde{\varepsilon}\} \tag{9.32a}$$

$$[\tilde{E}] = \frac{E}{(1-\nu^2)}\begin{bmatrix} 1 & \nu & 0 \\ \nu & 1 & 0 \\ 0 & 0 & \frac{(1-\nu)}{2} \end{bmatrix} \qquad \text{for plane strain} \tag{9.32b}$$

$$[\tilde{E}] = \frac{E}{(1-2\nu)(1+\nu)}\begin{bmatrix} (1-\nu) & \nu & 0 \\ \nu & (1-\nu) & 0 \\ 0 & 0 & \frac{(1-2\nu)}{2} \end{bmatrix} \qquad \text{for plane strain} \tag{9.32c}$$

Substituting Equation (9.32a) into Equation (9.31a), we obtain Equation (9.33).

$$U_0 = \frac{1}{2}\{\tilde{\sigma}\}^T\{\tilde{\varepsilon}\} = \frac{1}{2}\{\tilde{\varepsilon}\}^T[\tilde{E}]^T\{\tilde{\varepsilon}\} = \frac{1}{2}\{\tilde{\varepsilon}\}^T[\tilde{E}]\{\tilde{\varepsilon}\} \tag{9.33}$$

The strain–displacement relationships, Equations (8.1a), (8.1b), and (8.1d), can be written in matrix form as shown in Equation (9.34).

$$\{\varepsilon\} = \begin{bmatrix} \dfrac{\partial}{\partial x} & 0 \\[2mm] 0 & \dfrac{\partial}{\partial y} \\[2mm] \dfrac{\partial}{\partial y} & \dfrac{\partial}{\partial x} \end{bmatrix} \begin{Bmatrix} u \\ v \end{Bmatrix} \qquad (9.34)$$

The displacement at the element level can be approximated by using kinematically admissible functions as shown in Equations (9.35).

$$u^{(e)}(x) = \sum_{i=1}^{n} u_i^{(e)} f_i(x, y) \qquad \text{and} \qquad v^{(e)}(x) = \sum_{i=1}^{n} v_i^{(e)} f_i(x, y) \qquad (9.35)$$

Equation (9.35) can be written in matrix form as shown in Equations (9.36a) and (9.36b).

$$\begin{Bmatrix} u^{(e)} \\ v^{(e)} \end{Bmatrix} = \begin{bmatrix} f_1 & 0 & f_2 & 0 & \cdots & \cdots & f_n & 0 \\ 0 & f_1 & 0 & f_2 & \cdots & \cdots & 0 & f_n \end{bmatrix} \{d^{(e)}\} \qquad (9.36a)$$

$$\{d^{(e)}\}^T = \{u_1^{(e)} v_1^{(e)}, \quad u_2^{(e)} v_2^{(e)}, \quad \ldots, \quad u_n^{(e)} v_n^{(e)}\} \qquad (9.36b)$$

Substituting Equations (9.36a) and (9.36b) into Equation (9.34), we can write the strain in an element as shown in Equations (9.37a) and (9.37b).

$$\{\varepsilon^{(e)}\} = [B]\{d^{(e)}\} \qquad (9.37a)$$

$$[B] = \begin{bmatrix} \dfrac{\partial f_1}{\partial x} & 0 & \dfrac{\partial f_2}{\partial x} & 0 & \cdots & \cdots & \dfrac{\partial f_n}{\partial x} & 0 \\[3mm] 0 & \dfrac{\partial f_1}{\partial y} & 0 & \dfrac{\partial f_2}{\partial y} & \cdots & \cdots & 0 & \dfrac{\partial f_n}{\partial y} \\[3mm] \dfrac{\partial f_1}{\partial y} & \dfrac{\partial f_1}{\partial x} & \dfrac{\partial f_2}{\partial y} & \dfrac{\partial f_2}{\partial x} & \cdots & \cdots & \dfrac{\partial f_n}{\partial y} & \dfrac{\partial f_n}{\partial x} \end{bmatrix} \qquad (9.37b)$$

The matrix $[B]$ in Equation (9.37b) is called the strain–displacement matrix. Substituting Equation (9.37a) into Equation (9.33), we obtain the strain energy density for an element as shown in Equation (9.38).

$$U_0^{(e)} = \frac{1}{2}\{d^{(e)}\}^T \{B\}^T [\tilde{E}]\{B\}\{d^{(e)}\} \qquad (9.38)$$

The strain energy for the element can be written as shown in Equations (9.39a) and (9.39b).

$$U^{(e)} = \int_{V^{(e)}} U_0^{(e)} dV = \int_{V^{(e)}} \frac{1}{2}\{d^{(e)}\}^T \{B\}^T [\tilde{E}]\{B\}\{d^{(e)}\} dV = \frac{1}{2}\{d^{(e)}\}^T [K^{(e)}]\{d^{(e)}\} \qquad (9.39a)$$

$$\boxed{[K^{(e)}] = \int_{V^{(e)}} [B]^T [\tilde{E}][B] dV} \qquad (9.39b)$$

The matrix $[K^{(e)}]$ is the element stiffness matrix.

9.6.1 Constant Strain Triangle

The constant strain triangle (CST) is the simplest element in two dimensions and was one of the first to be used in finite element analysis. As the name suggests, the strain in the element is a constant, which implies that the displacements are linear functions of x and y as shown in Equations (9.40).

$$u^{(e)} = a_0 + a_1 x + a_2 y \quad \text{and} \quad v^{(e)}(x, y) = b_0 + b_1 x + b_2 y \tag{9.40}$$

There are three constants for u and v in Equation (9.40). To evaluate the three constants, we need nodal displacements at three nodes, which define a triangle shown in Figure 9.14. The strains in the element can be determined and are found to be constants, as shown in Equation (9.41).

$$\varepsilon_{xx}^{(e)} = \frac{\partial u^{(e)}}{\partial x} = a_1 \quad \varepsilon_{yy}^{(e)} = \frac{\partial v^{(e)}}{\partial y} = b_2 \quad \gamma_{xy}^{(e)} = \frac{\partial u^{(e)}}{\partial y} + \frac{\partial v^{(e)}}{\partial x} = a_2 + b_1 \tag{9.41}$$

The constants in Equation (9.40) can be solved[6] in terms of the nodal displacements of the three nodes shown in Figure 9.14 to obtain Equations (9.42).

$$u^{(e)}(x, y) = \sum_{i=1}^{3} f_i(x, y) u_i^{(e)} \quad v^{(e)}(x, y) = \sum_{i=1}^{3} f_i(x, y) v_i^{(e)} \tag{9.42}$$

Consider the u displacement at the jth node as given by Equation (9.43).

$$u^{(e)}(x_j, y_j) = \sum_{i=1}^{3} f_i(x_j, y_j) u_i^{(e)} = u_j^{(e)} \tag{9.43}$$

If Equation (9.43) is to be valid, the interpolation functions must satisfy the condition given by Equation (9.44), which is similar to the property of Lagrange polynomials given by Equation (9.3c).

$$f_i(x_j, y_j) = \begin{cases} 1 & i = j \\ 0 & i \neq j \end{cases} \tag{9.44}$$

The plots of the interpolation functions shown Figure 9.15 are constructed by using the observations that these functions are linear and must satisfy Equation (9.44).

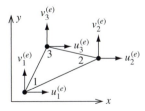

Figure 9.14 Constant strain triangle.

---

[6]See Example 9.4.

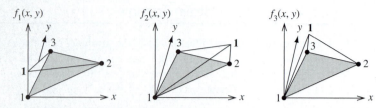

Figure 9.15 Interpolation functions for a CST.

The strain–displacement matrix $[B]$ can be found[7] as shown in Equation (9.45a).

$$[B] = \frac{1}{2A^{(e)}} \begin{bmatrix} y_2 - y_3 & 0 & y_3 - y_1 & 0 & y_1 - y_2 & 0 \\ 0 & x_3 - x_2 & 0 & x_1 - x_3 & 0 & x_2 - x_1 \\ x_3 - x_2 & y_2 - y_3 & x_1 - x_3 & y_3 - y_1 & x_2 - x_1 & y_1 - y_2 \end{bmatrix} \quad (9.45a)$$

$$A^{(e)} = \frac{1}{2}[(x_2 - x_1)(y_3 - y_1) - (y_1 - y_2)(x_1 - x_3)] \quad (9.45b)$$

The $A^{(e)}$ in Equation (9.45b) is the area of the triangle. The stiffness matrix can now be constructed from Equation (9.39b). Noting that the matrices $[B]$ and $[\tilde{E}]$ are constants in Equation (9.39b) and that the volume of the triangular element is area $A^{(e)}$ multiplied by the element thickness $t^{(e)}$, we obtain the stiffness matrix given by Equation (9.46).

$$[K^{(e)}] = [B]^T[\tilde{E}][B]\left[\int_{V^{(e)}} dV\right] = A^{(e)}t^{(e)}[B]^T[\tilde{E}][B] \quad (9.46)$$

The load vector can be constructed as discussed in Section 9.3.2, and the solution procedure would then proceed as discussed in Section 9.3.9. There are many issues and concepts in application of the equations described in this section, and the reader is referred to finite element textbooks[8] for additional details.

EXAMPLE 9.4

Obtain the strain–displacement matrix $[B]$ for the constant strain triangle given by Equation (9.45a).

PLAN

Equation (9.41) shows that the strain expression does not contain a_0 and b_0. The constants a_1 and a_2 in Equation (9.40) can be found in terms of the nodal displacement in the x direction. From these expressions, by replacing u with v and a's with b's, the constants b_1 and b_2 in Equation (9.40) can be determined. The strain expressions can be written in matrix form and the strain–displacement matrix $[B]$ determined.

---

[7] See Example 9.4

[8] See Zienkiewicz and Taylor [1989], or Bathe [1996], or Reddy [1993] in Appendix D for additional details.

SOLUTION

We use the coordinates of the three nodes substituted in Equation (9.40) and the displacements equated to the nodal displacements to obtain Equations (E1) through (E3).

$$a_0 + a_1 x_1 + a_2 y_1 = u_1^{(e)} \tag{E1}$$

$$a_0 + a_1 x_2 + a_2 y_2 = u_2^{(e)} \tag{E2}$$

$$a_0 + a_1 x_3 + a_2 y_3 = u_3^{(e)} \tag{E3}$$

We can use Cramer's rule (see Section B.6), to find the constants. The determinant $|D|$ of the matrix on the left-hand side of Equations (E1) through (E3) can be written as shown in Equation (E4)

$$|D| = \begin{vmatrix} 1 & x_1 & y_1 \\ 1 & x_2 & y_2 \\ 1 & x_3 & y_3 \end{vmatrix} \tag{E4}$$

Row 1 in Equation (E4) can be subtracted from rows 2 and 3 and the determinant evaluated from the first column, as shown in Equation (E5),

$$|D| = \begin{vmatrix} 1 & x_1 & y_1 \\ 0 & (x_2 - x_1) & (y_2 - y_1) \\ 0 & (x_3 - x_1) & (y_3 - y_1) \end{vmatrix} = (x_2 - x_1)(y_3 - y_1) - (y_2 - y_1)(x_3 - x_1) = 2A \tag{E5}$$

where A is the area of the triangle as given in Equation (9.45b).

By Cramer's rule, the constants a_1 and a_2 can be found as shown in Equations (E6) and (E7).

$$a_1 = \frac{1}{|D|} \begin{vmatrix} 1 & u_1^{(e)} & y_1 \\ 1 & u_2^{(e)} & y_2 \\ 1 & u_3^{(e)} & y_3 \end{vmatrix} = \frac{1}{2A}[-u_1^{(e)}(y_3 - y_2) + u_2^{(e)}(y_3 - y_1) - u_3^{(e)}(y_2 - y_1)] \tag{E6}$$

$$a_2 = \frac{1}{|D|} \begin{vmatrix} 1 & x_1 & u_1^{(e)} \\ 1 & x_2 & u_2^{(e)} \\ 1 & x_3 & u_3^{(e)} \end{vmatrix} = \frac{1}{2A}[u_1^{(e)}(x_3 - x_2) - u_2^{(e)}(x_3 - x_1) + u_3^{(e)}(x_2 - x_1)] \tag{E7}$$

The constants b_1 and b_2 can be written by replacing the u's with v's as shown in Equations (E8) and (E9).

$$b_1 = \frac{1}{2A}[-v_1^{(e)}(y_3 - y_2) + v_2^{(e)}(y_3 - y_1) - v_3^{(e)}(y_2 - y_1)] \tag{E8}$$

$$b_2 = \frac{1}{2A}[v_1^{(e)}(x_3 - x_2) - v_2^{(e)}(x_3 - x_1) + v_3^{(e)}(x_2 - x_1)] \tag{E9}$$

The strains in the element can be obtained by substituting Equations (E6) through (E9) into (9.41) to obtain Equations (E10) through (E12).

$$\varepsilon_{xx}^{(e)} = \frac{1}{2A}[u_1^{(e)}(y_2 - y_3) + u_2^{(e)}(y_3 - y_1) + u_3^{(e)}(y_1 - y_2)] \tag{E10}$$

$$\varepsilon_{yy}^{(e)} = \frac{1}{2A}[v_1^{(e)}(x_3 - x_2) + v_2^{(e)}(x_1 - x_3) + v_3^{(e)}(x_2 - x_1)] \tag{E11}$$

$$\gamma_{xy}^{(e)} = \frac{1}{2A}\begin{bmatrix} u_1^{(e)}(x_3 - x_2) + u_2^{(e)}(x_1 - x_3) + u_3^{(e)}(x_2 - x_1) \\ + v_1^{(e)}(y_2 - y_3) + v_2^{(e)}(y_3 - y_1) + v_3^{(e)}(y_1 - y_2) \end{bmatrix} \tag{E12}$$

Equations (E10) through (E12) can be written in matrix form as shown in Equation (E13).

$$\{\tilde{\varepsilon}^{(e)}\} = \begin{Bmatrix} \varepsilon_{xx}^{(e)} \\ \varepsilon_{yy}^{(e)} \\ \gamma_{xy}^{(e)} \end{Bmatrix} = \frac{1}{2A}\begin{bmatrix} y_2 - y_3 & 0 & y_3 - y_1 & 0 & y_1 - y_2 & 0 \\ 0 & x_3 - x_2 & 0 & x_1 - x_3 & 0 & x_2 - x_1 \\ x_3 - x_2 & y_2 - y_3 & x_1 - x_3 & y_3 - y_1 & x_2 - x_1 & y_1 - y_2 \end{bmatrix}\begin{Bmatrix} u_1^{(e)} \\ v_1^{(e)} \\ u_2^{(e)} \\ v_2^{(e)} \\ u_3^{(e)} \\ v_3^{(e)} \end{Bmatrix} = [B]\{d^{(e)}\} \tag{E13}$$

From Equation (E13), we see that the $[B]$ matrix is as given by Equation (9.45a).

COMMENT

The constants a_0 and b_0 in Equation (9.40) have no effect on the stiffness matrix and hence none on the nodal displacement values. However, if the displacement at any point inside the element were needed, the values of these constants in terms of the nodal displacements also would be needed (see Problems 9.25 and 9.27).

PROBLEM SET 9.1

FEM Problems

9.1 In a linear axial rod element, the nodal displacements were found to be $u_1^{(1)} = 0.05$ mm and $u_2^{(1)} = 0.25$ mm. The length of the element is 400 mm, the cross-sectional area $A = 50$ mm^2, and the modulus of elasticity $E = 200$ GPa. Determine (a) the displacement from node 1 at 100 mm, (b) the axial stress from node 1 at 100, and (c) the strain energy in the element.

9.2 In a quadratic axial rod element, the nodal displacements at the three equally spaced nodes were found to be

$$u_1^{(1)} = 0.0027 \text{ in} \quad u_2^{(1)} = 0.0098 \text{ in}$$
$$u_3^{(1)} = 0.017 \text{ in}$$

If the length of the element is 6 inches and the modulus of elasticity $E = 10,000$ ksi, determine (a) the displacement at 2 inches from node 1 and (b) the stress at 2 inches from node 1.

9.3 In a beam element of length 20 inches, the nodal displacement and slope at the two end nodes were found to be:

$$v_1^{(1)} = -0.407 \text{ in} \quad \theta_1^{(1)} = -0.012$$
$$v_2^{(1)} = -0.407 \text{ in} \quad \theta_2^{(1)} = 0.012$$

Find the deflection and slope at the midpoint of the element.

9.4 For the axial rod shown in Figure 9.8, find the stress at point B, the displacement at point C, and the strain energy in each element. Use the following FEM model: one linear element in AB and one quadratic element in BC. Compare your results with those shown in Table 9.1.

9.5 For the axial rod shown in Figure 9.8, find the stress at point B, the displacement at point C, and the strain energy in each element. Use the following FEM model: two equal linear elements for the entire rod AC. Compare your results with those shown in Table 9.1.

9.6 The axial rod shown in Figure P9.6 has an axial rigidity $EA = 15(10^6)$ lb. The rod is to be modeled by using a linear element for AB and a linear element for BC. Determine the displacement at point B and the reaction force at A.

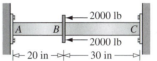

Figure P9.6

9.7 Two steel rods ($E_{st} = 200$ GPa) are securely fastened to a rigid plate that does not rotate during the application of load P, shown in Figure P9.7. The cross-sectional areas of AB and BC are $A_{AB} = 300$ mm^2 and $A_{BC} = 100$ mm^2. Model AB and BC with one linear element each. Determine (a) the displacement of the rigid plate B and (b) the reaction force at A.

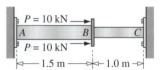

Figure P9.7

9.8 A force $F = 20$ kN is applied to a pin as shown in Figure P9.8. Both bars have a cross-sectional area $A = 100$ mm^2 and a modulus of elasticity $E = 200$ GPa. Bars AP and BP have lengths of $L_{AP} = 200$ mm and $L_{BP} = 250$ mm, respectively. Determine the displacement of pin P, using linear elements to represent each bar.

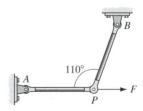

Figure P9.8

9.9 A force $F = 20$ kN is applied to the roller that slides inside a slot as shown in Figure P9.9. Both bars have cross-sectional area $A = 100$ mm² and modulus of elasticity $E = 200$ GPa. Bars AP and BP have lengths of $L_{AP} = 200$ mm and $L_{BP} = 250$ mm, respectively. Determine the displacement of the roller and the axial stress in bar A, using linear elements to represent each bar.

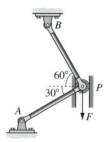

Figure P9.9

9.10 A force $F = 20$ kN is applied to a pin as shown in Figure P9.10. Both bars have a cross-sectional area $A = 100$ mm² and a modulus of elasticity $E = 200$ GPa. Bars AP and BP have lengths of $L_{AP} = 200$ mm and $L_{BP} = 250$ mm, respectively. Determine the displacement of pin P, using linear elements to represent each bar.

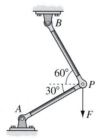

Figure P9.10

9.11 A force $F = 20$ kN is applied to a roller that slides inside a slot as shown in Figure P9.11. Both bars have cross-sectional area $A = 100$ mm² and modulus of elasticity $E = 200$ GPa. Bars AP and BP have lengths of $L_{AP} = 200$ mm and $L_{BP} = 250$ mm, respectively. Determine the displacement of the roller and the axial stress in bar A, using linear elements to represent each bar.

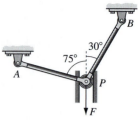

Figure P9.11

ANS. $v_P = -0.3$ mm
$R_{Px} = 2.89$ kN

9.12 A force $F = 20$ kN is applied to a pin as shown in Figure P9.12. Both bars have cross-sectional area $A = 100$ mm² and modulus of elasticity $E = 200$ GPa. Bars AP and BP have lengths of $L_{AP} = 200$ mm and $L_{BP} = 250$ mm, respectively. Determine the displacement of the pin, using linear elements to represent each bar.

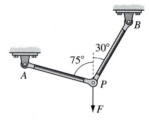

Figure P9.12

9.13 A steel ($G_{st} = 12,000$ ksi) shaft and a bronze ($G_{Cu/Sn} = 5600$ ksi) shaft are securely connected at B as shown in Figure P9.13. Determine the maximum torsional shear stress in the entire shaft and the rotation of the section at B, using one linear element to represent each segment, AB and BC.

Figure P9.13

ANS. $\phi_B = 0.0516$ rad
$T_A = -40.52$ in · kips

9.14 A solid circular steel (G_{st} = 12,000 ksi, E_{st} = 30,000 ksi) shaft of 4-inch diameter is loaded as shown in Figure P9.14. Determine the rotation of the sections at B and C and the reaction torque at A, using one linear element to represent each segment, AB, BC, and CD.

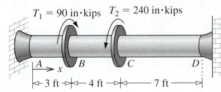

Figure P9.14

9.15 Starting with Equation (7.38c), obtain the first row of the element stiffness matrix given in Equation (9.29a).

9.16 Starting with Equation (7.38c), obtain the second row of the element stiffness matrix given in Equation (9.29a).

9.17 Starting with Equation (7.38c), obtain the third row of the element stiffness matrix given in Equation (9.29a).

9.18 Starting with Equation (7.38c), obtain the fourth row of the element stiffness matrix given in Equation (9.29a).

9.19 Starting with Equation (7.38d), obtain the element load vector given in Equation (9.29a).

9.20 Using just one beam element, determine the deflection and slope at the free end of the beam shown in Figure P9.20. Assume that EI is constant for the beam.

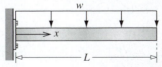

Figure P9.20

9.21 Using just one beam element, determine the deflection and slope at B (the midpoint) in the beam shown in Figure P9.21. Assume that EI is constant for the beam.

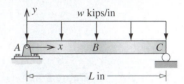

Figure P9.21

ANS. $v_B = -\left(\dfrac{wL^4}{96EI}\right) \qquad \theta_B = 0$

9.22 Using two beam elements, determine the deflection and slope at the free end and reaction force at B in the beam shown in Figure P9.22. Assume that EI is constant for the beam.

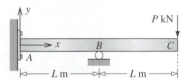

Figure P9.22

ANS. $v_C = -\left(\dfrac{7PL^3}{12EI}\right) \qquad \theta_C = -\left(\dfrac{3PL^2}{4EI}\right)$

$R_B = \dfrac{5P}{2}$

9.23 Using one element, determine the deflection and slope at midpoint of the beam shown in Figure P9.23. Assume that EI is constant for the beam.

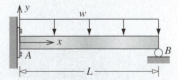

Figure P9.23

9.24 Using a single beam element for AB and a single beam element for BC in Figure P9.24, determine (a) the slope at A and B and (b) the reaction forces at A and B.

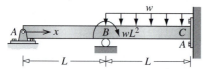

Figure P9.24

9.25 Determine the constants a_0 and b_0 in Equation (9.40).

9.26 In a constant strain triangle, the continuity of the displacement at the nodes also ensures continuity across the line joining the nodes. To prove this statement, show that the displacement[9] along the line joining nodes 1 and 2 depends only upon the nodal values of nodes 1 and 2.

9.27 A very convenient method of finding the interpolation functions for triangular elements is the use of nondimensional coordinates called area coordinates. Consider a point P in the triangle shown in Figure P9.27, and join it to the vertices of the triangle. The coordinates of point P can be defined by dividing the area opposite each vertex by the total area of the triangle, as shown in Figure P9.27 and Equation (9.47).

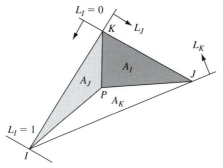

Figure P9.27

$$L_I = \frac{A_I}{A} \qquad L_J = \frac{A_J}{A} \qquad L_K = \frac{A_K}{A}$$

$$L_I + L_J + L_K = 1 \qquad (9.47)$$

Note that any point on a line parallel to the a side in Figure P9.26 would create triangles of equal area. Thus the area coordinate has zero value on side opposite the vertex, has a value of 1 on the vertex, and has a constant value on a line parallel to the side opposite the vertex. Determine the interpolation functions for the constant strain triangle in terms of the area coordinates L_I, L_J, and L_K.

9.28 The linear strain triangular element shown in Figure P9.28 will have displacements that will be quadratic in x and y, resulting in six nodes.[10] Determine the interpolation functions in terms of one area coordinates L_I, L_J, and L_K for nodes 1, 3, and 5.

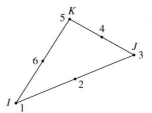

Figure P9.28

9.29 Determine the interpolation functions in terms of the area coordinates L_I, L_J, and L_K for nodes 2, 4, and 6 for the linear strain triangle shown in Figure P9.28.

ANS. $\mathcal{L}_2 = 4L_IL_J \qquad \mathcal{L}_4 = 4L_KL_J$

$\mathcal{L}_6 = 4L_IL_K$

---

[9] Nodal displacements are independent variables. If node 3 appeared in the equation for displacement along lines 1 and 2, it could be changed, and two adjoining elements could produce different displacement values along the line.

[10] A Pascal triangle is a very simple way of determining the number of nodes needed on a triangular element for polynomials of different orders. See Reddy [1993] in Appendix D for additional details.

9.7 CLOSURE

In this chapter, we used one-dimensional structural elements to elaborate the procedure for analysis by means of the finite element method. Familiarity with the following concepts will help in reading manuals and documents accompanying finite element software packages: nodes, elements, mesh, discretization, interpolation functions, Lagrange polynomials, Hermite polynomials, element stiffness matrix, global stiffness matrix, element load vector, nodal forces, nodal displacements, mesh refinement, h-method, p-method, and r-method.

The concepts introduced in this book can be further developed in solid mechanics courses covering subjects such as plates and shells, the finite element method, elasticity, plasticity, continuum mechanics, fracture mechanics, the mechanics of composites, and biomechanics.

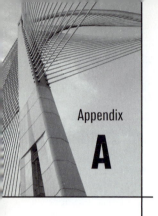

A

Statics and Mechanics of Materials Review

A.0 OVERVIEW

The book presupposes a course of statics and an introductory course in the mechanics of materials. The first three chapters briefly review many of the concepts from the introductory course on the mechanics of materials before extending the concepts to cover new material. Some concepts and details from the introductory course on the mechanics of materials, however, do not logically fit in the flow of presentation in the first three chapters. These concepts are briefly reviewed in this appendix. The presentation presupposes familiarity with the concepts. If you had the introductory course on the mechanics of materials some time ago, you may need to review your mechanics of materials textbook before attempting to assimilate the concepts presented briefly in this appendix.

A.1 TYPES OF FORCES AND MOMENTS

We can classify the forces and moments that we shall see in this book as external, internal, and reaction forces and moments.

A.1.1 External Forces and Moments

The forces and moments that are applied to the body and are often referred to as the *load* on the body are said to be external. These are assumed known in an analysis, though sometimes we carry external forces and moments as variables so that we may answer such questions as How much load can a structure support? or What loads are needed to produce a given deformation?

Surface forces (also called tractions) and moments are external forces and moments that act on the surface and are transmitted to the body by contact. Surface forces (moments) applied at a point are called concentrated forces (moments or couples). Surface forces (moments) applied along a line or over a surface are called distributed forces (moments).

Body forces are external forces that act at every point on the body. Body forces are not transmitted by contact. Gravitational and electromagnetic forces are two examples of body forces. A body force has units of force per unit volume.

A.1.2 Reaction Forces and Moments

The forces and moments that are developed at the supports of a body to resist movement due to the external forces and moments are called reaction forces and moments. Usually these are not known and must be calculated before further analysis can be conducted. The following principles are used to decide whether there is a reaction force or reaction moment at a support.

   (i) If a point cannot move in a given direction, then a reaction force opposite the direction is acting at that support point.

  (ii) If a line cannot rotate about an axis in a given direction, then a reaction moment opposite the direction is acting at that support.

 (iii) In making decisions about the movement of a point or rotation of a line at the support, we consider the support in isolation, not the entire body. Exceptions to the rule exist in three-dimensional problems such as bodies supported by balanced hinges or balanced bearings (rollers). Three-dimensional problems of these types are not covered in this book.

Appendix C shows several types of support that can be replaced by reaction forces and moments using the foregoing principles (see Section C.5).

A.1.3 Internal Forces and Moments

A body is held together by internal forces. Internal forces exist irrespective of whether we apply or do not apply external forces. A material resists changes due to applied forces and moments by increasing the internal forces. Our interest is in the resistance a material offers to an applied load (i.e., the internal forces). Internal forces always exist in pairs that are equal and opposite on the two surfaces produced by an imaginary cut. The internal forces and moments are shown in Figure A.1.

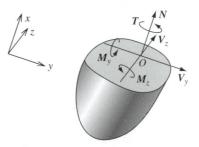

Figure A.1 Internal forces and moments.

Throughout the book we adhere to the following:

Convention: all internal forces and moments are given in ***bold italics*** (N = axial force, V_y = shear force, V_z = shear force, T = torque, M_y = bending moment, M_z = bending moment).

Definition 1 Forces that are normal to an imaginary cut surface are called *normal forces*. The normal force that points away from the surface (pulls the surface) is called a *tensile force*. The normal force that points into the surface (pushes the surface) is called a *compressive force*.

Definition 2 The normal force acting in the direction of the axis of the body is called the *axial force*.

Definition 3 Forces that are tangent to the imaginary cut surface are called *shear forces*.

Definition 4 An internal moment about an axis normal to an imaginary cut surface is called a torsional moment or torque.

Definition 5 Internal moments about an axis tangent to an imaginary cut are called *bending moments*.

A.2 FREE BODY DIAGRAMS

Newton's laws are applicable to free bodies only. We use "free" to designate a body that will move if it is not in equilibrium. If there are supports, they must be replaced by appropriate reaction forces and moments; the principles described in Section A.1.2 must be followed.

Definition 6 The diagram showing all the forces acting on a free body is called a *free body diagram* (FBD).

Additional free body diagrams may be created by making imaginary cuts for the calculation of internal quantities. Each imaginary cut will produce two additional free body diagrams; either one can be used for calculating internal forces and moments.

A body is in static equilibrium if the vector sum of all the forces acting on a free body is zero and the vector sum of all the moments about any point in space is zero. Mathematically, this is stated as Equation (A.1),

$$\sum \bar{F} = 0 \qquad \sum \bar{M} = 0 \tag{A.1}$$

where the symbol $\sum$ represents summation, and the overbar represents a vector quantity. In a three-dimensional Cartesian coordinate system, Equation (A.1) in scalar form is written as Equation (A.2).

$$
\begin{aligned}
\sum F_x &= 0 & \sum F_y &= 0 & \sum F_z &= 0 \\
\sum M_x &= 0 & \sum M_y &= 0 & \sum M_z &= 0
\end{aligned}
\tag{A.2}
$$

Equation (A.2) implies that there are *six independent* equations in three dimensions. In other words, we can at most solve for *six unknowns* from a free body diagram in three dimensions.

In two dimensions the sum of the forces in the z direction and the sum of the moments about the x and y axes are automatically satisfied because all forces must lie in the x-y plane. The remaining equilibrium equations in two dimensions that have to be satisfied are as given in Equation (A.3).

$$\sum F_x = 0 \qquad \sum F_y = 0 \qquad \sum M_z = 0 \qquad (A.3)$$

Equation (A.3) implies that there are *three independent* equations per free body diagram in two dimensions. In other words, we can at most solve for *three unknowns* from a two-dimensional free body diagram.

The following observations can be used to reduce computational effort.

- Balancing the moment at a point through which an unknown force (or forces) passes reduces the computational effort because such forces do not appear in the moment equation.

- Balancing the forces and/or moments perpendicular to the direction of an unknown force or moment reduces the computational effort because such forces do not appear in the equation.

Definition 7 A body on which there are more unknown reaction forces and moments than there are equilibrium equations (6 in three dimensions, and 3 in two dimensions) is called a statically indeterminate body.

Statically indeterminate problems arise when more supports than are needed are used to support a structure. Extra supports may be used for safety reasons or for purposes of increasing the stiffness of a structure. We define the following:

Definition 8 Degree of static redundancy = number of unknown reactions − number of equilibrium equations.

To solve a statically indeterminate problem, we generate equations on the displacement and/or rotation at the support points. Sometimes it is a mistake to try to generate enough equations for the unknowns by taking moments at many points. *A statically indeterminate problems cannot be solved from equilibrium equations alone.* There are only three independent equations of static equilibrium in two dimensions and six independent equations of static equilibrium in three dimensions. Additional equations must come from the displacements and/or rotation conditions at the support.

- The number of equations on the displacements and/or rotations needed to solve a statically indeterminate problem is equal to the degree of static redundancy.

A.3   TRUSSES

Definition 9 A truss is a structure made up of two-force members.

Definition 10 A two-force member is a structural member on which there is no moment couple and in which the forces act at two points only.

Two methods of calculating the internal forces in truss members are the method of joints and the method of sections.

In method of joints, a free body diagram is created by making imaginary cuts on all members joined at the pin. If a force is directed away from the pin, the two-force member is assumed to be in tension; if it is directed into the pin, the member is assumed to be in compression. By conducting a force balance in two (three) dimensions, two (three) equations per pin can be written.

In method of sections, an imaginary cut is made through the truss to produce a free body. The imaginary cut can be of any shape that will permit a quick calculation of the force in a member. Three equations in two dimensions and six equations in three dimensions can be written in accordance with the free body diagram produced from a single imaginary cut.

Definition 11 A zero-force member in a truss is a member that carries no internal force.

Identifying zero-force members can save significant computation time. Zero-force members can be identified by conducting the method of joints mentally. Usually if two members are collinear at a joint *and* there is no external force, the zero-force member is the member that is inclined to the collinear members.

A.4   CENTROIDS

The y and z coordinates of the centroid (y_c, z_c) of the two-dimensional area A shown in Figure A.2 are defined by Equation (A.4),

$$y_c = \frac{\int_A y \, dA}{\int_A dA} \qquad \text{and} \qquad z_c = \frac{\int_A z \, dA}{\int_A dA} \tag{A.4}$$

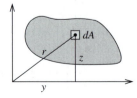

Figure A.2   Area moments.

Where the numerator in each case is referred to as the first moment of the area. If there is an axis of symmetry, then the area moment about the axis of symmetry from one part of the body is canceled by the moment from the symmetric part, and we conclude that *the centroid lies on the axis of symmetry.*

Consider a coordinate system fixed to the centroid of the area. If we now consider the first moment of the area in this coordinate system and it turns out to be nonzero, it implies that the centroid is not located at the origin, thus contradicting our starting assumption. We therefore conclude:

- The first moment of the area calculated in a coordinate system fixed to the centroid of the area is zero.

The centroid for a composite body in which the centroids of individual bodies are known can be calculated from the following equations,

$$
y_c = \frac{\sum\limits_{i=1}^{n} y_{c_i} A_i}{\sum\limits_{i=1}^{n} A_i}
\quad\text{and}\quad
z_c = \frac{\sum\limits_{i=1}^{n} z_{c_i} A_i}{\sum\limits_{i=1}^{n} A_i}
\tag{A.5}
$$

where y_{c_i} and z_{c_i} are the known coordinates of the centroids of the area A_i. Appendix C.6 shows the location of centroids of some common shapes that will be useful in solving the problems in this book.

A.5   STATICALLY EQUIVALENT LOAD SYSTEMS

> **Definition 12** Two systems of forces that generate the same resultant force and moment are called statically equivalent load systems.

If one system satisfies equilibrium, then the statically equivalent system also satisfies equilibrium, since the resultant force and the resultant moment must be zero in both systems. The concept of statically equivalent systems can simplify analysis significantly and is most often used in problems with distributed loads.

A.5.1   Distributed Force on a Line

Let $p(x)$ be a distributed force per unit length that varies with x. We can replace this distributed force by a force and moment acting at any point or by a single force acting at point x_c as shown in Figure A.3.

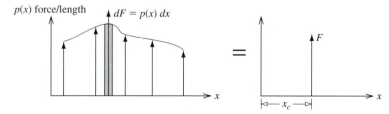

Figure A.3   Static equivalency for a distributed force on a line.

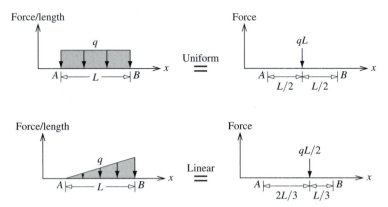

Figure A.4 Statically equivalent force for uniform and linear distributed forces on a line.

For the two system to be statically equivalent, the resultant force and the resultant moment about any point (origin) must be the same. This implies:

$$F = \int_L p(x)dx \qquad \text{and} \qquad x_c = \frac{\int_L x\,p(x)dx}{F} \qquad (A.6)$$

The force F is equal to the area under the curve, and x_c represents the location of the centroid of the distribution. This idea is used in replacing a uniform and linearly varying distribution by a statically equivalent force, as shown in Figure A.4.

The two statically equivalent systems are *not identical* systems. The deformation (change of shape of bodies) in a pair of statically equivalent systems is different. The distribution of the internal force and the internal moment of a pair of statically equivalent system is different. The following rule must be remembered:

> The imaginary cut needed for calculating internal forces and moments must be made on the original body, not on the statically equivalent body.

A.5.2 Distributed Force on a Surface

Let $\sigma(y, z)$ be a distributed force per unit area that varies in intensity with y and z. We would like to replace it by a single force, as shown in Figure A.5.

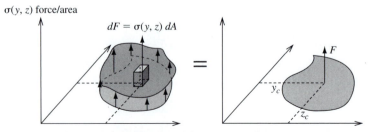

Figure A.5 Static equivalency for a distributed force on a surface.

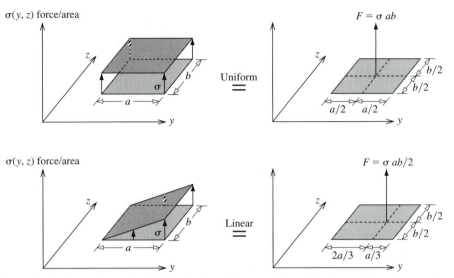

Figure A.6 Statically equivalent forces for uniform and linear distributed forces on a surface.

For the two systems shown in Figure A.5 to be statically equivalent load systems, the resultant force and the resultant moment y and z axes must be the same. This implies:

$$F = \iint_A \sigma(y, z)dy\,dz \qquad y_c = \frac{\iint_A y\sigma(y, z)dy\,dz}{F} \qquad z_c = \frac{\iint_A z\sigma(y, z)dy\,dz}{F} \qquad (A.7)$$

The force F is equal to the volume under the curve, whereas y_c and z_c represent the location of the centroid of the *distribution,* which can be different from the centroid of the area on which the distributed force acts. The centroid of the area depends only on the geometry of the area. The centroid of the distribution depends upon how the intensity of distributed load $\sigma(y, z)$ varies over the area.

Figure A.6 shows a uniform and a linearly varying distributed force that can be replaced by a single force at the centroid of the distribution. Notice that for a uniform distributed force, the centroid of the distributed force is the same as the centroid of the rectangular area, but for a linearly varying distributed force, the centroid of the distributed force is different from the centroid of the area. If we were to place the equivalent force at the centroid of the area, we would also need a moment at that point.

A.6   AREA MOMENTS OF INERTIA

The area moments of inertia, also referred to as second area moments, are defined as

$$I_{yy} = \int_A z^2 dA \qquad I_{zz} = \int_A y^2 dA \qquad I_{yz} = \int_A yz\,dA \qquad (A.8)$$

The polar moment of inertia is defined in Equation (A.9), with the relation to I_{yy} and I_{zz} deduced from Figure A.2.

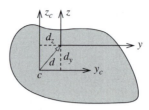

Figure A.7 Parallel axis theorem.

$$J = \int_A r^2 dA = I_{yy} + I_{zz} \tag{A.9}$$

If we know the area moment of inertia in a coordinate system fixed to the centroid, we can compute the area moments about an axis parallel to the coordinate axis by means of the parallel axis theorem given by Equation (A.10),

$$I_{yy} = I_{y_c y_c} + A d_y^2 \qquad I_{zz} = I_{z_c z_c} + A d_z^2 \qquad I_{yz} = I_{y_c z_c} + A d_y d_z \qquad J = J_c + A d^2 \tag{A.10}$$

where the subscript c refers to the axis fixed to the centroid of the body. The quantities y^2, z^2, r^2, A, d_y^2, d_z^2, and d^2 are always positive. Therefore, from Equations (A.8) through A.10, we conclude that I_{yy}, I_{zz}, and J are always positive and minimum about the axis passing through the centroid of the body. However, I_{yz} can be positive or negative, since y, z, d_y, and d_z can be positive or negative in Equation (A.8). If either y or z is an axis of symmetry, then the integral in I_{yz} on the positive side will cancel the integral on the negative side in Equation (A.8), and hence I_{yz} will be zero. We record the following observations.

- I_{yy}, I_{zz}, and J are always positive and minimum about the axis passing through the centroid of the body
- If either the y or the z axis is an axis of symmetry, then I_{yz} will be zero.

The moment of inertia of a composite body in which we know the moment of inertia of individual bodies about the centroid can be calculated from Equations (A.11a) through (A.11d),

$$I_{yy} = \sum_{i=1}^{n} (I_{y_{c_i} y_{c_i}} + A_i d_{y_i}^2) \tag{A.11a}$$

$$I_{zz} = \sum_{i=1}^{n} (I_{z_{c_i} z_{c_i}} + A_i d_{z_i}^2) \tag{A.11b}$$

$$I_{yz} = \sum_{i=1}^{n} (I_{y_{c_i} z_{c_i}} + A_i d_{y_i} d_{z_i}) \tag{A.11c}$$

$$J = \sum_{i=1}^{n} (J_{c_i} + A_i d_i^2) \tag{A.11d}$$

where $I_{y_{c_i} y_{c_i}}$, $I_{z_{c_i} z_{c_i}}$, $I_{y_{c_i} z_{c_i}}$, and J_{c_i} are the area moments of inertia about the axis passing through the centroid of the ith body. The Table in Section C.6 shows the area moments of inertia about an axis passing through the centroid of some common shapes that will be useful in solving problems in this book.

The *radius of gyration* $\hat{r}$ about an axis is defined by

$$\hat{r} = \sqrt{\frac{I}{A}} \qquad \text{or} \qquad I = A\hat{r}^2 \tag{A.12}$$

where I is the area moment of inertia about the same axis about which the radius of gyration $\hat{r}$ is being calculated.

A.7 PRINCIPAL MOMENTS OF INERTIA

In Section 6.1.6 we saw that significant simplification occurs in Equation (6.13) when $I_{yz} = 0$. Furthermore, we know that buckling occurs about the axis of the minimum moment of inertia. In this section we will see that the moments of inertia are components of a second-order tensor and, like stress, can be transformed in a similar manner. We have the following definitions and observations, which are analogous to those in stress transformation.

Definition 13 The coordinate system in which the cross moment of inertia is zero is called the principal coordinate system.

Definition 14 The moments of inertia in the principal coordinate system are called principal moments of inertia.

Consider the rotation of a coordinate system y-z by an angle θ counterclockwise to a coordinate system n-t as shown in Figure A.8, where the coordinates y and z can be resolved into components along the n and t directions to obtain the following coordinate transformation equations:

$$n = y \cos\theta + z \sin\theta \qquad t = -y \sin\theta + z \cos\theta \tag{A.13}$$

Substituting the coordinates into the definitions of area moments of inertia, we obtain the following transformation equations:

$$I_{nn} = \int_A t^2 dA = I_{yy}\cos^2\theta + I_{zz}\sin^2\theta - 2I_{yz}\cos\theta \, \sin\theta \tag{A.14a}$$

$$I_{tt} = \int_A n^2 dA = I_{yy}\sin^2\theta + I_{zz}\cos^2\theta + 2I_{yz}\cos\theta \, \sin\theta \tag{A.14b}$$

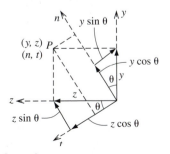

Figure A.8 Coordinate transformation.

$$I_{nt} = \int_A nt \, dA = (I_{yy} - I_{zz})\cos\theta\,\sin\theta + I_{yz}(\cos^2\theta - \sin^2\theta) \qquad \text{(A.14c)}$$

Equations (A.14a) through (A.14c) are similar to the stress transformation equations, and all the methods outlined for stress transformation can be used for finding the principal moments of inertia. We could set I_{nt} equal to zero in Equation (A.14c) and find the orientation of the principal axis as follows:

$$\tan 2\theta_p = \frac{-2I_{yz}}{(I_{yy} - I_{zz})} \qquad \text{(A.15)}$$

Once θ_p is known, it can be substituted into Equation (A.14a) and (A.14b) to obtain the principal moments of inertia as

$$I_{1,2} = \frac{I_{yy} + I_{zz}}{2} \pm \sqrt{\left(\frac{I_{yy} - I_{zz}}{2}\right)^2 + I_{yz}^2} \qquad \text{(A.16)}$$

where $I_{1,2}$ represents the two principal moment of inertias I_1 and I_2. The plus sign is to be taken with I_1 and the minus sign with I_2. Adding Equations (A.14a) and (A.14b) and I_1 and I_2, we obtain

$$I_{nn} + I_{tt} = I_{yy} + I_{zz} = I_1 + I_2 \qquad \text{(A.17)}$$

Equation (A.17) shows that the sum of the normal area moments of inertia in an orthogonal coordinate system does not depend upon the orientation of the coordinate system. As in stresses, it is not clear whether the principal angle found from Equation (A.15) is associated with I_1 or I_2. Once more, we resolve the problem by substituting θ_p into Equation (A.14a) [or (A.14b)] to obtain one of the principal area moments of inertia. We can then use Equation (A.14c) to find the other principal area moments of inertia. The greater of the two principal moments of inertia will be labeled I_1, and the angle corresponding to it will be reported as principal angle 1. Equation (A.16) can be used as a check on our results.

The bending normal stress equation in principal coordinates can be written as

$$\sigma_{xx} = -\left(\frac{M_n}{I_{nn}}\right)t - \left(\frac{M_t}{I_{tt}}\right)n \qquad \text{(A.18)}$$

where the internal moments in the n and t coordinates are related to the internal moments in the y and z coordinates as follows:

$$M_n = -\int_A t\sigma_{xx}dA = -M_z\sin\theta + M_y\cos\theta \qquad \text{(A.19a)}$$

$$M_t = -\int_A n\sigma_{xx}dA = M_z\cos\theta + M_y\sin\theta \qquad \text{(A.19b)}$$

The form of Equations (A.18) to (A.19b) is useful for standard sections, where the principal directions and the principal moments of inertia are tabulated. If, however, we need to calculate the area moments of inertia before proceeding to compute σ_{xx}, it is algebraically less tedious to use Equation (6.13).

EXAMPLE A.1

For the L-shaped beam shown in Figure A.9, the calculations show that

$$I_{yy} = 0.4125(10^6) \text{ mm}^4 \qquad I_{zz} = 1.5125(10^6) \text{ mm}^4$$

Determine the orientation of the principal axes and the principal moments of inertia about the axes passing through the centroid.

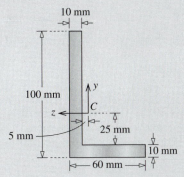

Figure A.9 Cross section for Example A.1.

PLAN

We can use Equation (A.11c) to find I_{yz}. We can then use Equation (A.15) to find the principal angle. Finally, we can use Equations (A.14a) and (A.14b) to find the principal moments of inertia.

SOLUTION

Figure A.10 shows L-shaped cross section as two rectangles with centroids C_1 and C_2. We can find the area of cross sections A_1 and A_2 and the values of d_{y1}, d_{z1} and d_{y2}, d_{z2}, the y and z components of the vectors from C_1 and C_2 to C, respectively, as follows:

$$A_1 = 500 \text{ mm}^2 \qquad A_2 = 1000 \text{ mm}^2$$

$$d_{y1} = 30 \text{ mm} \qquad d_{z1} = 20 \text{ mm} \qquad d_{y2} = -15 \text{ mm} \qquad d_{z2} = -10 \text{ mm}$$

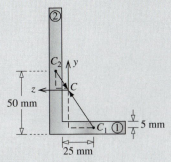

Figure A.10 Calculation of the cross moment of inertia.

The value of I_{yz} for a rectangle about its own centroid is zero. Thus I_{yz} about the centroid of the rectangles 1 and 2 shown in Figure A.10 is zero. From Equation (A.11c) we obtain.

$$I_{yz} = A_1 d_{y1} d_{z1} + A_2 d_{y2} d_{z2} = (1000)(-15)(-10) + (500)(30)(20) = 0.45(10^6) \text{ mm}^4 \quad \text{(E1)}$$

From Equation (A.15) we can find the principal angle as

$$\tan 2\theta_p = \frac{-2I_{yz}}{(I_{yy} - I_{zz})} = \frac{2(0.45)(10^6)}{(1.1)(10^6)} = 0.818 \quad \text{or} \quad \theta_p = 19.6° \quad \text{(E2)}$$

Substituting θ in Equations (A.14a), we obtain one of the principal moments of inertia as

$$I_p = [(0.4125)\cos^2 19.6 + (1.5125)\sin^2 19.6 - 2(0.45)\cos 19.6 \sin 19.6](10^6) \quad \text{or}$$

$$I_p = 0.2519(10^6) \text{ mm}^4 \quad \text{(E3)}$$

Noting that $I_{yy} + I_{zz} = 1.925(10^6)$, we obtain the other principal moment of inertia from Equation (A.17) as $(1.925 - 0.2519)(10^6) = 1.673(10^6)$. Thus the area moment of inertia in Equation (E3) is the second principal area moment of inertia, and 90° is added to (or subtracted from) the angle in Equation (E2) and reported as principal angle 1. The results are as follows:

ANS. $I_1 = 1.673(10^6) \text{ mm}^4 \qquad I_2 = 0.2519(10^6) \text{ mm}^4 \quad \theta_1 = 109.6° \text{ ccw, } (70.4° \text{ cw})$

Checking Results: From Equation (A.16) we have:

$$I_{1,2} = \left[\frac{0.4125 + 1.5125}{2} \pm \sqrt{\left(\frac{0.4125 - 1.5125}{2}\right)^2 + 0.45^2} \right](10^6) \quad \text{or}$$

$$I_1 = [0.9625 + 0.7106](10^6) = 1.673(10^6) \quad \text{Checks.}$$

$$I_2 = [0.9625 - 0.7106](10^6) = 0.2519(10^6) \quad \text{Checks.}$$

COMMENTS

In Figure A.11, which shows the orientation of the two solutions for principal coordinates, principal axis 2 is the minimum area moment of inertia and hence is the axis about which buckling will occur, and I_2 is the value that should be used in the Euler buckling formula for the calculation of the critical buckling load.

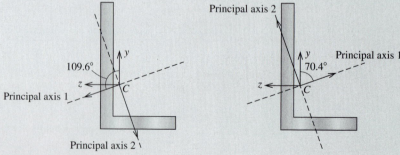

Figure A.11 Principal axis in Example A.1.

A.8 MOHR'S CIRCLE FOR STRESS

Mohr's circle for stress is represented by the following equation:

$$\left(\sigma_{nn} - \frac{\sigma_{xx} + \sigma_{yy}}{2}\right)^2 + \tau_{nt}^2 = \left(\frac{\sigma_{xx} - \sigma_{yy}}{2}\right)^2 + \tau_{xy}^2 \qquad (A.20)$$

The important observations about Mohr's circle for stress are as follows.

- Each point on the circle represents a unique plane that passes through the point at which the stresses are specified.
- The coordinates of a point on a Mohr's circle are (σ_{nn}, τ_{nt}), the normal and shear stresses on the plane represented by the point.
- The angles between planes on a stress cube are doubled when plotted on a Mohr's circle.

A.8.1 Construction of Mohr's Circle for Stress

The steps in the construction of Mohr's circle are as follows.

Step 1. Show the stresses σ_{xx}, σ_{yy}, and τ_{xy} on a stress cube and label the vertical plane as V and the horizontal plane as H as shown in Figure A.12.

Step 2. Write the coordinates of points V and H as

$$V(\sigma_{xx}, \tau_{xy}) \qquad \text{and} \qquad H(\sigma_{yy}, \tau_{yx})$$

The rotation arrow next to the shear stresses corresponds to the rotation of the cube caused by the set of shear stresses on planes V and H.

Step 3. Draw the horizontal axis with the tensile normal stress to the right and the compressive normal stress to the left as shown in Figure A.13. Draw the vertical axis with the clockwise direction of shear stress up and the counterclockwise direction of rotation down.

Step 4. Locate V and H and join these points by drawing a line. Label the point at which the line VH intersects the horizontal axis as C.

Step 5. With C as center and CV or CH as the radius, draw the Mohr circle.

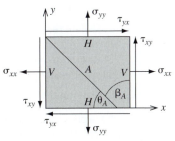

Figure A.12 Stress cube for the construction of a Mohr's circle.

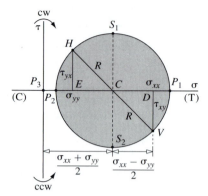

Figure A.13 Construction of Mohr's circle.

A.8.2 Principal Stresses and Maximum Shear Stress

Step 6. Calculate the principal stresses by finding the coordinates of points P_1 and P_2 in Figure A.13.

Step 7. Calculate the principal angle θ_p from either triangle VCD or triangle ECH. Find the angle between lines CV and CP_1 if θ_1 is different from θ_p.

Step 8. The *in-plane maximum shear stress* is the radius of the in-plane circle in Figure A.13 between P_1 and P_2. The in-plane maximum shear stress will exist on planes represented by S_1 and S_2 in Figure A.13.

Step 9. To find the absolute maximum shear stress, locate point P_3 at the value of the third principal stress (zero for plane stress and nonzero for plane strain) and draw two more circles: one between P_1 and P_3 and the other between P_2 and P_3. The maximum shear stress at a point is the radius of the biggest circle that can be drawn between P_1, P_2, and P_3.

A.8.3 Stresses on an Inclined Plane

The stresses on an inclined plane are found by first locating the point representing the plane on the Mohr circle and then determining the coordinate of the point. This is achieved as follows.

Step 10. Draw the inclined plane on the stress cube and label it A as in Figure A.12.

Step 11. Locate the inclined plane on the Mohr circle as described in step 13 and label it A as in Figure A.14(a).

Step 12. Calculate the coordinates of point A.
In accordance with Figure A.14, the coordinates of point A are $A\,(\sigma_A, \tau_A)$. The rotation corresponds to the rotation associated with the shear stress at point A.

Step 13. Determine the sign of shear stress.

To determine the sign of shear stress, we start by drawing the shear stress such that the inclined plane A rotates in the same direction as was recorded with the coordinates in step 12. A local coordinate system is established, and if the shear stress is in the positive

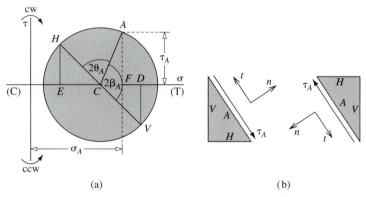

Figure A.14 Stresses on an inclined plane.

tangent direction, it is positive. The two possibilities are shown in Figure A.14(b). In both cases the shear stress is negative.

A.9 MOHR'S CIRCLE FOR STRAIN

The equation for Mohr's circle for strain is

$$\left(\varepsilon_{nn} - \frac{\varepsilon_{xx} + \varepsilon_{yy}}{2}\right)^2 + \left(\frac{\gamma_{nt}}{2}\right)^2 = \left(\frac{\varepsilon_{xx} - \varepsilon_{yy}}{2}\right)^2 + \left(\frac{\gamma_{xy}}{2}\right)^2 \tag{A.21}$$

The important observations in Mohr's circle for strain are as follows:

- Each point on the circle represents a unique direction passing through the point at which the strains are specified.
- The coordinates of each point on the circle are the strains (ε_{nn}, $\gamma_{nt}/2$). These represent the normal strain of a line in the n direction and half the shear strain that represents the rotation of the line passing through the point.
- Angles between lines are doubled when plotted on a Mohr circle.

A.9.1 Construction of Mohr's Circle for Strain

The steps in the construction of Mohr's circle for strain are as follows.

Step 1. Draw a square that has been deformed owing to shear strain γ_{xy}. Label the *intersection* of the vertical plane and x axis as V and the *intersection* of the horizontal plane and y axis as H as shown in Figure A.15.

Step 2. Write the coordinates of points V and H as

$$V(\varepsilon_{xx}, \gamma_{xy}/2)) \qquad \text{and} \qquad H(\varepsilon_{yy}, \gamma_{xy}/2)) \qquad \text{for} \qquad \gamma_{xy} > 0$$

The rotation arrow along the side of shear strains corresponds to the rotation of the line on which the point lies, as shown in Figure A.15.

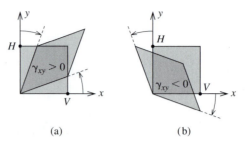

(a)   (b)

Figure A.15   Deformed cube for the construction of a Mohr circle.

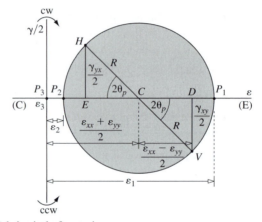

Figure A.16   Mohr's circle for strains.

Step 3. Draw the horizontal axis to represent the normal strain, with extension to the right and contractions to the left as shown in Figure A.16. Draw the vertical axis to represent *half the shear strain*, with clockwise rotation of a line in the upper plane, and counterclockwise rotation of a line in the lower plane.

Step 4. Locate points V and H and join the points by drawing a line. Label the point at which the line VH intersects the horizontal axis as C.

Step 5. The horizontal coordinate of point C is the average normal strain. The distance CE can be found from the coordinates of points E and C and the radius R, calculated from the Pythagorean theorem. With C as the center and CV or CH as radius, draw the Mohr circle.

A.9.2   Principal Strains and Maximum Shear Strain

Step 6. Calculate the principal strains by finding the coordinates of points P_1 and P_2 in Figure A.16.

Step 7. Calculate the principal angle θ_p from either triangle VCD or triangle ECH. Find the angle between lines CV and CP_1 if θ_1 is different from θ_p.

Step 8. The in-plane maximum shear strain $\gamma_p/2 = R$, the radius of the in-plane circle shown in Figure A.16. To find the absolute maximum shear strain, locate point P_3 at the value of the third principal strain (zero for plane strain and nonzero for plane stress) and draw two more circles: one between P_1 and P_3, the other between P_2 and P_3. The maximum shear strain (γ_{max}) at a point is found from the radius of the biggest circle.

A.9.3 Strains in a Specified Coordinate System

The strains in a specified coordinate system are found by first locating the coordinate directions on the Mohr circle and then determining the coordinate of the point representing the directions. This is achieved as follows.

Step 9. Draw Cartesian coordinate system as well as the specified coordinate system, placing a square in each coordinate system to represent the undeformed state. Label points V, H, N, and T to represent the four directions, as shown in Figure A.17(a).

Step 10. Points V and H on the Mohr circle are known. Point N is located by starting from point V and rotating by $2\theta_V$ in the same direction, as shown in Figure A.17(b). Similarly, starting from point H on the Mohr circle, point T is located as shown in Figure A.17(b).

Step 11. Calculate the coordinates of points N and T.

From Figure A.17(b), the coordinates of points N and T are as follows:

$$N(\varepsilon_{nn}, \gamma_{nt}/2)) \quad \text{and} \quad T(\varepsilon_{tt}, \gamma_{nt}/2))$$

The rotation of the line at point N is clockwise, since it is in the upper plane, while the rotation of the line at point T is counterclockwise, since it is in the lower plane in Figure A.17(b).

Step 12. Determine the sign of shear strain.

To draw the deformed shape, we rotate the n-coordinate line in the direction shown for point N in step 3. Similarly, we rotate the t-coordinate line in the direction shown for point: T in step 3 as shown in Figure A.17(c). The angle between the n and t directions increases; hence the shear strain γ_{nt} is negative.

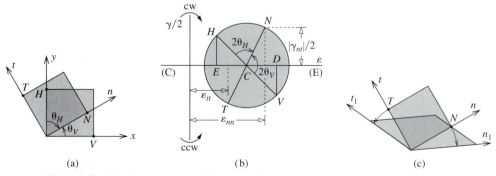

Figure A.17 Strains in a specified coordinate system.

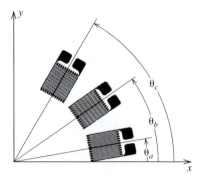

Figure A.18 Strain rosette.

A.10 STRAIN GAGES

The following are important observations about strain gages.

- Strain gages measure only normal strains directly.
- Strain gages are bonded to a free surface; that is, the strains are in a state of plane stress, not plane strain.
- Strain gages measure average strain at a point.
- An assembly of three strain gages used for finding three normal strains is called a *strain rosette*.

To determine the strains ε_{xx}, ε_{yy}, and γ_{xy} from strain gage readings, we use the strain transformation equation, as described in Equations (A.22),

$$\varepsilon_a = \varepsilon_{xx}\cos^2\theta_a + \varepsilon_{yy}\sin^2\theta_a + \gamma_{xy}\sin\theta_a\cos\theta_a \qquad \text{(A.22a)}$$

$$\varepsilon_b = \varepsilon_{xx}\cos^2\theta_b + \varepsilon_{yy}\sin^2\theta_b + \gamma_{xy}\sin\theta_b\cos\theta_b \qquad \text{(A.22b)}$$

$$\varepsilon_c = \varepsilon_{xx}\cos^2\theta_c + \varepsilon_{yy}\sin^2\theta_c + \gamma_{xy}\sin\theta_c\cos\theta_c \qquad \text{(A.22c)}$$

where ε_a, ε_b, and ε_c are the strain gage readings at angle θ_a, θ_{ab}, and θ_c, respectively, as shown in Figure A.18. The three equations can be solved for the three unknown strains ε_{xx}, ε_{yy}, and γ_{xy}.

A.11 THIN-WALLED PRESSURE VESSELS

The "thin wall" limitation implies that the ratio of the inner radius R to the wall thickness t is greater than 10.

Figure A.19a shows a thin-walled cylindrical pressure vessel with the nonzero stress components shown in polar coordinates. The stress σ_{xx} is called the *axial stress,* and $\sigma_{\theta\theta}$ is called the *hoop stress*. These stresses are related to the internal pressure as shown in Equations (A.23).

$$\sigma_{\theta\theta} = \frac{pR}{t} \qquad \text{and} \qquad \sigma_{xx} = \frac{pR}{2t} \qquad \text{(A.23)}$$

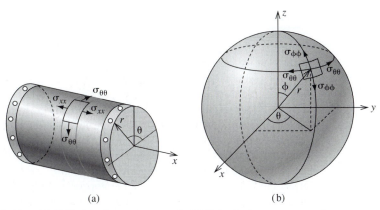

Figure A.19   (a) Cylindrical pressure vessel. (b) Spherical pressure vessel.

Figure A.19(b) shows a thin-walled spherical pressure vessel with the nonzero stress components drawn in spherical coordinates. The normal stresses are equal and are related to the internal pressure as shown in Equation (A.24).

$$\sigma_{\theta\theta} = \sigma_{\phi\phi} = \frac{pR}{2t} \qquad\qquad (A.24)$$

A.12   ADDITIONAL DETAILS ON ELEMENTARY STRUCTURAL THEORIES

This appendix describes details of the theoretical development of the formulas presented in Table 3.3. The equation and assumption numbers correspond to those in Table 3.3 [e.g., Equation (3.1-A), Assumption 1].

Our first assumption allows us to eliminate time as a variable and to develop the simplest possible theories.

Assumption 1 External forces and moments are not functions of time.[1]

If we relax Assumption 1, then time must be carried as an independent variable, and the derivatives with respect to x must be converted from ordinary to partial derivative status. The significant difference however is in the equilibrium equations.

For torsion and bending, we impose the following additional limitations.

Torsion: We limit ourself to circular cross sections.[2] This permits us to use the arguments of axisymmetry in deducing deformation.

Bending: In addition, there is a plane of symmetry, and the loading is in the plane of symmetry. This limitation ensures symmetric bending. We will assume that the x-y plane is the plane of symmetry and that bending is occurring about the z axis; that is, cross sections rotate about the z axis. In Section 6.1 we drop this limitation to develop the theory of unsymmetric bending of beams.

---

[1] See Problems 3.24 through 3.29 for dynamic problems.

[2] See Problems 3.32 and 3.33 on noncircular shafts.

A.12.1   Deformation

Figure A.20 shows grids as they deform in response to axial, torsional, and bending loads. The deformation of the grid is on the surface, but do the points in the interior of the cross section behave in a similar manner? An affirmative answer to the question is a reasonable approximation as long as the dimensions of the cross section are much smaller (factor of 10) than the length of the member, which is one of our limitations. We note that the following observations are true for all three cases. (1) The edges of the cross sections remain straight during deformation, suggesting that the plane sections remain plane. (2) The dimensional changes in the y direction are negligible (i.e., the normal strain in the y direction is negligible for *kinematic considerations*); however, the Poisson effect ensures that there will be some normal strain in the y direction, which we will be able to find once we know the normal strain in the x direction.

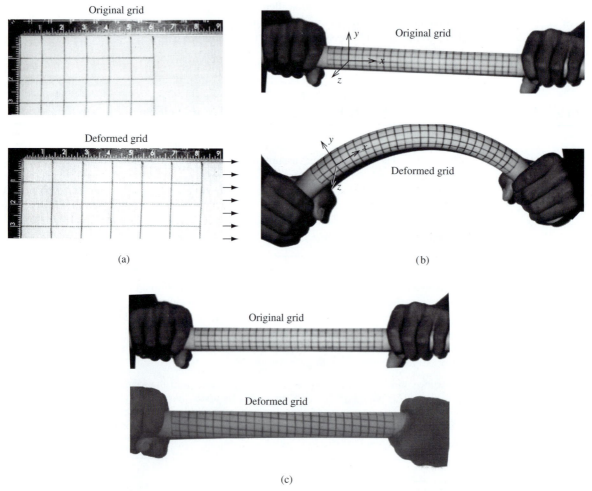

Figure A.20   Deformation of structural members in (a) axial, (b) bending, and (c) torsion modes. (*Courtesy Professor J.B. Ligon*)

Axial: Figure A.20(a) shows a grid on an elastic band that is pulled in the **axial** direction. The vertical lines remain approximately vertical, but the horizontal distance between the vertical lines changes. Thus, all points on a vertical displace by equal amounts. If this surface observation is true in the interior, then by implication we have the following assumption:

Assumption 2-A Plane sections remain plane and parallel.

The displacement in the x direction is measured by u. Assumption 2-A implies that u cannot be a function of y. This does not imply that all cross sections displace by equal amounts in the x direction. In fact, the strain would be zero if the displacement of cross sections in the x direction was constant at all values of x. The comments in this paragraph may be stated as follows:

$$u = u_0(x) \qquad\qquad (3.1\text{-}A)$$

Definition 1 The displacement u is considered to be positive in the positive x direction.

An alternative perspective is as follows. Because the cross section is significantly smaller than the length, we can approximate a function such as u by a constant (uniform) in the cross-sectional direction as shown in Figure A.21(a, b).

Bending: Figure A.20(b) shows a rubber beam with a grid on its surface that is being bent by hand. Notice that the lines in the y direction remain straight but rotate about the z axis. But the rotation is such that the original right angle between the x and y directions is *nearly* preserved during bending. The horizontal distance between the vertical lines does change, showing that there is normal strain in the x direction. Though there are some dimensional changes in the y direction, these appear to be significantly smaller than those along the x axis. The deformation observed in Figure A.20(b) is on the

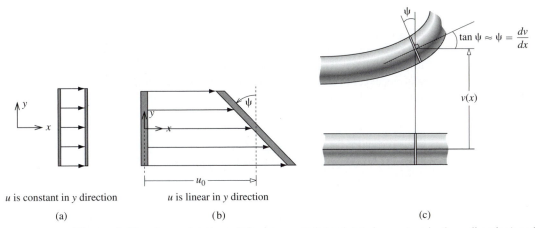

$\tan \psi \approx \psi = \dfrac{dv}{dx}$

$v(x)$

u_0

u is constant in y direction

u is linear in y direction

(a)　　　　　　　　(b)　　　　　　　　(c)

Figure A.21 Approximation of displacement: (a) axial (u is constant in the y direction) and (b) bending (u is linear in the y direction). (c) Elimination of a variable.

surface of the beam. To extrapolate the surface observations into the interior of the beam, we make the following assumptions:

Assumption 2a-B Squashing deformation is significantly smaller than deformation due to bending.

Assumption 2b-B Plane sections before deformation remain plane after deformation.

Assumption 2c-B A plane perpendicular to the beam axis remains *nearly* perpendicular after deformation.

Assumption 2a-B implies that the dimensional changes in the cross section are much smaller than the movement of the cross section as a whole. The longer the beam, the more valid Assumption 2a-B. Neglecting dimensional changes in the y direction implies that the normal strain in that direction can be assumed to be small,[3] hence can be neglected in kinematic calculations (i.e., $\varepsilon_{yy} = \partial v/\partial y \approx 0$). This assumption of small normal strain, in turn, implies that v cannot be a function of y as shown in Equation (3.1a-B).

$$v = v(x) \tag{3.1a-B}$$

Another way of viewing Assumption 2a-B is to think that the function v has been approximated as constant in the y direction. The implication is that if we know the *bending curve of one axial line on the beam, then we know the curves for all the lines on the beam.*

Assumption 2b-B *implies that the axial displacement u must vary linearly,* as shown in Figure A.21. In other words, the equation for u is $u = u_0 - \psi y$, where u_0 is the axial displacement at $y = 0$ and ψ is the slope of the plane. We can account for uniform axial displacement (u_0) separately as a problem in axial deformation. Thus, to study bending and axial problems independently,[4] we will assume $u_0 = 0$ in bending.

Assumption 2c-B *implies that the shear strain γ_{xy} is nearly zero,* for γ_{xy} is the measure of the change of angle between the axis of the beam and the perpendicular plane. We cannot use this assumption in building theoretical models of beam bending if shear is important,[5] as it is in sandwich beams.[6] But Assumption 2c-B helps simplify the theory by eliminating the variable ψ, which represents a measure of the rotation of a cross section by imposing the constraint that the angle between axial direction and the cross section is always 90°. This is accomplished by relating ψ to v as shown in Figure A.21(c). The displacement curve is defined by $v(x)$. The angle of the tangent to the curve $v(x)$ is equal to the rotation of the cross section if Assumption 2c-B is valid. For small strains, the tangent of

---

[3] It is accounted for as the Poisson effect, but the normal strain in the y direction is not an independent variable and hence is considered to be negligible in the kinematics.

[4] See Problem 3.22 in which u_0 is carried along with v.

[5] Such beams are called Timoshenko beams. See Example 3.7 for the procedure for accounting for shear.

[6] A sandwich beam cross section consists of two stiff plates bonded to a soft core. Sandwich beams are common in the design of lightweight structures such as are found in aircraft and boats.

an angle can be replaced by its argument (i.e., $\tan \psi \approx \psi = dv/dx$), which we can substitute into $u = u_0 - \psi y$, noting that $u_0 = 0$ to obtain

$$u = -y\frac{dv}{dx} \tag{3.1b-B}$$

Torsion: Figure A.20(c) shows a circular rubber shaft with a grid on surface that is being twisted by hand. The vertical lines, which are the edges of the circle, remain vertical during deformation. This *suggests* that plane sections remain plane during deformation. In other words, there is *no warping,* which is true only for circular cross sections. In Figure A.20(c), if we consider the various rectangles between two consecutive circles, we notice that the change of shape does not depend upon the angular position of the rectangle on the circle. This *suggests* that the radial lines from the center to the corners of the rectangle on the same circle must be rotating by the same amount, provided the lines are straight in the interior. Thus, for the interior, we make the following assumptions.

Assumption 2a-T: Plane sections perpendicular to the axis remain plane during deformation.

Assumption 2b-T: All radial lines rotate by the same angle during deformation on a cross section.

Assumption 2c-T: Radial lines remain straight during deformation.

These assumptions are analogous to thinking that each cross section in the shaft may be viewed as a rigid disk that rotates about its own axis, but the amount of rotation of each cross section can change along the length of the shaft, that is,

$$\phi = \phi(x) \tag{3.1-T}$$

Definition 2 The rotation ϕ is positive counterclockwise with respect to the x axis.

A.12.2 Strains

Assumption 3 The strains are small.[7]

Axial: Substituting Equation (3.1-A) into Equation 1.35a (i.e., $\varepsilon_{xx} = \partial u/\partial x$), we obtain

$$\varepsilon_{xx} = \frac{du_0}{dx}(x) \tag{3.2-A}$$

Equation (3.2-A) emphasizes that the *axial strain is uniform across the cross section* and is a function of x only. In deriving Equation (3.2-A), we made no statement regarding material behavior. In other words, *Equation (3.2-A) does not depend upon the material model* if Assumptions 2-A and 3 are valid. But clearly, if the material or the loading is such that Assumptions 2-A and 3 are not tenable, Equation (3.2-A) is not valid.

---

[7] See Problem 5.65 for large strains.

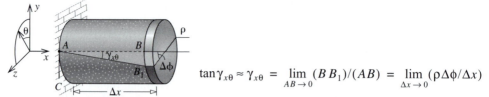

Figure A.22   Shear strain in torsion.

Bending: Substituting Equation (3.1b-B) into Equation 1.35a (i.e., $\varepsilon_{xx} = \partial u/\partial x$) we obtain

$$\varepsilon_{xx} = -y\frac{d^2v}{dx^2} \qquad (3.2\text{-B})$$

where the bending normal strain ε_{xx} varies linearly with y and has its maximum value at either the top or the bottom of the beam; d^2v/dx^2 is the curvature of the beam, and its magnitude is equal to the inverse of the radius of curvature.

Torsion: The shear strain that is of interest to us is the measure of the angle change between the axial direction and the tangent to the circle. If we use polar coordinates, then the change in angle we are interested in is between the x and θ directions: that is, $\gamma_{x\theta}$ as shown in Figure A.22. We consider a shaft with radius ρ and length Δx in which the right-hand section is rotated by an angle of $\Delta\phi$ with respect to the left, as shown in Figure A.22.

By letting Δx tend to zero, we obtain

$$\gamma_{x\theta} = \rho\frac{d\phi}{dx}(x) \qquad (3.2\text{-T})$$

where ρ is the radial coordinate of a point on the cross section. The quantity $d\phi/dx$ is called the *rate of twist,* and it is a function of x only because ϕ is a function of x only. Equation (3.2-T) shows that the *shear strain is a linear function of the radial coordinate* ρ and reaches a *maximum value at the outer surface* of the shaft.

A.12.3   Stress

In the introductory course on the mechanics of materials, our motivation was to develop a simple basic theory for structural members. Thus, we made assumptions regarding material behavior that permitted us to use the simplest material model, given by Hooke's law.

Assumption 4 The material is isotropic.

Assumption 5 There are no inelastic strains.

Assumption 6 The material is elastic.

Assumption 7 Stress and strains are linearly related.

With these assumptions, made, we can use Hooke's law, substituting Equations (3.2-A) and (3.2-B) into $\sigma_{xx} = E\varepsilon_{xx}$ and Equation (3.2-T) into $\tau_{x\theta} = G\gamma_{x\theta}$ to obtain the following stresses.

Axial   Bending   Torsion

$$\sigma_{xx} = E\frac{du_0}{dx}(x) \quad (3.3\text{-A}) \quad \sigma_{xx} = -Ey\frac{d^2v}{dx^2}(x) \quad (3.3\text{-B}) \quad \tau_{x\theta} = G\rho\frac{d\phi}{dx}(x) \quad (3.3\text{-T})$$

Inelastic strains may be due to temperature, humidity, plasticity, viscoelasticity, and so on. In Chapter 5, Inelastic Structural Behavior, we will drop Assumptions 5 through 7 and get different expressions for stresses, then redevelop the formulas in a manner similar to that shown in the sections that follow.

A.12.4 Internal Forces and Moments

Stresses, being an internal distributed force system, can be replaced by statically equivalent internal forces and moments as shown in Figure A.23 and Equations (3.4). These internal forces and moments are also referred to as *stress resultants*.

Axial   Bending   Torsion

$$N = \int_A \sigma_{xx}dA \qquad (3.4a\text{-A}) \quad N = \int_A \sigma_{xx}dA = 0 \qquad (3.4a\text{-B}) \quad T = \int_A \rho\tau_{x\theta}dA \qquad (3.4\text{-T})$$

$$M_z = -\int_A y\sigma_{xx}dA = 0 \qquad (3.4b\text{-A}) \quad M_z = -\int_A y\sigma_{xx}dA \qquad (3.4b\text{-B})$$

$$M_y = -\int_A z\sigma_{xx}dA = 0 \qquad (3.4c\text{-A}) \quad V_y = \int_A \tau_{xy}dA \qquad (3.4c\text{-B})$$

Axial: Equations (3.4b-A) and (3.4c-A) are requirements that the applied axial forces be such that there is no bending, hence no bending moments M_y and M_z. Another perspective is that the requirement, for zero bending moment decouples the axial and bending problems. We will use this condition to determine the origin of the coordinate system, at which time we will require that *the applied axial forces pass through the origin*.

Axial   Bending   Torsion

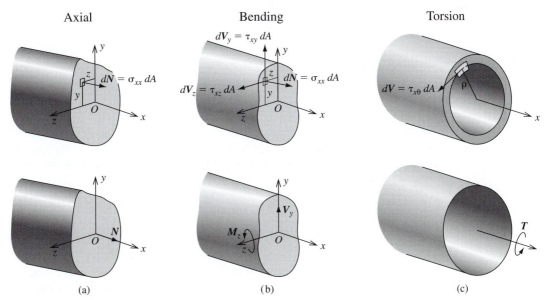

(a)   (b)   (c)

Figure A.23 Statically equivalent internal forces and moments.

Bending: M_y is implicitly accounted for in the requirements of symmetric bending. That is, σ_{xx} is symmetric about the y axis, and hence the equivalent internal moment about y axis is zero. Equation (3.4a-B) implies that the stress distribution across the cross section must be such that the compressive axial force equals the total tensile axial force on a cross section in bending. If stress is to change from compression to tension, this implies that there must be a line of zero normal stress in bending, which is called the neutral axis. *Determining the location of the neutral axis is critical in all bending problems.*

Definition 3 The line at which the bending normal stress is zero is called the *neutral axis*.

A.12.5 Location of the Origin

In bending, the origin of the y coordinate is located at the neutral axis irrespective of the material model. If Assumptions 5 through 7 are valid, then further simplification can be done in obtaining the location of the origin. Substituting Equations (3.3-A) and (3.3-B) into Equations (3.4b-A) and (3.4a-B), we note that the integration is with respect to y and z ($dA = dy\,dz$) while the quantities $(du_0/dx)(x)$ and $(d^2v/dx^2)(x)$ are functions of x only. Hence can be taken outside the integral signs as follows:

Axial	Bending

$$\int_A yE\frac{du_0}{dx}dA = \frac{du_0}{dx}\int_A yE\ dA = 0 \qquad\qquad -\int_A yE\frac{d^2v}{dx^2}dA = -\left(\frac{d^2v}{dx^2}\right)\int_A yE\ dA = 0$$

$$\int_A yE\ dA = 0 \qquad (3.5\text{-A}) \qquad \int_A yE\ dA = 0 \qquad (3.5\text{-B})$$

Equations (3.5-A) and (3.5-B) are identical because this is the condition for decoupling the axial and bending problems. Thus, it does not matter which problem is used to derive the equation. The origin of the coordinate system is chosen to satisfy Equation (3.5-A). Further simplification can be made by taking E outside the integral, as discussed in the next section.

A.12.6 Stress and Deformation Formulas

Substituting Equations (3.3-A), (3.3-B), and (3.3-T) into Equations (3.4a-A), (3.4b-B), and (3.4-T), we note that the integration is with respect to y and z ($dA = dy\,dz$), while the quantities (du_0/dx), (d^2v/dx^2), and $(d\phi/dx)$ are functions of x only, hence can be taken outside the integral signs as follows:

Axial		Bending		Torsion	

$$N = \int_A E\frac{du_0}{dx}dA \quad \text{or} \qquad M_z = \int_A Ey^2\frac{d^2v}{dx^2}dA \quad \text{or} \qquad T = \int_A G\rho^2\frac{d\phi}{dx}dA \quad \text{or}$$

$$N = \frac{du_0}{dx}\int_A E\ dA \quad (3.6\text{-A}) \qquad M_z = \frac{d^2v}{dx^2}\int_A Ey^2\ dA \quad (3.6\text{-B}) \qquad T = \frac{d\phi}{dx}\int_A G\rho^2\ dA \quad (3.6\text{-T})$$

To achieve further simplification, we would like to take E and G outside the integral; that is, E and G should not vary across the cross section.

Assumption 8 The material is homogeneous across the cross section.

We drop Assumption 8 in Chapter 4, Composite Structural Members, and derive new formulas following a similar sequence, as described next.

Based on Assumption 8, E can be taken outside the integral in Equation (3.5-A), resulting in the following condition:

$$\int_A y \, dA = 0 \qquad (3.7\text{-A})$$

Equation (3.7-A), implies that y should be measured from the centroid of the cross section. Thus, *for homogeneous, linear, elastic materials, the origin of the coordinate system is the centroid of the cross section.* All axial and bending loads must pass through the centroid if the axial problem is to be decoupled from the bending problem.[8]

With Assumption 8, E and G can be taken outside the integral. The remaining integrals represent the cross-sectional area $A = \int_A dA$, the second area moment of inertia $I_{zz} = \int_A y^2 \, dA$, and the polar area moment of inertia $J = \int_A \rho^2 \, dA$. We obtain the following formulas:

Axial	Bending	Torsion
$$\frac{du_0}{dx} = \frac{N}{EA} \qquad (3.8\text{-A})$$	$$\frac{d^2 v}{dx^2} = \frac{M_z}{EI_{zz}} \qquad (3.8\text{-B})$$	$$\frac{d\phi}{dx} = \frac{T}{GJ} \qquad (3.8\text{-T})$$
Axial rigidity $= EA$	Bending rigidity $= EI_{zz}$	Torsional rigidity $= GJ$

Substituting Equations (3.8-A), (3.8-B), and (3.8-T) into Equations (3.3-A), (3.3-B), and (3.3-T), we obtain the following stress formulas.

Axial	Bending	Torsion
$$\sigma_{xx} = \frac{N}{A} \qquad (3.9\text{-A})$$	$$\sigma_{xx} = -\left(\frac{M_z y}{I_{zz}}\right) \qquad (3.9\text{-B})$$	$$\tau_{x\theta} = \frac{T\rho}{J} \qquad (3.9\text{-T})$$

These stress formulas show that for a linear, elastic, homogeneous cross section,

- axial normal stress is uniform across the cross section,
- bending normal stress varies linearly across the cross section, being is zero at the centroid and maximum at point(s) furthest from the centroid,
- torsional shear stress varies linearly with the radial distance from the center and is maximum at the outermost circle on the material.

---

[8] Also see Problem 3.22 for derivation of this condition from another perspective.

Equations (3.8-A), (3.8-B), and (3.8-T) can be integrated to obtain deformation formulas. Integration in bending is discussed in detail in Section 3.4.3. Let u_1, u_2 and ϕ_1, ϕ_2 be the axial displacements and angles of rotation of the cross section, respectively, at points x_1 and x_2 on the structural member. To obtain simple formulas for axial and torsion deformation, we make the following assumptions:

Assumption 9 The material is homogeneous between x_1 and x_2.

Assumption 10 The shaft is not tapered.

Assumption 11 The external loads do not change with x between x_1 and x_2.

Assumption 9 implies that E and G are constants between x_1 and x_2.

Assumption 10, a shorter form of the assumption given in Table 3.3, implies that the cross-sectional dimensions are not changing with x: hence A and J do not change between x_1 and x_2. In Problems 3.7 through 3.9, A and J must be found as functions of x and Equations (3.8-A) and (3.8-T) integrated to obtain the deformation.

Assumption 11 implies that internal axial force and the internal torque are not changing with x. In Problems 3.34 and 3.36 Assumption 11 is not valid, and the internal force or moment as a function of x must be found as described in Section A.12.8, then Equations (3.8-A) and (3.8-T) can be integrated to find the deformation.

If Assumptions 9 through 11 are valid, then the following formulas can be used.

Axial   Bending   Torsional

$$u_2 - u_1 = \frac{N(x_2 - x_1)}{EA} \quad \text{(3.10-A)}$$

See Section 3.4.3 for deformation calculation in bending.

$$\phi_2 - \phi_1 = \frac{T(x_2 - x_1)}{GJ} \quad \text{(3.10-T)}$$

A.12.7   Sign Conventions

The internal forces and moments are calculated by making an imaginary cut through a structural member, drawing the free body diagram of one of the two parts produced from the cut, and writing the equilibrium equations. But in what direction should we draw the internal forces and moments on the imaginary cut?

One possibility is to draw the internal forces and moments in a direction that will serve to equilibrate the other forces. Thus, we will always obtain positive values for the internal forces and moments. The stress and deformation formulas can then be used only for determining the magnitude. The sign (direction) of the stresses and deformation then must be determined by inspection.

The second possibility is to draw the internal forces and moments at an imaginary cut according to the sign convention described in Figure A.24. The equilibrium equations would then give a positive or negative value for any internal quantity, and the formula would give the correct sign for the stresses and deformation.

Figure A.24 represents the stress distributions and statically equivalent internal forces and moments in a given coordinate system. From these sketches, we conclude that on an imaginary cut surface in a free body diagram,

- positive internal axial force N is shown as tensile,

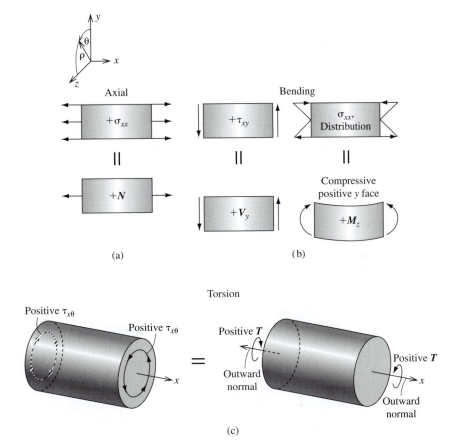

Figure A.24 Sign conventions for internal forces and moments.

- positive internal shear force V_y is shown in the same direction as positive τ_{xy},
- positive internal bending moment M_z is shown to put the positive y face in compression,
- positive internal torque T is shown counterclockwise with respect to the outward normal.

A.12.8 Equilibrium Equations

Consider an infinitesimal structural element that is created by making two imaginary cuts, separated by a distance dx, as shown in Figure A.25 and Equations (3.11), where $p_x(x)$ and $p_y(x)$ are the distributed forces in the x and y directions, respectively, having units of force per unit length, and $t(x)$ is the distributed torque, having units of moment per unit length.

Following are the sign conventions for the distributed loads:

- $p_x(x)$ is positive in the positive x direction.
- $p_y(x)$ is positive in the positive y direction.
- $t(x)$ is positive counterclockwise with respect to the x axis.

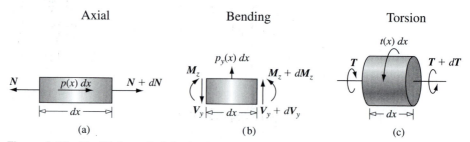

Axial   Bending   Torsion

Figure A.25 Equilibrium of infinitesimal structural elements: (a) axial, (b) bending, and (c) torsion.

Axial

$$\frac{dN}{dx} = -p_x(x) \qquad (3.11\text{-A})$$

Bending

$$\frac{dV_y}{dx} = -p_y(x) \qquad (3.11\text{a-B})$$

$$\frac{dM_z}{dx} = -V_y \qquad (3.11\text{b-B})$$

Torsional

$$\frac{dT}{dx} = -t(x) \qquad (3.11\text{-T})$$

Equations (3.11-A), (3.11a-B), (3.11b-B), and (3.11-T) can be integrated to obtain the internal forces and moments as functions of x. To obtain the integration constant, the value of the internal force or moment must be known at some point in the interval (length of the member, usually) of integration, generally the start or end of the interval.

A.12.9 Differential Equations

Substituting Equations (3.8-A), (3.8-B), and (3.8-T) into Equations (3.11-A), (3.11a-B), (3.11b-B), and (3.11-T), we obtain the following differential equations.

Axial

$$\frac{d}{dx}\left(EA\frac{du_0}{dx}\right) = -p_x(x) \quad (3.12\text{-A})$$

Bending

$$\frac{d^2}{dx^2}\left(EI_{zz}\frac{d^2v}{dx^2}\right) = p_y(x) \quad (3.12\text{-B})$$

Torsional

$$\frac{d}{dx}\left(GJ\frac{d\phi}{dx}\right) = -t(x) \quad (3.12\text{-T})$$

Appendix

B

Basic Matrix Algebra

This appendix briefly reviews basic matrix algebra from the perspective of this book. The presentation presupposes familiarity with the concepts. You may need to review a mathematics text for additional details.[1]

B.1 BASIC DEFINITIONS

A rectangular array of numbers is called a matrix. The matrix shown in Equation (B.1) has m rows and n columns. The size of the matrix is said to be $(m \times n)$. The element in the ith row and jth column is represented by a_{ij}.

$$[A] = \begin{bmatrix} a_{11} & a_{12} & \bullet & \bullet & a_{1n} \\ a_{21} & a_{22} & \bullet & \bullet & a_{2n} \\ \bullet & \bullet & \bullet & \bullet & \bullet \\ \bullet & \bullet & \bullet & \bullet & \bullet \\ a_{m1} & a_{m2} & \bullet & \bullet & a_{mn} \end{bmatrix} \tag{B.1}$$

B.2 ADDITION OF MATRICES

Addition of matrices can be performed only if the matrices have the same number of rows and columns. The sum of two matrices $[A]$ and $[B]$ of m rows and n columns results in a matrix $[C]$ of m rows and n columns and is represented by Equation (B.2a).

$$[C] = [A] + [B] \tag{B.2a}$$

---

[1] See Kreyszig [1979] in Appendix D.

The elements of the matrix $[C]$ can be found by using Equation (B.2b).

$$c_{ij} = a_{ij} + b_{ij} \qquad \begin{cases} i = 1, 2, \ldots, m \\ j = 1, 2, \ldots, n \end{cases} \tag{B.2b}$$

B.3 MULTIPLICATION OF MATRICES

Multiplication of a matrix by a number results in a matrix in which all elements are multiplied by the number, as shown in Equation (B.3).

$$q[A] = q \begin{bmatrix} a_{11} & a_{12} & \bullet & \bullet & a_{1n} \\ a_{21} & a_{22} & \bullet & \bullet & a_{2n} \\ \bullet & \bullet & \bullet & \bullet & \bullet \\ \bullet & \bullet & \bullet & \bullet & \bullet \\ a_{m1} & a_{m2} & \bullet & \bullet & a_{mn} \end{bmatrix} = \begin{bmatrix} qa_{11} & qa_{12} & \bullet & \bullet & qa_{1n} \\ qa_{21} & qa_{22} & \bullet & \bullet & qa_{2n} \\ \bullet & \bullet & \bullet & \bullet & \bullet \\ \bullet & \bullet & \bullet & \bullet & \bullet \\ qa_{m1} & a_{m2} & \bullet & \bullet & qa_{mn} \end{bmatrix} \tag{B.3}$$

The order of multiplication is important when two matrices are multiplied. In Equation (B.4a), matrix $[A]$ is said to premultiply matrix $[B]$ and matrix $[B]$ is said to postmultiply matrix $[A]$.

$$[C] = [A][B] \tag{B.4a}$$

In Equation (B.4a) the number of columns in matrix $[A]$ must equal the number of rows in matrix $[B]$. If matrix $[A]$ of size $(m \times n)$ premultiplies matrix $[B]$ of size $(n \times p)$, the result is matrix $[C]$, of size $(m \times p)$. The elements of matrix $[C]$ can be found from

$$c_{ij} = \sum_{k=1}^{n} a_{ik} b_{kj} \qquad \begin{cases} i = 1, 2, \ldots, m \\ j = 1, 2, \ldots, p \end{cases} \tag{B.4b}$$

B.4 A MATRIX TRANSPOSE

The transpose of a rectangular matrix $[A]$, consisting of m rows and n columns, is written as $[A]^T$ and related to $[A]$ as shown in Equation (B.5).

$$[A] = \begin{bmatrix} a_{11} & a_{12} & \bullet & \bullet & a_{1n} \\ a_{21} & a_{22} & \bullet & \bullet & a_{2n} \\ \bullet & \bullet & \bullet & \bullet & \bullet \\ \bullet & \bullet & \bullet & \bullet & \bullet \\ a_{m1} & a_{m2} & \bullet & \bullet & a_{mn} \end{bmatrix} \qquad [A]^T = \begin{bmatrix} a_{11} & a_{21} & \bullet & \bullet & a_{m1} \\ a_{12} & a_{22} & \bullet & \bullet & a_{m2} \\ \bullet & \bullet & \bullet & \bullet & \bullet \\ \bullet & \bullet & \bullet & \bullet & \bullet \\ a_{1n} & a_{2n} & \bullet & \bullet & a_{mn} \end{bmatrix} \tag{B.5}$$

The element a_{ij} of matrix $[A]$ becomes element a_{ji} in the transposed matrix $[A]^T$.

A *square* matrix (same number of rows and columns) is said to be symmetric if its transpose is the same as the original matrix, as shown in Equation (B.6).

$$\text{Symmetric matrix:} \quad [A]^T = [A] \tag{B.6}$$

The following rules apply to transposes of matrices during addition and multiplications:

$$([A] + [B])^T = [A]^T + [B]^T \quad \text{and} \quad ([A][B])^T = [B]^T[A]^T \tag{B.7}$$

B.5 DETERMINANT OF A MATRIX

A determinant is defined only for a square matrix and is represented as shown in Equation (B.8).

$$|A| = \det[A] = \begin{vmatrix} a_{11} & a_{12} & \bullet & \bullet & a_{1n} \\ a_{21} & a_{22} & \bullet & \bullet & a_{2n} \\ \bullet & \bullet & \bullet & \bullet & \bullet \\ \bullet & \bullet & \bullet & \bullet & \bullet \\ a_{n1} & a_{n2} & \bullet & \bullet & a_{nn} \end{vmatrix} \tag{B.8}$$

The minor M_{ij} associated with an element a_{ij} is the determinant of the matrix in which the ith row and jth column have been removed. The determinant of a matrix can be found by using Equation (B.9), where i is any row in the matrix, or by using Equation (B.10), where j is any column in the matrix.

$$|A| = \sum_{k=1}^{n} (-1)^{i+k} a_{ik} M_{ik} \tag{B.9}$$

$$|A| = \sum_{k=1}^{n} (-1)^{k+j} a_{kj} M_{kj} \tag{B.10}$$

If the determinant of a matrix is zero (i.e., $|A| = 0$), then the matrix $[A]$ is said to be singular. In a singular matrix, either all rows are not independent or all columns are not independent.

B.6 CRAMER'S RULE

Cramer's rule can be used for solving a set of linear algebraic equations. Consider the set of n linear algebraic equations in matrix form, as shown in Equation (B.11).

$$\begin{bmatrix} a_{11} & a_{12} & \bullet & a_{1j} & \bullet & a_{1n} \\ a_{21} & a_{22} & \bullet & a_{2j} & \bullet & a_{2n} \\ \bullet & \bullet & \bullet & & \bullet & \bullet \\ \bullet & \bullet & \bullet & a_{jj} & \bullet & \bullet \\ \bullet & \bullet & \bullet & \bullet & \bullet & \bullet \\ a_{n1} & a_{n2} & \bullet & a_{nj} & \bullet & a_{nn} \end{bmatrix} \begin{Bmatrix} x_1 \\ x_2 \\ \\ x_j \\ \\ x_n \end{Bmatrix} = \begin{Bmatrix} r_1 \\ r_2 \\ \\ r_j \\ \\ r_n \end{Bmatrix} \tag{B.11}$$

By Cramer's rule, the jth unknown x_j can be found by first replacing the jth column by the right-hand-side vector, taking the determinant of the resulting matrix, and then dividing by the determinant of the matrix $[A]$ as shown in Equation (B.12).

$$x_j = \frac{\begin{bmatrix} a_{11} & a_{12} & \bullet & r_1 & \bullet & a_{1n} \\ a_{21} & a_{22} & \bullet & r_2 & \bullet & a_{2n} \\ \bullet & \bullet & \bullet & & \bullet & \bullet \\ \bullet & \bullet & \bullet & r_j & \bullet & \bullet \\ \bullet & \bullet & \bullet & \bullet & \bullet & \bullet \\ a_{n1} & a_{n2} & \bullet & r_n & \bullet & a_{nn} \end{bmatrix}}{|A|} \qquad j = 1, 2, \ldots, n \tag{B.12}$$

B.7 INVERSE OF A MATRIX

The inverse of a matrix can be found only for a square matrix. The inverse of a matrix $[A]$ is denoted by $[A]^{-1}$. The product of a matrix and its inverse results in an identity matrix $[I]$ as shown in Equation (B.13). The identity matrix $[I]$ has 1 for the diagonal elements; all off-diagonal elements are 0.

$$[A]^{-1}[A] = [A][A]^{-1} = [I] \tag{B.13}$$

Equation (B.11) in matrix form can be written as Equation (B.14a),

$$[A]\{x\} = \{r\} \tag{B.14a}$$

where $\{x\}$ represents the unknown vector with components x_j, and $\{r\}$ represents the right-hand-side vector with components r_j. By premultiplying by $[A]^{-1}$, on both sides of Equation (B.14a) and using Equation (B.13), we obtain the unknown vector as shown in Equation (B.14b).

$$[A]^{-1}[A]\{x\} = [A]^{-1}\{r\} \quad \text{or} \quad [I]\{x\} = [A]^{-1}\{r\} \quad \text{or} \quad \{x\} = [A]^{-1}\{r\} \tag{B.14b}$$

Appendix

C

Information Charts and Tables

C.1 STRESS CONCENTRATION FACTOR FOR A FINITE PLATE WITH A CENTRAL HOLE[1]

Figure C.1 shows two stress concentration factors that differ owing to the cross-sectional area used in calculating the nominal stress. If the gross cross-sectional area $[Ht]$ of the plate is used, then we obtain the nominal stress $(\sigma_{nom})_{gross}$ as shown in Equation (C.1), and the upper curve of Figure C.1 should be used for the stress concentration factor. If the net area at the hole $[(H - d)t]$ is used, we obtain the nominal stress $(\sigma_{nom})_{net}$ as shown in Equation (C.1), and the lower curve of Figure C.1 should be used for the stress concentration factor. The two stress concentration factors are related as follows:

$$(\sigma_{nom})_{gross} = \frac{P}{Ht} \qquad (\sigma_{nom})_{net} = \frac{P}{(H - d)t} \qquad K_{net} = \left(1 - \frac{d}{H}\right)K_{gross} \qquad (C.1)$$

C.2 STRESS CONCENTRATION FACTOR FOR STEPPED AXIAL CIRCULAR BARS WITH FILLET[2]

The maximum axial stress in a stepped circular bar with a shoulder fillet will depend upon the values of the diameters (D and d) of the two circular bars and the radius of the fillet r. From these three variables, we can create two nondimensional variables (D/d and r/d) for showing the variation of stress concentration factor as in Figure C.2. The

---

[1] The stress concentration factor charts given here are approximate. For more accurate values, consult a handbook.

[2] The stress concentration factor charts given here are approximate. For more accurate values, consult a handbook.

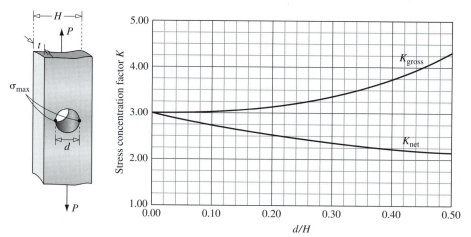

Figure C.1

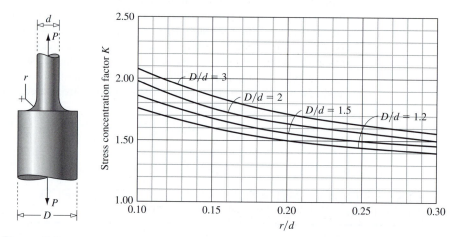

Figure C.2

maximum nominal axial stress will be in the smaller-diameter bar and is given by Equation (C.2).

$$\sigma_{nom} = \frac{4P}{\pi d^2} \qquad (C.2)$$

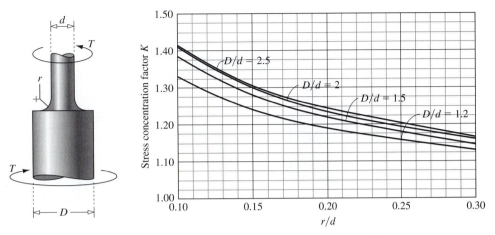

Figure C.3

C.3 STRESS CONCENTRATION FACTOR FOR STEPPED CIRCULAR SHAFTS WITH FILLET[3]

The maximum shear stress in a stepped circular shaft with a shoulder fillet will depend upon the values of the diameters (D and d) of the two circular shafts and the radius of the fillet r. From these three variables, we can create two nondimensional variables (D/d and r/d) for showing the variation of the stress concentration factor as in Figure C.3. The maximum nominal shear stress will be on the outer surface of the smaller-diameter bar and is given by Equation (C.3).

$$\tau_{nom} = \frac{16T}{\pi d^3} \qquad (C.3)$$

C.4 STRESS CONCENTRATION FACTOR FOR A STEPPED CIRCULAR BEAM WITH FILLET

The maximum bending normal stress in a stepped circular beam with a shoulder fillet will depend upon the values of the diameters (D and d) of the two circular shafts and the radius of the fillet r. From these three variables, we can create two nondimensional variables (D/d and r/d) for showing the variation of stress concentration factor as in Figure C.4. The maximum nominal bending normal stress will be on the outer surface in the smaller-diameter bar and is given by Equation (C.4).

$$\sigma_{nom} = \frac{32M}{\pi d^3} \qquad (C.4)$$

---

[3] The stress concentration factor charts given here are approximate. For more accurate values, consult a handbook.

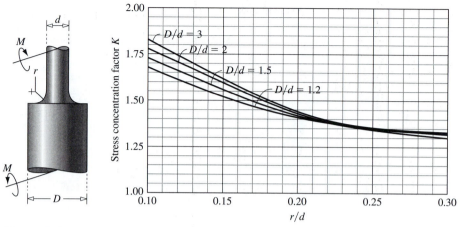

Figure C.4

C.5 REACTIONS AT THE SUPPORT

TABLE C.1

Type of Support	Reactions	Comments
Roller on smooth surface	R	Only the downward translation is prevented; hence the reaction force is upward as shown.
Smooth pin	R_x R_y	Translation in the horizontal and vertical directions is prevented; hence the reaction forces R_x and R_y can be in the direction shown *or opposite* to it.
Fixed support	R_x M_z R_y	Besides translation in the horizontal and vertical directions, rotation about the z axis is prevented. Hence the reactions R_x and R_y and $\boldsymbol{M}_z$ can be in the direction shown *or opposite* to it.
Roller in smooth slot	R	Translation perpendicular to the slot is prevented. The reaction force R can be in the direction shown *or opposite* to it.
Sliding support	R_x M_z	Translation perpendicular to the wall and rotation about the z axis are prevented. Hence the reactions R_x and $\boldsymbol{M}_z$ can be in the direction shown *or opposite* to it.
Ball and socket	R_x R_z R_y	Translation in all directions is prevented. The reaction forces can be in the direction shown *or opposite* to it.

TABLE C.1 (*Continued*)

Type of Support	Reactions	Comments
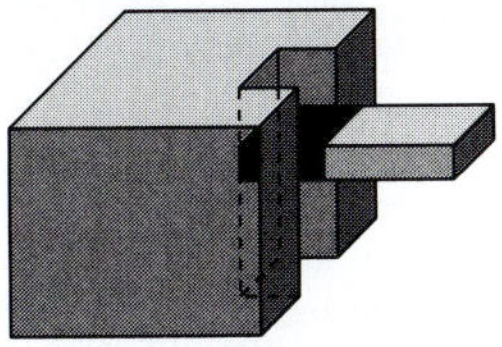 Smooth slot	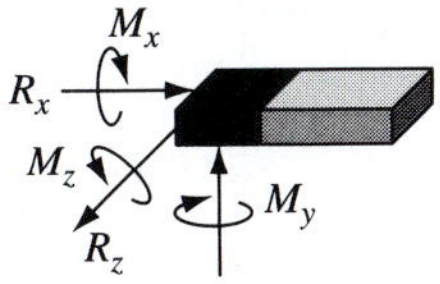	Translation in the z direction and rotation about any axis are prevented. Hence, the reaction force R_z and the reaction moments can be in the direction shown *or opposite* to it. Translation in the x direction into the slot is prevented but not translation out of it. Hence the reaction force R_x should be in the direction shown.

C.6 AREAS, CENTROIDS, AND SECOND AREA MOMENTS OF INERTIA

TABLE C.2

Shape	Location C of Centroid	Areas	Second Area Moments of Inertia
Rectangle	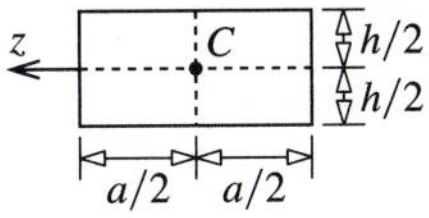	$A = ah$	$I_{zz} = \dfrac{1}{12}ah^3$
Circle	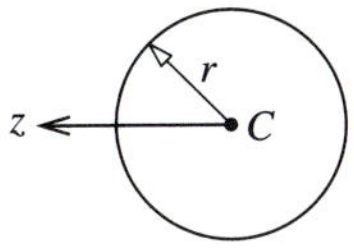	$A = \pi r^2$	$I_{zz} = \dfrac{1}{4}\pi r^4 \qquad J = \dfrac{1}{2}\pi r^4$
Ellipse	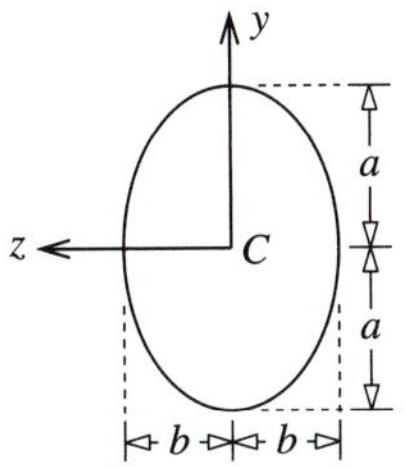	$A = \pi ab$	$I_{zz} = \dfrac{1}{4}\pi ba^3 \qquad I_{yy} = \dfrac{1}{4}\pi ab^3$ $J = \dfrac{1}{4}\pi ab(a^2 + b^2)$
Triangle	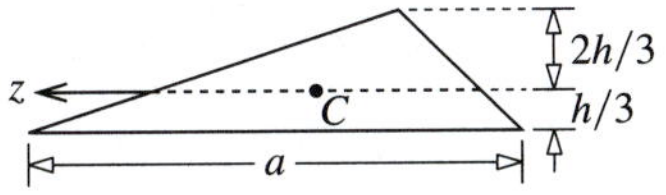	$A = ah/2$	$I_{zz} = \dfrac{1}{36}ah^3$
Semicircle	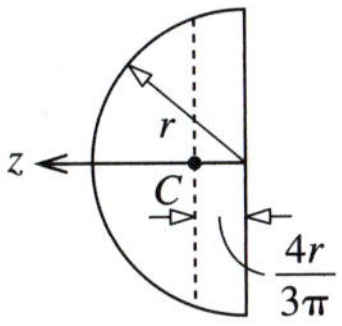	$A = (\pi r^2)/2$	$I_{zz} = \dfrac{1}{8}\pi r^4$
Trapezoid	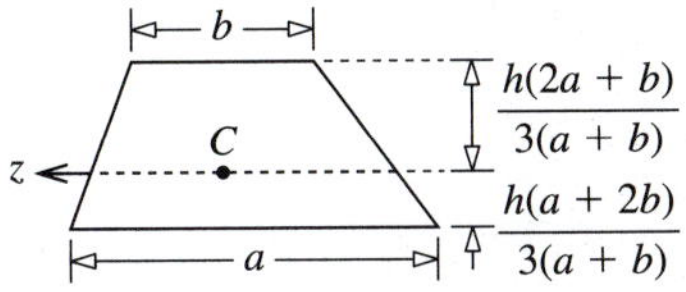	$A = h(a + b)/2$	

(*Continued*)

TABLE C.2 *(Continued)*

Shape	Location C of Centroid	Areas	Second Area Moments of Inertia
Quadratic curve	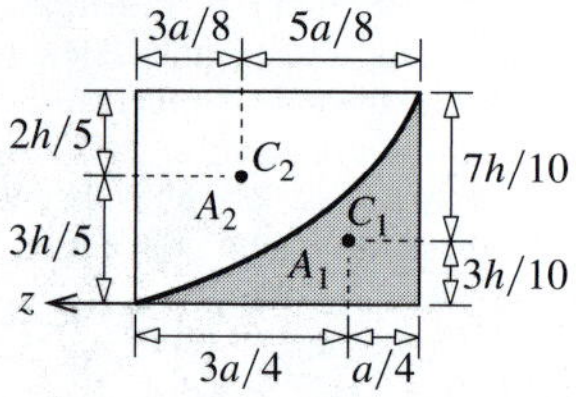	$A_1 = ah/3$ $A_2 = 2ah/3$	$(I_{zz})_1 = \dfrac{1}{21}ah^3 \qquad (I_{zz})_2 = \dfrac{2}{7}ah^3$
Cubic curve	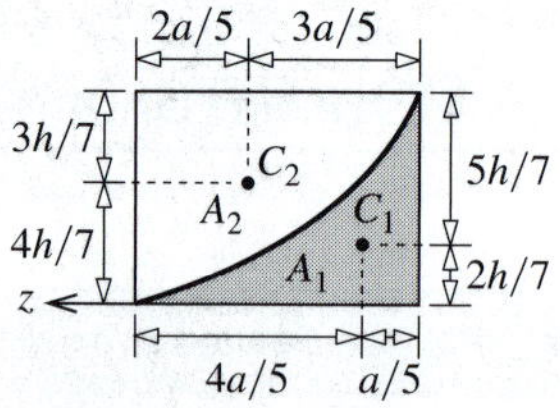	$A_1 = ah/4$ $A_2 = 3ah/4$	$(I_{zz})_1 = \dfrac{1}{30}ah^3 \qquad (I_{zz})_2 = \dfrac{3}{10}ah^3$

C.7   DEFLECTIONS AND SLOPES OF BEAMS

TABLE C.3

Case	Beam and Loading	Maximum Deflection and Slope[a]	Elastic Curve[a]
1.	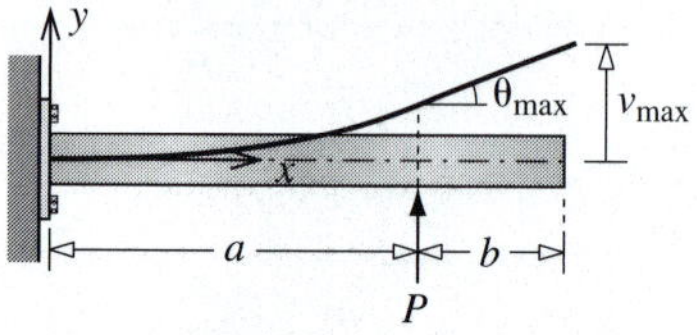	$v_{max} = \dfrac{Pa^2}{6EI}(2a + 3b)$ $\theta_{max} = \dfrac{Pa^2}{2EI}$	$v = \dfrac{Px^2}{6EI}(3a - x) \quad$ for $0 \leq x \leq a$ $v = \dfrac{Pa^2}{6EI}(3x - a) \quad$ for $x \geq a$
2.	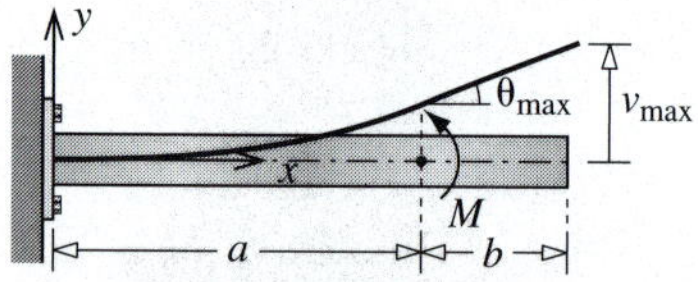	$v_{max} = \dfrac{Ma(a + 2b)}{2EI}$ $\theta_{max} = \dfrac{Ma}{EI}$	$v = \dfrac{Mx^2}{2EI} \quad$ for $0 \leq x \leq a$ $v = \dfrac{Ma}{2EI}(2x - a) \quad$ for $x \geq a$
3.	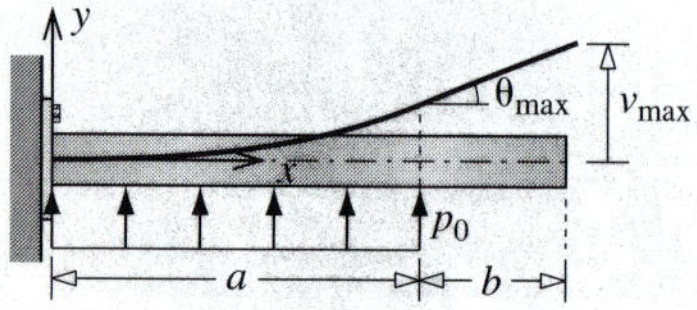	$v_{max} = \dfrac{p_o a^3(3a + 4b)}{24EI}$ $\theta_{max} = \dfrac{p_o a^3}{6EI}$	$v = \dfrac{p_o x^2}{24EI}(x^2 - 4ax + 6a^2)$ for $\quad 0 \leq x \leq a$ $v = \dfrac{p_o x^3}{24EI}(4x - a) \quad$ for $x \geq a$

TABLE C.3   *(Continued)*

Case	Beam and Loading	Maximum Deflection and Slope[a]	Elastic Curve[a]
4.		$v_{max} = \dfrac{PL^3}{48EI}$ $\theta_{max} = \dfrac{PL^2}{16EI}$	$v = \dfrac{Px}{48EI}(3L^2 - 4x^2)$ for $\quad 0 \le x \le L/2$
5.		$v_{max} = \dfrac{5p_0 L^4}{384EI}$ $\theta_{max} = \dfrac{p_0 L^3}{24EI}$	$v = \dfrac{p_0 x}{24EI}(x^3 - 2Lx^2 + L^3)$
6.		$v_{max} = \dfrac{ML^2}{9\sqrt{3}EI}$ @ $x = 0.4226L$ $\theta_1 = \dfrac{ML}{3EI} \quad \theta_2 = \left(\dfrac{ML}{6EI}\right)$	$v = \dfrac{Mx}{6EIL}(x^2 - 3Lx + 2L^2)$

[a] These equations can be used for composite beams by replacing the bending rigidity EI by the sum of the bending rigidities $\Sigma E_i I_i$.

C.8   PROPERTIES OF TRADITIONAL MATERIALS (FPS UNITS)

TABLE C.4

Material	Specific Weight[a] (lb/in³)	*E* Modulus of Elasticity[a] (ksi)	ν Poisson's Ratio[a] (ksi)	α Coefficient of Thermal Expansion[a] μ/°F	Elastic Strength[a] (ksi)	Ultimate Strength[a] (ksi)	Ductility (% elongation)[a]
Aluminum	0.100	10,000	0.25	12.5	40	45	17
Bronze	0.320	15,000	0.34	9.4	20	50	20
Concrete	0.087	4,000	0.15	6.0		2[b]	
Copper	0.316	15,000	0.35	9.8	12	35	35
Cast iron	0.266	25,000	0.25	6.0	25[b]	50[b]	
Glass	0.095	7,500	0.20	4.5		10	
Plastic	0.035	400	0.4	50		9	50
Rock	0.098	8,000	0.25	4	12[b]	78[b]	
Rubber	0.041	0.3	0.5	90	0.5	2	300
Steel	0.284	30,000	0.28	6.6	30	90	30
Titanium	0.162	14,000	0.33	5.3	135	155	13
Wood	0.02	1,800	0.30			5[b]	

[a] Material properties depend upon many variables and vary widely. The properties are approximate mean values. Elastic strength may be represented by yield stress, proportional limit, or offset yield stress. Both elastic strength and ultimate strength refer to tensile strength unless stated otherwise.

[b] Compressive strength.

C.9   PROPERTIES OF TRADITIONAL MATERIALS (METRIC UNITS)

TABLE C.5

Material	Density[a] (Mg./m^3)	E Modulus of Elasticity[a] (GPa)	ν Poisson's Ratio[a]	α Coefficient of Thermal Expansion[a] (μ/°C)	Elastic Strength[a] (MPa)	Ultimate Strength[a] (MPa)	Ductility (% elongation)[a]
Aluminum	2.77	70	0.25	12.5	280	315	17
Bronze	8.86	105	0.34	9.4	140	350	20
Concrete	2.41	28	0.15	6.0		14[b]	
Copper	8.75	105	0.35	9.8	84	245	35
Cast iron	7.37	175	0.25	6.0	175[b]	350[b]	
Glass	2.63	52.5	0.20	4.5		70	0
Plastic	0.97	2.8	0.4	50		63	50
Rock	2.72	56	0.25	4	84[b]	546[b]	
Rubber	1.14	2.1	0.5	90	3.5	14	300
Steel	7.87	210	0.28	6.6	210	630	30
Titanium	4.49	98	0.33	5.3	945	1185	13
Wood	0.55	12.6	0.30			35[b]	

[a] Material properties depend upon many variables and vary widely. The properties are approximate mean values. Elastic strength may be represented by yield stress, proportional limit, or offset yield stress. Both elastic strength and ultimate strength refer to tensile strength unless stated otherwise.
[b] Compressive strength.

C.10   PROPERTIES OF TYPICAL FIBER AND MATRIX MATERIALS

TABLE C.6

Material		Density[a]		Modulus of Elasticity[a]		Poisson's Ratio[a]	Tensile Strength[a]	
		lb/in^3	kg/m^3	GPa	ksi		MPa	ksi
Fiber	E-glass	0.092	2540	72	10,500	0.22	3450	500
	S-glass	0.090	2490	86	12,500	0.22	4800	700
	Carbon	0.064	1770	190	27,000	0.25	1700	250
	Kevlar	0.052	1440	62	18,850	0.35	2800	406
Matrix	Polyester	0.045	1250	3.0	435	0.38	65	9.5
	Epoxy	0.045	1250	4.0	580	0.34	63	9.1
	Polyimide	0.058	1600	4.0	580	0.35	95	13.8

[a] Material properties depend upon many variables and vary widely. The properties given are approximate mean values.

C.11   GEOMETRIC PROPERTIES OF WIDE-FLANGE SECTIONS (FPS UNITS)

TABLE C.7

Designation (in × lb/ft)	Depth d (in)	Area A (in²)	Web Thickness t_W (in)	Flange Width b_F (in)	Flange Thickness t_F (in)	I_{zz} (in⁴)	S_z (in³)	r_z (in)	I_{yy} (in⁴)	S_y (in³)	r_y (in)
W12 × 35	12.50	10.3	0.300	6.560	0.520	285.0	45.6	5.25	24.5	7.47	1.54
W12 × 30	12.34	8.79	0.260	6.520	0.440	238	38.6	5.21	20.3	6.24	1.52
W10 × 30	10.47	8.84	0.300	5.81	0.510	170	32.4	4.38	16.7	5.75	1.37
W10 × 22	10.17	6.49	0.240	5.75	0.360	118	23.2	4.27	11.4	3.97	1.33
W8 × 18	8.14	5.26	0.230	5.250	0.330	61.9	15.2	3.43	7.97	3.04	1.23
W8 × 15	8.11	4.44	0.245	4.015	0.315	48	11.8	3.29	3.41	1.70	0.876
W6 × 20	6.20	5.87	0.260	6.020	0.365	41.4	13.4	2.66	13.3	4.41	1.50
W6 × 16	6.28	4.74	0.260	4.03	0.405	32.1	10.2	2.60	4.43	2.20	0.967

C.12   GEOMETRIC PROPERTIES OF WIDE-FLANGE SECTIONS (METRIC UNITS)

TABLE C.8

Designation (mm × kg/m)	Depth d (mm)	Area A (mm²)	Web Thickness t_W (mm)	Flange Width b_F (mm)	Flange Thickness t_F (mm)	I_{zz} (10⁶mm⁴)	S_z (10³mm³)	r_z (mm)	I_{yy} (10⁶mm⁴)	S_y (10³mm³)	r_y (mm)
W310 × 52	317	6650	7.6	167	13.2	118.6	748	133.4	10.20	122.2	39.1
W310 × 44.5	313	5670	6.6	166	11.2	99.1	633	132.3	8.45	101.8	38.6
W250 × 44.8	266	5700	7.6	148	13.0	70.8	532	111.3	6.95	93.9	34.8
W250 × 32.7	258	4190	6.1	146	9.1	49.1	381	108.5	4.75	65.1	33.8
W200 × 26.6	207	3390	5.8	133	8.4	25.8	249	87.1	3.32	49.9	31.2
W200 × 22.5	206	2860	6.2	102	8.0	20.0	194.2	83.6	1.419	27.8	22.3
W150 × 29.8	157	3790	6.6	153	9.3	17.23	219	67.6	5.54	72.4	28.1
W150 × 24	160	3060	6.6	102	10.3	13.36	167	66	1.844	36.2	24.6

C.13   GEOMETRIC PROPERTIES OF S-SHAPED SECTIONS (FPS UNITS)

TABLE C.9

| Designation (in × lb/ft) | Depth d (in) | Area A (in²) | Web Thickness t_w (in) | Flange | | z Axis | | | y Axis | | |
				Width b_F (in)	Thickness t_F (in)	I_{zz} (in⁴)	S_z (in³)	r_z (in)	I_{yy} (in⁴)	S_y (in³)	r_y (in)
S12 × 35	12	10.3	0.428	5.078	0.544	229	38.4	4.72	9.87	3.89	0.98
S12 × 31.8	12	9.35	0.350	5.000	0.544	218	36.4	4.83	9.36	3.74	1.0
S10 × 35	10	10.3	0.594	4.944	0.491	147	29.4	3.78	8.36	3.38	0.901
S10 × 25.4	10	7.46	0.311	4.661	0.491	124	24.7	4.07	6.79	2.91	0.954
S8 × 23	8	6.77	0.411	4.171	0.426	64.9	16.2	3.10	4.31	2.07	0.798
S8 × 18.4	8	5.41	0.271	4.001	0.426	57.6	14.4	3.26	3.73	1.86	0.831
S7 × 20	7	5.88	0.450	3.860	0.392	42.4	12.1	2.69	3.17	1.64	0.734
S7 × 15.3	7	4.50	0.252	3.662	0.392	36.9	10.5	2.86	2.64	1.44	0.766

C.14   GEOMETRIC PROPERTIES OF S-SHAPED SECTIONS (METRIC UNITS)

TABLE C.10

| Designation (mm × kg/m) | Depth d (mm) | Area A (mm²) | Web Thickness t_w (mm) | Flange | | z Axis | | | y Axis | | |
				Width b_F (mm)	Thickness t_F (mm)	I_{zz} (10⁶mm⁴)	S_z (10³mm³)	r_z (mm)	I_{yy} (10⁶mm⁴)	S_y (10³mm³)	r_y (mm)
S310 × 52	305	6640	10.9	129	13.8	95.3	625	119.9	4.11	63.7	24.9
S310 × 47.3	305	6032	8.9	127	13.8	90.7	595	122.7	3.90	61.4	25.4
S250 × 52	254	6640	15.1	126	12.5	61.2	482	96.0	3.48	55.2	22.9
S250 × 37.8	254	4806	7.9	118	12.5	51.6	406	103.4	2.83	48.0	24.2
S200 × 34	203	4368	11.2	106	10.8	27.0	266	78.7	1.794	33.8	20.3
S200 × 27.4	203	3484	6.9	102	10.8	24	236	82.8	1.553	30.4	21.1
S180 × 30	178	3794	11.4	97	10.0	17.65	198.3	68.3	1.319	27.2	18.64
S180 × 22.8	178	2890	6.4	92	10.0	15.28	171.7	72.6	1.099	23.9	19.45

Bibliography

1. Barber, J. R. (1992) *Elasticity,* Kluwer Academic Publishers.

2. Bathe K. J. (1996) *Finite Element Procedures,* Prentice-Hall.

3. Boresi, A. P., and Chong, K. P. (1987) *Elasticity in Engineering Mechanics,* Elsevier.

4. Burger, C. P. (1993) Photoelasticity, pp. 165–266 in *Handbook on Experimental Mechanics,* A. S. Kobayashi, ed., Society of Experimental Mechanics.

5. Flugge, W. (1967) *Viscoelasticity,* Blaisdell Publishing.

6. Fung, Y. C. (1965) *Foundations of Solid Mechanics,* Prentice-Hall.

7. Gibson, R. F. (1994) *Principles of Composite Materials Mechanics,* McGraw-Hill.

8. Hyer, M. W. (1998) *Stress Analysis of Fiber-Reinforced Composite Materials,* WCB/McGraw-Hill.

9. Jones, R. M. (1975) *Mechanics of Composite Materials,* Hemisphere Publishing.

10. Kraus, H. (1967) *Thin Elastic Shells,* John Wiley & Sons.

11. Kreyszig, E. (1979) *Advanced Engineering Mathematics,* John Wiley & Sons.

12. Lanczos, C. (1986) *The Variational Principles of Mechanics,* Dover.

13. Mendelson, A. (1968) *Plasticity: Theory and Application,* Macmillan.

14. Morrison, F. A. (2001) *Understanding Rheology,* Oxford University Press.

15. Murakami, Y. (1987) *Stress Intensity Factors Handbook,* Pergamon Press.

16. Murray, N. W. (1984) *Introduction to the Theory of Thin-Walled Structures,* Oxford: Clarendon Press.

17. Muskhelishvili, N. I. (1963) *Some Basic Problems of the Mathematical Theory of Elasticity,* P. Noordhoff.

18. Pilkey, W. D. (1997) *Peterson's Stress Concentration Factors,* 2nd ed., John Wiley & Sons.

19. Post, D. (1993) Moiré interferometry, pp. 297–364 in *Handbook on Experimental Mechanics*, A. S. Kobayashi, ed., Society of Experimental Mechanics.

20. Reddy, J. N. (1993) *An Introduction to the Finite Element Method*, 2nd ed., McGraw-Hill.

21. Rolfe, S. T., and Barsom, J. M. (1977) *Fracture and Fatigue Control in Structures*, Prentice-Hall.

22. Shames, I. H., and Cozzarelli, F. A. (1997) *Elastic and Inelastic Stress Analysis*, Taylor & Francis.

23. Sneddon, I. N. (1980) *Fourier Transform*, McGraw-Hill.

24. Synge, J. L., and Schild, A. (1978) *Tensor Calculus*, Dover.

25. Szilard, R. (1974) *Theory and Analysis of Plates: Classical and Numerical Methods*, Prentice-Hall.

26. Timoshenko, S. P. (1983) *History of Strength of Materials*, Dover.

27. Timoshenko, S. P., and Goodier, J. N. (1951) *Theory of Elasticity*, McGraw-Hill.

28. Vinson, J. R. (1989) *The Behavior of Thin-Walled Structures: Beams, Plates, and Shells*, Kluwer Academic Publishers.

29. Zienkiewicz, O. C., and Taylor R. L. (1989) *The Finite Element Method*, 4th ed., McGraw-Hill.

INDEX

FORMULA SHEET

$$\sigma_{ij} = \lim_{\Delta A_i \to 0}\left(\frac{\Delta F_j}{\Delta A_i}\right) \qquad \sigma_{nn} = \sigma_{xx}\cos^2\theta + \sigma_{yy}\sin^2\theta + 2\tau_{xy}\sin\theta\cos\theta \qquad \tau_{nt} = -\sigma_{xx}\cos\theta\sin\theta + \sigma_{yy}\sin\theta\cos\theta + \tau_{xy}(\cos^2\theta - \sin^2\theta) \qquad \tan 2\theta_p = \frac{2\tau_{xy}}{(\sigma_{xx}-\sigma_{yy})}$$

$$\sigma_{1,2} = \frac{(\sigma_{xx}+\sigma_{yy})}{2} \pm \sqrt{\left(\frac{\sigma_{xx}-\sigma_{yy}}{2}\right)^2 + \tau_{xy}^2} \qquad \varepsilon_{nn} = \varepsilon_{xx}\cos^2\theta + \varepsilon_{yy}\sin^2\theta + \gamma_{xy}\sin\theta\cos\theta \qquad \gamma_{nt} = -2\varepsilon_{xx}\sin\theta\cos\theta + 2\varepsilon_{yy}\sin\theta\cos\theta + \gamma_{xy}(\cos^2\theta - \sin^2\theta)$$

$$\sigma_{nn} = \{n\}^T[\sigma]\{n\} \qquad \tau_{nt} = \{t\}^T[\sigma]\{n\} \qquad \sigma_{tt} = \{t\}^T[\sigma]\{t\} \qquad \{S\} = [\sigma]\{n\}$$

$$\sigma_p^3 - I_1\sigma_p^2 + I_2\sigma_p - I_3 = 0 \qquad I_1 = \sigma_{xx} + \sigma_{yy} + \sigma_{zz} \qquad I_2 = \begin{vmatrix}\sigma_{xx}&\tau_{xy}\\\tau_{yx}&\sigma_{yy}\end{vmatrix} + \begin{vmatrix}\sigma_{yy}&\tau_{yz}\\\tau_{zy}&\sigma_{zz}\end{vmatrix} + \begin{vmatrix}\sigma_{xx}&\tau_{xz}\\\tau_{zx}&\sigma_{zz}\end{vmatrix} \qquad I_3 = \begin{vmatrix}\sigma_{xx}&\tau_{xy}&\tau_{xz}\\\tau_{yx}&\sigma_{yy}&\tau_{yz}\\\tau_{zx}&\tau_{zy}&\sigma_{zz}\end{vmatrix}$$

$$x^3 - I_1 x^2 + I_2 x - I_3 = 0 \qquad x_1 = 2A\cos\alpha + I_1/3 \qquad x_{2,3} = -2A\cos(\alpha \pm 60^\circ) + I_1/3 \qquad A = \sqrt{(I_1/3)^2 - I_2/3} \qquad \cos 3\alpha = [2(I_1/3)^3 - (I_1/3)I_2 + I_3]/(2A^3)$$

$$\sigma_{oct} = (\sigma_1 + \sigma_2 + \sigma_3)/3 \qquad \tau_{oct} = \frac{1}{3}\sqrt{(\sigma_1-\sigma_2)^2 + (\sigma_2-\sigma_3)^2 + (\sigma_3-\sigma_1)^2} \qquad \varepsilon_{xx} = \frac{\partial u}{\partial x} \qquad \varepsilon_{yy} = \frac{\partial v}{\partial y} \qquad \varepsilon_{zz} = \frac{\partial w}{\partial z} \qquad \gamma_{xy} = \frac{\partial u}{\partial y} + \frac{\partial v}{\partial x} \qquad \gamma_{yz} = \frac{\partial v}{\partial z} + \frac{\partial w}{\partial y} \qquad \gamma_{zx} = \frac{\partial w}{\partial x} + \frac{\partial u}{\partial z}$$

$$\varepsilon_{xx} = [\sigma_{xx} - \nu(\sigma_{yy}+\sigma_{zz})]/E + \alpha\Delta T \qquad G = E/[2(1+\nu)] \qquad \gamma_{xy} = \tau_{xy}/G \qquad \varepsilon_{xx} = [\varepsilon_{xx} + \nu\varepsilon_{yy}]\frac{E}{(1-\nu^2)} \qquad \varepsilon_{zz} = -\left(\frac{\nu}{1-\nu}\right)(\varepsilon_{xx}+\varepsilon_{yy}) \qquad \sigma_{xx} = \frac{E[(1-\nu)\varepsilon_{xx} + \nu\varepsilon_{yy}]}{(1-2\nu)(1+\nu)}$$

$$\varepsilon_{xx} = \frac{\sigma_{xx}}{E_x} - \frac{\nu_{yx}}{E_y}\sigma_{yy} \qquad \gamma_{xy} = \frac{\tau_{xy}}{G_{xy}} \qquad \frac{\nu_{yx}}{E_y} = \frac{\nu_{xy}}{E_x} \qquad \sigma_{von} = \frac{1}{\sqrt{2}}\sqrt{(\sigma_1-\sigma_2)^2 + (\sigma_2-\sigma_3)^2 + (\sigma_3-\sigma_1)^2} \qquad \left|\frac{\sigma_2-\sigma_1}{\sigma_T}\right| \leq 1 \qquad \tau_{max} = \left|\max\left(\frac{\sigma_1-\sigma_2}{2}, \frac{\sigma_2-\sigma_3}{2}, \frac{\sigma_3-\sigma_1}{2}\right)\right|$$

$$K_1 = \sigma_{nom}\sqrt{\pi a} \qquad K_{II} = \tau_{nom}\sqrt{\pi a} \qquad K_{equiv} = \sqrt{K_I^2 + K_{II}^2} \qquad \left|\frac{\sigma_2-\sigma_1}{\sigma_C}\right| \leq 1$$

$$\sigma = \begin{cases}\sigma_{yield} & \varepsilon \geq \varepsilon_{yield}\\ E\varepsilon & -\varepsilon_{yield} \leq \varepsilon \leq \varepsilon_{yield}\\ -\sigma_{yield} & \varepsilon \leq -\varepsilon_{yield}\end{cases} \qquad \sigma = \begin{cases}\sigma_{yield} + E_2(\varepsilon - \varepsilon_{yield}) & \varepsilon \geq \varepsilon_{yield}\\ E_1\varepsilon & -\varepsilon_{yield} \leq \varepsilon \leq \varepsilon_{yield}\\ -\sigma_{yield} + E_2(\varepsilon + \varepsilon_{yield}) & \varepsilon \geq \varepsilon_{yield}\end{cases} \qquad \sigma = \begin{cases}E\varepsilon^n & \varepsilon \geq 0\\ -E(-\varepsilon)^n & \varepsilon < 0\end{cases}$$

Axial (Rods)

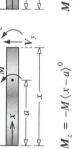

$$N = -F\langle x - a\rangle^0$$

$$P_x = F\langle x - a\rangle^{-1}$$

Torsion (Shafts)

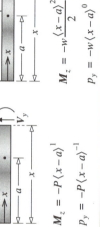

$$T = -T\langle x - a\rangle^0$$

$$t = T\langle x - a\rangle^{-1}$$

Bending (Beams)

$$M_z = -M\langle x - a\rangle^0 \qquad M_z = -P\langle x - a\rangle^1 \qquad M_z = -w\frac{\langle x - a\rangle^2}{2}$$

$$p_y = -M\langle x - a\rangle^{-2} \qquad p_y = -P\langle x - a\rangle^{-1} \qquad p_y = -w\langle x - a\rangle^0$$

$$U_0 = \frac{1}{2}\sigma\varepsilon \qquad U_0 = \frac{1}{2}[\sigma_{xx}\varepsilon_{xx} + \sigma_{yy}\varepsilon_{yy} + \sigma_{zz}\varepsilon_{zz} + \tau_{xy}\gamma_{xy} + \tau_{yz}\gamma_{yz} + \tau_{zx}\gamma_{zx}]$$

$$(F = 1)v_1(x_P) = \int_0^L \frac{M_2(x)M_1(x)}{EI}dx \qquad (M = 1)\frac{dv_1}{dx}(x_P) = \int_0^L \frac{M_2(x)M_1(x)}{EI}dx \qquad v_1(x_P) = \frac{\partial \overline{U}_B}{\partial F} \qquad \frac{dv_1(x_P)}{dx} = \frac{\partial \overline{U}_B}{\partial M}$$